Clinical Nutrition: Recent Advances and Remaining Challenges

Clinical Nutrition: Recent Advances and Remaining Challenges

Editor

Ina Bergheim

MDPI • Basel • Beijing • Wuhan • Barcelona • Belgrade • Manchester • Tokyo • Cluj • Tianjin

Editor
Ina Bergheim
University of Vienna
Vienna
Austria

Editorial Office
MDPI
St. Alban-Anlage 66
4052 Basel, Switzerland

This is a reprint of articles from the Special Issue published online in the open access journal *Nutrients* (ISSN 2072-6643) (available at: https://www.mdpi.com/journal/nutrients/special_issues/Clinical_Nutrition).

For citation purposes, cite each article independently as indicated on the article page online and as indicated below:

LastName, A.A.; LastName, B.B.; LastName, C.C. Article Title. *Journal Name* **Year**, *Volume Number*, Page Range.

ISBN 978-3-0365-4839-5 (Hbk)
ISBN 978-3-0365-4840-1 (PDF)

Contents

About the Editor

Ina Bergheim

Prof. Dr. Ina Bergheim is a Full Professor for Molecular Nutritional Sciences at the Department of Nutritional Sciences at the University of Vienna, Austria. She is a nutritional scientist by training. The main aim of her research is to determine molecular mechanisms underlying the development of alcohol-related and non-alcoholic fatty liver diseases as well as aging associated liver decline with a particular focus on the interplay of nutrition and the intestinal barrier herein. Furthermore, she aims to develop new strategies to target the molecular alterations underly the development of alcohol-related and non-alcoholic liver diseases as well as aging associated liver decline with nutrition including macronutrients but also secondary plant compounds.

Preface to "Clinical Nutrition: Recent Advances and Remaining Challenges"

In recent years, nutrition, often in combination with physical activity, has been acknowledged as a cornerstone in the prevention and, even more so, the treatment of many diseases. Indeed, nutritional intake and dietary pattern are often thought to be the main source of wellbeing, and often the impact of nutritional intake, dietary pattern and food-derived natural compounds in regard to their impact on human health are over- but also underestimated.

This book is a summary review article published within the Special Issue entitled 'Clinical Nutrition: Recent Advances and Remaining Challenges´ in the journal 'Nutrients´ and it aims to provide a broad overview and summary of recent findings and advances in various fields of clinical nutrition. A special focus is laid on new dietary approaches in the prevention and treatment of metabolic diseases, including obesity, non-alcoholic fatty liver disease and type 2 diabetes, but also on inflammatory bowel diseases as well as celiac disease and intestinal fibrosis in children and adults. Furthermore, new strategies in the prevention and treatment of malnutrition in elderly, adult patients with liver cirrhosis and children with cholestasis are presented and discussed. Additionally, recent findings and novel dietary approaches in the treatment of patients in intensive care and palliative care, respectively, as well as with gastrointestinal surgery are summarized. Moreover, the nutritional impact in the development and treatment as well as dietary approaches in the prevention of kidney stones, the Brugada Syndrome and prostatic hyperplasia are discussed. By summarizing novel findings regarding the relation of nutrition and the development of these disease, the present book aims to provide a better understanding of the impact but also the limitations of nutrition in disease development and treatment, but also highlight remaining questions and challenges in clinical nutrition.

Ina Bergheim
Editor

nutrients

Review

Nutrition Concepts for the Treatment of Obesity in Adults

Meike Wiechert and Christina Holzapfel *

Institute for Nutritional Medicine, School of Medicine, Technical University of Munich, 80992 Munich, Germany; meike.wiechert@tum.de
* Correspondence: christina.holzapfel@tum.de; Tel.: +49-89-289-249-23

Abstract: Obesity caused by a positive energy balance is a serious health burden. Studies have shown that obesity is the major risk factor for many diseases like type 2 diabetes mellitus, coronary heart diseases, or various types of cancer. Therefore, the prevention and treatment of increased body weight are key. Different evidence-based treatment approaches considering weight history, body mass index (BMI) category, and co-morbidities are available: lifestyle intervention, formula diet, drugs, and bariatric surgery. For all treatment approaches, behaviour change techniques, reduction in energy intake, and increasing energy expenditure are required. Self-monitoring of diet and physical activity provides an effective behaviour change technique for weight management. Digital tools increase engagement rates for self-monitoring and have the potential to improve weight management. The objective of this narrative review is to summarize current available treatment approaches for obesity, to provide a selective overview of nutrition trends, and to give a scientific viewpoint for various nutrition concepts for weight loss.

Keywords: dietary recommendation; weight loss; overweight; intermittent fasting; carbohydrate; fat; protein

Citation: Wiechert, M.; Holzapfel, C. Nutrition Concepts for the Treatment of Obesity in Adults. *Nutrients* **2022**, *14*, 169. https://doi.org/10.3390/nu14010169

Academic Editor: Ina Bergheim

Received: 25 November 2021
Accepted: 27 December 2021
Published: 30 December 2021

1. The Challenge Obesity

Obesity is a complex, multifactorial, and largely preventable chronic disease defined as abnormal or excessive fat accumulation [1]. Body mass index (BMI), calculated as weight in kilograms divided by the square of height in metres (kg/m^2), is the current most widely used criterion for classifying obesity [1]. People with a BMI $\geq$ 25 kg/m^2 are classified as overweight, and a BMI $\geq$ 30 kg/m^2 is categorized as obese [1–3] (Table 1). Although the BMI captures the degree of overweight and obesity, abdominal obesity measured mainly by waist circumference is additionally associated with health risks (Table 1). Clinical practice and medical guidelines focus on BMI and waist circumference as simple, objective, and reproducible tools to measure weight status and abdominal obesity. However, the diagnosis of obesity should not be based on BMI alone, rather together with other anthropometric and clinical parameters. Instrumental methods (e.g., bioimpedance analysis, Dual Energy X-ray Absorptiometry, and magnetic resonance imaging) to assess body composition and adipose tissue depots are available but are often time- and cost-intensive.

It is well known that overweight and obesity are the main risk factors for several diseases such as type 2 diabetes mellitus, hypertension, dyslipidaemia, cardiovascular disease, and several types of cancer [4]. Recently published data show that BMI is positively associated with severe coronavirus disease 2019 (COVID-19) outcomes [5]. Furthermore, an increased BMI might lead to a decline in quality of life and contribute to a decreased life expectancy [6–9].

Table 1. Body mass index and waist circumference: cut-offs and risk of co-morbidities.

	Body Mass Index		
	18.5–24.9 kg/m^2	**25–29.9 kg/m^2**	**≥30 kg/m^2**
Classification	Normal weight	Overweight	Obese
Risk of co-morbidities	Low	Increased	High
	Waist circumference		
	Men: <94 cm Women: <80 cm	Men: 94–101.9 cm Women: 80–87.9 cm	Men: ≥102 cm Women: ≥88 cm
Classification	Normal fat distribution	Moderate central fat accumulation	High central fat accumulation
Risk of co-morbidities	Low	Increased	High

The cause of obesity is a long-term energy imbalance caused by a combination of increased energy intake and reduced energy expenditure [1,6,10,11]. The National Health and Nutrition Examination Survey (NHANES) observed that the average daily energy intake increased from 1971 to 2000 in men by 168 kilocalories (kcal)/day and women by 335 kcal/day [12]. Without an active regulation or adaptation of energy balance, this increase theoretically could explain a weight gain per year of eight kilograms for men and 16 kg for women. Furthermore, energy expenditure has decreased over the last decades [12]. Basset and colleagues reported that in 2003, an American adult walked about 5000 steps/day. Compared to people living 300 years ago, it is a difference of 13,000 steps/day for men and 9000 steps/day for women [13]. Without any physiological adaption, this decline in physical activity could explain a yearly weight gain of 31 kg for men and 21 kg for women [10]. Daily occupation-related energy expenditure has decreased by more than 100 kcal/day in U.S. adults. That alone could explain a substantial weight gain in the population over the last five decades [14]. In addition to lifestyle factors, other contributing factors like food and environment have been identified [15].

Over the past 50 years, the prevalence of obesity has increased worldwide in pandemic dimensions [3,6,16–18]. The Global Burden of Disease study with data from 68.5 million persons demonstrated that in 2015 603.7 million adults were obese [19]. Since 1980, the prevalence of obesity has doubled in more than 70 countries and has continuously increased in most other countries [19]. If trends continue, by 2030, an estimated 38% of the world's adult population will be overweight, and another 20% will be obese [20]. The confinements by the COVID-19 pandemic have changed lifestyle behaviour and have promoted an obesogenic environment. Weight trajectories during the COVID-19 lockdown have been shown [21,22]. It is expected that the COVID-19 pandemic reinforces the obesity pandemic with long-term consequences on the prevalence of overweight and obesity [23]. Because of the rapid increase in the prevalence and disease burden of obesity, it is indispensable to focus on monitoring BMI and to identify, implement, and evaluate evidence-based interventions to address this health issue [19].

2. Treatment Approaches

Different evidence-based approaches considering weight history, BMI category, and co-morbidities to treat overweight and obesity are available: lifestyle intervention, formula diet, drugs, and bariatric surgery (Figure 1). The main aims of weight management are shown in Figure 2.

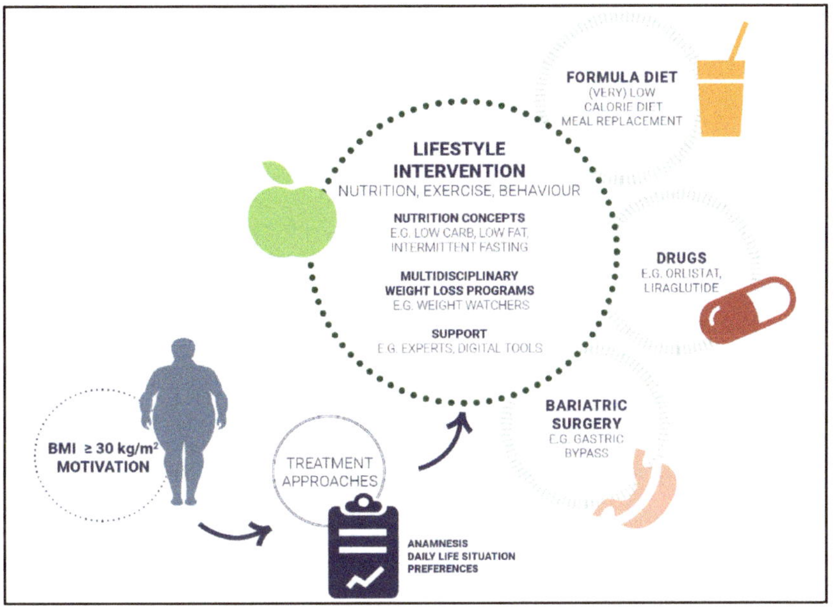

Figure 1. Treatment approaches.

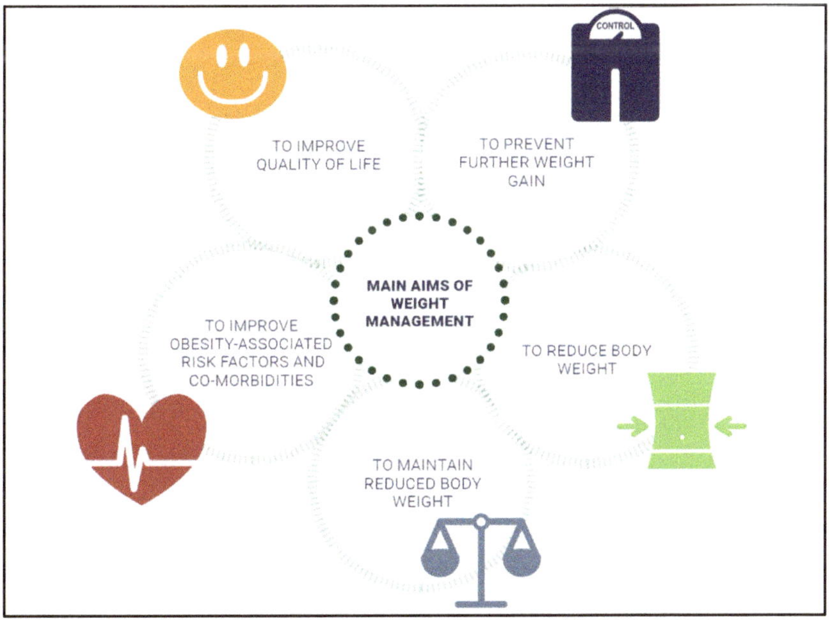

Figure 2. Overview of weight management goals.

2.1. Lifestyle Intervention

Lifestyle interventions inducing a negative energy balance provide the basis for the treatment of overweight and obesity and are part of the standard recommendation. Different lifestyle approaches exist, whereas nutrition, physical activity, and behaviour are the main components. By lowering energy intake and increasing physical activity accompanied by behavioural change techniques, a daily energy deficit of about 500 kcal is recommended for weight loss. This energy deficit can produce a moderate weight loss over one year. Energy balance changes with weight loss, making it necessary to adjust energy intake and expenditure during weight management. Energy balance is dynamic and weight loss leads to a new energy equilibrium on a lower level. Adherence to a lifestyle intervention is challenging for many people with overweight and obesity. In a systematic review and meta-analysis, three main factors were associated with improved adherence to weight loss interventions: supervision, social support, and focus on dietary intervention [24].

Nutrition is the main lifestyle factor. Therefore, nutrition aspects to decrease energy intake and to support weight management are highlighted in the following.

2.1.1. Energy Intake

Key components of energy balance include energy intake, expenditure, and storage. When energy intake exceeds energy expenditure, a state of positive energy balance increases body weight. The European Food Safety Authority (EFSA) recommends a daily dietary reference intake of 45% to 65% of total kilocalories from carbohydrates, 20% to 35% from fat [25], and 0.83 g protein/kg body weight [26]. For weight loss, a daily energy deficit of 500 kcal is recommended and can be reached by avoiding energy-dense food. Fat is a high-energy macronutrient and provides more than twice as high as the energy of carbohydrates or protein. Because of this, a reduction in daily fat intake supports to lower daily caloric intake. Fat intake can be reduced by using low fat dairy products like cheese, and yogurt, lean meat, and avoidance of hidden fats. In Table 2, recommendations and practical examples are shown for decreasing energy intake.

Table 2. Practical recommendations and examples to reduce energy intake.

Item	Recommendation	Implementation
Energy density	**Prefer low energy dense food.** Choose food with lower energy content per 100 g.	Salami → ham Rice → potatoes Cream cake → fruit tart
Dietary fat	**Watch out for hidden fats.** Choose food low in fat, or pick the fat-reduced alternative. **Prefer fat-reduced food preparing methods.** Choose methods without adding fat.	Potato chips → pretzel sticks Butter, cream cheese → low fat-cream cheese Frying → braising Roasting → cooking Chips → boiled potatoes
Plant-based food	**Take five portions a day of vegetables and fruits.** Choose plant-based food instead of meat.	Wheat noodles → zucchini pasta Mascarpone cream → fresh fruit for dessert Meat bolognese → vegetable bolognese
Dietary fiber	**Increase dietary fibre intake to 25 to 30 g a day.** Choose whole grain products, legumes, vegetables, fruits.	White noodles → whole grain noodles White bread → whole grain bread Rice → legumes
Beverages	**Drink caloric-free beverages.** Choose beverages without sugar and calories.	Water and tea Coffee without sugar Beverages with artificial sweeteners instead of sugar-sweetened beverages

2.1.2. Macronutrients

Studies have shown that not the macronutrient composition of the diet but the energy content is relevant for weight management [27,28].

Low carb diets often contain approximately 40% of carbohydrates per day. The lowest intake of carbohydrates is part of a ketogenic diet, where the aim is to minimize the carbohydrate intake as much as possible. Epidemiological data showed that a daily amount of carbohydrates of 50 to 55% correlates with the lowest mortality rate. Low carb diets, as well as high carb diets, increase the mortality risk (U-shaped association) [29]. A low carb diet includes a lower amount of plant-based food that has health-promoting effects. A meta-analysis with eight randomized controlled studies concluded that low carb diets are superior to diets with a low amount of fat regarding lipid metabolism in people with overweight and obesity [30]. However, data from NHANES indicate that realized decreases in the percentage of energy consumed from fat were associated with increased total energy intake caused by compensatory over-consumption of energy from sugars [31]. The stone age diet or paleo diet is a low carb diet, but also not clearly defined. In general, it is a diet with a high amount of meat and protein. The variety of foods is limited by the renouncing of grain. Smaller short-term intervention studies with methodological limitations examined the effects of a paleo-conform diet. Manheimer et al. evaluated data of four randomized controlled trials (RCTs) with 159 participants that compared the palaeolithic diet with any other dietary pattern. Results showed that palaeolithic nutrition resulted in stronger short-term improvements of cardiovascular risk factors like waist circumference and blood pressure than control diets [32]. In 70 post-menopausal women with obesity, it has been shown that there was no significant difference between a palaeolithic-type diet and the Nordic nutrition recommendations in anthropometric changes after 24 months [33].

An evaluation of almost 50 studies found that the participants, regardless of the macronutrient composition of the diet, lost the same amount of weight within 6 and 12 months of intervention [34]. In another study, 811 persons were randomized into four groups of diets with a different energy intake from fat, protein, and carbohydrates (20, 15, and 65%; 20, 25, and 55%; 40, 15, and 45%; 40, 25, and 35%, respectively). After two years of intervention, the weight loss was about 4 kg (completers-analysis), with no significant differences between the groups [28]. Furthermore, the comparison of three different diets (low fat/low energy diet; Mediterranean/low energy diet; low carb/non-energy reduced diet) resulted in similar findings. Mean weight loss after 2 years of intervention was 3.3, 4.6, and 5.5 kg, respectively (completers-analysis) [27]. In a study with 609 adults with BMI between 28 and 40 kg/m^2, the mean weight loss was about 6.0 kg (low carb diet) and 5.3 kg (low fat diet) after one year of intervention [35]. A systematic review of systematic reviews comparing low carb diets with control diets on weight loss concerned the low quality of studies, and concluded that better quality reviews and RCTs are needed for a clear recommendation of low carb diets as preferred to other energy-reduced diets [36]. Even with the plant-based form of Atkins diet or the Mediterranean diet, there is a moderate weight loss [27,37,38]. A meta-analysis indicates that a Mediterranean diet low in energy leads to moderate weight loss [39].

Finally, the macronutrient composition of a diet has no major impact on weight loss. Low carb, as well as low fat concepts, are effective for weight loss if a negative energy balance is provided.

A meal replacement is a high protein product used to replace at least one main meal per day. Those products are permitted to be marketed for weight management purposes and have specific regulatory requirements concerning supplementation with vitamins, minerals, and trace elements, as well as energy content, per portion. They are available e.g., as shakes, soups, or meal bars. Meal replacement strategy followed by a dietary change and behaviour modification strategy is popular among people trying to lose weight. One option is a very low calorie diet (VLCD) with < 800 kcal/day by total meal replacement. The other option is a low calorie diet (LCD) supplying > 800 kcal/day, generally in the range of 1200–1600 kcal/day. In a systematic review and meta-analysis on VLCDs, total weight loss ranged from 8.9 to 15 kg in persons with type 2 diabetes mellitus and 7.9 to 21 kg in persons without diabetes, over a treatment duration of 4 to 52 weeks. Study duration did not appear to influence overall weight loss. The average weight loss per week was about 0.5 kg [40]. Another review investigated the effect of weight loss interventions incorporating meal replacement compared with alternative interventions on weight change at 1 year in adults with overweight or obesity. In this review, studies with diets providing < 800 kcal/day, and with total diet replacement, were excluded. In general, all diets incorporating meal replacement resulted in a higher mean weight change at 1 year compared to the control groups or alternative diets [41]. The Diabetes Remission Clinical Trial (DiRECT) of 306 patients with type 2 diabetes mellitus demonstrated that diet-induced weight loss by total diet replacement (825–853 kcal/day formula diet for 3–5 months), followed by food reintroduction (2–8 weeks), and followed by structured support for long-term weight loss maintenance effectively reversed type 2 diabetes mellitus. At 12 months, 86% of the participants who achieved a weight loss >15 kg (24%) became drug-free and had remission of type 2 diabetes mellitus. An overall remission of type 2 diabetes mellitus in the intervention group was observed in 46% of patients after 1 year and in 36% of patients after 2 years [42,43].

2.1.3. Intermittent Fasting

There are several approaches of intermittent fasting. The 16:8 concept is a form of time-restricted eating where individuals eat within a time window of 8 h and fast for the remaining 16 h daily. The 5:2 concept consists of a normal eating routine on 5 days per week, without any specific recommendations or restrictions, and 2 days of fasting with an energy intake of about 500 kcal. Conley M. et al. compared the 5:2 diet (2 non-consecutive days with 600 kcal and 5 days of energy intake ad libitum per week) with an energy-reduced diet. After 6 months of intervention, both groups reduced their body weight similarly (5.3 kg (5:2) vs. 5.5 kg (standard)) with no significant difference [44]. A RCT compared the effects of alternate-day fasting with daily caloric restriction on body weight in participants with obesity. Findings demonstrate comparable weight loss after 6 (alternate day fasting: 6.8%, caloric restriction: 6.8%) and 12 months (6.0% versus 5.3%) [45]. In a systematic review and meta-analysis of RCTs intermittent versus continuous energy restriction on weight loss and cardio-metabolic outcomes have been investigated. The included eleven trials with a duration from 8 to 24 weeks resulted in a similar weight loss between the two intervention groups [46]. Compared to a continuous energy restriction, intermittent fasting leads to similar weight loss and similar improvement of cardio-metabolic parameters [47–49]. The recently published Cochrane review by Allaf et al. found that people lost more weight with intermittent fasting concepts than without a special dietary concept over three months (evidence from seven studies in 224 people). If intermittent fasting concepts were compared with energy-restricted diets for 3 months (10 studies; 719 people) or longer (3 to 12 months; 4 studies; 279 people), this difference in weight loss is lost [50]. The energy restriction causes the positive effect of intermittent fasting on weight loss, not fasting as a stand-alone intervention [51,52].

Besides weight loss, fasting-specific effects on metabolic regulation or cardiovascular health are discussed. In lean persons, there is no statistically significant difference between daily energy restriction and alternate-day fasting with or without energy restriction concerning postprandial indices of cardio-metabolic health, gut hormones, or the gene expression in subcutaneous adipose tissue [51].

A small study with eleven participants with overweight, early time-restricted fasting (eating between 8 a.m. and 2 p.m.) was investigated for acute effects on glucose metabolism and gene expression. In comparison to the control group (eating between 8 a.m. to 8 p.m.) 24 h glucose levels and glycaemic excursions decreased, and ketones, cholesterol, and the expression of the stress response and aging gene sirtuin 1 (SIRT1) and the autophagy gene microtube associated protein 1 light chain 3 alpha (LC3A) increased in the morning before breakfast. This was different to the gene expression pattern in the evening [53]. The early time-restricted feeding effect on cardio-metabolic health (insulin sensitivity, beta-cell responsiveness, blood pressure, oxidative stress, and appetite)—independent from weight loss—has further been observed by Sutton et al. in men with prediabetes [54]. In addition, a RCT with 17 participants with normal weight compared the metabolic effects of breakfast and dinner skipping. Compared to the three meals per day control group, skipping breakfast or dinner increased energy expenditure. Furthermore, fat oxidation increased on the breakfast skipping day [55]. In a prospective cohort study, it has been shown that breakfast skipping is associated with a 21% increased risk for the development of diabetes mellitus type 2 [56].

2.1.4. Personalized Nutrition

In the last years, concepts of personalized nutrition have been more focused, especially by commercial providers offering direct-to-consumer genetic testing. One of the main drivers for personalized dietary recommendations is the inter-individual variability of metabolic response on standardized meal challenges suggesting that personalized diets might successfully modify elevated postprandial blood glucose and its metabolic consequences [57]. In the Personalised Responses to Dietary Composition Trial (PREDICT1) with more than 1000 twins and unrelated healthy adults in the UK, large inter-individual variability in postprandial responses of blood triglycerides (103%), glucose (68%), and insulin (59%) following identical meals was observed. Various intrinsic and extrinsic factors could be identified as predictors of the inter-individual variability. In the following, scientific evidence of gene-based and microbiome-based dietary recommendations is summarized.

Gene-Based Dietary Recommendations

Gene-based dietary recommendations are based on individuals' genetic make-up. The fact that body weight is, inter alia, genetically determined, and more than hundreds of genetic loci are identified for a relationship with anthropometric parameters [58], is underlining the assumption that even the inter-individual varying degree of success in weight loss indicates a genetic component [59]. The *fat mass and obesity-associated* (*FTO*) gene is the gene with the largest effect on body weight. The function of the *FTO* gene is not yet fully understood, whereas it is shown that the *FTO* gene inhibits brown adipose tissue genesis [60]. Numerous companies offer genotyping and provide recommendations for a healthy diet or weight loss. Furthermore, commercial offerings for deoxyribonucleic acid (DNA) methylation profiling started to emerge. These commercially available direct-to-consumer tests are in contrast to the lacking scientific evidence that genotypes are associated with weight loss.

In a recently published pooled analysis of weight loss data, it has been shown that single nucleotide polymorphisms have a minor role in the inter-individual variation of weight loss [61]. The Food4Me study has shown that including genotype information for dietary recommendations had no beneficial effect on weight loss [62]. The American Society of Dietetics and Nutrition clearly states that "No significant differences in weight, body mass index (BMI; calculated as kg/m^2), or waist circumference were observed when results of genetic testing were incorporated into nutrition counselling compared with counselling or care that did not incorporate genetic results" [63], and the "use of nutrigenetic testing to provide dietary advice is not ready for routine dietetics practice" [64]. Present research cannot provide adequate evidence that individuals with a defined genetic make-up benefit especially from specific dietary recommendations concerning weight loss [65]. A systematic review on gene–diet interactions on weight change concluded that there is no evidence that gene–diet interactions are a main determinant for obesity treatment success [66]. For that reason, more future human studies are required to prove the clinical evidence of gene-based dietary recommendations on weight loss [59]. Furthermore, the investigation of single nucleotide polymorphisms will be replaced by the investigation of polygenetic scores to characterize humans according to their genetic predisposition. Khera AV et al. have developed and validated a genome-wide polygenetic score for five diseases (coronary heart disease, atrial fibrillation, diabetes mellitus type 2, inflammatory bowel disease, and breast cancer) [67]. In a further data analysis, a polygenetic predictor for body weight has been developed and validated [68].

Microbiome-Based Dietary Recommendations

Various correlations between the gut microbiota and individuals' nutrition, as well as the occurrence of diseases like obesity, have been shown [59]. These correlations indicate that personalized nutrition based on the microbiome is a further starting point for weight loss [59]. Increasing evidence suggests that changes in individuals' microbiome during dietary intervention are person-specific, and this heterogeneity, in addition to individuals' physiology, is due to a unique microbiome signature [69]. The integration of microbiome information in combination with other person-specific factors seems to have the potential for the understanding of complex interactions [70]. Von Schwarzenberg et al. demonstrated that a VLCD led to a decrease in bacterial abundance and restructuring of the gut microbiome. After this VLCD, the microbiota were transplanted to mice, which led to a decrease in body weight. This study reveals that diet–microbiome interactions modulate energy metabolism [71]. Independent from weight loss, there is no doubt that dietary intake influences gut microbiota structure [72]. In a literature review, the direct and indirect mechanisms behind the influence of the gut microbiome have been discussed. The composition of the microbiota, the presence of specific microbes, and their metabolic activity have to be considered in future human intervention studies to investigate the potential for targeting the microbiota for improving health [73]. Similar to the genetic direct-to-consumer tests, some companies offer the analysis of the microbiome for personalized dietary recommendations, whereas scientific evidence is still not given. The current knowledge about the microbiome's role in diet-mediated effects on health is too limited to provide evidence for microbiome-based dietary recommendations [74].

2.1.5. Weight Loss Programs

Multidisciplinary weight loss programs addressing nutrition, physical activity, and behaviour are available. Most of the programs are delivered in a single country or are regionally rolled out, e.g., by health insurances, health care providers, or companies. The largest global weight loss program is WW (formerly Weight Watchers). An international study has shown that the WW program resulted in a moderate weight loss at 12 and 24 months. Mean weight change at 12 months assessment was 6.65 kg (completers-analysis) [75] and mean weight loss after 2 years was 4.76 kg [76]. Furthermore, Johnston et al. have shown that persons using all provided intervention tools (weekly meetings, WW mobile application, and WW online tools) lose more weight than persons who selected not all three ways to access the treatment [77]. Another well-known program is OPTIFAST, which also provides scientific evidence for efficacy after 12 months. The 12 months OPTIFAST concept includes a total dietary replacement for three months followed by lifestyle recommendations and professional group sessions for further nine months. Comparing the effectiveness of the OPTIFAST program with a food-based dietary plan resulted in 10.5% versus 5.5% weight loss at 52 weeks [78]. In this study, there was an active weight maintenance phase, where meal replacement was allowed.

2.1.6. Support

Self-monitoring of diet and physical activity provides an effective behaviour change technique for weight management [79] and is a core component of behavioural obesity treatment [80]. It has been demonstrated that dietary self-monitoring itself and the frequency of self-monitoring is linked to weight loss [81]. Furthermore, self-monitoring tends to positively impact weight loss when combined with other self-regulation techniques, such as goal setting and feedback [82–84]. Engagement rates for self-monitoring diet were higher in digital than in paper-based self-monitoring [81,85].

Digital tools like online tools and applications (apps), tracking technologies, or even internet-based support have become attractive for teaching and supporting long-term behaviour change techniques [86]. Carter et al. examined in a RCT the acceptability and feasibility of a self-monitoring weight management intervention provided by a smartphone app compared to a website and paper diary. Results showed a mean weight change at

6 months of 4.6 kg in the smartphone app group, 2.9 kg in the paper diary group, and 1.3 kg in the website group and additionally, the app was highly appreciated in satisfaction and acceptability [85]. Weight management apps may have positive effects on weight-related outcomes; although, the methodological quality of many studies is low [87,88]. A meta-analysis by Villinger et al. with more than 6300 participants showed that app-based mobile interventions can be effective for changing nutrition behaviours and nutrition-related health outcomes [89]. Digital tools like apps are a time- and cost-effective method for the collection of health-related data with the potential of wide distribution and scalability [85,90,91]. Although many apps are available for weight loss, digital offerings for weight management rarely include evidence-based strategies for behaviour change [92,93].

3. Drugs

There are several agents for the pharmacologic therapy of obesity leading to decreased appetite, gastric emptying, nutrient absorption, or increased satiety. Some of these have gained marketing authorization during the last six years, and some others are still under development process [94,95]. Currently, the European Medicines Agency (EMA) authorized three drugs (orlistat, naltrexone/bupropion, and liraglutide), and the US Food and Drug Administration (FDA) has approved four drugs (orlistat, phentermine/topiramate, naltrexone/bupropion, and liraglutide) for obesity treatment [96]. The purpose of using pharmacotherapy to manage obesity is to increase patient adherence to lifestyle changes and to overcome the biological adaptations that occur with weight loss [97].

Increasing evidence has shown that behaviour-based interventions with one anti-obesity medication can result in greater weight loss than usual care conditions only [96]. The efficacy of available anti-obesity drugs is often limited to a reduction of 5–10% of body weight over a 1-year period and weight loss typically does not occur for more than 6–8 months [98]. In a systematic review and network meta-analysis, five weight loss medications (orlistat, lorcaserin, naltrexone-bupropion, phentermine-topiramate, or liraglutide) were compared regarding efficacy on weight loss. In total, 28 RCTs with 29,018 patients were included. Study participants in the placebo group had a statistically significant lower odds ratio for achieving 5% weight loss after one year than participants taking drugs. Excess weight loss was 8.8 kg for phentermine-topiramate, 5.3 kg for liraglutide, 5.0 kg for naltrexone-bupropion, 3.2 kg for lorcaserin, and 2.6 kg for orlistat [99].

4. Bariatric Surgery

Bariatric surgery is appropriate for persons with severe obesity. Indications of bariatric surgery vary across countries. In most countries, the access to bariatric surgery is, e.g., restricted to persons for whom other weight loss measures have failed. There are two primary mechanisms: restriction and malabsorption of ingested food. A restrictive approach is physically limiting the quantity of food that can be ingested by limiting the size and capacity of the stomach while leaving the remainder of the gastrointestinal tract intact. Malabsorption of calories and nutrients occurs when a portion is bypassed or removed. Kilocalories and nutrients are less able to be absorbed because ingested food remains in the gut to a shorter distance. The International Federation for Surgery of Obesity and Metabolic Disorders (IFSO) has published an annual report of all bariatric surgery provided to the Global Registry [100]. Data from 51 countries with documented surgery between 2014 and 2018 were evaluated. The surgery procedure carried out most frequently was sleeve gastrectomy (46.0%) followed by Roux en Y gastric bypass (38.2%), one anastomosis gastric bypass (7.6%), and gastric banding (5.0%) [100]. Within the systematic review by Puzziferry et al., 29 clinical studies with 7971 patients were evaluated. The main finding was that gastric bypass resulted in greater weight loss than the gastric band [101]. A further systematic review and meta-analysis showed that all current bariatric procedures are associated with significant weight loss, but more long-term data are needed for one-anastomosis gastric bypass and sleeve gastrectomy [102]. Particularly worthy to mention is that patients with higher adherence rates to visits and behaviour changes before surgery are more likely to lose more weight after surgery [103]. In any case, good preparation before and after surgery is indispensable to ensure the best outcomes [104]. Several steps during the preoperative evaluation are necessary. These include the individual's psychological fitness to undergo bariatric surgery, the professional nutritional assessment, and patient education to guide the patient towards dietary modifications that are necessary after surgery [104]. Nutritional deficiencies are a long-term clinical issue in patients through modifications to the gastrointestinal tract [105]. The Clinical Practice Guidelines of the European Association for Endoscopic Surgery (EAES) recommend postoperative nutritional and behavioural advice for patients undergoing bariatric surgery [106]. Nutritional monitoring is an essential component of weight management after bariatric surgery to increase the patients' adherence to healthy dietary habits and to appropriate supplementation measures [105]. In addition, monitoring prevents the risk of weight regain, makes it easier to detect possible nutritional deficiencies, and contributes to the preservation of a good quality of life [105].

5. Conclusions

This scientific viewpoint is a narrative review and not comparable with a systematic review but gives an overview of various treatment approaches, which should be used and combined considering the individuals' needs, preferences, weight status, and cardiometabolic risk factors. All treatment approaches have to result in a negative energy balance. Independent of the weight loss concept (e.g., intermittent fasting, low carb, low fat, drugs or, bariatric surgery), weight loss failed without a negative energy balance. Many trends like gene-based or microbiome-based dietary recommendations still lack conclusive scientific evidence. In general, weight loss studies often have methodological limitations (e.g., study design or duration), leading to results not being comparable, and they therefore should be interpreted with caution. With lifestyle changes, a moderate weight loss after one year is possible. Other approaches, such as bariatric surgery, lead to greater weight loss, but are proven only for specific target groups. More research, especially by long-term intervention studies, is needed to evaluate weight loss concepts and to obtain evidence-based tailored recommendations.

Author Contributions: M.W. and C.H. wrote the manuscript. All authors have read and agreed to the published version of the manuscript.

Funding: This work was supported by the *enable* competence cluster, an interdisciplinary cluster of nutrition science, funded by the German Federal Ministry of Education and Research (Bundesministerium für Bildung und Forschung, BMBF). The manuscript was written by the research group "Personalized nutrition and eHealth" (grant number: 01EA1709). The *enable* publication number is 79.

Acknowledgments: The authors thank Teresa Hölzl for designing the figures.

Conflicts of Interest: M.W. declares no conflict of interest. C.H. is a member of the scientific advisory board of the 4sigma GmbH Oberhaching, Germany.

References

1. World Health Organization. *Obesity: Preventing and Managing the Global Epidemic: Report of a WHO Consultation. WHO Technical Report Series 894*; World Health Organization: Geneva, Switzerland, 2000; ISBN 9241208945.
2. Waxman, A. Why a global strategy on diet, physical activity and health? *World Rev. Nutr. Diet.* **2005**, *95*, 162–166. [CrossRef] [PubMed]
3. World Health Organization. Obesity and Overweight. Available online: https://www.who.int/en/news-room/fact-sheets/detail/obesity-and-overweight (accessed on 20 June 2021).
4. Guh, D.P.; Zhang, W.; Bansback, N.; Amarsi, Z.; Birmingham, C.L.; Anis, A.H. The incidence of co-morbidities related to obesity and overweight: A systematic review and meta-analysis. *BMC Public Health* **2009**, *9*, 88. [CrossRef] [PubMed]
5. Gao, M.; Piernas, C.; Astbury, N.M.; Hippisley-Cox, J.; O'Rahilly, S.; Aveyard, P.; Jebb, S.A. Associations between body-mass index and COVID-19 severity in 6·9 million people in England: A prospective, community-based, cohort study. *Lancet Diabetes Endocrinol.* **2021**, *9*, 350–359. [CrossRef]
6. Blüher, M. Obesity: Global epidemiology and pathogenesis. *Nat. Rev. Endocrinol.* **2019**, *15*, 288–298. [CrossRef] [PubMed]
7. Fontaine, K.R.; Redden, D.T.; Wang, C.; Westfall, A.O.; Allison, D.B. Years of life lost due to obesity. *JAMA* **2003**, *289*, 187–193. [CrossRef]

8. Whitlock, G.; Lewington, S.; Sherliker, P.; Clarke, R.; Emberson, J.; Halsey, J.; Qizilbash, N.; Collins, R.; Peto, R. Body-mass index and cause-specific mortality in 900 000 adults: Collaborative analyses of 57 prospective studies. *Lancet* **2009**, *373*, 1083–1096. [CrossRef]
9. Yumuk, V.; Tsigos, C.; Fried, M.; Schindler, K.; Busetto, L.; Micic, D.; Toplak, H. European Guidelines for Obesity Management in Adults. *Obes. Facts* **2015**, *8*, 402–424. [CrossRef] [PubMed]
10. Hill, J.O.; Wyatt, H.R.; Peters, J.C. Energy balance and obesity. *Circulation* **2012**, *126*, 126–132. [CrossRef] [PubMed]
11. Camacho, S.; Ruppel, A. Is the calorie concept a real solution to the obesity epidemic? *Glob. Health Action* **2017**, *10*, 1289650. [CrossRef]
12. Wright, J.D.; Kennedy-Stephenson, J.; Wang, C.Y.; McDowell, M.A.; Johnson, C.L. Trends in Intake of Energy and Macronutrients—United States, 1971–2000. *JAMA* **2004**, *53*, 80–82. [CrossRef]
13. Bassett, D.R.; Wyatt, H.R.; Thompson, H.; Peters, J.C.; Hill, J.O. Pedometer-measured physical activity and health behaviors in U.S. adults. *Med. Sci. Sports Exerc.* **2010**, *42*, 1819–1825. [CrossRef]
14. Church, T.S.; Thomas, D.M.; Tudor-Locke, C.; Katzmarzyk, P.T.; Earnest, C.P.; Rodarte, R.Q.; Martin, C.K.; Blair, S.N.; Bouchard, C. Trends over 5 decades in U.S. occupation-related physical activity and their associations with obesity. *PLoS ONE* **2011**, *6*, e19657. [CrossRef]
15. Rolls, B.J. The Supersizing of America: Portion Size and the Obesity Epidemic. *Nutr. Today* **2003**, *38*, 42–53. [CrossRef]
16. Yanovski, J.A. Obesity: Trends in underweight and obesity—Scale of the problem. *Nat. Rev. Endocrinol.* **2018**, *14*, 5–6. [CrossRef]
17. Abarca-Gómez, L.; Abdeen, Z.A.; Hamid, Z.A.; Abu-Rmeileh, N.M.; Acosta-Cazares, B.; Acuin, C.; Adams, R.J.; Aekplakorn, W.; Afsana, K.; Aguilar-Salinas, C.A.; et al. Worldwide trends in body-mass index, underweight, overweight, and obesity from 1975 to 2016: A pooled analysis of 2416 population-based measurement studies in 128·9 million children, adolescents, and adults. *Lancet* **2017**, *390*, 2627–2642. [CrossRef]
18. NCD Risk Factor Collaboration. Trends in adult body-mass index in 200 countries from 1975 to 2014: A pooled analysis of 1698 population-based measurement studies with 19·2 million participants. *Lancet* **2016**, *387*, 1377–1396. [CrossRef]
19. Afshin, A.; Forouzanfar, M.H.; Reitsma, M.B.; Sur, P.; Estep, K.; Lee, A.; Marczak, L.; Mokdad, A.H.; Moradi-Lakeh, M.; Naghavi, M.; et al. Health Effects of Overweight and Obesity in 195 Countries over 25 Years. *N. Engl. J. Med.* **2017**, *377*, 13–27. [CrossRef]
20. Smith, K.B.; Smith, M.S. Obesity Statistics. *Prim. Care* **2016**, *43*, 121–135, ix. [CrossRef]
21. Dicken, S.J.; Mitchell, J.J.; Le Newberry Vay, J.; Beard, E.; Kale, D.; Herbec, A.; Shahab, L. Impact of COVID-19 Pandemic on Weight and BMI among UK Adults: A Longitudinal Analysis of Data from the HEBECO Study. *Nutrients* **2021**, *13*, 2911. [CrossRef] [PubMed]
22. Seal, A.; Schaffner, A.; Phelan, S.; Brunner-Gaydos, H.; Tseng, M.; Keadle, S.; Alber, J.; Kiteck, I.; Hagobian, T. COVID-19 pandemic and stay-at-home mandates promote weight gain in US adults. *Obesity* **2021**, *30*, 240–248. [CrossRef] [PubMed]
23. Stefan, N.; Birkenfeld, A.L.; Schulze, M.B. Global pandemics interconnected—Obesity, impaired metabolic health and COVID-19. *Nat. Rev. Endocrinol.* **2021**, *17*, 135–149. [CrossRef]
24. Lemstra, M.; Bird, Y.; Nwankwo, C.; Rogers, M.; Moraros, J. Weight loss intervention adherence and factors promoting adherence: A meta-analysis. *Patient Prefer. Adherence* **2016**, *10*, 1547–1559. [CrossRef] [PubMed]
25. European Food Safety Authority. EFSA Sets European Dietary Reference Values for Nutrient Intakes. Available online: https://www.efsa.europa.eu/en/press/news/nda100326 (accessed on 20 June 2021).
26. European Food Safety Authority. Scientific Opinion on Dietary Reference Values for protein. *EFSA J.* **2012**, *10*, 2557. [CrossRef]
27. Shai, I.; Schwarzfuchs, D.; Henkin, Y.; Shahar, D.R.; Witkow, S.; Greenberg, I.; Golan, R.; Fraser, D.; Bolotin, A.; Vardi, H.; et al. Weight loss with a low-carbohydrate, Mediterranean, or low-fat diet. *N. Engl. J. Med.* **2008**, *359*, 229–241. [CrossRef] [PubMed]
28. Sacks, F.M.; Bray, G.A.; Carey, V.J.; Smith, S.R.; Ryan, D.H.; Anton, S.D.; McManus, K.; Champagne, C.M.; Bishop, L.M.; Laranjo, N.; et al. Comparison of weight-loss diets with different compositions of fat, protein, and carbohydrates. *N. Engl. J. Med.* **2009**, *360*, 859–873. [CrossRef] [PubMed]
29. Seidelmann, S.B.; Claggett, B.; Cheng, S.; Henglin, M.; Shah, A.; Steffen, L.M.; Folsom, A.R.; Rimm, E.B.; Willett, W.C.; Solomon, S.D. Dietary carbohydrate intake and mortality: A prospective cohort study and meta-analysis. *Lancet Public Health* **2018**, *3*, e419–e428. [CrossRef]
30. Gjuladin-Hellon, T.; Davies, I.G.; Penson, P.; Amiri Baghbadorani, R. Effects of carbohydrate-restricted diets on low-density lipoprotein cholesterol levels in overweight and obese adults: A systematic review and meta-analysis. *Nutr. Rev.* **2019**, *77*, 161–180. [CrossRef] [PubMed]
31. Austin, G.L.; Ogden, L.G.; Hill, J.O. Trends in carbohydrate, fat, and protein intakes and association with energy intake in normal-weight, overweight, and obese individuals: 1971-2006. *Am. J. Clin. Nutr.* **2011**, *93*, 836–843. [CrossRef] [PubMed]
32. Manheimer, E.W.; van Zuuren, E.J.; Fedorowicz, Z.; Pijl, H. Paleolithic nutrition for metabolic syndrome: Systematic review and meta-analysis. *Am. J. Clin. Nutr.* **2015**, *102*, 922–932. [CrossRef]
33. Mellberg, C.; Sandberg, S.; Ryberg, M.; Eriksson, M.; Brage, S.; Larsson, C.; Olsson, T.; Lindahl, B. Long-term effects of a Palaeolithic-type diet in obese postmenopausal women: A 2-year randomized trial. *Eur. J. Clin. Nutr.* **2014**, *68*, 350–357. [CrossRef]
34. Johnston, B.C.; Kanters, S.; Bandayrel, K.; Wu, P.; Naji, F.; Siemieniuk, R.A.; Ball, G.D.C.; Busse, J.W.; Thorlund, K.; Guyatt, G.; et al. Comparison of weight loss among named diet programs in overweight and obese adults: A meta-analysis. *JAMA* **2014**, *312*, 923–933. [CrossRef]

35. Gardner, C.D.; Trepanowski, J.F.; Del Gobbo, L.C.; Hauser, M.E.; Rigdon, J.; Ioannidis, J.P.A.; Desai, M.; King, A.C. Effect of Low-Fat vs Low-Carbohydrate Diet on 12-Month Weight Loss in Overweight Adults and the Association With Genotype Pattern or Insulin Secretion: The DIETFITS Randomized Clinical Trial. *JAMA* **2018**, *319*, 667–679. [CrossRef] [PubMed]
36. Churuangsuk, C.; Kherouf, M.; Combet, E.; Lean, M. Low-carbohydrate diets for overweight and obesity: A systematic review of the systematic reviews. *Obes. Rev.* **2018**, *19*, 1700–1718. [CrossRef] [PubMed]
37. Jenkins, D.J.A.; Wong, J.M.W.; Kendall, C.W.C.; Esfahani, A.; Ng, V.W.Y.; Leong, T.C.K.; Faulkner, D.A.; Vidgen, E.; Greaves, K.A.; Paul, G.; et al. The effect of a plant-based low-carbohydrate ("Eco-Atkins") diet on body weight and blood lipid concentrations in hyperlipidemic subjects. *Arch. Intern. Med.* **2009**, *169*, 1046–1054. [CrossRef] [PubMed]
38. Jenkins, D.J.A.; Wong, J.M.W.; Kendall, C.W.C.; Esfahani, A.; Ng, V.W.Y.; Leong, T.C.K.; Faulkner, D.A.; Vidgen, E.; Paul, G.; Mukherjea, R.; et al. Effect of a 6-month vegan low-carbohydrate ('Eco-Atkins') diet on cardiovascular risk factors and body weight in hyperlipidaemic adults: A randomised controlled trial. *BMJ Open* **2014**, *4*, e003505. [CrossRef] [PubMed]
39. Esposito, K.; Kastorini, C.-M.; Panagiotakos, D.B.; Giugliano, D. Mediterranean diet and weight loss: Meta-analysis of randomized controlled trials. *Metab. Syndr. Relat. Disord.* **2011**, *9*, 1–12. [CrossRef] [PubMed]
40. Leslie, W.S.; Taylor, R.; Harris, L.; Lean, M.E.J. Weight losses with low-energy formula diets in obese patients with and without type 2 diabetes: Systematic review and meta-analysis. *Int. J. Obes.* **2016**, *41*, 96–101. [CrossRef] [PubMed]
41. Astbury, N.M.; Piernas, C.; Hartmann-Boyce, J.; Lapworth, S.; Aveyard, P.; Jebb, S.A. A systematic review and meta-analysis of the effectiveness of meal replacements for weight loss. *Obes. Rev.* **2019**, *20*, 569–587. [CrossRef]
42. Lean, M.E.J.; Leslie, W.S.; Barnes, A.C.; Brosnahan, N.; Thom, G.; McCombie, L.; Peters, C.; Zhyzhneuskaya, S.; Al-Mrabeh, A.; Hollingsworth, K.G.; et al. Primary care-led weight management for remission of type 2 diabetes (DiRECT): An open-label, cluster-randomised trial. *Lancet* **2018**, *391*, 541–551. [CrossRef]
43. Lean, M.E.J.; Leslie, W.S.; Barnes, A.C.; Brosnahan, N.; Thom, G.; McCombie, L.; Peters, C.; Zhyzhneuskaya, S.; Al-Mrabeh, A.; Hollingsworth, K.G.; et al. Durability of a primary care-led weight-management intervention for remission of type 2 diabetes: 2-year results of the DiRECT open-label, cluster-randomised trial. *Lancet Diabetes Endocrinol.* **2019**, *7*, 344–355. [CrossRef]
44. Conley, M.; Le Fevre, L.; Haywood, C.; Proietto, J. Is two days of intermittent energy restriction per week a feasible weight loss approach in obese males? A randomised pilot study. *Nutr. Diet.* **2018**, *75*, 65–72. [CrossRef]
45. Trepanowski, J.F.; Kroeger, C.M.; Barnosky, A.; Klempel, M.C.; Bhutani, S.; Hoddy, K.K.; Gabel, K.; Freels, S.; Rigdon, J.; Rood, J.; et al. Effect of Alternate-Day Fasting on Weight Loss, Weight Maintenance, and Cardioprotection Among Metabolically Healthy Obese Adults: A Randomized Clinical Trial. *JAMA Intern. Med.* **2017**, *177*, 930–938. [CrossRef] [PubMed]
46. Cioffi, I.; Evangelista, A.; Ponzo, V.; Ciccone, G.; Soldati, L.; Santarpia, L.; Contaldo, F.; Pasanisi, F.; Ghigo, E.; Bo, S. Intermittent versus continuous energy restriction on weight loss and cardiometabolic outcomes: A systematic review and meta-analysis of randomized controlled trials. *J. Transl. Med.* **2018**, *16*, 371. [CrossRef]
47. Harvie, M.N.; Pegington, M.; Mattson, M.P.; Frystyk, J.; Dillon, B.; Evans, G.; Cuzick, J.; Jebb, S.A.; Martin, B.; Cutler, R.G.; et al. The effects of intermittent or continuous energy restriction on weight loss and metabolic disease risk markers: A randomized trial in young overweight women. *Int. J. Obes.* **2011**, *35*, 714–727. [CrossRef] [PubMed]
48. Sundfør, T.M.; Svendsen, M.; Tonstad, S. Effect of intermittent versus continuous energy restriction on weight loss, maintenance and cardiometabolic risk: A randomized 1-year trial. *Nutr. Metab. Cardiovasc. Dis.* **2018**, *28*, 698–706. [CrossRef]
49. Harris, L.; Hamilton, S.; Azevedo, L.B.; Olajide, J.; de Brún, C.; Waller, G.; Whittaker, V.; Sharp, T.; Lean, M.; Hankey, C.; et al. Intermittent fasting interventions for treatment of overweight and obesity in adults: A systematic review and meta-analysis. *JBI Database System. Rev. Implement. Rep.* **2018**, *16*, 507–547. [CrossRef]
50. Allaf, M.; Elghazaly, H.; Mohamed, O.G.; Fareen, M.F.K.; Zaman, S.; Salmasi, A.-M.; Tsilidis, K.; Dehghan, A. Intermittent fasting for the prevention of cardiovascular disease. *Cochrane Database Syst. Rev.* **2021**, *1*, CD013496. [CrossRef] [PubMed]
51. Templeman, I.; Smith, H.A.; Chowdhury, E.; Chen, Y.-C.; Carroll, H.; Johnson-Bonson, D.; Hengist, A.; Smith, R.; Creighton, J.; Clayton, D.; et al. A randomized controlled trial to isolate the effects of fasting and energy restriction on weight loss and metabolic health in lean adults. *Sci. Transl. Med.* **2021**, *13*, eabd8034. [CrossRef] [PubMed]
52. Lowe, D.A.; Wu, N.; Rohdin-Bibby, L.; Moore, A.H.; Kelly, N.; Liu, Y.E.; Philip, E.; Vittinghoff, E.; Heymsfield, S.B.; Olgin, J.E.; et al. Effects of Time-Restricted Eating on Weight Loss and Other Metabolic Parameters in Women and Men With Overweight and Obesity: The TREAT Randomized Clinical Trial. *JAMA Intern. Med.* **2020**, *180*, 1491–1499. [CrossRef]
53. Jamshed, H.; Beyl, R.A.; Della Manna, D.L.; Yang, E.S.; Ravussin, E.; Peterson, C.M. Early Time-Restricted Feeding Improves 24-h Glucose Levels and Affects Markers of the Circadian Clock, Aging, and Autophagy in Humans. *Nutrients* **2019**, *11*, 1234. [CrossRef]
54. Sutton, E.F.; Beyl, R.; Early, K.S.; Cefalu, W.T.; Ravussin, E.; Peterson, C.M. Early Time-Restricted Feeding Improves Insulin Sensitivity, Blood Pressure, and Oxidative Stress Even without Weight Loss in Men with Prediabetes. *Cell Metab.* **2018**, *27*, 1212–1221.e3. [CrossRef] [PubMed]
55. Nas, A.; Mirza, N.; Hägele, F.; Kahlhöfer, J.; Keller, J.; Rising, R.; Kufer, T.A.; Bosy-Westphal, A. Impact of breakfast skipping compared with dinner skipping on regulation of energy balance and metabolic risk. *Am. J. Clin. Nutr.* **2017**, *105*, 1351–1361. [CrossRef] [PubMed]
56. Mekary, R.A.; Giovannucci, E.; Willett, W.C.; van Dam, R.M.; Hu, F.B. Eating patterns and type 2 diabetes risk in men: Breakfast omission, eating frequency, and snacking. *Am. J. Clin. Nutr.* **2012**, *95*, 1182–1189. [CrossRef]

57. Zeevi, D.; Korem, T.; Zmora, N.; Israeli, D.; Rothschild, D.; Weinberger, A.; Ben-Yacov, O.; Lador, D.; Avnit-Sagi, T.; Lotan-Pompan, M.; et al. Personalized Nutrition by Prediction of Glycemic Responses. *Cell* **2015**, *163*, 1079–1094. [CrossRef]
58. Loos, R.J. The genetics of adiposity. *Curr. Opin. Genet. Dev.* **2018**, *50*, 86–95. [CrossRef]
59. Holzapfel, C.; Dawczynski, C.; Henze, A.; Simon, M.C. Personalized dietary recommendations for weight loss. A scientific perspective from various angles. *Ernahr. Umsch.* **2021**, *68*, 26–35.
60. Claussnitzer, M.; Dankel, S.N.; Kim, K.-H.; Quon, G.; Meuleman, W.; Haugen, C.; Glunk, V.; Sousa, I.S.; Beaudry, J.L.; Puviindran, V.; et al. FTO Obesity Variant Circuitry and Adipocyte Browning in Humans. *N. Engl. J. Med.* **2015**, *373*, 895–907. [CrossRef]
61. Holzapfel, C.; Sag, S.; Graf-Schindler, J.; Fischer, M.; Drabsch, T.; Illig, T.; Grallert, H.; Stecher, L.; Strack, C.; Caterson, I.D.; et al. Association between Single Nucleotide Polymorphisms and Weight Reduction in Behavioural Interventions-A Pooled Analysis. *Nutrients* **2021**, *13*, 819. [CrossRef] [PubMed]
62. Celis-Morales, C.; Marsaux, C.F.; Livingstone, K.M.; Navas-Carretero, S.; San-Cristobal, R.; Fallaize, R.; Macready, A.L.; O'Donovan, C.; Woolhead, C.; Forster, H.; et al. Can genetic-based advice help you lose weight? Findings from the Food4Me European randomized controlled trial. *Am. J. Clin. Nutr.* **2017**, *105*, 1204–1213. [CrossRef]
63. Ellis, A.; Rozga, M.; Braakhuis, A.; Monnard, C.R.; Robinson, K.; Sinley, R.; Wanner, A.; Vargas, A.J. Effect of Incorporating Genetic Testing Results into Nutrition Counseling and Care on Health Outcomes: An Evidence Analysis Center Systematic Review-Part II. *J. Acad. Nutr. Diet.* **2021**, *121*, 582–605.e17. [CrossRef]
64. Camp, K.M.; Trujillo, E. Position of the Academy of Nutrition and Dietetics: Nutritional genomics. *J. Acad. Nutr. Diet.* **2014**, *114*, 299–312. [CrossRef] [PubMed]
65. Drabsch, T.; Holzapfel, C. A Scientific Perspective of Personalised Gene-Based Dietary Recommendations for Weight Management. *Nutrients* **2019**, *11*, 617. [CrossRef]
66. Bayer, S.; Winkler, V.; Hauner, H.; Holzapfel, C. Associations between Genotype-Diet Interactions and Weight Loss-A Systematic Review. *Nutrients* **2020**, *12*, 2891. [CrossRef] [PubMed]
67. Khera, A.V.; Chaffin, M.; Aragam, K.G.; Haas, M.E.; Roselli, C.; Choi, S.H.; Natarajan, P.; Lander, E.S.; Lubitz, S.A.; Ellinor, P.T.; et al. Genome-wide polygenic scores for common diseases identify individuals with risk equivalent to monogenic mutations. *Nat. Genet.* **2018**, *50*, 1219–1224. [CrossRef] [PubMed]
68. Khera, A.V.; Chaffin, M.; Wade, K.H.; Zahid, S.; Brancale, J.; Xia, R.; Distefano, M.; Senol-Cosar, O.; Haas, M.E.; Bick, A.; et al. Polygenic Prediction of Weight and Obesity Trajectories from Birth to Adulthood. *Cell* **2019**, *177*, 587–596.e9. [CrossRef]
69. Kolodziejczyk, A.A.; Zheng, D.; Elinav, E. Diet-microbiota interactions and personalized nutrition. *Nat. Rev. Microbiol.* **2019**, *17*, 742–753. [CrossRef]
70. Magni, P.; Bier, D.M.; Pecorelli, S.; Agostoni, C.; Astrup, A.; Brighenti, F.; Cook, R.; Folco, E.; Fontana, L.; Gibson, R.A.; et al. Perspective: Improving Nutritional Guidelines for Sustainable Health Policies: Current Status and Perspectives. *Adv. Nutr.* **2017**, *8*, 532–545. [CrossRef]
71. von Schwartzenberg, R.J.; Bisanz, J.E.; Lyalina, S.; Spanogiannopoulos, P.; Ang, Q.Y.; Cai, J.; Dickmann, S.; Friedrich, M.; Liu, S.-Y.; Collins, S.L.; et al. Caloric restriction disrupts the microbiota and colonization resistance. *Nature* **2021**, *595*, 272–277. [CrossRef]
72. Breuninger, T.A.; Wawro, N.; Breuninger, J.; Reitmeier, S.; Clavel, T.; Six-Merker, J.; Pestoni, G.; Rohrmann, S.; Rathmann, W.; Peters, A.; et al. Associations between habitual diet, metabolic disease, and the gut microbiota using latent Dirichlet allocation. *Microbiome* **2021**, *9*, 61. [CrossRef]
73. Cani, P.D.; van Hul, M.; Lefort, C.; Depommier, C.; Rastelli, M.; Everard, A. Microbial regulation of organismal energy homeostasis. *Nat. Metab.* **2019**, *1*, 34–46. [CrossRef]
74. DGAC. Scientific Report of the 2015 Dietary Guidelines Advisory Committee. Available online: https://health.gov/sites/default/files/2019-09/Scientific-Report-of-the-2015-Dietary-Guidelines-Advisory-Committee.pdf (accessed on 20 June 2021).
75. Jebb, S.A.; Ahern, A.L.; Olson, A.D.; Aston, L.M.; Holzapfel, C.; Stoll, J.; Amann-Gassner, U.; Simpson, A.E.; Fuller, N.R.; Pearson, S.; et al. Primary care referral to a commercial provider for weight loss treatment versus standard care: A randomised controlled trial. *Lancet* **2011**, *378*, 1485–1492. [CrossRef]
76. Holzapfel, C.; Cresswell, L.; Ahern, A.L.; Fuller, N.R.; Eberhard, M.; Stoll, J.; Mander, A.P.; Jebb, S.A.; Caterson, I.D.; Hauner, H. The challenge of a 2-year follow-up after intervention for weight loss in primary care. *Int. J. Obes.* **2014**, *38*, 806–811. [CrossRef]
77. Johnston, C.A.; Rost, S.; Miller-Kovach, K.; Moreno, J.P.; Foreyt, J.P. A randomized controlled trial of a community-based behavioral counseling program. *Am. J. Med.* **2013**, *126*, 1143.e19–1143.e24. [CrossRef]
78. Ard, J.D.; Lewis, K.H.; Rothberg, A.; Auriemma, A.; Coburn, S.L.; Cohen, S.S.; Loper, J.; Matarese, L.; Pories, W.J.; Periman, S. Effectiveness of a Total Meal Replacement Program (OPTIFAST Program) on Weight Loss: Results from the OPTIWIN Study. *Obesity* **2019**, *27*, 22–29. [CrossRef] [PubMed]
79. Harvey, J.; Krukowski, R.; Priest, J.; West, D. Log Often, Lose More: Electronic Dietary Self-Monitoring for Weight Loss. *Obesity* **2019**, *27*, 380–384. [CrossRef]
80. Jensen, M.D.; Ryan, D.H.; Apovian, C.M.; Ard, J.D.; Comuzzie, A.G.; Donato, K.A.; Hu, F.B.; van Hubbard, S.; Jakicic, J.M.; Kushner, R.F.; et al. 2013 AHA/ACC/TOS Guideline for the Management of Overweight and Obesity in Adults. *Circulation* **2014**, *129*, S102–S138. [CrossRef]
81. Patel, M.L.; Wakayama, L.N.; Bennett, G.G. Self-Monitoring via Digital Health in Weight Loss Interventions: A Systematic Review Among Adults with Overweight or Obesity. *Obesity* **2021**, *29*, 478–499. [CrossRef]

82. Michie, S.; Abraham, C.; Whittington, C.; McAteer, J.; Gupta, S. Effective techniques in healthy eating and physical activity interventions: A meta-regression. *Health Psychol.* **2009**, *28*, 690–701. [CrossRef]
83. Harkin, B.; Webb, T.L.; Chang, B.P.I.; Prestwich, A.; Conner, M.; Kellar, I.; Benn, Y.; Sheeran, P. Does monitoring goal progress promote goal attainment? A meta-analysis of the experimental evidence. *Psychol. Bull.* **2016**, *142*, 198–229. [CrossRef] [PubMed]
84. Spring, B.; Champion, K.E.; Acabchuk, R.; Hennessy, E.A. Self-regulatory behaviour change techniques in interventions to promote healthy eating, physical activity, or weight loss: A meta-review. *Health Psychol. Rev.* **2021**, *15*, 508–539. [CrossRef] [PubMed]
85. Carter, M.C.; Burley, V.J.; Nykjaer, C.; Cade, J.E. Adherence to a smartphone application for weight loss compared to website and paper diary: Pilot randomized controlled trial. *J. Med. Internet Res.* **2013**, *15*, e32. [CrossRef]
86. Stubbs, R.J.; Duarte, C.; Palmeira, A.L.; Sniehotta, F.F.; Horgan, G.; Larsen, S.C.; Marques, M.M.; Evans, E.H.; Ermes, M.; Harjumaa, M.; et al. Evidence-Based Digital Tools for Weight Loss Maintenance: The NoHoW Project. *Obes. Facts* **2021**, *14*, 320–333. [CrossRef] [PubMed]
87. Turner-McGrievy, G.; Tate, D. Tweets, Apps, and Pods: Results of the 6-month Mobile Pounds Off Digitally (Mobile POD) randomized weight-loss intervention among adults. *J. Med. Internet Res.* **2011**, *13*, e120. [CrossRef] [PubMed]
88. Wharton, C.M.; Johnston, C.S.; Cunningham, B.K.; Sterner, D. Dietary self-monitoring, but not dietary quality, improves with use of smartphone app technology in an 8-week weight loss trial. *J. Nutr. Educ. Behav.* **2014**, *46*, 440–444. [CrossRef]
89. Villinger, K.; Wahl, D.R.; Boeing, H.; Schupp, H.T.; Renner, B. The effectiveness of app-based mobile interventions on nutrition behaviours and nutrition-related health outcomes: A systematic review and meta-analysis. *Obes. Rev.* **2019**, *20*, 1465–1484. [CrossRef]
90. Roess, A. The Promise, Growth, and Reality of Mobile Health—Another Data-free Zone. *N. Engl. J. Med.* **2017**, *377*, 2010–2011. [CrossRef] [PubMed]
91. Svetkey, L.P.; Batch, B.C.; Lin, P.-H.; Intille, S.S.; Corsino, L.; Tyson, C.C.; Bosworth, H.B.; Grambow, S.C.; Voils, C.; Loria, C.; et al. Cell phone intervention for you (CITY): A randomized, controlled trial of behavioral weight loss intervention for young adults using mobile technology. *Obesity* **2015**, *23*, 2133–2141. [CrossRef] [PubMed]
92. Bardus, M.; van Beurden, S.B.; Smith, J.R.; Abraham, C. A review and content analysis of engagement, functionality, aesthetics, information quality, and change techniques in the most popular commercial apps for weight management. *Int. J. Behav. Nutr. Phys. Act.* **2016**, *13*, 35. [CrossRef]
93. Rivera, J.; McPherson, A.; Hamilton, J.; Birken, C.; Coons, M.; Iyer, S.; Agarwal, A.; Lalloo, C.; Stinson, J. Mobile Apps for Weight Management: A Scoping Review. *JMIR Mhealth Uhealth* **2016**, *4*, e87. [CrossRef]
94. Pilitsi, E.; Farr, O.M.; Polyzos, S.A.; Perakakis, N.; Nolen-Doerr, E.; Papathanasiou, A.-E.; Mantzoros, C.S. Pharmacotherapy of obesity: Available medications and drugs under investigation. *Metabolism* **2019**, *92*, 170–192. [CrossRef]
95. Daneschvar, H.L.; Aronson, M.D.; Smetana, G.W. FDA-Approved Anti-Obesity Drugs in the United States. *Am. J. Med.* **2016**, *129*, 879.e1–879.e6. [CrossRef] [PubMed]
96. Tak, Y.J.; Lee, S.Y. Anti-Obesity Drugs: Long-Term Efficacy and Safety: An Updated Review. *World J. Mens. Health* **2021**, *39*, 208–221. [CrossRef]
97. Apovian, C.M.; Aronne, L.J.; Bessesen, D.H.; McDonnell, M.E.; Murad, M.H.; Pagotto, U.; Ryan, D.H.; Still, C.D. Pharmacological management of obesity: An endocrine Society clinical practice guideline. *J. Clin. Endocrinol. Metab.* **2015**, *100*, 342–362. [CrossRef] [PubMed]
98. May, M.; Schindler, C.; Engeli, S. Modern pharmacological treatment of obese patients. *Ther. Adv. Endocrinol. Metab.* **2020**, *11*, 2042018819897527. [CrossRef] [PubMed]
99. Khera, R.; Murad, M.H.; Chandar, A.K.; Dulai, P.S.; Wang, Z.; Prokop, L.J.; Loomba, R.; Camilleri, M.; Singh, S. Association of Pharmacological Treatments for Obesity With Weight Loss and Adverse Events: A Systematic Review and Meta-analysis. *JAMA* **2016**, *315*, 2424–2434. [CrossRef] [PubMed]
100. Welbourn, R.; Hollyman, M.; Kinsman, R.; Dixon, J.; Liem, R.; Ottosson, J.; Ramos, A.; Våge, V.; Al-Sabah, S.; Brown, W.; et al. Bariatric Surgery Worldwide: Baseline Demographic Description and One-Year Outcomes from the Fourth IFSO Global Registry Report 2018. *Obes. Surg.* **2019**, *29*, 782–795. [CrossRef]
101. Puzziferri, N.; Roshek, T.B.; Mayo, H.G.; Gallagher, R.; Belle, S.H.; Livingston, E.H. Long-term follow-up after bariatric surgery: A systematic review. *JAMA* **2014**, *312*, 934–942. [CrossRef]
102. O'Brien, P.E.; Hindle, A.; Brennan, L.; Skinner, S.; Burton, P.; Smith, A.; Crosthwaite, G.; Brown, W. Long-Term Outcomes After Bariatric Surgery: A Systematic Review and Meta-analysis of Weight Loss at 10 or More Years for All Bariatric Procedures and a Single-Centre Review of 20-Year Outcomes After Adjustable Gastric Banding. *Obes. Surg.* **2019**, *29*, 3–14. [CrossRef]
103. Toussi, R.; Fujioka, K.; Coleman, K.J. Pre- and postsurgery behavioral compliance, patient health, and postbariatric surgical weight loss. *Obesity* **2009**, *17*, 996–1002. [CrossRef]
104. Benalcazar, D.A.; Cascella, M. *StatPearls: Obesity Surgery Pre-Op Assessment And Preparation*; StatPearls Publishing: Florida, FL, USA, 2021.

105. Lupoli, R.; Lembo, E.; Saldalamacchia, G.; Avola, C.K.; Angrisani, L.; Capaldo, B. Bariatric surgery and long-term nutritional issues. *World J. Diabetes* **2017**, *8*, 464–474. [CrossRef]
106. Di Lorenzo, N.; Antoniou, S.A.; Batterham, R.L.; Busetto, L.; Godoroja, D.; Iossa, A.; Carrano, F.M.; Agresta, F.; Alarçon, I.; Azran, C.; et al. Clinical practice guidelines of the European Association for Endoscopic Surgery (EAES) on bariatric surgery: Update 2020 endorsed by IFSO-EC, EASO and ESPCOP. *Surg. Endosc.* **2020**, *34*, 2332–2358. [CrossRef] [PubMed]

nutrients MDPI

Review

Endovascular Bariatric Surgery as Novel Minimally Invasive Technique for Weight Management in the Morbidly Obese: Review of the Literatur

Giuseppe Massimo Sangiorgi [1,*], Alberto Cereda [2], Nicola Porchetta [1], Daniela Benedetto [1], Andrea Matteucci [1], Michela Bonanni [1], Gaetano Chiricolo [1] and Antonino De Lorenzo [1]

[1] Department of Biomedicine and Prevention, Institute of Cardiology, Cardiac Cath Lab, University of Rome Tor Vergata, 00133 Rome, Italy; nicola.porchetta@gmail.com (N.P.); dania.benedetto@gmail.com (D.B.); andrea.matteucci2@gmail.com (A.M.); michelabonanni91@gmail.com (M.B.); gaetano.chiricolo@ptvonline.it (G.C.); delorenzo@uniroma2.it (A.D.L.)

[2] Department of Cardiology, Cardiac Cath Lab, San Gaudenzio Clinic, 28100 Novara, Italy; tskcer@hotmail.it

* Correspondence: gsangiorgi@gmail.com

Abstract: Nowadays, obesity represents one of the most unresolved global pandemics, posing a critical health issue in developed countries. According to the World Health Organization, its prevalence has tripled since 1975, reaching a prevalence of 13% of the world population in 2016. Indeed, as obesity increases worldwide, novel strategies to fight this condition are of the utmost importance to reduce obese-related morbidity and overall mortality related to its complications. Early experimental and initial clinical data have suggested that endovascular bariatric surgery (EBS) may be a promising technique to reduce weight and hormonal imbalance in the obese population. Compared to open bariatric surgery and minimally invasive surgery (MIS), EBS is much less invasive, well tolerated, with a shorter recovery time, and is probably cost-saving. However, there are still several technical aspects to investigate before EBS can be routinely offered to all obese patients. Further prospective studies and eventually a randomized trial comparing open bariatric surgery vs. EBS are needed, powered for clinically relevant outcomes, and with adequate follow-up. Yet, EBS may already appear as an appealing alternative treatment for weight management and cardiovascular prevention in morbidly obese patients at high surgical risk.

Keywords: bariatric surgery; cardiovascular disease; endovascular bariatric surgery; obesity; prevention

Citation: Sangiorgi, G.M.; Cereda, A.; Porchetta, N.; Benedetto, D.; Matteucci, A.; Bonanni, M.; Chiricolo, G.; De Lorenzo, A. Endovascular Bariatric Surgery as Novel Minimally Invasive Technique for Weight Management in the Morbidly Obese: Review of the Literature. *Nutrients* **2021**, *13*, 2541. https://doi.org/10.3390/nu13082541

Academic Editor: Ina Bergheim

Received: 30 May 2021
Accepted: 21 July 2021
Published: 25 July 2021

1. Introduction

Nowadays, obesity represents one of the most unresolved global pandemics, posing a critical health issue in developed countries. According to the World Health Organization, its prevalence has tripled since 1975, reaching 13% of the world population in 2016. Obesity is defined as a body mass index (BMI) greater than 30 kg/m^2, while morbid obesity is defined as BMI > 40 kg/m^2. Numerous comorbidities such as major stroke, acute myocardial infarction, hypertension, type 2 diabetes, hyperlipidemia, obstructive sleep apnea, and all-cause mortality are strongly associated with this disease [1]. As a result, approximately 2.8 million deaths per year may occur in adult populations affected by obesity.

Given the tremendous impact on public health with an approximate attributable cost of nearly 150 billion dollars per year in the United States, obesity treatments have become of critical importance for the healthcare system, the medical community, and policymakers [2].

The cornerstone of obesity treatment is represented by behavioral modifications (i.e., diet and physical exercise), ideally in a highly motivated patient that should be followed by a multidisciplinary team of healthcare professionals. If successful, this strategy consents modest and durable weight loss reduction of 5% to 10% [3]. The long-term efficacy of all behavioral therapies is limited. For those unable to reach these goals, few drugs (orlistat, lorcaserin, phentermine/topiramate, nartrexone/bupropion, semaglutide and liraglutide) are available as adjuvant therapy, but in general, are not free of side effects, usually dose-dependent, have limited adherence (frequently due to arbitrary withdrawal of the drug), and with suboptimal outcome in obtaining the goal of weight-reduction. Recently, non-surgical endoscopic bariatric therapies such as intragastric balloons, endoscopic gastric plication, and endoluminal duodenal-jejunal sleeve have been implemented in patients not willing to undergo conventional bariatric surgery. However, potentially severe complications have been reported with these techniques (gastric perforation, bowel obstruction, and gastrointestinal bleeding), and for these reasons they are currently performed only in highly experienced centers.

Surgical approaches, mainly represented by Roux-en-Y gastric bypass, adjustable gastric banding, sleeve gastrectomy, and biliopancreatic division (Figure 1) are reserved to morbidly obese individuals or obese individuals with one or more obesity-related comorbidities (or even lower for uncontrolled diabetes) who have not been able to reach the aforementioned goals with behavioral modifications and drug therapy.

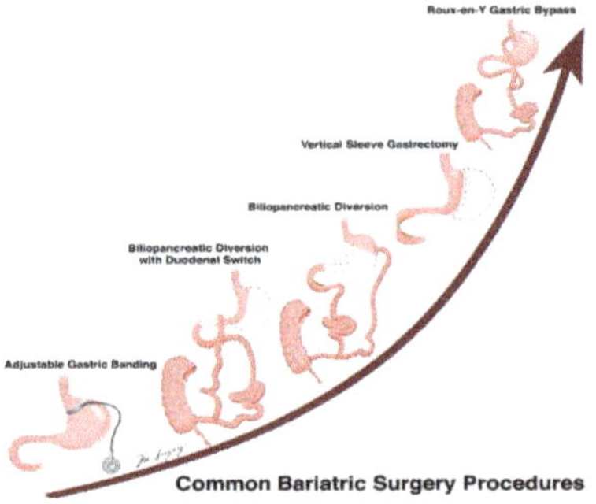

Figure 1. Overview of different bariatric surgery options. The most effective options in weight reduction are the Roux-en-Y gastric bypass and sleeve gastrectomy. Arrow indicates an increase in efficacy.

Well-known short- and long-term complications, even if uncommon, including bleeding, infections, deep venous thrombosis, gastric dumping syndrome, and internal hernia have been reported with different surgical techniques.

Another frequent eventuality is post-bariatric surgery anemia: it is in most cases due to iron deficiency, along with vitamin B12 deficiency as a secondary cause. Iron deficiency is expressed by low serum ferritin and it occurs because of its lower absorption secondary to hypocloridria and the bypassing of the duodenum and proximal jejunum [4]. In addition to anemia, vitamin B12 deficiency (resulting from inadequate secretion of intrinsic factor, limited gastric acidity and the bypassing of the duodenum) can lead to neurological disorders [4]. In the absence of adequate vitamin B12 supplement, up to 30% of patients will be unable to maintain normal levels of plasma B12 at 1 year [5].

Bariatric surgery results in calcium/vitamin D malabsorption (results from bypassing the duodenum and proximal jejunum) with secondary hyperparathyroidism, changes in fat mass and alterations in fat- and gut-derived hormone; the final effect is an accelerated bone loss [4]. Patients affected by secondary hyperparathyroidism should obtain bone benefits from oral supplementation of vitamin D [5,6]. In fact, the European Association for Endoscopic Surgery (EAES) Clinical practice guidelines on bariatric surgery strongly recommend vitamin D supplement post-surgery because the anticipated benefits outweigh the potential risks of vitamin therapy [7].

Poor protein digestion and absorption, secondary to altered biliary and pancreatic function, is involve in protein malnutrition and can be observed after bariatric surgery [4]; albumin levels can be considered as marker of protein deficiency [6].

Low serum levels of fat-soluble vitamins (vitamin A, K and E) usually occur after bariatric procedure [4,5].

As a "treatment gap" option, a novel minimally invasive procedure, i.e., endovascular embolization of the gastric fundus arterial supply (endovascular bariatric surgery -EBS), has been proposed as a supplementary technique with successful results in animal models and a few recently published small clinical trials [8–13].

We report herein an updated review on the rational, pathophysiology, experimental, and clinical outcomes with this procedure in obese patients.

2. Pathophysiological Basis for EBS

Gastric fundus is mainly supplied by the left gastric artery (LGA) and sometimes by the gastroepiploic artery. The stomach has a neurohumoral role on hunger regulation through ghrelin: this is the rationale for gastric fundus embolization. Ghrelin is a ligand of the growth hormone secretagogue receptor (GHS-R) in neuropeptide Y (NPY) and agouti-related peptide (AgRP) in the arcuate nucleus of the hypothalamus with a downstream effect to inhibit the release of the α-melanocyte-stimulating hormone. Therefore, ghrelin acts to increase appetite and food intake, increasing weight gain (Figure 2). Practically, ghrelin plasma level rises sharply shortly before meals, which correlates with hunger sensation that occurs before consuming food. Conversely, ghrelin falls immediately after eating, which correlates with the sense of satiation after eating [14,15]. In addition, ghrelin downregulates anorexigenic hormone receptors for PYY, GLP-1, and cholecystokinin and reduces the sensitivity of gastric distension by selectively inhibition of gastric subpopulation of mechanically sensitive vagal afferent nerves [16,17]. Nearly 90% of body ghrelin is produced in the ghrelin-secreting cells of the stomach, mainly in the gastric fundus with fewer amounts produced in the small intestine, brain, and pancreas. Therefore, the main goal of the EBS procedure is to reduce ghrelin production by the stomach.

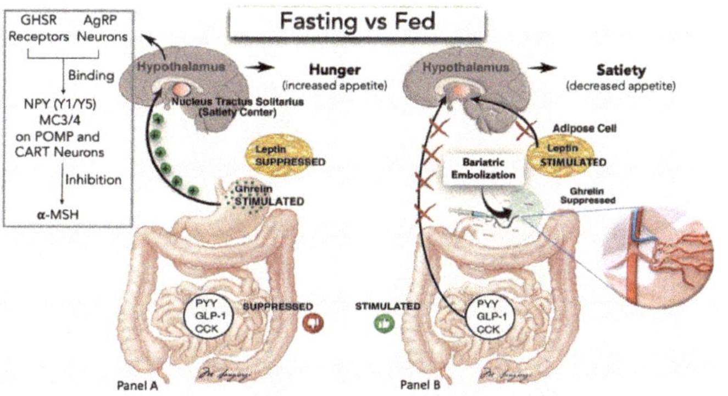

Figure 2. Hormonal changes and diagram of the ghrelin signal pathway. (**A**) The "hunger hormone" ghrelin is secreted by the gastric fundus, whereas peptide YY (PPY), cholecystokinin (CCK), and glucagon-like peptide (GLP-1) are secreted in the gut from L cells. Adipocytes produce leptin (LPT). In the fasting state, decreased food intake suppresses the release of PPY, GLP-1, CCK, and LPT, while stimulates ghrelin production from the stomach. Ghrelin binds in the hypothalamic arcuate nucleus to growth hormone secretagogue receptor (GHSR) in neuropeptide Y (NPY) and agouti-related peptide (AgRP) neu-rons. NPY and AgRP bind subsequently to NPY subtype 1 and 5 (NPH Y1/Y5) and melacortin-3 and-4 (MC3/4) receptors on proopiomelanocortin (POMP) and cocaine-amphetamine-regulated transcript neurons (CART), inhibiting the release of α- melanocyte-stimulating hormone (α-MSH). By inhibiting α-MSH, ghrelin acts to increase hunger and food intake. (**B**) BES procedure reducing ghrelin production in the stomach fundus area, mimics a fed state characterized by PYY, GLP-1, CCK, and LPT hormone increases. As a result, appetite decreases and an increase in the feeling of satiety occurs.

It has been previously reported in small retrospective studies [18,19] that patients with gastrointestinal bleeding treated with LGA embolization had significant weight loss after the procedure. Although these studies are difficult to interpret due to the small sample size, associated comorbidities, high-risk profile of the population included, and variable follow-up, they proved the concept and set the basis for gastric fundus embolization as a possible treatment for obesity.

3. EBS Procedure

The celiac trunk branches from the aorta at the level of the twelfth thoracic vertebra (T12). The LGA is the first and smaller branch of the celiac trunk, even if there are less common possibilities of independent origins from the aorta, splenic artery, common hepatic artery, gastroduodenal artery and superior mesenteric artery. It runs along the superior portion of the lesser gastric curvature and anastomoses with the right gastric artery that arises from the common hepatic artery. The left gastroepiploic artery (GEA) is the largest branch of the splenic artery and gives gastric branches to both surfaces of the stomach. It anastomoses with the right GEA that arises from the gastroduodenal artery (Figure 3). Normal anatomic variants are frequent and can be present in up to 30% of patients.

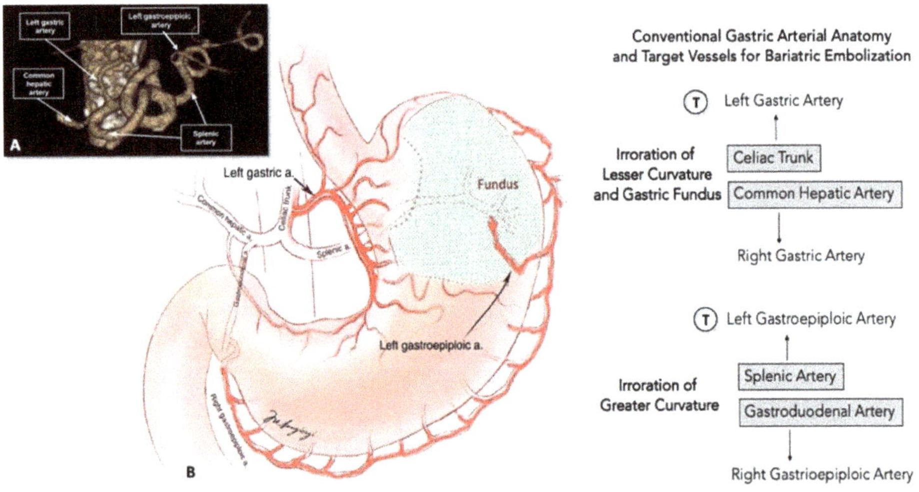

Figure 3. Voxel Gradient Angio-CT 3D reconstruction (**A**) and schematic drawn (**B**) of left gastric artery and left gastroepiploic artery normal anatomy. Most commonly the left gastric artery originates from the celiac trunk. Less frequently, the artery may arise directly from the aorta, splenic artery, common hepatic artery, and superior mesenteric artery. The superior part of the greater curvature of the stomach is supplied by the left gastroepiploic artery while the inferior part of the greater curvature by the right gastroepiploic artery. T indicates target for BES procedure.

Although there is no standard bariatric embolization procedure, similar approaches have been used in the majority of studies done so far. Through the femoral or radial artery, a selective digital subtraction angiography (usually in AP, LAO 60°, or LAO 90° projections) of the celiac trunk is performed to identify the LGA and other potentials embolization targets (all arteries supplying the gastric fundus and potential accessory gastric arteries); the decision to embolize other vessels, especially the gastroepiploic artery, is based on their contribution to fundal blood supply (assessed angiographically). Different types of diagnostic catheters can be used for this purpose (JR, Simons, Cobra, SOS, among others) depending on the operator's preference and experience. Cone beam CT has also been used by some authors for best determining fundal perfusion and the eventual need to embolize other arterial territories. After targets have been identified, selective cannulation and angiography using the same catheter or a microcatheter are performed. Embolic material choice has been variable throughout trials with 300–500 μm and 500–700 μm microspheres and 300–500 μm or 500–700 μm polyvinyl alcohol (PVA) particles used. LGA embolization can also be performed with an occlusion balloon microcatheter (OBC) advanced into the target artery over a standard guidewire: a subsequent balloon inflation at the OBC tip can be used to prevent retrograde reflux, with tip pressure/resistance monitored to prevent overembolization and antegrade reflux. Embolization is taken to stasis, which was defined as the visual absence of the flow of contrast after five heartbeats; postembolization DSA is usually acquired to confirm the success of embolization (Figure 4). Unfortunately there are no consensus statements to standardize this procedure.

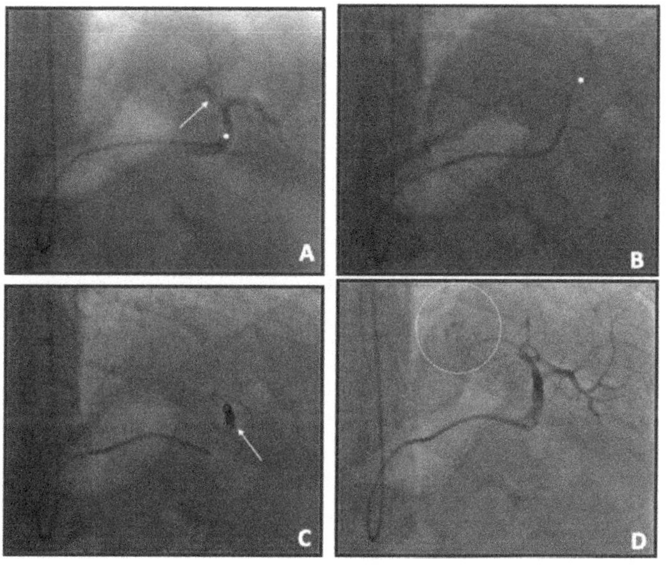

Figure 4. Case example of a left gastroepiploic artery (arrow) embolization in a 55 years old male with a BMI of 43.2. (**A**) Selective angiography by multipurpose 6F 125 cm into the splenic artery (asterisk). (**B**): A Rebar 0.27′ microcatheter (Medtronic, Santa Rosa, CA, USA—asterisk) was advanced into the multipurpose catheter selectively engaging the gastroepiploic artery. (**C**) Three contour spirals (Medtronic, Santa Rosa, CA, USA) of different size and length (4 × 40 mm; 4 × 40 mm; 5 × 30 mm) were subsequently released (arrow). (**D**) Final angiography with target fundus zone indicated by dashed circle.

4. EBS Preclinical Evidence

In 2007, Arepally and coauthors published the first experience of catheter embolization of the LGA to reduce systemic plasma levels of ghrelin in a swine model [20]. The study showed a reduction in ghrelin but no significant weight change. Micro-ulcers at the gastroesophageal junction were also observed in the euthanized animals. A follow-up study with a sham procedure in the control group evaluated ghrelin levels and natural weight gain in 4 weeks follow-up. A significant decrease in weight (7.8%) compared to the control group (15.1%) was shown [21].

In a canine model, Bawudun et al. [22] showed a significant decrease in plasma levels of ghrelin, abdominal subcutaneous fat, and body weight in Lipiodol-embolized (or a combination of Lipiodol and polyvinyl-alcohol) animals compared to control group. Peak effect was obtained between weeks three and four and compensatory rise in plasma levels of ghrelin was observed at seven weeks after embolization. According to the author, this result may suggest the compensatory production of ghrelin that occurred in the gastric fundus after the embolization.

Paxton and colleagues obtained 55% of ghrelin reduction associated with significant weight loss at eight weeks in a swine model using 40 μm microsphere as embolizing agent [23]. The same author evaluated by histology the gastric mucosa of swine after the procedure and demonstrated healed or healing ulcers in 50% of the gastric body of treated animals [24].

In 2016, to assess whether the number of fundal arteries embolized could impact the ghrelin reduction and gastric ulceration rate, the same authors utilized embolic microspheres into four arteries supplying the gastric fundus, vs. two arteries vs. one artery. Only the group of swine undergoing complete embolization demonstrated significant ghrelin reduction compared to sham control animals. Gastric ulcers were present in 50% of animals that embolized four arteries compared to 40% in the other groups [25].

In 2017, Kim et al. [26] performed EBS in five swine by selectively infusing 50–150 or 150–250 μm PVA particles into the gastric arteries while five animals were treated with saline infusion as a sham procedure. Endoscopy was performed three weeks after EBS to see whether any gastric ulcer occurred. Celiac trunk angiography was performed eight weeks after EBS [27]. No statistically significant differences in ghrelin levels were observed after eight weeks despite a reduction compared to the control group. In the embolized group, ulcerations were identified in three animals (60%). Re-canalization of the embolized arteries was identified on follow-up angiography in three animals (60%), respectively, suggesting that EBS procedure with PVA particles can transiently suppress ghrelin levels in embolized animals. However, ulcerations of gastric fundus and recanalization of the embolized arteries are present in the long-term follow-up.

A comparison of the studies involving animal models is reported in Table 1.

Table 1. Experimental studies evaluating BES procedure for weight reduction.

Author/Year/REF.	Animal Model	N of Animals	Follow-Up	Absolute Weight Change (%)	Absolute Ghrelin Change (%)
Arepally 2007 [20]	Swine	5 2 CTRL 4 LDSM 4 HDSM	4 weeks	8.6 increase 1.4 increase 1.4 increase	15% increase 130% increase 60% decrease
Arepally 2008 [21]	Swine	10 5 CTRL 5 SM	4 weeks	15.8 increase 7.8 increase	10.1 increase 12.9 decrease
Bawudun 2012 [22]	Dog	15 5 CTRL 5 EO 5 PVA	8 weeks	5 increase 5 decrease 10 decrease	13.6 increase 15.8 decrease 30.1 decrease
Paxton 2013 [23]	Swine	12 6 CTRL 6 MS	8 weeks	28 increase 10 increase	30 increase 30 decrease
Paxton 2014 [24]	Swine	12 6 CTRL 6 EO	8 weeks	Not Reported Not Reported	Not Reported Not Reported
Paxton 2016 [25]	Swine	22 6 CTRL 6 EA [8]5 EA [2]5 EA [1]	4–8 weeks	26.5 increase 8.7 increase 19.7 increase 23.0 increase	19.9 increase 29.9 decrease 18.2 increase 234 increase
Kim 2017 [26]	Swine	10 5 CTRL 5 PVA	2 mos	30% increase 27% increase	No change 40% decrease

CTRL = control; LDSM = low dose sodium morrhuate; HDSM = high dose sodium morrhuate; EO = ethiodized oil (Lipiodol); PVA= polyvinyl-alcohol; MS = microspheres; EA (*n*) = number of embolized arteries.

5. EBS Human Clinical Evidence

The available clinical evidence on bariatric embolization has been thoroughly described in previous reports [10,27,28]. The first human experience was published in 2013 by Kipshidze and colleagues [11] with its two years follow-up published in 2015. In this five-patient series (mean age 44.7 ± 7.4), the average weight decreased from 128.1 ± 24.4 to 106 ± 21 at two years. All procedures were performed through femoral access using 6F JR 4 catheters for angiography and Excelsior 1018 Microcatheter (Boston Scientific Corp., Cork, Ireland) for selective cannulation. Embolization was done with 300 to 500 μm embospheres (Biocompatibles UK Limited, Surrey, United Kingdom). There were no major procedural complications. Three of the five patients complained of abdominal discomfort after the procedure, all with unremarkable follow-up gastroscopies.

The GET LEAN trial [13] (Gastric Artery Embolization Trial for the Lessening of Appetite Nonsurgically) published in 2016 analyzed the safety and efficacy of LGA embolization at six months. With four patients included (mean age 41 years, range, 30–54), average weight loss was 9.2 kg (range 2.7–17.2) at six months follow-up. Ultrasound (US) guided right femoral or left radial access and 4–5 F Simmons 1 catheter for angiography were use and embolization was performed with 300 to 500 μm microspheres through a microcatheter. No major complications were reported. Three of the five patients complained of mild nausea, occasional vomiting, and mild epigastric discomfort immediately following the procedure that were resolved within 24 h for two patients and within 3–4 days for the other.

In another series published by Bai [8] in 2018 that included five patients with a mean age of 42.8 ± 13.9, weight loss was on average 12.9 ± 14.66 kg at nine months follow-up. Operators used US guided femoral access in all patients, 5F standard catheter for angiography, Progreat 2, 7 F microcatheter (Terumo) for selective cannulation, and embolization was done with 500–710 μm PVA particles (COOK Incorporated, Bloomington). No major complications were reported: four patients experienced slight epigastric discomfort in the first hours after the procedure, which resolved within 48 h. One patient developed a small ulceration (3 mm in length) below the cardia (grade II according to CTCAE v4.0), which endoscopically healed 30 days later after treatment with omeprazole (20 mg).

Pirlet in 2019 [12] studied seven morbid-obese patients (mean age 48 ± 7 years, mean BMI of 52 ± 8 kg/m^2) who were referred for coronary angiography. Weight loss was on average was 13 ± 17 kg (median loss: −11 kg [0, −25]) at up to 12 months after the procedure. The procedures were done through 5–6F right radial access, and selective cannulation of the LGA was done by 5F JR catheter in all but one patient where a 3F Renegade Microcatheter (Boston Scientific Corp., Cork, Ireland) was used. Embolization with 300–500 μm PV particles (Cook Medical, Bloomington, Indiana, USA) was performed. No major complications were observed in this series. Six patients had transient epigastric discomfort resolved with pro-ton pump inhibitors (PPIs).

In a retrospective series published by Elens and colleagues [9] in 2019 that included 16 patients who underwent embolization of the LGA the mean weight loss was 8 ± 5.12 kg. The procedure was successful in all but one patient. Four patients (25%) were lost in follow-up. Femoral access was used in all patients; celiac trunk angiogram was done with a 5F cobra catheter (Cook Medical, Bloomington, Indiana, USA), LGA angiography with 5F MP catheter, selective cannulation with Progreat 2.7F (Terumo, Tokyo, Japan) microcatheter and embolization was performed with 300–500 μm embospheres (Merit Medical) in the first patient and 500–700 μm embospheres in the remaining patients. The patient treated with 300–500 μm embosphere had gastric ulceration on control endoscopy that resolved at three months. One major complication was reported in this series of a patient that ended in the intensive care unit for pancreatitis, splenic infarct and late gastric perforation.

The BEAT trial [29] (Bariatric Embolization of Arteries for the Treatment of Obesity) 1-year results included 20 patients (mean age of 44 years 6 +/− 11 years) and the procedure was successful in all of them. The mean excess weight loss at one year was 11.5% (95% CI: 6.8%–16%; $p < 0.001$). In this series, patients underwent two-weeks pre-treatment with omeprazole and sucralfate that continued up to 6 weeks post-procedure. Operators used femoral access, celiac trunk angiography plus cone-beam CT to evaluate the gastric arterial distribution. Catheterization was performed with a 5-F SOS guiding catheter (Angiodynamics, Latham, NY), and a 2.9-F high-flow microcatheter (Maestro; Merit Medical) was utilized for selective cannulation. After administration of 200 mg of nitroglycerin and 2.5 mg of verapamil (only into the LGA), embolization was performed with 300–500 µm embospheres (Merit Medical). No major complications were reported on this trial. Eleven minor adverse events occurred in eight participants. One participant had subclinical pancreatitis, evident by transient elevation of lipase levels during the hospital stay that was treated with supportive care and discharged within 48 h, with no further clinical sequels.

Zaitoun and collaborators recently published a pilot study of EBS in 10 patients with obesity (BMI > 30 kg/m^2) and prediabetes (hemoglobin A1c (HbA1c) level 5.7%–6.4%) [30]. A statistically significant reductions in HbA1c (from 6.1% $\pm$ 0.2 to 4.7% $\pm$ 0.6; $p < 0.0001$), mean body weight (from 107.4 kg $\pm$ 12.8 to 98 kg $\pm$ 11.6; $p < 0.0001$), and mean BMI (from 37.4 kg/m^2 $\pm$ 3.3 to 34.1 kg/m^2 $\pm$ 3; $p < 0.0001$) was observed at 6 months follow-up. Embolization was performed with 300–500 µm microspheres. The authors also reported a significant positive correlation between BMI and HbA1c levels (r = 0.91; CI, 0.66–0.98; $p = 0.0002$).

In 2020 the LOSEIT group, led by Vivek Reddy of the Icahn School of Medicine at Mount Sinai in New York City, published the first in-human, sham-controlled, randomized clinical trial about EBS [31]. In this trial 40 patients (BMI of 35.0 to 55.0 kg/m^2 and age 21 to 60 years) underwent randomization in 1:1 to either sham or transcatheter bariatric embolotherapy (TBE) targeting the left gastric artery; patients randomized to sham were unblinded at 6 months and crossover to TBE was allowed. Patients of both groups received a lifestyle support (diet and behavioral education for weight loss). TBE has been performed via femoral arterial access using standard 6-F guiding catheters for celiac artery angiography. In a case the gastroduodenal artery was the origin of the left gastric artery which supplied the gastric fundus and in one patient the embolization target vessel was the left hepatic artery. An occlusion balloon microcatheter (Endo-bar Solutions LLC, Orangeburg, New York) was advanced into the target artery and a balloon at the tip was inflated to prevent retrograde reflux. The embolization was achieved by using 300- to 500-mm microspheres (BeadBlock, Biocompatibles Ltd., Farnham, United Kingdom) into the LGA. The procedure was repeated until adequate angiographic stasis was achieved with the balloon deflated over 5 cardiac cycles. Patients randomized to sham received propofol only, without arterial access. At 6 months, in both the intention-to-treat and per-protocol populations, the total body weight loss was greater with TBE (7.4 kg/6.4% and 9.4 kg/8.3% loss, respectively) than sham (3.0 kg/2.8% and 1.9 kg/1.8%, respectively; p ¼ 0.034/0.052 and p ¼ 0.0002/0.0011, respectively); the total body weight loss was maintained with TBE at 12 months (intention-to-treat 7.8 kg/6.5% loss, per-protocol 9.3 kg/9.3% loss; p ¼ 0.0011/0.0008, p ¼ 0.0005/0.0005, respectively). After 1 week, all the patients underwent endoscopic examination: 5 cases of asymptomatic ulceration in the treatment group occurred (4 small superficial ulcers in the sub-cardiac region of the stomach and 1 superficial ulcer in the greater curvature).

Bariatric Embolization of Arteries With Imaging Visible Embolics (BEATLES) (BAE2) is an on-going, randomized, sham-controlled study sponsored by Johns Hopkins University [32]. The aim of the trial, with an estimated enrollment of 59 participants, is to evaluate the change in body weight 12 months after randomization in the bariatric embolization procedure arm versus the control (sham) arm; the estimated study completion date is December 2023.

A comparison of the studies involving humans is reported in Table 2.

Table 2. Clinical Studies Evaluating BES procedure for Weight Reduction.

Author/Year/REF	Study Design	N of pts	Follow-UpLength	Total Weight Loss (%)	Embolic Material
Kipshidze 2015 [11]	Prospective	5	2 years	22 kg (17.1%)	300–500 μm BeadBlock embospheres
GET-LEAN 2016 [13]	Prospective	4	6 months	9.2 kg (8.5%)	300–500 μm BeadBlock embospheres
Bai 2018 [8]	Prospective	5	9 months	12.9 kg (12.6%)	500–710 μm PVA particles
Pirlet 2019 [12]	Prospective	7	12 months	13 kg (4.7%)	300–500 μm PVA particles
Elens 2019 [9]	Prospective	16	12 months	8 kg (10%)	500–700 μm Merit Medical embospheres
BEAT 2019 [29]	Prospective	20	12 months	7.6 kg (11.5%)	300–500 μm Merit Medical embospheres
Zaitun 2019 [30]	Prospective (pre-diabetic)	10	6 months	9 kg (8.9%)	300–500 μm Merit Medical embospheres
LOSEIT 2020 [31]	Randomized, sham-controlled	40	12 months	9.3 Kg (9.3%)	300–500 μm BeadBlock microspheres
BEATLES 2023 [32]On-going Trial	Randomized, sham-controlled	56	12 months	On-going trial	100–200 μm radiopaque microspheres

6. Current and Future Perspectives

Considering the huge impact of obesity in public health worldwide, novel treatments are of utmost importance to reduce associated morbidity and mortality related to this condition and its complications. Early experimental and initial clinical data have suggested that BES may be a promising technique to reduce weight and hormonal imbalance in the obese population. Compared to bariatric surgery, BSE seems to be less invasive, well-tolerated, with shorter recovery time, and is probably cost-saving; further randomized clinical trials are needed to confirm this hypothesis and to evaluate eventual consequences regarding metabolic and nutritional status of patients underwent EBS. The duration of BES procedure varies from 80 to 100 min, with a fluoroscopy time of 32 $\pm$ 14 min. Current human clinical evidence shows that bariatric LGA embolization is an effective treatment that is associated with statistically significant weight loss during follow-up. In addition to weight change, some clinical series have also demonstrated a reduction in serum level of ghrelin and/or leptin [8,11,29,31] and in the level of hemoglobin A1c [29,30] and mean total cholesterol [29]; a statistically significant mean decrease in diastolic blood pressure in the embolization arm was observed in one trial [31]. Furthermore, the procedure is associated with improvements in quality-of-life after embolization, especially regarding the domains of physical function, self-esteem, sexual life, and public distress. The most frequent side effects after the procedure are nausea, vomiting and epigastric pain, and minor complications such as transient superficial mucosal ulcers are common after LGA embolization (usually treated with proton pump inhibitors and spontaneous healing within 4 weeks to 3 months without the needed of further hospitalization); only one case of major complications after the procedure (pancreatitis, splenic infarct and late gastric perforation) was reported by Elens et al. [9].

Distal embolizing agents (mostly embospheres and PVA particles) have been the main embolic material used in recent trials. Although bariatric embolization seems to be safe in this initial phase and results in terms of weight loss are promising, we need further data on patients subsequently treated with conventional bariatric surgery (given that the gastroesophageal junction is de-vascularized) to establish if BES can be utilized as bridge-therapy to surgery in this population. The ideal embolic material remains to be elucidated to achieve what seems to be a perfect safety-efficacy balance mostly on the long term. A larger well designed RCT ideally with a sham-controlled group is eagerly expected to establish the future role of BES in favoring weight loss and hormone homeostasis and for prevention of obese-related morbidity and overall mortality. There is also the needed of a trial of comparison between BES and optimal medical therapy, including new drugs like Semaglutide [33]. At last there are not data of comparison between BES and surgical techniques to evaluate if BES should be considered as an alternative in those patients who are not passible of surgery (multimorbid patients).

Author Contributions: Conceptualization, G.M.S. and A.C.; Methodology, G.M.S.; Validation, G.M.S., A.C. and N.P.; Formal Analysis, G.M.S., A.C. and G.C.; Investigation, G.M.S., A.C. and G.C.; Data Curation, A.M. and M.B.; Writing—Original Draft Preparation, G.M.S. and N.P.; Writing—Review & Editing, all authors. Visualization, all authors. Supervision, G.M.S. and N.P.; Project Administration, G.M.S.; Funding Acquisition, G.M.S. All authors have read and agreed to the published version of the manuscript.

Funding: This research received no external funding.

Acknowledgments: Authors wish to thank Sara Sguotti from Medtronic Corporation for her invaluable technical assistance during the procedures and in manuscript preparation. All the authors have consented to the acknowledgement.

Conflicts of Interest: The authors declare no conflict of interest.

References

1. Flegal, K.M.; Kit, B.K.; Orpana, H.; Graubard, B.I. Association of all-cause mortality with overweight and obesity using standard body mass index categories: A systematic review and meta-analysis. *JAMA* **2013**, *309*, 71–82. [CrossRef]
2. Kim, D.D.; Basu, A. Estimating the medical care costs of obesity in the united states: Systematic review, meta-analysis, and empirical analysis. *Value Health* **2016**, *19*, 602–613. [CrossRef]
3. Jensen, M.D.; Ryan, D.H.; Apovian, C.M.; Ard, J.D.; Comuzzie, A.G.; Donato, K.A.; Hu, F.B.; Hubbard, V.S.; Jakicic, J.M.; Kushner, R.F.; et al. 2013 AHA/ACC/TOS guideline for the management of overweight and obesity in adults: A report of the American College of Cardiology/American Heart Association Task Force on Practice Guidelines and the Obesity Society. *Circulation* **2014**, *129*, S102–S138. [CrossRef]
4. Lupoli, R.; Lembo, E.; Saldalamacchia, G.; Avola, C.K.; Angrisani, L.; Capaldo, B. Bariatric surgery and long-term nutritional issues. *World J. Diabetes* **2017**, *8*, 464–474. [CrossRef]
5. Fujioka, K. Follow-up of nutritional and metabolic problems after bariatric surgery. *Diabetes Care* **2005**, *28*, 481–484. [CrossRef] [PubMed]
6. Mangan, A.; Le Roux, C.W.; Miller, N.G.; Docherty, N.G. Iron and Vitamin D/Calcium Deficiency after Gastric Bypass: Mechanisms Involved and Strategies to Improve Oral Supplement Disposition. *Curr. Drug Metab.* **2019**, *20*, 244–252. [CrossRef]
7. Di Lorenzo, N.; Antoniou, S.A.; Batterham, R.L.; Busetto, L.; Godoroja, D.; Iossa, A.; Carrano, F.M.; Agresta, F.; Alarçon, I.; Azran, C.; et al. Clinical practice guidelines of the European Association for Endoscopic Surgery (EAES) on bariatric surgery: Update 2020 endorsed by IFSO-EC, EASO and ESPCOP. *Surg. Endosc.* **2020**, *34*, 2332–2358. [CrossRef] [PubMed]
8. Bai, Z.-B.; Qin, Y.-L.; Deng, G.; Zhao, G.-F.; Zhong, B.-Y.; Teng, G.-J. Bariatric embolization of the left gastric arteries for the treatment of obesity: 9-month data in 5 patients. *Obes. Surg.* **2018**, *28*, 907–915. [CrossRef] [PubMed]
9. Elens, S.; Roger, T.; Elens, M.; Rommens, J.; Sarafidis, A.; Capelluto, E.; Delcour, C. Gastric embolization as treatment for overweight patients; Ef-ficacy and safety. *Cardiovasc. Interv. Radiol.* **2019**, *42*, 513–519. [CrossRef]
10. Hafezi-Nejad, N.; Bailey, C.R.; Weiss, C.R. Bariatric embolization: A narrative review of clinical data from human trials. *Tech. Vasc. Interv. Radiol.* **2020**, *23*, 100658. [CrossRef]
11. Kipshidze, N.; Archvadze, A.; Bertog, S.; Leon, M.B.; Sievert, H. Endovascular bariatrics: First in humans study of gastric artery embolization for weight loss. *JACC Cardiovasc. Interv.* **2015**, *8*, 1641–1644. [CrossRef]
12. Pirlet, C.; Ruzsa, Z.; Costerousse, O.; Nemes, B.; Merkely, B.; Poirier, P.; Bertrand, O.F. Transradial left gastric artery embolization to treat severe obesity: A pilot study. *Catheter. Cardiovasc. Interv.* **2019**, *93*, 365–370. [CrossRef]
13. Syed, M.I.; Morar, K.; Shaikh, A.; Craig, P.; Khan, O.; Patel, S.; Khabiri, H. Gastric artery embolization trial for the lessening of appetite nonsurgically (GET LEAN): Six-month preliminary data. *J. Vasc. Interv. Radiol.* **2016**, *27*, 1502–1508. [CrossRef]
14. Korbonits, M.; Goldstone, A.P.; Gueorguiev, M.; Grossman, A.B. Ghrelin—A hormone with multiple functions. *Front. Neuroendocrinol.* **2004**, *25*, 27–68. [CrossRef] [PubMed]
15. Nakazato, M.; Murakami, N.; Date, Y.; Kojima, M.; Matsuo, H.; Kangawa, K.; Matsukura, S. A role for ghrelin in the central regulation of feeding. *Nat. Cell Biol.* **2001**, *409*, 194–198. [CrossRef]
16. Duca, F.A.; Covasa, M. Current and emerging concepts on the role of periperal signals in the control of food intake and development of obesity. *Br. J. Nutr.* **2012**, *108*, 778–793. [CrossRef]
17. Page, A.J.; Slattery, J.A.; Milte, C.; Laker, R.; O'Donnell, T.; Dorian, C.; Brierley, S.; Blackshaw, A. Ghrelin selectively reduces mechanosensitivity of upper gastrointestinal vagal afferents. *Am. J. Physiol. Liver Physiol.* **2007**, *292*, G1376–G1384. [CrossRef]
18. Gunn, A.J.; Oklu, R. A preliminary observation of weight loss following left gastric artery embolization in humans. *J. Obes.* **2014**, *2014*, 185349. [CrossRef] [PubMed]

19. Kim, D.J.; Raman, H.S.; Salter, A.; Ramaswamy, R.; Gunn, A.J.; Weiss, C.R.; Akinwande, O. Analysis of weight changes after left gastric artery embolization in a cancer-naive population. *Diagn. Interv. Radiol.* **2018**, *24*, 94–97. [CrossRef] [PubMed]
20. Arepally, A.; Barnett, B.; Montgomery, E.; Patel, T.H. Catheter-directed gastric artery chemical embolization for modulation of systemic ghrelin levels in a porcine model: Initial experience. *Radiology* **2007**, *244*, 138–143. [CrossRef]
21. Arepally, A.; Barnett, B.; Patel, T.T.; Howland, V.; Boston, R.C.; Kraitchman, D.L.; Malayeri, A.A. Catheter-directed gastric artery chemical embolization suppresses systemic ghrelin levels in porcine model. *Radiology* **2008**, *249*, 127–133. [CrossRef] [PubMed]
22. Bawudun, D.; Xing, Y.; Liu, W.-Y.; Huang, Y.-J.; Ren, W.-X.; Ma, M.; Xu, X.-D.; Teng, G.-J. Ghrelin suppression and fat loss after left gastric artery em-bolization in canine model. *Cardiovasc. Interv. Radiol.* **2012**, *35*, 1460–1466. [CrossRef] [PubMed]
23. Paxton, B.E.; Kim, C.Y.; Alley, C.L.; Crow, J.H.; Balmadrid, B.; Keith, C.G.; Kankotia, R.J.; Stinnett, S.; Arepally, A. Bariatric embolization for suppression of the hunger hormone ghrelin in a porcine model. *Radiology* **2013**, *266*, 471–479. [CrossRef] [PubMed]
24. Paxton, B.E.; Alley, C.L.; Crow, J.H.; Burchette, J.; Weiss, C.R.; Kraitchman, D.L.; Arepally, A.; Kim, C.Y. Histopathologic and immunohistochemical sequelae of bar-iatric embolization in a porcine model. *J. Vasc. Interv. Radiol.* **2014**, *25*, 455–461. [CrossRef] [PubMed]
25. Paxton, B.E.; Arepally, A.; Alley, C.L.; Kim, C.Y. Bariatric embolization: Pilot study on the impact of gastroprotective agents and arterial distribution on ulceration risk and efficacy in a porcine model. *J. Vasc. Interv. Radiol.* **2016**, *27*, 1923–1928. [CrossRef]
26. Kim, J.M.; Kim, M.D.; Han, K.; Muqmiroh, L.; Kim, S.U.; Kim, G.M.; Kwon, J.; Pari, S., Jr.; Won, J.Y.; Lee, D.Y. Bariatric arterial embolization with non-spherical polyvinyl alcohol particles for ghrelin suppression in a swine model. *Cardiovasc. Interv. Radiol.* **2017**, *40*, 744–749. [CrossRef]
27. Hafezi-Nejad, N.; Bailey, C.R.; Gunn, A.J.; Weiss, C.R. Weight loss after left gastric artery embolization: A systematic review and meta-analysis. *J. Vasc. Interv. Radiol.* **2019**, *30*, 1593–1603.e3. [CrossRef]
28. Fu, Y.; Kraitchman, D.L. Rationale and preclinical data supporting bariatric arterial embolization. *Tech. Vasc. Interv. Radiol.* **2020**, *23*, 100656. [CrossRef]
29. Weiss, C.R.; Abiola, G.O.; Fischman, A.M.; Cheskin, L.J.; Vairavamurthy, J.; Holly, B.P.; Akinwande, O.; Nwoke, F.; Paudel, K.; Belmustakov, S.; et al. Bariatric embolization of arteries for the treatment of obesity (BEAT Obesity) trial: Results at 1 year. *Radiology* **2019**, *291*, 792–800. [CrossRef]
30. Zaitoun, M.M.; Basha, M.A.A.; Hassan, F.; Elsayed, S.B.; Farag, A.A.; Amer, M.; Aly, S.A.; Zaitoun, N. Left gastric artery embolization in obese, prediabetic patients: A pilot study. *J. Vasc. Interv. Radiol.* **2019**, *30*, 790–796. [CrossRef] [PubMed]
31. Reddy, V.Y.; Neužil, P.; Musikantow, D.; Sramkova, P.; Rosen, R.; Kipshidze, N.; Kipshidze, N.; Fried, M. Transcatheter bariatric embolotherapy for weight reduction in obesity. *JACC* **2020**, *76*, 2305–2317. [CrossRef] [PubMed]
32. Weiss, C.R.; Abiola, G.O.; Fischman, A.M.; Cheskin, L.J.; Vairavamurthy, J.; Holly, B.P.; Akinwande, O.; Nwoke, F.; Paudel, K.; Belmustakov, S.; et al. Bariatric embolization of arteries with imaging visible embolics (BEATLES) (BAE2). *Clin. Trials.*
33. Davies, M.; Færch, L.; Jeppesen, O.K.; Pakseresht, A.; Pedersen, S.D.; Perreault, L.; Rosenstock, J.; Shimomura, I.; Viljoen, A.; Wadden, T.A.; et al. Semaglutide 2·4 mg once a week in adults with overweight or obesity, and type 2 diabetes (STEP 2): A randomised, double-blind, double-dummy, placebo-controlled, phase 3 trial. *Lancet* **2021**, *397*, 971–984. [CrossRef]

nutrients

MDPI

Review

Nutrition in Patients with Type 2 Diabetes: Present Knowledge and Remaining Challenges

Maria Letizia Petroni [1,2], Lucia Brodosi [1,2], Francesca Marchignoli [1], Anna Simona Sasdelli [1], Paolo Caraceni [1,2], Giulio Marchesini [2,*] and Federico Ravaioli [1,2]

[1] IRCCS-Azienda Ospedaliera di Bologna Sant'Orsola-Malpighi, I-40138 Bologna, Italy; marialetizia.petroni@unibo.it (M.L.P.); lucia.brodosi2@unibo.it (L.B.); francesca.marchignoli@gmail.com (F.M.); annasimona.sasdelli@aosp.bo.it (A.S.S.); paolo.caraceni@unibo.it (P.C.); f.ravaioli@unibo.it (F.R.)

[2] Department of Medical and Surgical Sciences, Alma Mater University of Bologna, I-40138 Bologna, Italy

* Correspondence: giulio.marchesini@unibo.it; Tel.: +39-051-2144889

Abstract: Unhealthy behaviours, including diet and physical activity, coupled with genetic predisposition, drive type 2 diabetes (T2D) occurrence and severity; the present review aims to summarise the most recent nutritional approaches in T2D, outlining unmet needs. Guidelines consistently suggest reducing energy intake to counteract the obesity epidemic, frequently resulting in sarcopenic obesity, a condition associated with poorer metabolic control and cardiovascular disease. Various dietary approaches have been proposed with largely similar results, with a preference for the Mediterranean diet and the best practice being the diet that patients feel confident of maintaining in the long term based on individual preferences. Patient adherence is indeed the pivotal factor for weight loss and long-term maintenance, requiring intensive lifestyle intervention. The consumption of nutritional supplements continues to increase even if international societies do not support their systematic use. Inositols and vitamin D supplementation, as well as micronutrients (zinc, chromium, magnesium) and pre/probiotics, result in modest improvement in insulin sensitivity, but their use is not systematically suggested. To reach the desired goals, patients should be actively involved in the collaborative development of a personalised meal plan associated with habitual physical activity, aiming at normal body weight and metabolic control.

Keywords: behaviour; diet; lifestyle; nutrition supplements; sarcopenia; type 2 diabetes

Citation: Petroni, M.L.; Brodosi, L.; Marchignoli, F.; Sasdelli, A.S.; Caraceni, P.; Marchesini, G.; Ravaioli, F. Nutrition in Patients with Type 2 Diabetes: Present Knowledge and Remaining Challenges. *Nutrients* **2021**, *13*, 2748. https://doi.org/10.3390/nu13082748

Academic Editor: Ina Bergheim

Received: 12 July 2021
Accepted: 6 August 2021
Published: 10 August 2021

1. Introduction

Diabetes mellitus, namely type 2 diabetes (T2D), constitutes a significant challenge for health systems worldwide. According to the 2019 Diabetes Atlas of the International Diabetes Federation [1], 463 million adults are currently living with diabetes (1 on 11 individuals worldwide, but 1 in 5 are aged over 65). The total number is expected to increase further by 700 million in 2045. The economic impact is huge—driven by the direct costs of treatment and complications, the indirect costs of disability and premature death, and the intangible costs of poor quality of life.

Despite its characterizations as a disease of affluence, nutritional problems are frequent in T2D. Unhealthy lifestyles expressed by overnutrition and/or scarce physical activity, leading to overweight and obesity, add to genetic defects in the pathogenesis of the disease. Dietary restrictions are prescribed to reduce the incidence of T2D as well as to improve metabolic control [2], but weight loss is burdened by the loss of muscle mass [3] and sarcopenia adds to age-dependent muscle wasting [4], increasing frailty [5]. These two opposite needs make a correct nutritional approach mandatory to reduce disease burden, improve metabolic control, limit pharmacologic treatment and reduce the risk of impending cardiovascular disease.

National and international guidelines for nutritional and lifestyle recommendations are available [5–9], together with protocols to guide weight loss to produce long-term T2D remission [10]. The proposed strategies (dietary prescription, lifestyle counselling, cognitive behaviour therapy), although all-inclusive of nutritional components, are markedly different in their approach and goals and should be known by clinicians approaching patients with T2D (Table 1) [11]. The present review is intended to summarize the most recent nutritional approaches in T2D, also outlining unmet needs.

Table 1. Comparison of strategies and goals of different dietary interventions.

	Dietary Prescription	Dietary Counseling	Behaviour Therapy
1-4 Dietary program	• Based on rigid meal prescription (food weight, substitution schedule)	• Proposed food choices with templates of daily meals	• Food choices proposed within the frame of a healthy diet
Patient role	• Passive adherence to the prescriptive plans	• Acquires competence in healthy diet strategies	• Meals and physical activity planned according to personal preferences
Role of therapist	• Active–gives the solution • The planned calorie intake is mandatory for patients	• Provides education on healthy lifestyles • Helps in identifying best practices according to patients' preferences and individual status	• Communicates empathically • Supports patients' activities, success and failures to stimulate self-efficacy • Helps in identifying obstacles and presents possible solutions
Treatment goals	• Strong focus on weight loss or HbA1c targets	• Set realistic expectations and acceptable body weight	• Behavioural changes are the main targets, independent of the amount of weight loss
Temporal terms	• Usually limited to weeks, with frequent changes	• Life-long adherence to healthy lifestyles	• Life-long adherence to healthy lifestyles
Additional components	———	• Integration of dietary and physical activity counselling	• Generation of a mindset favouring lifestyle targets
Psychological support	———	• Support by family, significant others, both on food choices and in habitual physical activity	• Implementation of a pro-active problem solving • Stimulus control aimed at modifying the environment • Strategies of cognitive restructuring to address dysfunctional thoughts

Note that enrolment into counselling and behaviour therapy may be facilitated by motivational interviewing. Treatment may be provided either in individual or in group settings; group strategies are likely to enhance the coping skills of the participants, via relational and interpersonal communication with people experiencing similar difficulties.

2. Methods and Areas of Research

2.1. Literature Search

The literature on T2D is immense. A PubMed search of June 2021, limited to the period 2016–2021 using the string "Type 2 diabetes" [MeSH Terms] AND "nutrition" [All Fields] AND "human" [MeSH Terms], retrieved 4865 references, including 887 review articles (234 systematic reviews), 255 meta-analyses and 760 clinical trials. The authors used the search to enucleate the most relevant data and unmet treatment needs. The reference lists of selected articles were used to retrieve older documents in order to provide a complete overview of present problems.

2.2. Diabetes, Obesity and Sarcopenia

The association between T2D and obesity is so strict that the term "diabesity" was originally used to indicate the dreadful association of the two conditions in a JAMA editorial in 1980 [12]. The term was finally proposed by Astrup and Finer [13], as well as by Zimmet et al. [14] and it is largely accepted inside the metabolic community. The accumulation of body fat characterizes obesity, but it is measured by a formula (the body mass index, i.e., weight (kg)/height^2 (m)), not at all considering body fat. Muscle mass is frequently increased in obesity but might be relatively scarce in quantity and quality compared to body fat.

Sarcopenia is particularly common in older patients, synergistically driven by age and obesity; body fat increases until the seventh decade of life (the median age of patients with diabetes attending diabetes centres) and decreases thereafter [15]. At the same time, sedentariness progressively reduces muscle mass, finally resulting in sarcopenic obesity [16], frequently associated with cardiometabolic disorders [17].

By definition, sarcopenia implies a quantitatively reduced muscle mass, as measured by dual-energy X-ray absorptiometry (DXA), the commonly accepted gold standard. Several studies have validated the use of bioelectrical impedance analysis (BIA), an easy, time-saving, and cost-effective bedside technique for assessing regional muscle mass and body composition [18,19]. BIA-assessed sarcopenia is defined by the skeletal muscle mass index (SMI), calculated as total appendicular skeletal mass (ASM, kg) divided by body weight (kg) $\times$ 100. These measurements do not consider qualitative muscle mass, and most recent guidelines suggest that functional measurements (e.g., low muscle strength by handgrip) should be primarily used to characterize sarcopenia, with quantitative data as supportive measures [20].

The prevalence of sarcopenia in diabetes has been extensively investigated. In a recent narrative review, the prevalence of sarcopenia varied between 7% and 29% [21], according to age and metabolic control, but higher figures are frequently reported. A systematic review with meta-analysis including 15 studies confirmed a prevalence varying up to 50% [22], again driven by age and metabolic control. A study with BIA concluded that patients with T2D have an enlarged ectopic fat at the expense of skeletal muscle, i.e., relative sarcopenia [23], and lower muscle mass is coupled with decreased muscle strength [24], also predicting diabetes in the general population [25]. The contribution of diabetes duration remains controversial [21,22], but older patients with T2D, with an expected longer duration of disease, show a larger decline in appendicular lean mass, muscle strength, and functional capacity compared with normoglycemic controls [26]. Notably, when compared with matched control populations, the risk of sarcopenia increased systematically in the presence of T2D (odds ratio (OR) 1.55; 95% confidence interval (CI) 1.25–1.91; $p < 0.001$ [22] and OR 1.63; 95% CI 1.20–2.22; $p = 0.002$ [27]). This indicates a need for preventive measures to limit quantitative and qualitative muscle defects by effective nutritional treatments.

2.3. Metabolic Control

The primary defect in T2D is insulin resistance, a condition where normal insulin levels are associated with lower metabolic effects or where higher than normal insulin levels are needed to elicit a normal metabolic response. Insulin resistance accounts for diffuse impairment in whole body, as well as in selective defects in different organs and tissues (liver, muscle, adipose tissue).

Whole-body insulin resistance mainly reflects muscle insulin resistance [28], reducing glucose and amino acid uptake in the postprandial phase, as well accelerating glycogen and amino acid release in the post-absorptive state, also accelerated by glucagon release [29]. Glucagon constitutes the link between muscle and liver in substrate disposal; by stimulating hepatic glucose production and ketogenesis, glucagon favours the utilization of substrates released in the periphery, whereas high insulin concentrations favour hepatic fat deposition. In both obese and nonobese subjects, higher plasma insulin levels have been associated with a linear increase in the rates of hepatic de novo lipogenesis [30], as supported by the hypoglycaemic effects of glucagon suppression of glucagon-receptor antagonists [31,32]. In the hepatocytes, fatty acids may be derived from de novo lipogenesis, uptake of non-esterified fatty acids and low-density lipoproteins, or lipolysis of intracellular triacylglycerol. Their accumulation due to higher synthesis and decreased export in the presence of high insulin concentrations in the portal vein is the likely cause of fatty liver disease, occurring in up to 73% of patients with T2D [33].

The link between muscle tissue and the liver is exerted by amino acids (Figure 1) [34]. Branched-chain amino acids, bypassing the liver in the post-prandial state, serve as nitrogen carriers to the periphery, whereas alanine and glutamine are used to carry nitrogen from the periphery to the liver, intestine and kidney. In insulin-resistant states, including obesity [35], the post-load uptake of branched-chain amino acids is impaired, possibly leading to defective amino acid supply to the muscle tissue and sarcopenia. In summary, the complex trafficking of glucose, lipid and amino acid in response to insulin resistance should be considered in the treatment of diabetes.

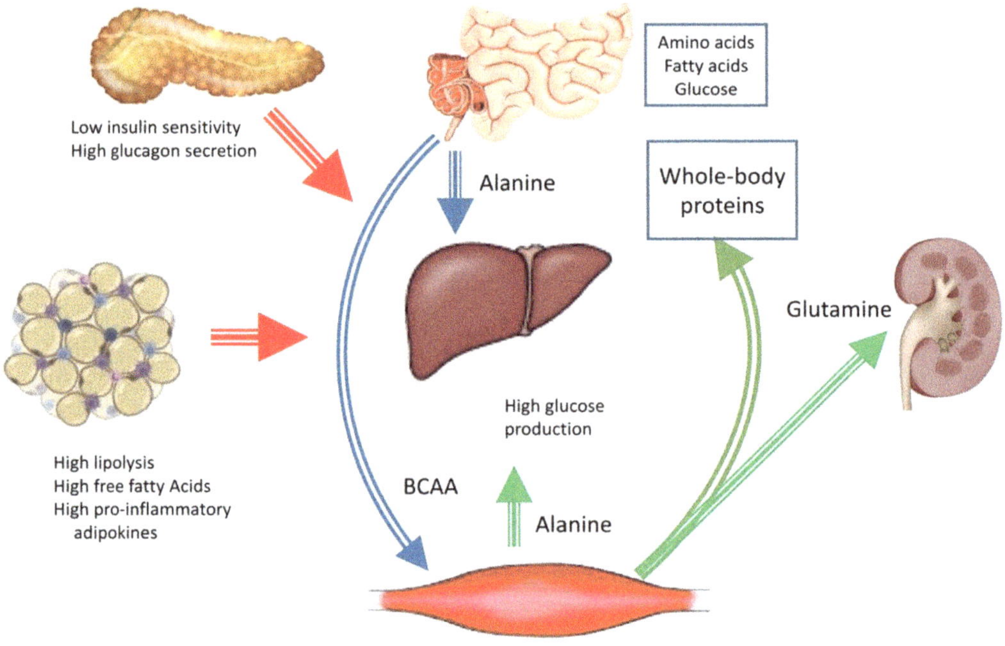

Figure 1. Interorgan amino acid exchange in the postabsorptive state and after meals in diabetes. Note the importance of BCAAs (valine, isoleucine and leucine) as nitrogen carriers to the muscle tissue (lean mass) in the post-prandial period (blue arrows) and the reverse importance of alanine and glutamine as nitrogen carriers to central organs in the post-absorptive state (liver, kidney, intestine) (green arrows). In this context, the regulatory role of the pancreas (altered secretion of insulin and glucagon) and the adipose tissue (lipolysis, release of free fatty acids and inflammatory adipokines in the general circulation, particularly in the post-absorptive state) is pivotal for the regulation of hepatic and whole-body homeostasis (red arrows).

3. Medical Nutrition Therapy for Type 2 Diabetes

The foundation of medical nutrition therapy (MNT) of T2D is to achieve glucose, lipids, and blood pressure within the target range to prevent, delay or manage microvascular and macrovascular complications [36,37].

MNT plays a pivotal role in the overall management of diabetes, and patients with T2D should be actively involved with their healthcare team for the collaborative development of a personalized meal plan. If these patients are referred to a registered dietitian or a nutritionist proficient in providing diabetes-specific treatment, an absolute reduction of glycated A1C haemoglobin of up to 1.9% may be observed [8]. Continuous dietary counselling integrated with mobile apps and wearable devices has also been advocated to facilitate the real-time assessment of dietary intake, to strengthen adherence, and support motivation and self-efficacy [38].

3.1. Comparison between Different Guidelines

Table 2 summarizes the main nutritional recommendations for patients with T2D, derived from guidelines, and the dietary patterns with a high degree of evidence [5–9]. All proposed interventions are designed to reduce energy intake and promote 5–10% loss of initial body weight, leading to improved insulin sensitivity, blood glucose and blood pressure control, and reduced lipid levels [39]. Regular mealtimes and a healthy diet should be combined with increased physical activity [4].

Table 2. Summary of nutritional recommendations for type 2 diabetes, as derived from international guidelines.

Nutrients	Recommendations
1-2 Calorie intake	• Reduce energy intake in all individuals with overweight/obesity (calorie deficit of 250–500 kcal/day) to promote weight loss (0.5–1.0 kg/week) to a final body weight within the normal range
1-2 Macronutrient distribution	• There is insufficient evidence to recommend specific macronutrient distribution, but a moderate carbohydrate reduction might favour glucose control and promote a moderate weight loss
1-2 Carbohydrates	• Prefer low glycaemic index foods (whole grains, fruits, legumes, green salad with olive oil dressing and most vegetables). Limit refined carbohydrates (pasta, white bread, rice, white potatoes, etc.)
1-2 Sugars	• Limit intake of sucrose-containing foods and sugary drinks • Prefer non-nutritive sweeteners as substitutes of sugar • Low calorie or unsweetened beverages should be preferred, but their consumption limited
1-2 Fibers	• 30–50 g/day of dietary fibres (at least one-third of soluble origin)
1-2 Proteins	• As in the general population, 1.0–1.5 g/kg ideal body weight • Reduce protein intake to 0.8 g kg/b.w. or lower in patients with chronic kidney disease
1-2 Fats	• As in the general population, 20–35% of total kcal/day • Avoid trans-fatty acids and limit saturated fatty acids (SFAs) to 7–9%. Increase foods enriched in long-chain omega-3 polyunsaturated fatty acids (PUFAs) and monounsaturated fatty acids (MUFAs).
1-2 Micronutrients & Vitamins	• Correct micronutrient and vitamin deficiencies • Consider vitamin supplementation (B-group vitamins or folic acid) in metformin-treated patients
1-2 Sodium	• Limited as in the general population; consider additional limitations in those with hypertension
1-2 Alcohol	• Limited as in the general population
1-2 Dietary pattern	• Favour a dietary model based on Mediterranean-style

The optimal distribution of macronutrients as a percentage of total energy is highly variable, from 45 to 60% for carbohydrates, from 15 to 20% for proteins and 20 to 35% for fats, suggesting no ideal percentage of calories from macronutrients [7]. As to carbohydrates, high-fibre sources (30–50 g/day of dietary fibre, ≥30% as soluble fibres) and minimally processed, low-glycaemic index carbohydrates should be preferred to improve glycaemic control, LDL-cholesterol and cardiovascular (CV) risk. Overall, reducing carbohydrate intake for individuals with T2D has been shown to improve blood glucose [6]; a systematic review and meta-analysis (9 studies with 734 patients) confirmed a beneficial effect of low-carb diets vs. normal-or high-carb diets on HbA1c and on short-term weight loss, not on long-term weight loss [40]. Food plans should emphasize the consumption of non-starchy vegetables, with minimal added sugars, fruits, whole grains, and dairy products [41]. Using non-nutritive sweeteners as substitutes for added sugar (sucrose, high fructose corn syrup, fructose, glucose) can reduce daily calories and total carbohydrates. For those who regularly consume sugary drinks, consuming a low calorie or unsweetened drink can be an alternative, but both should be consumed with caution.

Additionally, recommendations on protein intake do not differ from the general population (1.0–1.2 g/kg body weight or corrected body weight for patients with overweight/obese); protein intake should be reduced to 0.8 g/kg body weight in subjects with chronic diabetic nephropathy [36]. At present, there is some inconsistency across guidelines from different countries as to protein sources (some do not limit animal proteins) and as to allowed maximal amount of protein intake (1.2–1.5 g/kg/day) [42]. A recent meta-analysis of 54 RCTs (4344 participants) showed a significant effect of moderate high-protein diets (20–45% of total energy) vs. low-protein diets (10–23%) on weight loss and weight loss maintenance, total fat mass reduction and cardiometabolic risk [43]. The authors suggest that the effects might also be due to the blood-pressure-lowering effect of bioactive peptides that inhibit the angiotensin-converting enzyme activity observed in protein isolates [44].

Among dietary fats, it is recommended to avoid trans-fatty acids as much as possible and to consume less than 7–9% of the total daily energy from saturated fatty acids (SFAs). SFAs should be replaced with polyunsaturated fatty acids (PUFAs), mainly mixed sources of omega-3/omega-6, and with monounsaturated fatty acids (MUFAs) of vegetable origin whole grains, nuts and seed (rich in alpha-linolenic fatty acid) [36,45].

The recommendations have largely focused on the quality of the diet and the importance of a healthy eating pattern that contains nutrient-rich foods, with less attention to the percentage of specific nutrients, with a reduction in daily caloric intake (250–500 kcal) for subjects with overweight and obesity [6]. Several dietary patterns have been studied and proposed, but no single dietary pattern should be preferred [8]. Individual preferences and treatment goals will determine the long-term use of these models; systematic reviews and meta-analyses have shown that a Mediterranean-style dietary pattern significantly improves hard outcomes such as glycaemic control, systolic blood pressure, total cholesterol, HDL-cholesterol and triglycerides [46]. The Mediterranean diet is characterised by a moderate-to-low carbohydrate intake, entirely covering the micronutrient needs [47]. Additionally, a low fat diet, i.e., the DASH-diet, promoted in the prevention of cardiovascular disease and the treatment of high blood pressure [48], has also reached consensus [49]. In a review comparing low-carbohydrate and ketogenic diets, the vegan diet, and the Mediterranean diet, all diets improved glycaemic control and weight loss, but patient adherence and long-term manageability were pivotal factors for the efficacy of each diet [50].

3.2. Intensive Lifestyle Intervention

Intensive lifestyle intervention (ILI) that supports behaviour changes, as initially experienced in the Finnish Diabetes Prevention Study and the U.S. Diabetes Prevention Program [51,52], represents the recommended approach to prevent and/or delay the onset of T2D in prediabetic patients [5]. The ILI behaviour approach combines diet and physical activity interventions with the goal to achieve and maintain a 7% loss of initial body weight and to increase moderate-intensity physical activity to at least 150 min/week. The effect of ILI has also been investigated in the treatment of T2D. The Look AHEAD study randomized 5145 individuals with T2D and associated overweight or obesity to either ILI or diabetes support and education (as control group), having cardiovascular outcomes as primary goal. Weight loss was achieved by reducing caloric intake to 1200–1800 kcal/day depending on baseline weight using portion-controlled meal plans, calorie-counting techniques, and meal replacements combined to moderate physical activity to ≥175 min/week. ILI was delivered as individual and group sessions over the first year, with a median follow-up of 9.6 years. [53]. After one year, the average weight loss in the ILI group was 8.6%, compared with 0.7% in the control group, with 55% of ILI participants having lost ≥7% of their initial b.w. vs. 7% of controls. This led to remission of T2D in 11.2% of ILI participants vs. 2.0% in controls. However, by the fifth year of follow-up, ILI participants had regained half of their initial weight loss, and the study was closed at the end of the follow-up (10-years) after an interim analysis had shown that the intervention had failed its primary outcome [54]. Thus, the critical point becomes how to achieve long-term weight loss maintenance, a difficult task in the general population [55], and a core problem in T2D treatment with approaches based on lifestyle changes. Although more effective than behaviour change in inducing and sustaining remission of T2D, bariatric surgery also suffers from reduced durability over time [56].

A novel approach was tested in the DIRECT trial, a primary care-led management intervention in patients with T2D diagnosed by less than 6 years and not receiving insulin. The ILI strategy was preceded by a commercial very-low-calorie diet followed by stepwise food reintroduction. Primary outcomes were weight loss $\geq$15 kg and T2D remission. At 12 months, almost half of participants achieved T2D remission off all glucose-lowering medications [57]; this percentage dropped to 36% at 24 months [58]. Notably, the maintenance of diabetes remission paralleled weight loss maintenance and particularly fat removal from the liver and pancreas, suggesting recovered insulin secretion [59]. With the limits of durability, all these data support the use of ILI, including dietary interventions, as an effective adjuvant treatment to improve glycaemic control [60].

Another approach is the so-called intermittent fasting, which has gained increased popularity for treating T2D based on very limited literature [61]. This term encompasses various eating behaviours that avoid (or limit) nutrient and energy intake for a significant amount of time (a full day or a time-restricted feeding between 6 to 8 h) on a regular intermittent schedule. Intermittent fasting is claimed to improve glucose control, insulin resistance and to induce weight loss by generating a 'metabolic switch', i.e., a sort of rejuvenation of the metabolic homeostasis, leading to increased health span and longevity [62], but no advantage over conventional caloric restriction has been proven. Moreover, this regimen could carry the risk of hypoglycaemia even when following a medication dose-change protocol and should only be used under strict medical control and/or continuous glucose monitoring [63].

Finally, the use of mobile apps and wearable devices has recently gained consensus to facilitate weight loss. The use of these devices allows a direct analysis of daily calorie intake and physical activity (daily steps), translated into calorie consumption [64]. This provides immediate feedback and is likely to support long-term adherence to well-defined goals [38]. Several commercial apps are available, and have been tested in the prevention and treatment of diabetes in trials mimicking the U.S. Diabetes Prevention trial [52]. Toro-Ramos et al. confirmed a modest efficacy of weight loss for app users after 6 and 12 months of systematic use in subjects with prediabetes compared with usual care [65], and similar studies are available with the most recent apps that also support by tailored messages interactivity [66]. Although all these supports are expected to improve long-term weight loss, and a few patients may really reach impressive results [67], their use is biased by higher attrition rates [68]. Nonetheless, the possibility to reach a larger audience makes this approach a useful opportunity.

4. Nutritional Supplements for Metabolic Control

International diabetes societies do not support the use of nutritional supplements in diabetes, but their use continues to increase in several countries, despite lack of evidence and uncertainty on safety [36]. A complete analysis of available products (combinations may account for several hundreds) is outside the scope of this review, but a few of them are of interest. Their putative mechanism(s) of action are summarized in Table 3 [69–89]. They are not expected to replace diet and glucose-lowering drugs but might be confidently used, provided their safety is proven.

Table 3. Putative mechanism(s) responsible for the beneficial effects of nutrition supplements and micronutrients on diabetes risk and glycaemic control.

Product	Mechanism of Action
1-2 Inositols [69–71]	Myo-inositol (MI) and D-chiro-inositol (DCI) act as insulin second messengers. MI takes part in cellular glucose uptake and is high in the brain and heart, where high rates of glucose utilization occur MI prevents the release of free fatty acids from adipose tissues; on the contrary, DCI is involved in glycogen storage, being elevated in the liver, muscle, and fat tissue DCI may be preferred to MI to restore insulin sensitivity and glycogen synthesis because it bypasses the defective epimerization of MI to DCI in the presence of insulin resistance
1-2 Vitamin D [72–75]	Serum levels of 25(OH)D are significantly lower in patients with T2D compared with values measured in healthy people, with a negative correlation with HOMA-IR and adipokines Direct effect on insulin secretion, mediated by nuclear vitamin D receptors also present in pancreatic β-cells, but the effects on insulin sensitivity have long been conflicting Vitamin D deficiency is associated with vascular inflammatory responses by promoting the secretion of inflammatory cytokines
1-2 Niacin [76]	This compound mediates hundreds of oxidation-reduction redox reactions, which are essential sources of energy for a myriad of cellular functions, finally known to improve lipid profile and to reduce cardiovascular risk Restoration of C-peptide release, but unexplained negative results on glycaemic control
1-2 Resveratrol [77,78]	Activator of the sirtuin pathway, regulating several cellular functions related to metabolism, oxidation, and aging Anti-oxidant activity
1-2 Pre/probiotics [79–83]	Effects on insulin sensitivity by modification of gut microbiota
1-2 Zinc [84,85]	Participation in insulin synthesis, storage, crystallization, and secretion in the pancreatic β-cell, as in well as in insulin action and translocation inside the cells Stimulation of insulin sensitivity through the activation of the phosphoinositol-3-kinase/protein kinase B cascade. Stimulation of glucose uptake in insulin-independent tissues (insulin-mimetic action) Suppression of proinflammatory cytokines (interleukin-1β and nuclear factor kβ), thus avoiding β-cells death and protecting insulin
1-2 Chromium [86,87]	Effects on insulin signalling Insulin sensitising activity in experimental animals
1-2 Magnesium [88,89]	Possible effects of Mg^{2+} deficiency on the tricarboxylic acid cycle, increasing the risk of hyperinsulinemia and insulin resistance Modulation of insulin action and oxidative glucose metabolism Alteration of lipid metabolism and the antioxidant system

Abbreviations: HOMA-IR, homeostasis model assessment of insulin resistance; IL, interleukin; TNF, tumor-necrosis factor.

4.1. Inositols

Several reviews and meta-analyses have been published on the treatment of gestational diabetes with myo-inositol (MI) or D-chiro-inositol (DCI) [70,90–93]. A Cochrane review was inconclusive [94]; MI supplementation did not reduce the need for insulin or produce any significant effect on blood glucose. Conflicting data have also been reported using DCI or the combination of MI and DCI, and the optimum dosage to achieve a significant effect on glucose metabolism remains unsettled [91]. A position statement of the two largest Italian diabetes societies concluded that MI (at the dose of 4 g/day) might be safely used for the prevention and treatment of gestational diabetes [95], but the level of evidence and the strength of recommendations are low. No data are available on the use of MI or DCI to treat insulin resistance outside gestational diabetes. Studies are in progress on the combined use of MI and myo-inositol hexa-phosphate (IP6), or phytic acid, showing more effective anti-oxidant and glucose-lowering activity in experimental animals [96], but no clinical data are available.

The use of inositol(s) in polycystic ovary syndrome is not considered in the present review; in that setting, specific hormonal activity is likely to produce clinical effects [97].

4.2. Vitamin D

Vitamin D levels are frequently suboptimal in T2D, probably driven by overweight/obesity, and specifically by visceral adiposity [98], and have been associated with chronic inflammation and insulin resistance, as well as impaired insulin release [99]. Epidemiological studies support the existence of a relationship between low vitamin D levels and the presence of T2D, metabolic syndrome [100,101], nonalcoholic fatty liver disease (NAFLD) [102], cardiovascular risk factors [103] and insulin resistance, also tested by glucose clamp [75]. However, a clear association between vitamin D levels, insulin and glucose metabolism has not been systematically confirmed by intervention studies, and a causal association has never been established [104]. In a subset of the RECORD trial, a placebo-controlled trial of oral vitamin D_3 and/or calcium supplementation for the secondary prevention of osteoporotic fractures in older people, vitamin D_3 at the daily dose of 800 IU with or without 1000 mg of calcium did not prevent the development of T2D and did not reduce the need for glucose-lowering drugs in T2D patients [105]. Although the effects on insulin sensitivity have long been conflicting [73], a recent systematic review with metanalysis confirmed that vitamin D supplementation resulted in a significant improvement in HOMA-IR (standardized mean difference = −0.57; 95% CI: −1.09 to −0.04), particularly when vitamin D was administered in large doses and for a short period of time to nonobese, vitamin D deficient patients, or to individuals with optimal glucose control at baseline [106]. Data have been confirmed in another recent study in vitamin D-deficient adults randomized to high dose vitamin D supplementation. The HOMA value of insulin resistance was significantly reduced, and a lower rate of progression toward diabetes was observed vs. the control group (3% vs. 22%; p = 0.002) [107].

Of note, vitamin D has been extensively used also to treat sarcopenia, considering the role of insulin resistance extending from glucose metabolism to protein and amino acid metabolism, as discussed below.

4.3. Niacin

Niacin is a water-soluble derivative of pyridine, present in several forms (namely as nicotinic acid or nicotinamide), also named as vitamin B3. It is a derivative of vitamin B, frequently associated with inositols as inositol hexanicotinate. The effects on insulin release from islet β-cells have been extensively investigated in T2D with secondary failure of sulfonylureas, where niacin at the daily dose of 1.5 g significantly restored C-peptide release [108]. However, a meta-analysis of eight trials where niacin was used to treat hyperlipidemia in 2110 T2D patients showed no significant effects on plasma glucose (weighted mean difference (WMD), 0.18 mmol/L; 95% CI, −0.14 to 0.50) and HbA1c levels (WMD, 0.39%; 95% CI, −0.15 to 0.94) [109]. Niacin appeared to cause a deterioration of glucose control, in keeping with data observed in a meta-analysis of 11 trials in patients without diabetes at entry, where niacin was used to treat dyslipidaemia and prevent cardiovascular events [110] (relative risk of de novo T2D: 1.34 (95% CI 1.21–1.49)). Similar results were provided by the large trial of combination treatment with niacin plus laropiprant [111], where niacin treatment (2 g/day for a median of 3.9 years) was associated with an increased incidence of de novo T2D (rate ratio, 1.32; 95% CI 1.16–1.51) and deterioration in metabolic control in subjects with diabetes (1.55; 1.34–1.78) [112]. This deleterious effect is similar to the well-known, mild negative effect of statins on glucose metabolism.

It adds to the well-known poor tolerability of niacin because of flushing, occurring at pharmacologic doses.

4.4. Nutraceuticals

Natural compounds derived from plant extracts, spices, herbs, and essential oils have been tested for alleged benefits in managing patients with metabolic syndrome [77,113]. They include Mediterranean diet components, olive oil and its anti-oxidant components, natural legumes and cereals, as well as specific compounds, alone or in combination. Curcumin [114], cinnamon [115,116], berberine [117,118], citrus flavonoids [119,120], quercetin [121,122], the bioactive compounds of garlic [123,124], red yeast rice [125] and neem extracts [126] have all demonstrated some activity on insulin sensitivity, but studies are usually of poor quality and very few received extensive validation, although supported by systematic reviews [119]. They may be included in dietary recommendations but should never replace pharmacologic treatment.

Resveratrol, a polyphenol present in plants such as grapes and nuts and mainly in derivatives (wine), merits a specific citation [127–129]. A recent Cochrane review identified three RCTs with a total of 50 participants who received graded doses of daily oral resveratrol for 4–5 weeks vs. placebo. Studies had a low risk of bias, but the analysis did not demonstrate any significant effect on glucose and HbA1c levels, with the limit of a short observation period. The authors found eight more ongoing RCTs with approximately 800 participants, likely to contribute more solid results [128]. Clinical studies in patients with insulin resistance and NAFLD have shown promising results [130], but even moderate alcohol intake is questioned in these patients due to the negative effects of alcohol on hepatic and extrahepatic cancers, which outweigh the possible beneficial effects on the cardiovascular system, largely derived from retrospective studies [131]. Finally, alcohol provides extra calories that should be considered in patients on dietary restriction, the pivotal intervention to reduce body weight and NAFLD burden.

Probiotics and/or prebiotics could be a promising approach to improve insulin sensitivity by modification of gut microbiota. Clinical data are specifically referred to gestational diabetes [132,133]; in these women four high-quality RCTs (288 participants) showed that treatment was associated with a significant reduction in insulin resistance (HOMA-IR: −0.69%; 95% CI −1.24, −0.14, $p = 0.01$), not in fasting glucose (−0.13 mmol/L; 95% CI −0.32, 0.06, $p = 0.18$) or LDL-cholesterol (−0.16 mmol/L; 95% CI −0.45, 0.13, $p = 0.67$) [133]. In the general diabetes population, the most recent review identified 38 studies totalling 2086 participants fitting pre-defined criteria to be included in a meta-analysis [134]. Overall, the use of prebiotics, probiotics or synbiotics reduced fasting glucose (−0.58 mmol/L; 95% CI −0.86, −0.30; $p < 0.01$), total cholesterol (−0.14 mmol/L; 95% CI −0.26, −0.02, $p = 0.02$) and triglyceride levels (−0.11 mmol/L; 95% CI −0.20, −0.02, $p = 0.01$) and increased HDL-cholesterol (0.04 mmol/L; 95% CI 0.01, 0.07, $p < 0.01$), but failed to reach the significance threshold in HbA1c (−2.17 mmol/mol; 95% CI, −4.37 to 0.03; $p = 0.05$) and had no effect on LDL-cholesterol [134].

Fructans are compounds acting as prebiotics, i.e., non-digestible food ingredients neither metabolized nor absorbed while passing through the upper gastrointestinal tract and fermented by bacteria in the colon. They include fructo-oligosaccharides, galacto-oligosaccharides, lactulose and large polysaccharides (inulin, resistant starches, cellulose, hemicellulose, pectin and gum) [135,136]. Diets rich in fructans might improve glucose metabolism in T2D also via decreased intake and intestinal absorption of food, adding to modifications of gut microbiota [137,138]. A systematic review with meta-analysis of 25 studies did not provide evidence for a beneficial effect on BMI, but inulin-type carbohydrate supplementation reduced fasting glucose (−16.4 mg/dL; 95% CI, −17.6 to −15.2), HbA1c (−0.58%; 95% CI, −0.78 to −0.39), and HOMA-IR (−0.99%; 95% CI, −1.76 to −0.2). However, a large heterogeneity was demonstrated, raising doubts on data validity [139].

4.5. Other Micronutrients

4.5.1. Zinc

Zinc deficiency is common in T2D [140], likely as an effect of both hyperzincuria [141] and reduced intestinal absorption [142], resulting in insulin resistance [143]. Its antioxidant role further strengthens the importance of zinc levels for diabetes control and the prevention of microvascular complications [144].

In the clinical setting, a systematic review with meta-analysis of 12 studies in T2D patients showed that zinc supplementation resulted in a significant reduction of fasting blood glucose (pooled mean difference, −18.1 mg/dL; 95% CI −33.8 to −2.41) and HbA1c (−0.54 %; 95%CI, -0.86 to -0.21), accompanied by a systematic reduction of total and LDL-cholesterol levels [145]. Among diabetes-related complications, zinc supplementation was shown to reduce lipoperoxidation [146] and to decrease urinary albumin excretion, independently of glucose control [147,148]. However, a few studies failed to demonstrate any positive effect of zinc supplementation in the metabolic control of T2D patients [146], also in the presence of long-term supplementation and low zinc levels at baseline [149]. Zinc supplementation might prove useful only in specific settings. In zinc-deficient patients with cirrhosis, independently of diabetes status, zinc treatment (zinc sulfate, 200 mg three times per day) was associated with improved non-insulin-mediated glucose disposal (so-called glucose effectiveness) [150], as well as improved alanine stimulated urea synthesis rate, a measure of amino acid utilization in tissues [151], also resulting in decreased ammonia levels and improved mental state. All these complementary effects might be important in subjects with T2D progressed to NAFLD-cirrhosis [152].

No relevant side effects of zinc supplements have been reported in chronic diseases [153].

4.5.2. Chromium

A possible role of deficient chromium levels as risk factor T2D has long been suggested based on its insulin-sensitising activity, but the effects on human disease remain uncertain. In a large case-control study involving 4443 Chinese individuals (nearly half with either newly diagnosed T2D or newly diagnosed pre-diabetes), plasma chromium levels were approximately 10% lower in the T2D and pre-diabetes groups vs. controls, and the risk of T2D and pre-diabetes decreased across quartiles of chromium [154]. This evidence fits with smaller studies reporting decreased chromium levels and/or increased chromium excretion in T2D [141,155].

The effects of chromium supplementation have been tested in multiple review articles with pooled analysis or metanalysis [156–159]. Based on 25 RCTs of chromium supplementation, Suksomboon et al., concluded for positive effects of chromium supplementation on glucose control in patients with diabetes, with no increased risks of adverse events compared with placebo [156]. On the contrary, Yin et al., in a meta-analysis of 14 trials (875 participants, mean age range: 30 to 83 years old, 8 to 24 weeks of follow-up) did not demonstrate any significant effect of chromium, either as Cr chloride, or Cr picolinate, or Cr yeast) on HbA1c levels [157]. In a review limited to patients with T2D, very few studies reached clinically meaningful goals, defined as fasting plasma glucose (FPG) ≤7.2 mmol/dL, a decline in HbA1c to values ≤7%, or a decrease of ≥0.5% in baseline levels [158]. Finally, in the most recent and largest analysis in T2D (28 studies, 1295 participants, heterogeneous chromium supplements with daily intake ranging up to 3000 μg for 6–24 weeks), the authors concluded for a positive effect of Cr supplements on glucose metabolism [159] and include chromium supplements into the treatment of T2D [159], despite uncertainty about long-term use. Treatment reduced fasting glucose (WMD, −0.99 mmol/L; 95% CI, −1.72 to −0.25), HbA1c (WMD, −0.54 %; 95% CI, −0.82 to −0.25), triglycerides and increased HDL-cholesterol. The effects were mainly reported using both chloride and picolinate formulations and were independent of treatment duration.

4.5.3. Magnesium

Insulin modulates the shift of magnesium from extracellular to intracellular space; in turn, intracellular Mg^{2+} concentration modulates insulin action, as well as blood pressure [160]; thus, low magnesium induces insulin resistance, and insulin resistance further decreases magnesium levels [161]. In the past 20 years, several epidemiological and clinical studies have demonstrated the protective role of magnesium on the risk of diabetes. In U.S. women aged ≥45 years (Women's Health Study) with no previous history of T2D, an inverse association was found between dietary magnesium and incident T2D, which was significant among women with increasing grades of overweight/obesity (P for trend, 0.02). It was associated with a progressive decline of insulin levels (P for trend, 0.03) [162]. Data were confirmed in 1122 individuals (20–65 years of age) enrolled between 1996 and 1997 and re-examined about 10 years later. The relative risk of new-onset prediabetes and T2D were increased in the presence of low magnesium levels at baseline [163].

Oral magnesium supplementation in subjects with T2D and low magnesium levels have been reported to improve insulin sensitivity and metabolic control [164–166]. In a meta-analysis of 40 prospective cohort studies enrolling more than 1 million participants and follow-up periods ranging from 4 to 30 years, dietary magnesium intake was associated with a 19% reduction in the relative risk of T2D (RR 0.81; 95% CI, 0.77–0.86 per 100 mg/day increment) [167]. In a different analysis of 28 studies involving 1694 subjects (834 in the treatment arm and 860 in the placebo arm), magnesium supplementation was demonstrated to produce favourable effects on blood glucose (WMD, −4.64 mg dL, 95% CI −7.60 to −1.68), as well as on HDL- and LDL-cholesterol, triglycerides and systolic blood pressure, also reducing cardiovascular risk [168].

Additionally, for magnesium supplements, no safety concerns have been raised; Verma and coll. argue that large trials should be performed to validate the use of magnesium supplements to prevent and treat T2D [168], but no consensus exists in the community [169].

5. Prevention and Treatment of Diabetes-Related Sarcopenia

Optimal energy intake, healthy food choices and sufficient protein intake, coupled with habitual physical activity, especially resistance training, are the cornerstones for metabolic control and the prevention of frailty in T2D. Despite the mounting evidence of the negative impact of sarcopenia on the natural history [170] and quality of life of T2D patients [171], there is a surprising dearth of intervention studies addressing T2D-related sarcopenia. Therefore, we must rely on findings from general intervention studies on sarcopenia and/or sarcopenic obesity.

Resistance training represents the most effective intervention for prevention and treatment and can be safely carried out even in fragile patients [172]. High protein (1.2–1.4 g/kg) hypocaloric diets—either exclusively food-based or including protein supplements, both as an adjunct to resistance training—have proven effective for preventing muscle mass loss during weight-reduction diets in women with obesity [173]. To reach the anabolic threshold, the protein supplement should be provided at meals rather than between meals in the elderly. The optimal protein dose (including food protein and proteins from supplements) should be 30–45 g of proteins per serving in the elderly [174]. However, high protein load cannot be recommended to T2D patients with chronic kidney disease (CKD) [175].

Whey proteins, rich in the anabolic amino acid leucine, represent the most frequently used protein supplements. Additionally, BCAA supplement or the leucine metabolite β-hydroxy-β-methyl butyrate have been proposed. These supplements are generally ineffective as sole treatment in patients without diabetes [173,176,177] and must be added to resistance training to improve already-established sarcopenia (associated or not to obesity). Leucine has strong insulinotropic properties, and leucine-rich supplements may increase the availability of amino acids for protein synthesis and reduce protein breakdown in the muscle, at the same time enhancing glucose disposal and glycaemic control, but solid data are lacking [178]. A noteworthy issue is that BCAA treatment has proven effective both in preventing and in improving sarcopenia in patients with liver cirrhosis, also independently of physical exercise/resistance training [179,180].

Finally, vitamin D was also proposed as a nutritional supplement to control sarcopenia. The activation of the vitamin D receptor present in muscle cells promotes their differentiation, proliferation and hypertrophy. Vitamin D deficiency is associated with reduced muscle mass and strength in the elderly [181], and vitamin D supplementation increased muscle strength, particularly in vitamin D-deficient cases and in the elderly [181]. Data were not confirmed by a Cochrane review in patients with liver disease; no data are available in T2D [182] and trials are eagerly warranted.

6. Management of Other Comorbidity in Patients with T2D

6.1. Cirrhosis

Nutrition therapy in cirrhosis has already been discussed in this Special Issue of Nutrients. Nonetheless, its association with T2D deserves a special focus considering the high prevalence—up to two-thirds of patients with cirrhosis listed for liver transplantation have T2D [183]—and its importance as a risk factor for the development of complications (ascites, hepatic encephalopathy, bacterial infections, renal insufficiency, hepatocellular carcinoma) [184]. Nutrition treatment becomes extremely challenging since additional determinants of malnutrition may be present, including reduced food intake and/or defective absorption of nutrients and impaired albumin synthesis. Sarcopenia—accelerated by upregulation of myostatin due to hyperammonaemia—becomes a predictor of morbidity and mortality, aggravated by obesity (sarcopenic obesity) [185,186], and is difficult to treat. Bariatric surgery is frequently contraindicated [187]; also pharmacologic treatment with GLP-1 agonists favouring weight loss [188], such as liraglutide, may be contraindicated by the presence of varices at risk of bleeding [189], and dietary treatment remains the sole possibility.

Unfortunately, there are no specific guidelines for the nutritional treatment of T2D associated with cirrhosis, and individualized, structured nutritional programs are suggested to accomplish the need for restriction of sodium and fluids [190]. Due to the accelerated depletion of glycogen stores, it is important to provide frequent (3 to 5) meals containing carbohydrates, plus a late evening carbohydrate snack to prevent muscle protein catabolism [191,192].

Protein restriction is not systematically advocated, as these patients usually tolerate a normal protein intake. Besides hypoalbuminemia, potentially requiring a higher protein intake, albumin glycation is present in T2D [193]. The structurally damaged albumin molecule is also dysfunctional, and albumin administration may be required to reduce ascites. Although the specific indications for use are clearly defined by international guidelines [194], albumin is frequently administered outside evidence-based indications, including nutritional support [195]. At present, no studies showed a direct link between albumin administration and nutritional correction in decompensated cirrhosis; it can only be hypothesized that the clinical improvement seen with long-term albumin treatment could indirectly improve the nutritional status through different mechanisms, which include the control/resolution of ascites and whole body edema, or the reduction of systemic inflammation [196].

6.2. Renal Failure

In T2D patients with CKD, protein restriction may be advised; low protein diets (daily protein intake reduced to 0.8 g/kg b.w.) showed a beneficial impact on the trajectory of renal function leading to an attenuation in the progression of CKD and delayed initiation of dialysis treatment, an important goal for patients [197–199]. However, protein restriction may worsen sarcopenia and should be limited as long as possible. According to the National Kidney Foundation/Kidney Disease Outcomes Quality Initiative (NKF-KDOQI) Guidelines, protein intake must actually be increased up to 1.2 g/kg body in patients undergoing maintenance dialysis due to important additional amino acid losses occurring in dialysate [200,201].

Different sources of dietary protein may have a different impact on CKD-related complications; meat intake increases the production of nitrogenous end products, worsens uraemia and may increase the risk of constipation with consequent hyperkalaemia associated with the low fibre intake [199]. A predominantly plant-based diet, fibre-rich and low in protein content (0.6–0.8 g/kg/day), can produce favourable changes in the intestinal microbiome, thus modulating the generation of uremic toxins and slowing down the progression of CKD, finally reducing cardiovascular risk [202]. Carbohydrates from sugars should be limited to less than 10% of the energy intake [203], and saturated fatty acids, trans fats, and cholesterol should be replaced by polyunsaturated and monounsaturated fats, associated with more favourable outcomes [204]. Dietary sodium restriction should be considered, but a deficient sodium intake (to less than 1.5–2.0 g/day) carries the risk of hyponatremia, leading to reduced insulin sensitivity and prediabetes [205]. T2D patients with advanced CKD progressing to end-stage renal disease may be prone to the "burnt-out diabetes" phenomenon (i.e., spontaneous resolution of hyperglycaemia and frequent hypoglycaemic episodes); further studies in this frail population in chronic hemodialysis treatment are particularly needed to determine the safety and the effectiveness of dietary manipulations [206].

7. Conclusions

T2D is the paradigm of conditions where genetic, behavioural and individual factors drive disease occurrence and severity. Despite decades of epidemiological studies and randomized trials, several unmet needs remain (Table 4). The goal of optimal nutritional approach is to maintain or regain a body weight within the normal range, providing adequate intake of macronutrients and micronutrients to reduce the risk of sarcopenia. Various dietary approaches have been proposed to improve outcome, with the Mediterranean diet supported by solid evidence. However, as long-term adherence is the main goal to be achieved, the dietary plan and the calorie restriction that patients feel confident to maintain life-long should always be preferred. At present, supplementation with inositols, vitamin D and micronutrients (zinc, chromium, magnesium) is not systematically suggested, but might be considered in individual patients.

Table 4. Principal unmet needs for optimal nutritional treatment of patients with type 2 diabetes.

Target	Unmet Needs
1-2 Weight control	Define the best dietary plan to support weight loss and weight loss maintenance Define the role of psychological support in individual, difficult cases Define the role of e-health technology and individual apps to improve long-term adherence to dietary recommendations
Prevention and treatment of sarcopenia	Define the optimal protein intake to prevent sarcopenia How to integrate physical activity in the daily life of frail patients Assess the relative role of resistance vs. aerobic exercise
Vitamins and nutritional supplements	Are they really needed (if, when, to whom)? Who should be screened for micronutrient deficit Define the role of nutritional supplements in the prevention of sarcopenia Define the possible interaction of vitamins and nutrients with drug treatment

Although advances in nutrigenomics and metabolomics offer the rationale for tailored precision medicine, a personalized meal plan, supported by continuous dietary counselling by registered dietitians remains at present the key strategy for long-term success in weight and glycaemic control [37], particularly in individual high-risk cases [38].

Author Contributions: Conceptualization, M.L.P., G.M. and F.R.; literature search, M.L.P., L.B., F.M., A.S.S., P.C., G.M. and F.R.; writing—original draft preparation, M.L.P., G.M. and F.R.; writing—review and editing, M.L.P., L.B., F.M., A.S.S., P.C., G.M. and F.R. All authors have read and agreed to the published version of the manuscript.

Funding: This research received no external funding.

Institutional Review Board Statement: Not applicable.

Informed Consent Statement: Not applicable.

Data Availability Statement: Not applicable.

Acknowledgments: F.M. is supported by a contract financed by the Italian Ministry of Health and Italian Regions (NET-2016-02364191).

Conflicts of Interest: The authors declare no conflict of interest.

References

1. International Diabetes Federation. *IDF Diabetes Atlas*, 9th ed.; International Diabetes Federation: Brussels, Belgium, 2019; Volume 1. Available online: https://www.diabetesatlas.org/en/ (accessed on 8 June 2021).
2. American Diabetes Association. 3. Prevention or delay of type 2 diabetes: Standards of medical care in Diabetes-2021. *Diabetes Care* **2021**, *44* (Suppl. S1), S34–S39. [CrossRef]
3. Cruz-Jentoft, A.J.; Landi, F.; Schneider, S.M.; Zuniga, C.; Arai, H.; Boirie, Y.; Chen, L.K.; Fielding, R.A.; Martin, F.C.; Michel, J.P.; et al. Prevalence of and interventions for sarcopenia in ageing adults: A systematic review. Report of the International Sarcopenia Initiative (EWGSOP and IWGS). *Age Ageing* **2014**, *43*, 748–759. [CrossRef] [PubMed]
4. Kalyani, R.R.; Corriere, M.; Ferrucci, L. Age-related and disease-related muscle loss: The effect of diabetes, obesity, and other diseases. *Lancet Diabetes Endocrinol.* **2014**, *2*, 819–829. [CrossRef]
5. American Diabetes Association. 5. Facilitating behavior change and well-being to improve health outcomes: Standards of medical care in Diabetes-2021. *Diabetes Care* **2021**, *44* (Suppl. S1), S53–S72. [CrossRef] [PubMed]
6. American Diabetes Association. 8. Obesity management for the treatment of type 2 diabetes: Standards of medical care in Diabetes-2021. *Diabetes Care* **2021**, *44* (Suppl. S1), S100–S110. [CrossRef]
7. Sievenpiper, J.L.; Chan, C.B.; Dworatzek, P.D.; Freeze, C.; Williams, S.L. Diabetes Canada Clinical Practice Guidelines Expert Committee. Nutrition therapy. *Can. J. Diabetes* **2018**, *42* (Suppl. S1), S64–S79. [CrossRef] [PubMed]
8. Franz, M.J.; MacLeod, J.; Evert, A.; Brown, C.; Gradwell, E.; Handu, D.; Reppert, A.; Robinson, M. Academy of Nutrition and Dietetics Nutrition Practice Guideline for type 1 and type 2 diabetes in adults: Systematic review of evidence for medical nutrition therapy effectiveness and recommendations for integration into the nutrition care process. *J. Acad. Nutr. Diet.* **2017**, *117*, 1659–1679. [CrossRef] [PubMed]
9. Associazione Medici Diabetologi (AMD); Società Italiana di Diabetologia (SID). *Standard Italiani per la Cura del Diabete*; AMD, SID: Rome, Italic, 2018.
10. Leslie, W.S.; Ford, I.; Sattar, N.; Hollingsworth, K.G.; Adamson, A.; Sniehotta, F.F.; McCombie, L.; Brosnahan, N.; Ross, H.; Mathers, J.C.; et al. The Diabetes Remission Clinical Trial (DiRECT): Protocol for a cluster randomised trial. *BMC Fam. Pract.* **2016**, *17*, 20. [CrossRef]
11. Dalle Grave, R.; Calugi, S.; Centis, E.; Marzocchi, R.; El Ghoch, M.; Marchesini, G. Lifestyle modification in the management of the metabolic syndrome: Achievements and challenges. *Diabetes Metab. Syndr. Obes. Targets Ther.* **2010**, *3*, 373–385. [CrossRef]
12. National Institutes of Health. Successful diet and exercise therapy is conducted in Vermont for "diabesity". *JAMA* **1980**, *243*, 519–520. [CrossRef]
13. Astrup, A.; Finer, N. Redefining type 2 diabetes: 'Diabesity' or 'obesity dependent diabetes mellitus'? *Obes. Rev.* **2000**, *1*, 57–59. [CrossRef] [PubMed]
14. Zimmet, P. The Negev desert, migration, islands and diabesity: Genes and environment clashing. In Proceedings of the Cohen Memorial Lecture 36th Annual Meeting of the European Association for the Study of Diabetes, Jerusalem, Israel, 17–21 September 2000.
15. Flegal, K.M.; Shepherd, J.A.; Looker, A.C.; Graubard, B.I.; Borrud, L.G.; Ogden, C.L.; Harris, T.B.; Everhart, J.E.; Schenker, N. Comparisons of percentage body fat, body mass index, waist circumference, and waist-stature ratio in adults. *Am. J. Clin. Nutr.* **2009**, *89*, 500–508. [CrossRef] [PubMed]
16. Wilson, M.M.; Morley, J.E. Invited review: Aging and energy balance. *J. Appl. Physiol.* **2003**, *95*, 1728–1736. [CrossRef] [PubMed]
17. The Lancet Diabetes Endocrinology. Sarcopenia: A fate worth challenging. *Lancet Diabetes Endocrinol.* **2014**, *2*, 183. [CrossRef]
18. Marini, E.; Buffa, R.; Saragat, B.; Coin, A.; Toffanello, E.D.; Berton, L.; Manzato, E.; Sergi, G. The potential of classic and specific bioelectrical impedance vector analysis for the assessment of sarcopenia and sarcopenic obesity. *Clin. Interv. Aging* **2012**, *7*, 585–591. [CrossRef]
19. Buffa, R.; Saragat, B.; Cabras, S.; Rinaldi, A.C.; Marini, E. Accuracy of specific BIVA for the assessment of body composition in the United States population. *PLoS ONE* **2013**, *8*, e58533. [CrossRef]

20. Cruz-Jentoft, A.J.; Bahat, G.; Bauer, J.; Boirie, Y.; Bruyere, O.; Cederholm, T.; Cooper, C.; Landi, F.; Rolland, Y.; Sayer, A.A.; et al. Sarcopenia: Revised European consensus on definition and diagnosis. *Age Ageing* **2019**, *48*, 16–31. [CrossRef]
21. Izzo, A.; Massimino, E.; Riccardi, G.; Della Pepa, G. A narrative review on sarcopenia in type 2 diabetes mellitus: Prevalence and associated factors. *Nutrients* **2021**, *13*, 13. [CrossRef]
22. Anagnostis, P.; Gkekas, N.K.; Achilla, C.; Pananastasiou, G.; Taouxidou, P.; Mitsiou, M.; Kenanidis, E.; Potoupnis, M.; Tsiridis, E.; Goulis, D.G. Type 2 diabetes mellitus is associated with increased risk of sarcopenia: A systematic review and meta-analysis. *Calcif. Tissue Int.* **2020**, *107*, 453–463. [CrossRef]
23. Solanki, J.D.; Makwana, A.H.; Mehta, H.B.; Gokhale, P.A.; Shah, C.J. Body composition in type 2 diabetes: Change in quality and not just quantity that matters. *Int. J. Prev. Med.* **2015**, *6*, 122. [CrossRef]
24. Bawadi, H.; Alkhatib, D.; Abu-Hijleh, H.; Alalwani, J.; Majed, L.; Shi, Z. Muscle strength and glycaemic control among patients with type 2 diabetes. *Nutrients* **2020**, *12*, 771. [CrossRef]
25. Lombardo, M.; Padua, E.; Campoli, F.; Panzarino, M.; Mindrescu, V.; Annino, G.; Iellamo, F.; Bellia, A. Relative handgrip strength is inversely associated with the presence of type 2 diabetes in overweight elderly women with varying nutritional status. *Acta Diabetol.* **2021**, *58*, 25–32. [CrossRef] [PubMed]
26. Leenders, M.; Verdijk, L.B.; van der Hoeven, L.; Adam, J.J.; van Kranenburg, J.; Nilwik, R.; van Loon, L.J. Patients with type 2 diabetes show a greater decline in muscle mass, muscle strength, and functional capacity with aging. *J. Am. Med. Dir. Assoc.* **2013**, *14*, 585–592. [CrossRef]
27. Veronese, N.; Stubbs, B.; Punzi, L.; Soysal, P.; Incalzi, R.A.; Saller, A.; Maggi, S. Effect of nutritional supplementations on physical performance and muscle strength parameters in older people: A systematic review and meta-analysis. *Ageing Res. Rev.* **2019**, *51*, 48–54. [CrossRef] [PubMed]
28. Petersen, K.F.; Dufour, S.; Savage, D.B.; Bilz, S.; Solomon, G.; Yonemitsu, S.; Cline, G.W.; Befroy, D.; Zemany, L.; Kahn, B.B.; et al. The role of skeletal muscle insulin resistance in the pathogenesis of the metabolic syndrome. *Proc. Natl. Acad. Sci. USA* **2007**, *104*, 12587–12594. [CrossRef]
29. Haedersdal, S.; Lund, A.; Knop, F.K.; Vilsboll, T. The role of glucagon in the pathophysiology and treatment of type 2 diabetes. *Mayo Clin. Proc.* **2018**, *93*, 217–239. [CrossRef] [PubMed]
30. Taylor, R. Type 2 diabetes: Etiology and reversibility. *Diabetes Care* **2013**, *36*, 1047–1055. [CrossRef] [PubMed]
31. Galsgaard, K.D.; Winther-Sorensen, M.; Orskov, C.; Kissow, H.; Poulsen, S.S.; Vilstrup, H.; Prehn, C.; Adamski, J.; Jepsen, S.L.; Hartmann, B.; et al. Disruption of glucagon receptor signaling causes hyperaminoacidemia exposing a possible liver-alpha-cell axis. *Am. J. Physiol. Endocrinol. Metab.* **2018**, *314*, E93–E103. [CrossRef] [PubMed]
32. Wewer Albrechtsen, N.J.; Faerch, K.; Jensen, T.M.; Witte, D.R.; Pedersen, J.; Mahendran, Y.; Jonsson, A.E.; Galsgaard, K.D.; Winther-Sorensen, M.; Torekov, S.S.; et al. Evidence of a liver-alpha cell axis in humans: Hepatic insulin resistance attenuates relationship between fasting plasma glucagon and glucagonotropic amino acids. *Diabetologia* **2018**, *61*, 671–680. [CrossRef] [PubMed]
33. Younossi, Z.M.; Golabi, P.; de Avila, L.; Paik, J.M.; Srishord, M.; Fukui, N.; Qiu, Y.; Burns, L.; Afendy, A.; Nader, F. The global epidemiology of NAFLD and NASH in patients with type 2 diabetes: A systematic review and meta-analysis. *J. Hepatol.* **2019**, *71*, 793–801. [CrossRef] [PubMed]
34. Felig, P. Amino acid metabolism in man. *Annu. Rev. Biochem.* **1975**, *44*, 933–955. [CrossRef] [PubMed]
35. Forlani, G.; Vannini, P.; Marchesini, G.; Zoli, M.; Ciavarella, A.; Pisi, E. Insulin-dependent metabolism of branched-chain amino acids in obesity. *Metabolism* **1984**, *33*, 147–150. [CrossRef]
36. Evert, A.B.; Dennison, M.; Gardner, C.D.; Garvey, W.T.; Lau, K.H.K.; MacLeod, J.; Mitri, J.; Pereira, R.F.; Rawlings, K.; Robinson, S.; et al. Nutrition therapy for adults with diabetes or prediabetes: A consensus report. *Diabetes Care* **2019**, *42*, 731–754. [CrossRef] [PubMed]
37. Forouhi, N.G.; Misra, A.; Mohan, V.; Taylor, R.; Yancy, W. Dietary and nutritional approaches for prevention and management of type 2 diabetes. *BMJ* **2018**, *361*, k2234. [CrossRef] [PubMed]
38. Wang, D.D.; Hu, F.B. Precision nutrition for prevention and management of type 2 diabetes. *Lancet Diabetes Endocrinol.* **2018**, *6*, 416–426. [CrossRef]
39. Ojo, O. Dietary intake and type 2 diabetes. *Nutrients* **2019**, *11*, 2177. [CrossRef]
40. Meng, Y.; Bai, H.; Wang, S.; Li, Z.; Wang, Q.; Chen, L. Efficacy of low carbohydrate diet for type 2 diabetes mellitus management: A systematic review and meta-analysis of randomized controlled trials. *Diabetes Res. Clin. Pract.* **2017**, *131*, 124–131. [CrossRef]
41. Ojo, O.; Ojo, O.O.; Adebowale, F.; Wang, X.H. The effect of dietary glycaemic index on glycaemia in patients with type 2 diabetes: A systematic review and meta-analysis of randomized controlled trials. *Nutrients* **2018**, *10*, 373. [CrossRef]
42. Hamdy, O.; Barakatun-Nisak, M.Y. Nutrition in Diabetes. *Endocrinol. Metab. Clin. N. Am.* **2016**, *45*, 799–817. [CrossRef]
43. Vogtschmidt, Y.D.; Raben, A.; Faber, I.; de Wilde, C.; Lovegrove, J.A.; Givens, D.I.; Pfeiffer, A.F.H.; Soedamah-Muthu, S.S. Is protein the forgotten ingredient: Effects of higher compared to lower protein diets on cardiometabolic risk factors. A systematic review and meta-analysis of randomised controlled trials. *Atherosclerosis* **2021**, *328*, 124–135. [CrossRef]
44. Xue, L.; Yin, R.; Howell, K.; Zhang, P. Activity and bioavailability of food protein-derived angiotensin-I-converting enzyme-inhibitory peptides. *Compr. Rev. Food Sci. Food Saf.* **2021**, *20*, 1150–1187. [CrossRef]

45. Qian, F.; Korat, A.A.; Malik, V.; Hu, F.B. Metabolic effects of monounsaturated fatty acid-enriched diets compared with carbohydrate or polyunsaturated fatty acid-enriched diets in patients with type 2 diabetes: A systematic review and meta-analysis of randomized controlled trials. *Diabetes Care* **2016**, *39*, 1448–1457. [CrossRef]
46. Georgoulis, M.; Kontogianni, M.D.; Yiannakouris, N. Mediterranean diet and diabetes: Prevention and treatment. *Nutrients* **2014**, *6*, 1406–1423. [CrossRef]
47. Mesías, M.; Seiquer, I.; Delgado-Andrade, C. The Mediterranean diet and mineral composition. In *The Mediterranean Diet: An Evidence-Based Approach*, 2nd ed.; Preedy, V., Watson, R., Eds.; Academic Press Books: Cambridge, MS, USA, 2020; pp. 151–163.
48. U.S. Department of Health and Human Services; U.S. Department of Agriculture. 2015–2020 Dietary Guidelines for Americans, 8th ed. 2015. Available online: http://health.gov/dietaryguidelines/2015/guidelines (accessed on 8 June 2021).
49. Campbell, A.P. DASH eating plan: An eating pattern for diabetes management. *Diabetes Spectr.* **2017**, *30*, 76–81. [CrossRef] [PubMed]
50. Chester, B.; Babu, J.R.; Greene, M.W.; Geetha, T. The effects of popular diets on type 2 diabetes management. *Diabetes Metab. Res. Rev.* **2019**, *35*, e3188. [CrossRef] [PubMed]
51. Tuomilehto, J.; Lindstrom, J.; Eriksson, J.G.; Valle, T.T.; Hamalainen, H.; Ilanne-Parikka, P.; Keinanen-Kiukaanniemi, S.; Laakso, M.; Louheranta, A.; Rastas, M.; et al. Prevention of type 2 diabetes mellitus by changes in lifestyle among subjects with impaired glucose tolerance. *N. Engl. J. Med.* **2001**, *344*, 1343–1350. [CrossRef] [PubMed]
52. The Diabetes Prevention Program Research Group. The Diabetes Prevention Program (DPP): Description of lifestyle intervention. *Diabetes Care* **2002**, *25*, 2165–2171. [CrossRef]
53. The Look Ahead Research Group; Wadden, T.A.; West, D.S.; Delahanty, L.; Jakicic, J.; Rejeski, J.; Williamson, D.; Berkowitz, R.I.; Kelley, D.E.; Tomchee, C.; et al. The Look AHEAD study: A description of the lifestyle intervention and the evidence supporting it. *Obesity* **2006**, *14*, 737–752. [CrossRef]
54. Wing, R.R.; Bolin, P.; Brancati, F.L.; Bray, G.A.; Clark, J.M.; Coday, M.; Crow, R.S.; Curtis, J.M.; Egan, C.M.; Espeland, M.A.; et al. Cardiovascular effects of intensive lifestyle intervention in type 2 diabetes. *N. Engl. J. Med.* **2013**, *369*, 145–154. [CrossRef] [PubMed]
55. Petroni, M.L.; Caletti, M.T.; Calugi, S.; Dalle Grave, R.; Marchesini, G. Long-term treatment of severe obesity: Are lifestyle interventions still an option? *Expert Rev. Endocrinol. Metab.* **2017**, *12*, 391–400. [CrossRef]
56. Ikramuddin, S.; Korner, J.; Lee, W.J.; Thomas, A.J.; Connett, J.E.; Bantle, J.P.; Leslie, D.B.; Wang, Q.; Inabnet, W.B., 3rd; Jeffery, R.W.; et al. Lifestyle Intervention and medical management with vs without Roux-en-Y gastric bypass and control of hemoglobin A1c, LDL cholesterol, and systolic blood pressure at 5 years in the Diabetes Surgery Study. *JAMA* **2018**, *319*, 266–278. [CrossRef]
57. Lean, M.E.; Leslie, W.S.; Barnes, A.C.; Brosnahan, N.; Thom, G.; McCombie, L.; Peters, C.; Zhyzhneuskaya, S.; Al-Mrabeh, A.; Hollingsworth, K.G.; et al. Primary care-led weight management for remission of type 2 diabetes (DiRECT): An open-label, cluster-randomised trial. *Lancet* **2018**, *391*, 541–551. [CrossRef]
58. Lean, M.E.J.; Leslie, W.S.; Barnes, A.C.; Brosnahan, N.; Thom, G.; McCombie, L.; Peters, C.; Zhyzhneuskaya, S.; Al-Mrabeh, A.; Hollingsworth, K.G.; et al. Durability of a primary care-led weight-management intervention for remission of type 2 diabetes: 2-year results of the DiRECT open-label, cluster-randomised trial. *Lancet Diabetes Endocrinol.* **2019**, *7*, 344–355. [CrossRef]
59. Taylor, R.; Al-Mrabeh, A.; Zhyzhneuskaya, S.; Peters, C.; Barnes, A.C.; Aribisala, B.S.; Hollingsworth, K.G.; Mathers, J.C.; Sattar, N.; Lean, M.E.J. Remission of human type 2 diabetes requires decrease in liver and pancreas fat content but Is dependent upon capacity for beta cell recovery. *Cell Metab.* **2018**, *28*, 547–556. [CrossRef] [PubMed]
60. Garcia-Molina, L.; Lewis-Mikhael, A.M.; Riquelme-Gallego, B.; Cano-Ibanez, N.; Oliveras-Lopez, M.J.; Bueno-Cavanillas, A. Improving type 2 diabetes mellitus glycaemic control through lifestyle modification implementing diet intervention: A systematic review and meta-analysis. *Eur. J. Nutr.* **2020**, *59*, 1313–1328. [CrossRef]
61. Grajower, M.M.; Horne, B.D. Clinical management of intermittent fasting in patients with diabetes mellitus. *Nutrients* **2019**, *11*, 873. [CrossRef] [PubMed]
62. Rajpal, A.; Ismail-Beigi, F. Intermittent fasting and 'metabolic switch': Effects on metabolic syndrome, prediabetes and type 2 diabetes. *Diabetes Obes. Metab.* **2020**, *22*, 1496–1510. [CrossRef]
63. Horne, B.D.; Grajower, M.M.; Anderson, J.L. Limited evidence for the health effects and safety of intermittent fasting among patients with type 2 diabetes. *JAMA* **2020**, *324*, 341–342. [CrossRef]
64. Higgins, J.P. Smartphone applications for patients' health and fitness. *Am. J. Med.* **2016**, *129*, 11–19. [CrossRef]
65. Toro-Ramos, T.; Michaelides, A.; Anton, M.; Karim, Z.; Kang-Oh, L.; Argyrou, C.; Loukaidou, E.; Charitou, M.M.; Sze, W.; Miller, J.D. Mobile delivery of the Diabetes Prevention Program in people with prediabetes: Randomized controlled trial. *JMIR mHealth uHealth* **2020**, *8*, e17842. [CrossRef]
66. Stein, N.; Brooks, K. A fully automated conversational artificial intelligence for weight loss: Longitudinal cbservational study among overweight and obese ddults. *JMIR Diabetes* **2017**, *2*, e28. [CrossRef] [PubMed]
67. Chin, S.O.; Keum, C.; Woo, J.; Park, J.; Choi, H.J.; Woo, J.T.; Rhee, S.Y. Successful weight reduction and maintenance by using a smartphone application in those with overweight and obesity. *Sci. Rep.* **2016**, *6*, 34563. [CrossRef] [PubMed]
68. Mazzotti, A.; Caletti, M.T.; Brodosi, L.; Di Domizio, S.; Forchielli, M.L.; Petta, S.; Bugianesi, E.; Bianchi, G.; Marchesini, G. An internet-based approach for lifestyle changes in patients with NAFLD: Two-year effects on weight loss and surrogate markers. *J. Hepatol.* **2018**, *69*, 1155–1163. [CrossRef] [PubMed]

69. Nestler, J.E.; Unfer, V. Reflections on inositol(s) for PCOS therapy: Steps toward success. *Gynecol. Endocrinol.* **2015**, *31*, 501–505. [CrossRef]
70. Unfer, V.; Nestler, J.E.; Kamenov, Z.A.; Prapas, N.; Facchinetti, F. Effects of inositol(s) in women with PCOS: A systematic review of randomized controlled trials. *Int. J. Endocrinol.* **2016**, *2016*, 1849162. [CrossRef]
71. Larner, J. D-chiro-inositol–its functional role in insulin action and its deficit in insulin resistance. *Int. J. Exp. Diabetes Res.* **2002**, *3*, 47–60. [CrossRef]
72. Walters, M.R. Newly identified actions of the vitamin D endocrine system. *Endocr. Rev.* **1992**, *13*, 719–764. [CrossRef]
73. Pittas, A.G.; Harris, S.S.; Stark, P.C.; Dawson-Hughes, B. The effects of calcium and vitamin D supplementation on blood glucose and markers of inflammation in nondiabetic adults. *Diabetes Care* **2007**, *30*, 980–986. [CrossRef]
74. Wang, W.; Zhang, J.; Wang, H.; Wang, X.; Liu, S. Vitamin D deficiency enhances insulin resistance by promoting inflammation in type 2 diabetes. *Int J. Clin. Exp. Pathol* **2019**, *12*, 1859–1867.
75. Chiu, K.C.; Chu, A.; Go, V.L.; Saad, M.F. Hypovitaminosis D is associated with insulin resistance and beta cell dysfunction. *Am. J. Clin. Nutr.* **2004**, *79*, 820–825. [CrossRef]
76. Montserrat-de la Paz, S.; Bermudez, B.; Naranjo, M.C.; Lopez, S.; Abia, R.; Muriana, F.J. Pharmacological effects of niacin on acute hyperlipemia. *Curr. Med. Chem.* **2016**, *23*, 2826–2835. [CrossRef]
77. Cicero, A.F.; Tartagni, E.; Ertek, S. Nutraceuticals for metabolic syndrome management: From laboratory to benchside. *Curr. Vasc. Pharmacol.* **2014**, *12*, 565–571. [CrossRef]
78. Xia, N.; Daiber, A.; Forstermann, U.; Li, H. Antioxidant effects of resveratrol in the cardiovascular system. *Br. J. Pharmacol.* **2017**, *174*, 1633–1646. [CrossRef] [PubMed]
79. Al Bander, Z.; Nitert, M.D.; Mousa, A.; Naderpoor, N. The gut microbiota and inflammation: An overview. *Int. J. Environ. Res. Public Health* **2020**, *17*, 7618. [CrossRef] [PubMed]
80. Salazar, J.; Angarita, L.; Morillo, V.; Navarro, C.; Martinez, M.S.; Chacin, M.; Torres, W.; Rajotia, A.; Rojas, M.; Cano, C.; et al. Microbiota and diabetes mellitus: Role of lipid mediators. *Nutrients* **2020**, *12*, 3039. [CrossRef] [PubMed]
81. Tanase, D.M.; Gosav, E.M.; Neculae, E.; Costea, C.F.; Ciocoiu, M.; Hurjui, L.L.; Tarniceriu, C.C.; Maranduca, M.A.; Lacatusu, C.M.; Floria, M.; et al. Role of gut microbiota on onset and progression of microvascular complications of type 2 diabetes (T2DM). *Nutrients* **2020**, *12*, 3719. [CrossRef]
82. Tiderencel, K.A.; Hutcheon, D.A.; Ziegler, J. Probiotics for the treatment of type 2 diabetes: A review of randomized controlled trials. *Diabetes Metab. Res. Rev.* **2020**, *36*, e3213. [CrossRef]
83. Woldeamlak, B.; Yirdaw, K.; Biadgo, B. Role of gut microbiota in type 2 diabetes mellitus and its complications: Novel insights and potential intervention strategies. *Korean J. Gastroenterol.* **2019**, *74*, 314–320. [CrossRef]
84. Shan, Z.; Bao, W.; Zhang, Y.; Rong, Y.; Wang, X.; Jin, Y.; Song, Y.; Yao, P.; Sun, C.; Hu, F.B.; et al. Interactions between zinc transporter-8 gene (SLC30A8) and plasma zinc concentrations for impaired glucose regulation and type 2 diabetes. *Diabetes* **2014**, *63*, 1796–1803. [CrossRef]
85. Fukunaka, A.; Fujitani, Y. Role of zinc homeostasis in the pathogenesis of diabetes and obesity. *Int. J. Mol. Sci.* **2018**, *19*, 476. [CrossRef]
86. Maret, W. Chromium supplementation in human health, metabolic syndrome, and diabetes. *Met. Ions Life Sci.* **2019**, *19*, 231–252. [CrossRef]
87. Yang, X.; Li, S.Y.; Dong, F.; Ren, J.; Sreejayan, N. Insulin-sensitizing and cholesterol-lowering effects of chromium (D-Phenylalanine)3. *J. Inorg. Biochem.* **2006**, *100*, 1187–1193. [CrossRef]
88. Feng, J.; Wang, H.; Jing, Z.; Wang, Y.; Cheng, Y.; Wang, W.; Sun, W. Role of magnesium in type 2 diabetes mellitus. *Biol. Trace Elem. Res.* **2020**, *196*, 74–85. [CrossRef]
89. Paolisso, G.; Barbagallo, M. Hypertension, diabetes mellitus, and insulin resistance: The role of intracellular magnesium. *Am. J. Hypertens.* **1997**, *10*, 346–355. [CrossRef]
90. D'Anna, R.; Santamaria, A.; Alibrandi, A.; Corrado, F.; DI Benedetto, A.; Facchinetti, F. Myo-inositol for the prevention of gestational diabetes mellitus. A brief review. *J. Nutr. Sci. Vitaminol.* **2019**, *65*, S59–S61. [CrossRef] [PubMed]
91. Formoso, G.; Baldassarre, M.P.A.; Ginestra, F.; Carlucci, M.A.; Bucci, I.; Consoli, A. Inositol and antioxidant supplementation: Safety and efficacy in pregnancy. *Diabetes Metab. Res. Rev.* **2019**, *35*, e3154. [CrossRef]
92. Guardo, F.D.; Curro, J.M.; Valenti, G.; Rossetti, P.; Di Gregorio, L.M.; Conway, F.; Chiofalo, B.; Garzon, S.; Bruni, S.; Rizzo, G. Non-pharmacological management of gestational diabetes: The role of myo-inositol. *J. Complement. Integr. Med.* **2019**, *17*, 20190111. [CrossRef]
93. Tahir, F.; Majid, Z. Inositol supplementation in the prevention of gestational diabetes mellitus. *Cureus* **2019**, *11*, e5671. [CrossRef] [PubMed]
94. Brown, J.; Crawford, T.J.; Alsweiler, J.; Crowther, C.A. Dietary supplementation with myo-inositol in women during pregnancy for treating gestational diabetes. *Cochrane Database Syst. Rev.* **2016**, *9*, CD012048. [CrossRef] [PubMed]
95. Associazione Medici Diabetologi; Società Italiana di Diabetologia. Position Statement: Integratori Vitaminici, Inositolo e Probiotici nelle donne con Iperglicemia in Gravidanza. 2018. Available online: https://www.siditalia.it/news/2103-17-10-2018-position-statement-amd-sid-integratori-vitaminici-inositolo-e-probiotici-nelle-donne-con-iperglicemia-in-gravidanza.edn (accessed on 8 June 2021).

96. Omoruyi, F.O.; Stennett, D.; Foster, S.; Dilworth, L. New frontiers for the use of IP6 and inositol combination in treating diabetes mellitus: A review. *Molecules* **2020**, *25*, 1720. [CrossRef]
97. Davinelli, S.; Nicolosi, D.; Di Cesare, C.; Scapagnini, G.; Di Marco, R. Targeting metabolic consequences of insulin resistance in polycystic ovary syndrome by D-chiro-inositol and emerging nutraceuticals: A focused review. *J. Clin. Med.* **2020**, *9*, 987. [CrossRef]
98. Cheng, S.; Massaro, J.M.; Fox, C.S.; Larson, M.G.; Keyes, M.J.; McCabe, E.L.; Robins, S.J.; O'Donnell, C.J.; Hoffmann, U.; Jacques, P.F.; et al. Adiposity, cardiometabolic risk, and vitamin D status: The Framingham Heart Study. *Diabetes* **2010**, *59*, 242–248. [CrossRef]
99. Mitri, J.; Muraru, M.D.; Pittas, A.G. Vitamin D and type 2 diabetes: A systematic review. *Eur. J. Clin. Nutr.* **2011**, *65*, 1005–1015. [CrossRef] [PubMed]
100. Ford, E.S.; Ajani, U.A.; McGuire, L.C.; Liu, S. Concentrations of serum vitamin D and the metabolic syndrome among U.S. adults. *Diabetes Care* **2005**, *28*, 1228–1230. [CrossRef] [PubMed]
101. Scragg, R.; Sowers, M.; Bell, C. Serum 25-hydroxyvitamin D, diabetes, and ethnicity in the Third National Health and Nutrition Examination Survey. *Diabetes Care* **2004**, *27*, 2813–2818. [CrossRef]
102. Eliades, M.; Spyrou, E.; Agrawal, N.; Lazo, M.; Brancati, F.L.; Potter, J.J.; Koteish, A.A.; Clark, J.M.; Guallar, E.; Hernaez, R. Meta-analysis: Vitamin D and non-alcoholic fatty liver disease. *Aliment. Pharmacol. Ther.* **2013**, *38*, 246–254. [CrossRef] [PubMed]
103. Lee, J.H.; O'Keefe, J.H.; Bell, D.; Hensrud, D.D.; Holick, M.F. Vitamin D deficiency an important, common, and easily treatable cardiovascular risk factor? *J. Am. Coll. Cardiol.* **2008**, *52*, 1949–1956. [CrossRef]
104. Tai, K.; Need, A.G.; Horowitz, M.; Chapman, I.M. Vitamin D, glucose, insulin, and insulin sensitivity. *Nutrition* **2008**, *24*, 279–285. [CrossRef]
105. Avenell, A.; Cook, J.A.; MacLennan, G.S.; McPherson, G.C. Vitamin D supplementation and type 2 diabetes: A substudy of a randomised placebo-controlled trial in older people (RECORD trial, ISRCTN 51647438). *Age Ageing* **2009**, *38*, 606–609. [CrossRef] [PubMed]
106. Li, X.; Liu, Y.; Zheng, Y.; Wang, P.; Zhang, Y. The effect of vitamin D supplementation on glycemic control in type 2 diabetes patients: A systematic review and meta-analysis. *Nutrients* **2018**, *10*, 375. [CrossRef]
107. Niroomand, M.; Fotouhi, A.; Irannejad, N.; Hosseinpanah, F. Does high-dose vitamin D supplementation impact insulin resistance and risk of development of diabetes in patients with pre-diabetes? A double-blind randomized clinical trial. *Diabetes Res. Clin. Pract.* **2019**, *148*, 1–9. [CrossRef]
108. Polo, V.; Saibene, A.; Pontiroli, A.E. Nicotinamide improves insulin secretion and metabolic control in lean type 2 diabetic patients with secondary failure to sulphonylureas. *Acta Diabetol.* **1998**, *35*, 61–64. [CrossRef] [PubMed]
109. Xiang, D.; Zhang, Q.; Wang, Y.T. Effectiveness of niacin supplementation for patients with type 2 diabetes: A meta-analysis of randomized controlled trials. *Medicine* **2020**, *99*, e21235. [CrossRef] [PubMed]
110. Goldie, C.; Taylor, A.J.; Nguyen, P.; McCoy, C.; Zhao, X.Q.; Preiss, D. Niacin therapy and the risk of new-onset diabetes: A meta-analysis of randomised controlled trials. *Heart* **2016**, *102*, 198–203. [CrossRef] [PubMed]
111. Landray, M.J.; Haynes, R.; Hopewell, J.C.; Parish, S.; Aung, T.; Tomson, J.; Wallendszus, K.; Craig, M.; Jiang, L.; Collins, R.; et al. Effects of extended-release niacin with laropiprant in high-risk patients. *N. Engl. J. Med.* **2014**, *371*, 203–212. [CrossRef] [PubMed]
112. Swerdlow, D.I.; Preiss, D.; Kuchenbaecker, K.B.; Holmes, M.V.; Engmann, J.E.; Shah, T.; Sofat, R.; Stender, S.; Johnson, P.C.; Scott, R.A.; et al. HMG-coenzyme A reductase inhibition, type 2 diabetes, and bodyweight: Evidence from genetic analysis and randomised trials. *Lancet* **2015**, *385*, 351–361. [CrossRef]
113. Rochlani, Y.; Pothineni, N.V.; Kovelamudi, S.; Mehta, J.L. Metabolic syndrome: Pathophysiology, management, and modulation by natural compounds. *Ther. Adv. Cardiovasc. Dis.* **2017**, *11*, 215–225. [CrossRef]
114. Jin, T.; Song, Z.; Weng, J.; Fantus, I.G. Curcumin and other dietary polyphenols: Potential mechanisms of metabolic actions and therapy for diabetes and obesity. *Am. J. Physiol. Endocrinol. Metab.* **2018**, *314*, E201–E205. [CrossRef]
115. Deyno, S.; Eneyew, K.; Seyfe, S.; Tuyiringire, N.; Peter, E.L.; Muluye, R.A.; Tolo, C.U.; Ogwang, P.E. Efficacy and safety of cinnamon in type 2 diabetes mellitus and pre-diabetes patients: A meta-analysis and meta-regression. *Diabetes Res. Clin. Pract.* **2019**, *156*, 107815. [CrossRef]
116. Yilmaz, Z.; Piracha, F.; Anderson, L.; Mazzola, N. Supplements for diabetes mellitus: A review of the literature. *J. Pharm. Pract.* **2017**, *30*, 631–638. [CrossRef]
117. Perez-Rubio, K.G.; Gonzalez-Ortiz, M.; Martinez-Abundis, E.; Robles-Cervantes, J.A.; Espinel-Bermudez, M.C. Effect of berberine administration on metabolic syndrome, insulin sensitivity, and insulin secretion. *Metab. Syndr. Relat. Disord.* **2013**, *11*, 366–369. [CrossRef]
118. Yang, J.; Yin, J.; Gao, H.; Xu, L.; Wang, Y.; Xu, L.; Li, M. Berberine improves insulin sensitivity by inhibiting fat store and adjusting adipokines profile in human preadipocytes and metabolic syndrome patients. *Evid. Based Complement. Alternat. Med.* **2012**, *2012*, 363845. [CrossRef]
119. Gandhi, G.R.; Vasconcelos, A.B.S.; Wu, D.T.; Li, H.B.; Antony, P.J.; Li, H.; Geng, F.; Gurgel, R.Q.; Narain, N.; Gan, R.Y. Citrus flavonoids as promising phytochemicals targeting diabetes and related complications: A systematic review of in vitro and in vivo studies. *Nutrients* **2020**, *12*, 2907. [CrossRef] [PubMed]

120. Al-Aubaidy, H.A.; Dayan, A.; Deseo, M.A.; Itsiopoulos, C.; Jamil, D.; Hadi, N.R.; Thomas, C.J. Twelve-week mediterranean diet intervention increases citrus bioflavonoid levels and reduces inflammation in people with type 2 diabetes mellitus. *Nutrients* **2021**, *13*, 1133. [CrossRef] [PubMed]
121. Chen, S.; Jiang, H.; Wu, X.; Fang, J. Therapeutic effects of quercetin on inflammation, obesity, and type 2 diabetes. *Mediat. Inflamm.* **2016**, *2016*, 9340637. [CrossRef]
122. Eid, H.M.; Haddad, P.S. The antidiabetic potential of quercetin: Underlying mechanisms. *Curr. Med. Chem.* **2017**, *24*, 355–364. [CrossRef] [PubMed]
123. Gomez-Arbelaez, D.; Lahera, V.; Oubina, P.; Valero-Munoz, M.; de Las Heras, N.; Rodriguez, Y.; Garcia, R.G.; Camacho, P.A.; Lopez-Jaramillo, P. Aged garlic extract improves adiponectin levels in subjects with metabolic syndrome: A double-blind, placebo-controlled, randomized, crossover study. *Mediat. Inflamm.* **2013**, *2013*, 285795. [CrossRef] [PubMed]
124. Shang, A.; Cao, S.Y.; Xu, X.Y.; Gan, R.Y.; Tang, G.Y.; Corke, H.; Mavumengwana, V.; Li, H.B. Bioactive compounds and biological functions of garlic (*Allium sativum* L.). *Foods* **2019**, *8*, 246. [CrossRef]
125. Zhu, B.; Qi, F.; Wu, J.; Yin, G.; Hua, J.; Zhang, Q.; Qin, L. Red yeast rice: A systematic review of the traditional uses, chemistry, pharmacology, and quality control of an important chinese folk medicine. *Front. Pharmacol.* **2019**, *10*, 1449. [CrossRef]
126. Saleem, S.; Muhammad, G.; Hussain, M.A.; Bukhari, S.N.A. A comprehensive review of phytochemical profile, bioactives for pharmaceuticals, and pharmacological attributes of Azadirachta indica. *Phytother. Res.* **2018**, *32*, 1241–1272. [CrossRef]
127. Huang, D.D.; Shi, G.; Jiang, Y.; Yao, C.; Zhu, C. A review on the potential of resveratrol in prevention and therapy of diabetes and diabetic complications. *Biomed. Pharmacother.* **2020**, *125*, 109767. [CrossRef]
128. Jeyaraman, M.M.; Al-Yousif, N.S.H.; Singh Mann, A.; Dolinsky, V.W.; Rabbani, R.; Zarychanski, R.; Abou-Setta, A.M. Resveratrol for adults with type 2 diabetes mellitus. *Cochrane Database Syst. Rev.* **2020**, *1*, CD011919. [CrossRef] [PubMed]
129. Ozturk, E.; Arslan, A.K.K.; Yerer, M.B.; Bishayee, A. Resveratrol and diabetes: A critical review of clinical studies. *Biomed. Pharmacother.* **2017**, *95*, 230–234. [CrossRef] [PubMed]
130. Chen, S.; Zhao, X.; Ran, L.; Wan, J.; Wang, X.; Qin, Y.; Shu, F.; Gao, Y.; Yuan, L.; Zhang, Q.; et al. Resveratrol improves insulin resistance, glucose and lipid metabolism in patients with non-alcoholic fatty liver disease: A randomized controlled trial. *Dig. Liver Dis.* **2015**, *47*, 226–232. [CrossRef] [PubMed]
131. Petroni, M.L.; Brodosi, L.; Marchignoli, F.; Musio, A.; Marchesini, G. Moderate alcohol intake in non-alcoholic fatty liver disease: To drink or not to drink? *Nutrients* **2019**, *11*, 3048. [CrossRef]
132. Plows, J.F.; Reynolds, C.M.; Vickers, M.H.; Baker, P.N.; Stanley, J.L. Nutritional supplementation for the prevention and/or treatment of gestational diabetes mellitus. *Curr. Diabetes Rep.* **2019**, *19*, 73. [CrossRef]
133. Taylor, B.L.; Woodfall, G.E.; Sheedy, K.E.; O'Riley, M.L.; Rainbow, K.A.; Bramwell, E.L.; Kellow, N.J. Effect of probiotics on metabolic outcomes in pregnant women with gestational diabetes: A systematic review and meta-analysis of randomized controlled trials. *Nutrients* **2017**, *9*, 461. [CrossRef] [PubMed]
134. Bock, P.M.; Telo, G.H.; Ramalho, R.; Sbaraini, M.; Leivas, G.; Martins, A.F.; Schaan, B.D. The effect of probiotics, prebiotics or synbiotics on metabolic outcomes in individuals with diabetes: A systematic review and meta-analysis. *Diabetologia* **2021**, *64*, 26–41. [CrossRef] [PubMed]
135. Aoun, A.; Darwish, F.; Hamod, N. The influence of the gut microbiome on obesity in adults and the role of probiotics, prebiotics, and synbiotics for weight loss. *Prev. Nutr. Food Sci.* **2020**, *25*, 113–123. [CrossRef] [PubMed]
136. Greger, M. A whole food plant-based diet is effective for weight loss: The evidence. *Am. J. Lifestyle Med.* **2020**, *14*, 500–510. [CrossRef] [PubMed]
137. Wang, L.; Yang, H.; Huang, H.; Zhang, C.; Zuo, H.X.; Xu, P.; Niu, Y.M.; Wu, S.S. Inulin-type fructans supplementation improves glycemic control for the prediabetes and type 2 diabetes populations: Results from a GRADE-assessed systematic review and dose-response meta-analysis of 33 randomized controlled trials. *J. Transl. Med.* **2019**, *17*, 410. [CrossRef] [PubMed]
138. Gao, C.; Rao, M.; Huang, W.; Wan, Q.; Yan, P.; Long, Y.; Guo, M.; Xu, Y.; Xu, Y. Resistant starch ameliorated insulin resistant in patients of type 2 diabetes with obesity: A systematic review and meta-analysis. *Lipids Health Dis.* **2019**, *18*, 205. [CrossRef] [PubMed]
139. Rao, M.; Gao, C.; Xu, L.; Jiang, L.; Zhu, J.; Chen, G.; Law, B.Y.K.; Xu, Y. Effect of inulin-type carbohydrates on insulin resistance in patients with type 2 diabetes and obesity: A systematic review and meta-analysis. *J. Diabetes Res.* **2019**, *2019*, 5101423. [CrossRef] [PubMed]
140. Kaur, B.; Henry, J. Micronutrient status in type 2 diabetes: A review. *Adv. Food Nutr. Res.* **2014**, *71*, 55–100. [CrossRef] [PubMed]
141. Kazi, T.G.; Afridi, H.I.; Kazi, N.; Jamali, M.K.; Arain, M.B.; Jalbani, N.; Kandhro, G.A. Copper, chromium, manganese, iron, nickel, and zinc levels in biological samples of diabetes mellitus patients. *Biol. Trace Elem. Res.* **2008**, *122*, 1–18. [CrossRef] [PubMed]
142. Overbeck, S.; Rink, L.; Haase, H. Modulating the immune response by oral zinc supplementation: A single approach for multiple diseases. *Arch. Immunol. Ther. Exp.* **2008**, *56*, 15–30. [CrossRef]
143. Jansen, J.; Karges, W.; Rink, L. Zinc and diabetes—Clinical links and molecular mechanisms. *J. Nutr. Biochem.* **2009**, *20*, 399–417. [CrossRef]
144. Ruz, M.; Carrasco, F.; Rojas, P.; Basfi-Fer, K.; Hernandez, M.C.; Perez, A. Nutritional effects of zinc on metabolic syndrome and type 2 diabetes: Mechanisms and main findings in human studies. *Biol. Trace Elem. Res.* **2019**, *188*, 177–188. [CrossRef]
145. Jayawardena, R.; Ranasinghe, P.; Galappatthy, P.; Malkanthi, R.; Constantine, G.; Katulanda, P. Effects of zinc supplementation on diabetes mellitus: A systematic review and meta-analysis. *Diabetol Metab. Syndr.* **2012**, *4*, 13. [CrossRef]

146. Anderson, R.A.; Roussel, A.M.; Zouari, N.; Mahjoub, S.; Matheau, J.M.; Kerkeni, A. Potential antioxidant effects of zinc and chromium supplementation in people with type 2 diabetes mellitus. *J. Am. Coll. Nutr.* **2001**, *20*, 212–218. [CrossRef]
147. Farvid, M.S.; Jalali, M.; Siassi, F.; Hosseini, M. Comparison of the effects of vitamins and/or mineral supplementation on glomerular and tubular dysfunction in type 2 diabetes. *Diabetes Care* **2005**, *28*, 2458–2464. [CrossRef]
148. Parham, M.; Amini, M.; Aminorroaya, A.; Heidarian, E. Effect of zinc supplementation on microalbuminuria in patients with type 2 diabetes: A double blind, randomized, placebo-controlled, cross-over trial. *Rev. Diabet. Stud.* **2008**, *5*, 102–109. [CrossRef]
149. Roussel, A.M.; Kerkeni, A.; Zouari, N.; Mahjoub, S.; Matheau, J.M.; Anderson, R.A. Antioxidant effects of zinc supplementation in Tunisians with type 2 diabetes mellitus. *J. Am. Coll. Nutr.* **2003**, *22*, 316–321. [CrossRef]
150. Marchesini, G.; Bugianesi, E.; Ronchi, M.; Flamia, R.; Thomaseth, K.; Pacini, G. Zinc supplementation improves glucose disposal in patients with cirrhosis. *Metabolism* **1998**, *47*, 792–798. [CrossRef]
151. Marchesini, G.; Fabbri, A.; Bianchi, G.; Brizi, M.; Zoli, M. Zinc supplementation and amino acid-nitrogen metabolism in patients with advanced cirrhosis. *Hepatology* **1996**, *23*, 1084–1092. [CrossRef]
152. Morgan, M.Y.; Blei, A.; Grungreiff, K.; Jalan, R.; Kircheis, G.; Marchesini, G.; Riggio, O.; Weissenborn, K. The treatment of hepatic encephalopathy. *Metab. Brain Dis.* **2007**, *22*, 389–405. [CrossRef]
153. Huang, H.Y.; Caballero, B.; Chang, S.; Alberg, A.J.; Semba, R.D.; Schneyer, C.R.; Wilson, R.F.; Cheng, T.Y.; Vassy, J.; Prokopowicz, G.; et al. The efficacy and safety of multivitamin and mineral supplement use to prevent cancer and chronic disease in adults: A systematic review for a National Institutes of Health state-of-the-science conference. *Ann. Intern. Med.* **2006**, *145*, 372–385. [CrossRef]
154. Chen, S.; Jin, X.; Shan, Z.; Li, S.; Yin, J.; Sun, T.; Luo, C.; Yang, W.; Yao, P.; Yu, K.; et al. Inverse association of plasma chromium levels with newly diagnosed type 2 diabetes: A case-control study. *Nutrients* **2017**, *9*, 294. [CrossRef]
155. Basaki, M.; Saeb, M.; Nazifi, S.; Shamsaei, H.A. Zinc, copper, iron, and chromium concentrations in young patients with type 2 diabetes mellitus. *Biol. Trace. Elem. Res.* **2012**, *148*, 161–164. [CrossRef]
156. Suksomboon, N.; Poolsup, N.; Yuwanakorn, A. Systematic review and meta-analysis of the efficacy and safety of chromium supplementation in diabetes. *J. Clin. Pharm. Ther.* **2014**, *39*, 292–306. [CrossRef]
157. Yin, R.V.; Phung, O.J. Effect of chromium supplementation on glycated hemoglobin and fasting plasma glucose in patients with diabetes mellitus. *Nutr. J.* **2015**, *14*, 14. [CrossRef]
158. Costello, R.B.; Dwyer, J.T.; Bailey, R.L. Chromium supplements for glycemic control in type 2 diabetes: Limited evidence of effectiveness. *Nutr. Rev.* **2016**, *74*, 455–468. [CrossRef]
159. Huang, H.; Chen, G.; Dong, Y.; Zhu, Y.; Chen, H. Chromium supplementation for adjuvant treatment of type 2 diabetes mellitus: Results from a pooled analysis. *Mol. Nutr. Food Res.* **2018**, *62*, 1700438. [CrossRef]
160. De Valk, H.W.; Verkaaik, R.; van Rijn, H.J.; Geerdink, R.A.; Struyvenberg, A. Oral magnesium supplementation in insulin-requiring Type 2 diabetic patients. *Diabet. Med.* **1998**, *15*, 503–507. [CrossRef]
161. Kostov, K. Effects of magnesium deficiency on mechanisms of insulin resistance in type 2 diabetes: Focusing on the processes of insulin secretion and signaling. *Int. J. Mol. Sci.* **2019**, *20*, 1351. [CrossRef]
162. Song, Y.; Manson, J.E.; Buring, J.E.; Liu, S. Dietary magnesium intake in relation to plasma insulin levels and risk of type 2 diabetes in women. *Diabetes Care* **2004**, *27*, 59–65. [CrossRef]
163. Guerrero-Romero, F.; Rascon-Pacheco, R.A.; Rodriguez-Moran, M.; de la Pena, J.E.; Wacher, N. Hypomagnesaemia and risk for metabolic glucose disorders: A 10-year follow-up study. *Eur. J. Clin. Investig.* **2008**, *38*, 389–396. [CrossRef]
164. Rodriguez-Moran, M.; Guerrero-Romero, F. Oral magnesium supplementation improves insulin sensitivity and metabolic control in type 2 diabetic subjects: A randomized double-blind controlled trial. *Diabetes Care* **2003**, *26*, 1147–1152. [CrossRef]
165. Mooren, F.C.; Kruger, K.; Volker, K.; Golf, S.W.; Wadepuhl, M.; Kraus, A. Oral magnesium supplementation reduces insulin resistance in non-diabetic subjects—A double-blind, placebo-controlled, randomized trial. *Diabetes Obes. Metab.* **2011**, *13*, 281–284. [CrossRef] [PubMed]
166. El Derawi, W.A.; Naser, I.A.; Taleb, M.H.; Abutair, A.S. The effects of oral magnesium supplementation on glycemic response among type 2 diabetes patients. *Nutrients* **2018**, *11*, 44. [CrossRef]
167. Fang, X.; Wang, K.; Han, D.; He, X.; Wei, J.; Zhao, L.; Imam, M.U.; Ping, Z.; Li, Y.; Xu, Y.; et al. Dietary magnesium intake and the risk of cardiovascular disease, type 2 diabetes, and all-cause mortality: A dose-response meta-analysis of prospective cohort studies. *BMC Med.* **2016**, *14*, 210. [CrossRef] [PubMed]
168. Verma, H.; Garg, R. Effect of magnesium supplementation on type 2 diabetes associated cardiovascular risk factors: A systematic review and meta-analysis. *J. Hum. Nutr. Diet.* **2017**, *30*, 621–633. [CrossRef] [PubMed]
169. Hannon, B.A.; Fairfield, W.D.; Adams, B.; Kyle, T.; Crow, M.; Thomas, D.M. Use and abuse of dietary supplements in persons with diabetes. *Nutr. Diabetes* **2020**, *10*, 14. [CrossRef]
170. Low, S.; Pek, S.; Moh, A.; Khin, C.Y.A.; Lim, C.L.; Ang, S.F.; Wang, J.; Ang, K.; Tang, W.E.; Lim, Z.; et al. Low muscle mass is associated with progression of chronic kidney disease and albuminuria—An 8-year longitudinal study in Asians with Type 2 Diabetes. *Diabetes Res. Clin. Pract.* **2021**, *174*, 108777. [CrossRef] [PubMed]
171. Ross, E.; Wright, H.; Villani, A. Lower body extremity function is associated with health-related quality of life: A cross-sectional analysis of overweight and obese older adults with and without type 2 diabetes mellitus. *Qual. Life Res.* **2021**, *30*, 2265–2273. [CrossRef] [PubMed]

172. Fragala, M.S.; Cadore, E.L.; Dorgo, S.; Izquierdo, M.; Kraemer, W.J.; Peterson, M.D.; Ryan, E.D. Resistance training for older adults: Position statement from the National Strength and Conditioning Association. *J. Strength Cond. Res.* **2019**, *33*, 2019–2052. [CrossRef] [PubMed]
173. Petroni, M.L.; Caletti, M.T.; Dalle Grave, R.; Bazzocchi, A.; Aparisi Gomez, M.P.; Marchesini, G. Prevention and treatment of sarcopenic obesity in women. *Nutrients* **2019**, *11*, 1302. [CrossRef] [PubMed]
174. Chapman, I.; Oberoi, A.; Giezenaar, C.; Soenen, S. Rational use of protein supplements in the elderly-Relevance of gastrointestinal mechanisms. *Nutrients* **2021**, *13*, 1227. [CrossRef]
175. Li, Q.; Wen, F.; Wang, Y.; Li, S.; Lin, S.; Qi, C.; Chen, Z.; Qiu, X.; Zhang, Y.; Zhang, S.; et al. Diabetic kidney disease benefits from intensive low-protein diet: Updated systematic review and meta-analysis. *Diabetes Ther.* **2021**, *12*, 21–36. [CrossRef]
176. Tu, D.Y.; Kao, F.M.; Tsai, S.T.; Tung, T.H. Sarcopenia among the elderly population: A systematic review and meta-analysis of randomized controlled trials. *Healthcare* **2021**, *9*, 650. [CrossRef]
177. Osuka, Y.; Kojima, N.; Sasai, H.; Wakaba, K.; Miyauchi, D.; Tanaka, K.; Kim, H. Effects of exercise and/or beta-hydroxy-beta-methylbutyrate supplementation on muscle mass, muscle strength, and physical performance in older women with low muscle mass: A randomized, double-blind, placebo-controlled trial. *Am. J. Clin. Nutr.* **2021**. [CrossRef]
178. Manders, R.J.; Little, J.P.; Forbes, S.C.; Candow, D.G. Insulinotropic and muscle protein synthetic effects of branched-chain amino acids: Potential therapy for type 2 diabetes and sarcopenia. *Nutrients* **2012**, *4*, 1664–1678. [CrossRef] [PubMed]
179. Hey, P.; Gow, P.; Testro, A.G.; Apostolov, R.; Chapman, B.; Sinclair, M. Nutraceuticals for the treatment of sarcopenia in chronic liver disease. *Clin. Nutr. ESPEN* **2021**, *41*, 13–22. [CrossRef]
180. Kim, Y. Emerging treatment options for sarcopenia in chronic liver disease. *Life* **2021**, *11*, 250. [CrossRef] [PubMed]
181. Beaudart, C.; Buckinx, F.; Rabenda, V.; Gillain, S.; Cavalier, E.; Slomian, J.; Petermans, J.; Reginster, J.Y.; Bruyere, O. The effects of vitamin D on skeletal muscle strength, muscle mass, and muscle power: A systematic review and meta-analysis of randomized controlled trials. *J. Clin. Endocrinol. Metab.* **2014**, *99*, 4336–4345. [CrossRef] [PubMed]
182. Bjelakovic, G.; Nikolova, D.; Bjelakovic, M.; Gluud, C. Vitamin D supplementation for chronic liver diseases in adults. *Cochrane Database Syst. Rev.* **2017**, *11*, CD011564. [CrossRef]
183. Braganca, A.C.; Alvares-da-Silva, M.R. Prevalence of diabetes mellitus and impaired glucose tolerance in patients with decompensated cirrhosis being evaluated for liver transplantation: The utility of oral glucose tolerance test. *Arq. Gastroenterol.* **2010**, *47*, 22–27. [CrossRef] [PubMed]
184. Elkrief, L.; Rautou, P.E.; Sarin, S.; Valla, D.; Paradis, V.; Moreau, R. Diabetes mellitus in patients with cirrhosis: Clinical implications and management. *Liver Int.* **2016**, *36*, 936–948. [CrossRef] [PubMed]
185. Kallwitz, E.R. Sarcopenia and liver transplant: The relevance of too little muscle mass. *World J. Gastroenterol.* **2015**, *21*, 10982–10993. [CrossRef]
186. Schiavo, L.; Busetto, L.; Cesaretti, M.; Zelber-Sagi, S.; Deutsch, L.; Iannelli, A. Nutritional issues in patients with obesity and cirrhosis. *World J. Gastroenterol.* **2018**, *24*, 3330–3346. [CrossRef]
187. Brodosi, L.; Petta, S.; Petroni, M.L.; Marchesini, G.; Morelli, M.C. Management of diabetes in candidates for liver transplantation and in transplant recipients. *Transplantation* **2021**, in press. [CrossRef] [PubMed]
188. Colosimo, S.; Ravaioli, F.; Petroni, M.L.; Brodosi, L.; Marchignoli, F.; Barbanti, F.A.; Sasdelli, A.S.; Marchesini, G.; Pironi, L. Effects of antidiabetic agents on steatosis and fibrosis biomarkers in type 2 diabetes: A real-world data analysis. *Liver Int.* **2021**, *41*, 731–742. [CrossRef] [PubMed]
189. Vukotic, R.; Raimondi, F.; Brodosi, L.; Vitale, G.; Petroni, M.L.; Marchesini, G.; Andreone, P. The effect of liraglutide on beta-blockade for preventing variceal bleeding: A case series. *Ann. Intern. Med.* **2020**, *173*, 404–405. [CrossRef] [PubMed]
190. Bischoff, S.C.; Bernal, W.; Dasarathy, S.; Merli, M.; Plank, L.D.; Schutz, T.; Plauth, M. ESPEN practical guideline: Clinical nutrition in liver disease. *Clin. Nutr.* **2020**, *39*, 3533–3562. [CrossRef]
191. Plank, L.D.; Gane, E.J.; Peng, S.; Muthu, C.; Mathur, S.; Gillanders, L.; McIlroy, K.; Donaghy, A.J.; McCall, J.L. Nocturnal nutritional supplementation improves total body protein status of patients with liver cirrhosis: A randomized 12-month trial. *Hepatology* **2008**, *48*, 557–566. [CrossRef]
192. Verboeket-van de Venne, W.P.; Westerterp, K.R.; van Hoek, B.; Swart, G.R. Energy expenditure and substrate metabolism in patients with cirrhosis of the liver: Effects of the pattern of food intake. *Gut* **1995**, *36*, 110–116. [CrossRef]
193. Baldassarre, M.; Naldi, M.; Zaccherini, G.; Bartoletti, M.; Antognoli, A.; Laggetta, M.; Gagliardi, M.; Tufoni, M.; Domenicali, M.; Waterstradt, K.; et al. Determination of effective albumin in patients with decompensated cirrhosis: Clinical and prognostic implications. *Hepatology* **2021**. [CrossRef]
194. European Association for the Study of the Liver. EASL Clinical Practice Guidelines for the management of patients with decompensated cirrhosis. *J. Hepatol.* **2018**, *69*, 406–460. [CrossRef] [PubMed]
195. Caraceni, P.; Pavesi, M.; Baldassarre, M.; Bernardi, M.; Arroyo, V. The use of human albumin in patients with cirrhosis: A European survey. *Expert Rev. Gastroenterol. Hepatol.* **2018**, *12*, 625–632. [CrossRef] [PubMed]
196. Caraceni, P.; Riggio, O.; Angeli, P.; Alessandria, C.; Neri, S.; Foschi, F.G.; Levantesi, F.; Airoldi, A.; Boccia, S.; Svegliati-Baroni, G.; et al. Long-term albumin administration in decompensated cirrhosis (ANSWER): An open-label randomised trial. *Lancet* **2018**, *391*, 2417–2429. [CrossRef]
197. Goldstein-Fuchs, J.; Kalantar-Zadeh, K. Nutrition Intervention for advanced stages of diabetic kiidney disease. *Diabetes Spectr.* **2015**, *28*, 181–186. [CrossRef] [PubMed]

198. Syauqy, A.; Hsu, C.Y.; Lee, H.A.; Rau, H.H.; Chao, J.C. Association between dietary patterns and kidney function parameters in adults with metabolic syndrome: A cross-sectional study. *Nutrients* **2020**, *13*, 40. [CrossRef]
199. Ko, G.J.; Kalantar-Zadeh, K.; Goldstein-Fuchs, J.; Rhee, C.M. Dietary approaches in the management of diabetic patients with kidney disease. *Nutrients* **2017**, *9*, 824. [CrossRef] [PubMed]
200. Isaka, Y. Optimal protein intake in pre-dialysis chronic kidney disease patients with sarcopenia: An overview. *Nutrients* **2021**, *13*, 1205. [CrossRef]
201. Kidney Disease Ourcomes Quality Initiative. KDOQI clinical practice guidelines and clinical practice recommendations for diabetes and chronic kidney disease. *Am. J. Kidney Dis.* **2007**, *49* (Suppl. S2), S12–S154. [CrossRef] [PubMed]
202. Kalantar-Zadeh, K.; Joshi, S.; Schlueter, R.; Cooke, J.; Brown-Tortorici, A.; Donnelly, M.; Schulman, S.; Lau, W.L.; Rhee, C.M.; Streja, E.; et al. Plant-dominant low-protein diet for conservative management of chronic kidney disease. *Nutrients* **2020**, *12*, 1931. [CrossRef]
203. Tuttle, K.R.; Bakris, G.L.; Bilous, R.W.; Chiang, J.L.; de Boer, I.H.; Goldstein-Fuchs, J.; Hirsch, I.B.; Kalantar-Zadeh, K.; Narva, A.S.; Navaneethan, S.D.; et al. Diabetic kidney disease: A report from an ADA Consensus Conference. *Am. J. Kidney Dis.* **2014**, *64*, 510–533. [CrossRef]
204. Shapiro, H.; Theilla, M.; Attal-Singer, J.; Singer, P. Effects of polyunsaturated fatty acid consumption in diabetic nephropathy. *Nat. Rev. Nephrol.* **2011**, *7*, 110–121. [CrossRef]
205. Kong, Y.W.; Baqar, S.; Jerums, G.; Ekinci, E.I. Sodium and its role in cardiovascular disease—The debate continues. *Front. Endocrinol.* **2016**, *7*, 164. [CrossRef]
206. Abe, M.; Hamano, T.; Hoshino, J.; Wada, A.; Inaba, M.; Nakai, S.; Masakane, I. Is there a "burnt-out diabetes" phenomenon in patients on hemodialysis? *Diabetes Res. Clin. Pract.* **2017**, *130*, 211–220. [CrossRef]

nutrients

Review

Beyond the Paradigm of Weight Loss in Non-Alcoholic Fatty Liver Disease: From Pathophysiology to Novel Dietary Approaches

Angelo Armandi [1,2] and Jörn M. Schattenberg [2,3,*]

[1] Department of Medical Sciences, University of Turin, 10126 Turin, Italy; angelo.armandi@unito.it
[2] Department of Internal Medicine I, University Medical Center, Johannes Gutenberg University, 55131 Mainz, Germany
[3] Metabolic Liver Research Program, Department of Internal Medicine I, University Medical Center, Johannes Gutenberg University, 55131 Mainz, Germany
* Correspondence: joern.schattenberg@unimedizin-mainz.de; Tel.: +49-(0)-6131-17-6074

Abstract: Current treatment recommendations for non-alcoholic fatty liver disease (NAFLD) rely heavily on lifestyle interventions. The Mediterranean diet and physical activity, aiming at weight loss, have shown good results in achieving an improvement of this liver disease. However, concerns related to compliance and food accessibility limit the feasibility of this approach, and data on the long-term effects on liver-related outcomes are lacking. Insulin resistance is a central aspect in the pathophysiology of NAFLD; therefore, interventions aiming at the improvement of insulin sensitivity may be preferable. In this literature review, we provide a comprehensive summary of the available evidence on nutritional approaches in the management of NAFLD, involving low-calorie diets, isocaloric diets, and the novel schemes of intermittent fasting. In addition, we explore the harmful role of single nutrients on liver-specific key metabolic pathways, the role of gene susceptibility and microbiota, and behavioral aspects that may impact liver disease and are often underreported in clinical setting. At present, the high variability in terms of study populations and liver-specific outcomes within nutritional studies limits the generalizability of the results and highlights the urgent need of a tailored and standardized approach, as seen in regulatory trials in Non-Alcoholic Steatohepatitis (NASH).

Keywords: insulin; lifestyle; non-alcoholic; steatohepatitis; fibrosis; metabolic syndrome; weight loss; time-restricted feeding; intermittent fasting; low-carb diet; liver disease

Citation: Armandi, A.; Schattenberg, J.M. Beyond the Paradigm of Weight Loss in Non-Alcoholic Fatty Liver Disease: From Pathophysiology to Novel Dietary Approaches. *Nutrients* **2021**, *13*, 1977. https://doi.org/10.3390/nu13061977

Academic Editor: James King

Received: 16 May 2021
Accepted: 7 June 2021
Published: 8 June 2021

1. Introduction: Rationale for Lifestyle Interventions in NAFLD

Non-alcoholic fatty liver disease (NAFLD) represents the leading form of chronic liver disease, with a worldwide prevalence of 25% [1] and a disease burden which is projected to dramatically increase by 2030 [2], in parallel with the pandemic of metabolic-related affections, mainly obesity and type 2 diabetes (T2DM). Overall, the economic and social burden of NAFLD in Europe is high [3].

NAFLD is a multi-systemic disease which involves multi-directional metabolic derangements [4]. The term NAFLD covers different disease activities and disease stages, ranging from simple steatosis (non-alcoholic fatty liver, NAFL), to the histological evidence of lobular inflammation and ballooning (non-alcoholic steatohepatitis, NASH), which represents the progressive form of the metabolic-related liver damage that can potentially lead to cirrhosis and further complications, including hepatocellular carcinoma (HCC) [5]. Progression along this disease continuum is driven by alternating inflammatory bouts that foster excessive fibrogenesis and promote scars inside the liver parenchyma. Progression through the histological fibrosis stages to cirrhosis has been recognized as the most relevant prognostic factor in NAFLD [6]. In addition, resolution of NASH is more difficult to achieve when compared to NAFL, where no inflammatory activity is detected upon histology. Presently, liver biopsy is the reference standard to reliably grade NASH and stage fibrosis. Hence, in the absence of accepted non-invasive markers, an accurate assessment of treatment efficacy requires liver biopsy. This is frequently not performed in lifestyle intervention trials and thus results are not directly comparable with current registration studies exploring pharmacotherapy options.

Currently, no liver-directed pharmacological therapy is approved for the treatment of NAFLD, and the management relies on lifestyle interventions. The majority of individuals with NAFLD are overweight or obese and suffer from T2DM [7]. Therefore, weight loss, aiming at ameliorating the liver injury and decreasing fibrosis, is suggested by all guidelines. However, NAFLD can also occur in lean individuals, where other factors, such as an unfavorable genetic background, may lead to a comparable liver phenotype. The patients considered as lean NAFLD can develop all disease stages of NASH, supporting the concept that even in the absence of overt obesity, steatohepatitis and fibrosis can occur. [8]. In all NASH patients, a balance between harmful (metabolic) co-factors, intrinsic genetic factors, and extrinsic lifestyle choices impact on the clinical phenotype. This variability also adds to the inhomogeneity of responses observed in lifestyle interventions.

The seminal studies conducted by Vilar-Gomez et al. on 293 NAFLD individuals undergoing intensified lifestyle changes (low-fat hypocaloric diet) with psychological coaching for 52 weeks highlighted the results that nutrition and physical activity can have on liver histology: weight loss of more than 10% of the initial body weight led to a resolution of NASH in 90% of cases, and a regression of fibrosis in 45% of cases [9]. Unfortunately, only 10% of all intensively guided participants could reach this endpoint. Similar data resulted from another randomized, controlled trial conducted on 31 overweight or obese patients with biopsy-proven NASH undergoing lifestyle intervention for 48 weeks, with significant improvements in histological inflammatory activity in those who could reach at least a 7% reduction in weight loss [10]. In a routine clinical setting which offers less intensive counseling, the effects of lifestyle modifications are clearly lower, which is linked to the lower rate of adherence.

Evidence coming from morbidly obese patients undergoing bariatric surgery have provided insightful results on the impact of rapid weight loss on liver damage. Despite differences of NAFLD in bariatric cohorts with typically lower disease activity and stages, the pronounced and sustained weight loss that occurs highlights the degree of effect that can be achieved in a short time frame. Studies conducted by Lassailly et al. showed that NASH resolves in 85% of cases at 1 year after bariatric surgery [11], and an additional beneficial effect on fibrosis can be observed across 5 years following surgery [12].

Histological improvements resulting from weight loss are also observed across clinical trials with investigational drugs. In a phase II trial conducted on patients with NASH, liraglutide, a synthetic agonist of glucagon-like peptide-1 (GLP-1) approved for the treatment of T2DM, showed a resolution of inflammatory activity in 39% of cases along with a weight reduction of 15% from the initial value [13]. Empagliflozin, a sodium–glucose co-transporter-2 (SGLT2) inhibitor used for the treatment of T2DM, showed a 20% resolution of liver fat content as measured by magnetic resonance spectroscopy, accompanied by a reduction in a placebo-corrected 2.5 kg of weight [14]. On the contrary, pioglitazone, a peroxisome proliferator-activated receptor (PPAR)-γ agonist, has been shown to improve NASH in up to 34% of cases with a mean weight gain of 4.7 kg [15].

In addition, data extrapolated by clinical trials have also demonstrated the significant influence of lifestyle interventions in modulating the course of NAFLD. Dietary recommendations and physical activity are suggested for all participants in experimental studies, in which proper, systematic designs may help to identify strong endpoints, limit heterogeneity, and predict the size of placebo responses [16]. Therefore, results emerging from placebo arms mirror the effective action of lifestyle modifications, which are strengthened by higher compliance in the setting of close clinical visits. One recent meta-analysis of the placebo groups from 39 randomized controlled trials of adults with NASH showed a significant response: up to 25% of patients given a placebo had improvements in all the histology scores, including fibrosis, with moderate heterogeneity among the different studies. Moreover, patients given the placebo had a significant reduction in steatosis, as measured by proton magnetic resonance spectroscopy, and a significant decrease in serum transaminases [17].

2. Response to Lifestyle Intervention between Genes and Environment

Diverse genetic susceptibility, as expressed by different single nucleotide polymorphisms (SNPs) in targeted genes, as well as the additional impact of metabolic co-morbidities and lifestyle habits, are together responsible for the high variability in the efficacy of lifestyle interventions.

Genome-wide association studies (GWAS) have identified vulnerable sites in many gene loci, such as Patatin-like phospholipase domain-containing protein 3 (PNPLA3) [18], Membrane-bound O-acyltransferase domain-containing protein 7 (MBOAT7) [19] and Transmembrane 6 Superfamily Member 2 (TM6SF2) [20], which are related to a more aggressive disease phenotype, leading to a major risk for developing advanced liver disease. However, post hoc analyses conducted in a randomized controlled trial of lifestyle modifications in NAFLD populations have shown that variants in PNPLA3 were associated with better improvements in weight loss, better impacts on dyslipidemia, and greater reductions in intrahepatic fat, as evaluated by proton magnetic resonance spectroscopy [21]. These data suggest that the presence of the PNPLA3 variant may induce a better response to lifestyle intervention, and favorable results also come from a study from Sevastianova et al., where the same variant did not prevent the decrease in liver fat upon undertaking a hypocaloric low-carbohydrate diet for 6 days [22].

This evidence may lead to a reduced need for drug therapy in patients with PNPLA3 variants, and further studies investigating the role of other variants would potentially bring stronger results. Currently, polygenic risk scores are being evaluated, putting together different gene variants, in order to assess their potential role in preventing hard endpoints [23], and response to lifestyle interventions might be included in this novel approach.

In addition, gene expression is continuously shaped by environmental agents, which act as epigenetic modulators of specific protein translation and synthesis. In one study conducted on mice, a high-cholesterol diet was associated with a down-regulation of key genes involved in cholesterol metabolism, such as farnesoid-X receptor (FXR), which exerts ani-inflammatory and anti-fibrotic effects upon the liver [24]. Hence, the unhealthy dietary environment may predispose the liver to unfavorable gene expression, with relevant implications in clinical settings.

The close connection between the liver and the gut supports the concept that gut microbiota (GM) correlate with the liver disease. However, these interactions are multidimensional and complex. Alterations of the gut vascular barrier and intestinal epithelial barrier are in reciprocal interaction with the GM composition and contribute to liver injury in murine models [25]. Gut integrity—a function of intestinal epithelium and dietary composition—also affects the balance between bacterial species and microbial gene richness, which may give rise to hepatic inflammation [26]. Obese people present with low microbial gene richness, and dietary intervention has been shown to improve this pattern [27]. Additionally, reductions in GM heterogeneity lead to a decrease in short-chain fatty acids and increased lipopolysaccharide (LPS)—both factors are associated with insulin resistance [28]. To add to this evidence, one longitudinal study conducted in 307 males observed a reduced risk of cardio-metabolic disease in those individuals who adhered to a Mediterranean-style dietary pattern, and this was associated with a specific GM taxonomy [29].

When considering the diverse response on a lifestyle treatment, ethnicity plays a relevant role. A recent meta-analysis of over two million Chinese individuals highlighted a dramatic increase in the prevalence of NAFLD, reaching 29% in the 2010s. A greater predominance of PNPLA3 variants was observed, partly explaining the 10% of NAFLD diagnosed in non-obese individuals. This might also impact on the response from lifestyle treatment. On the other hand, a major prevalence of NAFLD was observed in Westernized areas of Asia, underlying the combining impact of environmental factors [30]. The balance between genes and environment with regard to response to lifestyle treatment is still underreported in NAFLD. Randomized control trials of lifestyle interventions in Asian NAFLD populations have shown that both aerobic and resistance training, alone or in combination with dietary changes, have a positive impact on weight loss and reductions in liver fat and inflammation, however with substantial heterogeneity among studies involving different ethnicities [31]. Therefore, ethnic differences may be responsible for different outcomes for lifestyle interventions and may affect the interpretation of data.

3. Improving Insulin Resistance as Metabolic Endpoint for Lifestyle Intervention

Insulin resistance has been widely assessed as the major trigger of liver damage in NAFLD [32,33]. Defective action of insulin in peripheral tissues causes a reduced insulin-mediated glucose uptake in skeletal muscle, resulting in persistent hyperglycemia, and enhances lipolysis in adipose tissue, with increased levels of free fatty acids. The liver is responsible for the capture of the overflow of free fatty acids, accumulating esterified fats in lipid droplets inside the cytoplasm of hepatocytes. Insulin resistance inside the liver, linked to the alterations in post-receptor insulin signaling, leads to increased gluconeogenesis, with further harmful impacts on glycemic homeostasis.

A mechanistic link between insulin resistance and chronic liver inflammation has been explored in studies conducted on animal models. In mice, overexpression of cytochrome P450 2E1 (CYP2E1) as a result of oxidative stress-derived liver inflammation, impairs intrahepatic insulin signaling, by decreasing the tyrosine phosphorylation of insulin receptor substrates (IRS), one key passage in the insulin metabolic pathway [34]. Hence, this interference prevents the liver from implementing a proper response to insulin. Interestingly, CYP2E1-derived inhibition of insulin signaling was partially mediated by downstream c-Jun N-terminal kinase (JNK), tightly connected to the activation and apoptosis of signal-regulating kinase 1 (ASK1). These are two major components of the hepatic inflammasome that promote apoptosis, inflammation and fibrosis, and their inhibition has been shown to improve liver damage [35], potentially acting as a target to treat insulin resistance [36].

Moreover, it has been shown that the crosstalk between adipose tissue and the liver, mediated by free fatty acids, results in the activation of Kuppfer cells, promoting inflammation and fibrogenesis. This pathophysiological milieu develops in the absence of obesity and T2DM, as an independent mechanism of disease [37]. One study conducted on non-obese, non-diabetic NASH individuals showed a direct association between saturated fat intake, derived indices of insulin resistance, and the postprandial rise of triglycerides, suggesting a shared pathological ground between nutrition, fats, and insulin activity, in the absence of overt metabolic-related morbidities [38].

Moreover, a high heterogeneity was observed in the obese population with NAFLD. Obesity is not a unique phenotype and the different compartments of adipose tissue, e.g., subcutaneous tissue versus abdominal/visceral adipose tissue, contribute differently to the disease. Visceral adipose tissue is characterized by proinflammatory activity and contributes to insulin resistance in peripheral tissues, in particular, in skeletal muscle. Muscle insulin resistance is linked to a worse metabolic phenotype. Overall, these individuals have a different clinical phenotype compared to obese patients that do not exhibit visceral adiposity and accompanying insulin-resistance. These patients are currently considered metabolically healthy obese patients. This difference may also have important implications in terms of prognosis, as well in the response to lifestyle treatment.

High insulin levels are therefore the results of multiple drivers, involving both environmental and genetic factors, of which balance determines the phenotype and the natural history of liver disease. A diet rich in saturated fats, sucrose-enriched beverages, refined carbohydrates, high glycemic index foods, high fructose intake, high caloric foods, inserted in the picture of the Western diet, as well as harmful eating habits and sedentary lifestyles, promote hyperinsulinemia and NAFLD [39]. In particular, the detrimental effect of fructose in the context of a hypercaloric diet leads to increased de novo lipogenesis and lipotoxicity.

These two major factors are involved in the progression of NAFLD to fibrosing NASH [40]. Hence, the improvement of insulin resistance would be the preferable endpoint for lifestyle interventions, with consequent beneficial effects on lipid profiles, cardio-metabolic parameters, and anthropometric measures. This complex crosstalk may be influenced by genes, as well as the co-presence of metabolic affections that would likely confer variable responses to the dietary approach.

The importance of insulin resistance in the clinical setting is supported by lifestyle interventions examining the improvement of NAFLD. Secondary or co-primary endpoints of most interventional studies are linked to the evaluation in changes of insulin resistance indices. In one randomized crossover trial conducted on obese non-diabetic biopsy-proven NAFLD individuals, lifestyle interventions using a Mediterranean diet led to a significant improvement in insulin sensitivity, as determined by a euglycemic clamp, in parallel to a reduced liver fat content, without relevant weight loss [41]. Another study conducted on adolescents with NAFLD undergoing a low-fructose diet led to a reduction in liver transaminases, accompanied by a significant improvement in systolic blood pressure and the Homeostatic Model Assessment for Insulin Resistance (HOMA-IR) [42]. The emerging role of the tight connection between liver outcomes and improvements in insulin resistance across the clinical studies suggest that insulin sensitivity should be addressed as a solid endpoint of lifestyle interventions.

4. Quantitative and Qualitative Aspects of Nutrition

Although association studies have highlighted a link of nutritional components, and in particular, hypercaloric diets, to liver injury in NAFLD, the specific dietary patterns or, more generally, the type of lifestyle intervention, that will reverse the disease phenotype beyond adherence to a hypocaloric diet is less clear. Resolution of steatohepatitis and regression of fibrosis are viewed as the most relevant endpoint in clinical trials. However, data on long-term outcomes, including the different incidence of metabolic concomitant affections and overall mortality, are lacking. Hence, most data are extrapolated from the efficacy of some dietary patterns in clinical trials, as well as the diverse harmful impact of different micro or macronutrients in animal models or surrogate endpoints in humans.

Recently, the evaluation of these interconnected pathways has been explored through the approach of the Geometric Framework of Nutrition (GFN), a dimensional model that graphically integrates key aspects of nutritional systems and maps the relationship between nutrient intake and health outcomes [43]. One experimental study conducted in mice showed, with the GFN model, that a carbohydrate intake of less than 25 kj/day and protein intake of more than 10 kj/day was associated with a lower probability of developing fatty liver disease, while increased fat content was positively associated with a fatty liver [44]. This evidence suggests that not only the reduction in calories, but rather the quality and the energy content of nutritional components, can add to a successful dietary intervention.

4.1. Mediterranean Diet versus Western Diet

Dietary patterns that approximate the Mediterranean diet have been repeatedly assessed in patients with metabolic diseases. In a large meta-analysis of 50 studies comprising 534,906 individuals, adherence to the Mediterranean diet was associated with a reduced risk of metabolic syndrome and, importantly, overall mortality. This was accompanied by a lower waist circumference, higher glucose tolerance, higher levels of high-density lipoprotein (HDL) cholesterol, and better levels of systolic and diastolic blood pressure [45].

In treating NAFLD, it is likely that easy and simple dietary patterns will be most beneficial for patients. One recent Italian randomized controlled trial, conducted by Franco et al. in 144 non-diabetic NAFLD patients, showed that a low glycemic index Mediterranean diet significantly improved hepatic steatosis, assessed by a control attenuation parameter (CAP), as well as markers of insulin resistance, as determined by HOMA-IR [46]. Interestingly, this pattern of diet ameliorates insulin resistance even in the absence of weight loss, acting on a pathophysiological level [41], and improves blood levels of transaminases (alanine aminotransferase, ALT) as surrogate markers of hepatic necroinflammatory activity, but also liver stiffness at elastography after 6 months of treatment [47].

One European prospective population study conducted on 2288 Swiss individuals without baseline hepatic steatosis for a mean time of 5.3 years has confirmed the previous evidence on a larger scale, because higher adherence to Mediterranean diet was associated with a reduced risk of developing NAFLD [48].

The Mediterranean food pattern comprises the high consumption of vegetables, fruits, mainly unrefined grains, low-fat milk, nuts; low glycemic index carbohydrates; higher proportions of monounsaturated fatty acids (MUFA) or polyunsaturated fatty acids (PUFA) with minimal saturated fats; a weekly consumption of fish, legumes, poultry and eggs; daily consumption of olive oil and a moderate consumption of red wine with meals as sources of polyphenols; and a more sporadic consumption of potatoes, red meat, sweets [49]. The model of the Mediterranean diet is based on low-glycemic index foods, which favor low levels of insulin and, more generally, a lower risk of developing insulin resistance. Therefore, the beneficial effect of the Mediterranean diet on NAFLD is likely to be dependent on the improvement of insulin sensitivity. However, any excessive intake of nutrients should be avoided, regardless of the beneficial impact on health. For instance, the anti-inflammatory benefits derived from PUFA are linked to extra-virgin olive oil, which is widely used in this dietary pattern. At the same time, an excessive intake of fatty acids (including PUFA) is associated with increased fat deposition, with harmful effects on health, highlighting the importance of balance in a dietary pattern.

Notably, the spread of the Western diet has counterbalanced the original paradigms of the Mediterranean diet, leading to less strict adherence in recent decades. In one study conducted on an Italian population, a greater use of animal proteins, processed and sugary foods, and higher intake of simple sugars and saturated fats was observed among young individuals, with respect to old subjects that continued on the original pattern of the Mediterranean diet [50].

In general, a lower intake of fibers and a higher intake of carbohydrates, saturated fats, fructose, and animal proteins favor the onset of NAFLD, in particular, in the picture of the Western diet. One prospective study conducted on 14-year-old adolescents reported a higher incidence of fatty livers at the age of 17 in those who followed a Western diet pattern, rich in take-away foods, refined cereals, and processed meats [51].

4.2. Behavioral Aspects That Contribute to Liver Damage

Regardless of the type of diet, behavioral aspects need to be considered. When approaching food intake, several mechanisms involve the frequency and the number of meals, and in many cases the psychological underlying background is not fully investigated. Conceiving food either as a reward, or as one way to adapt to recurrent frustrations, can result in altered nutritional behavior: craving of carbohydrates, sweet-eating, night-eating, and emotional eating are some of the labels used to identify these patterns. In particular, attitude to snacking increases intrahepatic and abdominal fat, independently from caloric contents of meals [52]. Eating before bedtime seems to be associated with a higher risk of developing NAFLD in otherwise healthy individuals [53], as does the highly perceived stressfulness [54]. Fast eating leads to an increase in total calorie intake, in particular with regard to carbohydrates, impacting on the incidence of NAFLD [52,55], and should be investigated as a potential driver of liver disease in lean individuals as well [56]. Moreover, sleep disorders are common in NAFLD population, with delayed sleep onset, poor sleep quality, and shortened sleep duration, with subsequent daily sleepiness and diminished quality of life. In one study, food intake times were switched towards the night, and was partially responsible for the sleep disturbances [57]. On the contrary, binge eating disorder was not associated with the severity of liver disease, with respect to NASH and fibrosis [58], although a higher prevalence of binge eating has been observed in NAFLD populations [59].

As for alcohol reporting, of which correct estimations in clinical setting are often challenging, dietary misbehavior is frequently underreported among obese individuals. This may be a reason of concern when approaching a failure in lifestyle intervention, possibly caused by the discrepancy between effective and reported calorie intake [60]. In recent years, mobile technology has offered novel approaches to reduce underreporting, in particular with the help of image-assisted methods that can improve the accuracy of dietary assessment [61]. With this system, food information is directly extracted from the images and the calorie content is automatically calculated through portion size estimation [62]. In one study, eating pattern among healthy individuals was monitored with the use of a mobile app. Most subjects ate frequently and irregularly for more than 14 h during the day. Regulation of eating pattern and restriction of eating duration, assisted by a novel system of data visualization ("feedogram"), resulted in reduced body weight and increased wellness [63]. These approaches might help clinicians to gather comprehensive reports of dietary patterns and habits, with the possibility of defining more precise lifestyle interventions.

5. Diverse Impact of Nutrients on NAFLD

Dietary patterns are highly conditioned by the geographical area, with subsequent differences in nutrient availability, but also cultural heritage, socio-economic status, and food accessibility as determined by local institutions. Moreover, the variability of the outcomes across the clinical studies, as well as the unknown effect of single nutrients on liver histology, make it hard to assess the harmful size of one specific reported dietary pattern. This aspect seems of crucial relevance in clinical practice, because it would lead to a tailored dietary pattern after addressing potentially harmful nutrients (Figure 1).

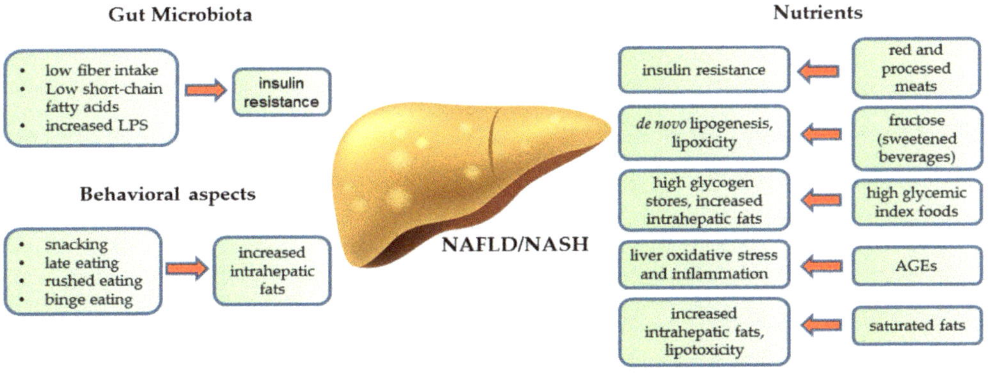

Figure 1. Impact of gut microbiota, behavioral aspects and specific nutrients on features of liver damage leading to Non Alcoholic Fatty Liver Disease/Non Alcoholic Steatohepatitis (NAFLD/NASH). Abbreviations: AGEs, advanced glycation end-products; LPS, lipopolysaccharide.

5.1. The Harmful Effect of Fructose Intake

For instance, in one cross-sectional study conducted on 789 individuals undergoing screening colonoscopy, the consumption of red and processed meat was associated with both NAFLD and insulin resistance [64]. Similarly, a high amount of fructose or high glycemic index foods lead to enhanced hepatic fat synthesis, while a high intake of saturated fats predisposes the liver to fat accumulation. The subsequent lipid-driven toxic damage represents the main driver of hepatic insulin resistance that arises along with NAFLD. Exogenous and endogenous advanced glycation end-products (AGEs) represent a further source of oxidative stress that might interfere with hepatic glucose metabolism in the onset of liver inflammation. Likewise, low fiber intake negatively shapes intestinal bacterial species to dysbiosis, increasing proinflammatory bacterial products that directly impact on liver metabolism.

Fructose is among the main drivers of liver disease, because it feeds into hepatic de novo lipogenesis, thus increasing the amount of steatosis and consequent lipotoxicity, and moreover promoting insulin resistance in the context of a hypercaloric diet. One study conducted on NAFLD adolescents showed how a 6 month reduction in overall fructose intake, together with low glycemic index diet, improved both metabolic parameters (systolic blood pressure and HOMA-IR) and liver biochemistry (ALT) [42]. However, specific patterns of fructose intake have been investigated; not fructose ingestion per se, but rather its addition in sweet beverages, which has been attributed to metabolic derangements. Indeed, one recent randomized, double blind, placebo-controlled trial evaluated the impact of beverage consumption: containing fructose, sucrose (glucose-fructose disaccharide) or glucose. Interestingly, fructose- and sucrose-, but not glucose-containing beverages increased the hepatic synthesis of fatty acids, even in basal state and at a stable average total energy intake. These findings support the hypothesis of long-term fructose-induced intrahepatic metabolic changes, leading to adaptive pathways and increased basal lipogenic activity. During the 7 weeks of intervention, the authors did not detect any differences in metabolic outcomes (fasting plasma triglycerides, glucose and insulin concentrations, HOMA-IR modifications, arterial hypertension rates), concluding that chronic exposure to fructose-containing beverages impacts only on liver lipogenesis, the first hallmark in the onset of metabolic liver disease [65].

5.2. *Exploration of High-Glycemic Index Nutrients*

The detrimental effect of high glycemic index nutrients on NAFLD has been shown in multiple studies. High glycemic index foods (potatoes, white rice, white bread, honey) rapidly increase insulin blood levels, with a subsequent harsh decrease in blood glucose levels. This metabolic pattern is linked to two main aspects: the hyperinsulinemia involved in the long-term deleterious adaptation in insulin-sensitive tissues, and the quick onset of starvation caused by hypoglycemia, causing an increase in caloric intake and potentially leading to eating disorders. This eating pattern leads to hepatic fat accumulation and higher glycogen stores [66]. As with fructose consumption, a high glycemic index food pattern acts on an early step of the liver disease, without impacts on other metabolic parameters (namely, blood glucose levels, triglycerides, high-density lipoprotein cholesterol) [67]. However, longitudinal studies are needed to assess the long-term impacts of this eating pattern on the multiple components of the metabolic syndrome, as well as on the evolution of NAFLD.

In a diabetic population, high glycemic index food patterns are particularly unfavorable, and a long-term low glycemic index diet has been shown to improve glycated hemoglobin and fasting blood glucose, with respect to controls [68]. This nutritional strategy may therefore be tailored to diabetic NAFLD patients, in order to improve overall metabolic disruption.

5.3. *AGEs and Oxidative Stress*

Oxidative stress, induced by reactive oxygen species (ROS), is part of the pathophysiology in NAFLD [69]. Overproduction of ROS occurs when the mitochondrial oxidative capacity is exhausted by an excess of substrates, including saturated fats undergoing β-oxidation. This leads to impaired mitochondrial phosphorylation, which is associated with reduced ATP synthesis and caspase-mediated apoptosis. Excessive ROS can also cause lipid peroxidation, which is a strong driver of lipid-derived liver inflammation.

The susceptibility to ROS-driven hepatic oxidative stress is also influenced by the reduced availability of antioxidant molecules, such as glutathione (GSH). GSH is an ROS scavenger that enables activity of the key antioxidant enzyme GSH peroxidase. In mice fed with a high fat diet, a reduction in GSH and GSH peroxidase was detected, with reduced ability to control the inflammation and mitochondrial-derived oxidative stress, perpetuating the damage. Diet-induced weight loss reduced hepatic fat content and restored the regular transcription and synthesis of antioxidant enzymes [70]. Similar results have been observed in NAFLD patients undergoing bariatric surgery. One year after the intervention, significant differences in plasma and liver markers of oxidative stress were detected, in comparison with baseline evaluations [71].

In addition, one source of oxidative stress is AGEs. These are a biologically active group of molecules that are formed through no-enzymatic reactions between reducing sugars and proteins, lipids and nucleic acids [72]. Their synthesis greatly depends on the concentration of reactants and the half time of the proteins involved; the longer the time, the more prone to develop AGEs. Low rates of AGE formation are normal in metabolism in healthy individuals, but increased production occurs in the presence of carbohydrate overload. Hence, in the context of T2DM, these products are increasingly synthesized. In addition, a Western dietary pattern, in particular with high consumption of red meat and refined grains, sustains the onset of pro-inflammatory activity [73] and hence the development of oxidative stress, which favors AGE formation. In addition, AGEs can be synthesized at an exogenous level, in processed foods (heat treatment, caramel production, bread baking) which are prevalent in Western diets [74].

Accumulation of AGEs results in the activation of pro-inflammatory and pro-fibrotic pathways. In fact, the liver is responsible for the clearance of AGEs, through the interaction between specific receptors (RAGEs) expressed by Kuppfer cells and endothelial cells [75]. This link activates intracellular signaling involved in oxidative stress, thus perpetuating the damage.

Interestingly, AGEs have been shown to potentially discriminate between healthy individuals and NAFLD [76], as well as to distinguish between minimal steatosis versus moderate steatosis; therefore, they may likely be non-invasive biomarkers in high-risk populations [77]. In one randomized, placebo-controlled trial, the simple restriction of oral AGE intake was shown to ameliorate insulin resistance in obese individuals, suggesting that this could be a valuable nutritional intervention acting on a pathophysiological level [78].

5.4. Contribution of Lipids to Liver Damage

The exploration of the impact of lipid profiles in NAFLD has provided notable results. As discussed above, carbohydrates mainly act by increasing hepatic de novo lipogenesis, whereas the excessive intake of saturated fats increases intrahepatic triglyceride content, as compared to unsaturated fats, and increases the rate of lipolysis. Moreover, saturated fats seem to have the highest impact on insulin resistance and stimulate the synthesis of ceramides, which are mainly involved in the process of lipotoxicity and oxidative stress [79]. Similarly, diacylglycerols (DAGs) are intermediates of dietary fat oxidation directly implied in the disrupted hepatic glucose metabolism. By interfering with downstream regulating factors of the insulin pathway, DAGs promote insulin resistance and lipid-mediated hepatocellular damage [80].

In patients with NASH, long-term supplementation of PUFA was shown to decrease blood triglyceride levels and to improve or stabilize intrahepatic inflammatory activity [81], as well as to increase circulating levels of adiponectin, which is an adipose tissue-derived anti-inflammatory hormone [82]. The effect of non-saturated fats is mostly evident in the setting of the Mediterranean diet; nonetheless, one study conducted on an Asian Indian population showed an improvement in the grading of fatty livers and HOMA-IR in NAFLD patients consuming MUFA from vegetable oil, regardless of the overall alimentary pattern [83]. In addition, fish oil, which has been widely used as a source of unsaturated fats for the modulation of dyslipidemia and inflammation, has shown benefits in the setting of liver disease. In a randomized, double blind, placebo-controlled trial of patients with NAFLD and dyslipidemia, fish oil intake was associated with a significant improvement in glucose, cholesterol and triglyceride levels, as well the normalization of transaminases and increases in adiponectin levels. Moreover, a further beneficial effect on inflammation was reported from reductions in prostaglandins and tumor necrosis factor-α (TNF) [84].

5.5. Current Evidence on the Role of Fiber Supplementation

One further aspect that needs to be highlighted is the contribution of fiber intake to metabolic equilibrium. Fibers modulate gut microbiota, which is a substrate for producing short-chain fatty acids (acetate, propionate and butyrate), improving intestinal motility and enterocyte function [85]. Moreover, their positive effects on satiety lead to a better control of body weight [86]. The Western diet is characterized by a lower intake of fibers, and the subsequent reduced modulation of gut microbiota may lead to altered metabolic host status and impaired regulating pathways of enteric hormones, favoring the onset of obesity and T2DM [87]. Despite these two conditions being the key metabolic factors inducing NAFLD, evidence of the direct impact of fibers of NAFLD are lacking, due to the low level of scientific evidence [88]. Fiber supplementation seems to improve NAFLD-related surrogate outcomes, namely, Body Mass Index (BMI), HOMA-IR and transaminases [89], although prospective studies are needed to evaluate the long-term impact of fibers on hard NAFLD outcomes.

5.6. Alcohol and NAFLD: What Are the Proper Recommendations?

Alcohol consumption is a co-factor that exerts an injurious action on liver health. A small intake of red wine with meals (with a threshold of 30 g/day for men and 20 g/day for women) has been considered safe, with regard to substantial harm for chronic liver disease. Additionally, the regular intake of red wine is promoted, due to its beneficial effect related to the antioxidant action of resveratrol. However, when considering individuals presenting NAFLD, the beneficial action of red wine becomes challenging. Cross-sectional studies have suggested that modest consumption of alcohol is associated with a reduced degree of severity (comprising histological necroinflammation and fibrosis), with respect to abstinence [90,91]. However, in the complex metabolic scenario in which NAFLD is imbricated, the compound effect of the concomitant conditions has to be considered, because some features of alcohol metabolism (e.g., oxidative stress, hypertriglyceridemia) may have impacts in a considerable and unique manner. One large prospective population study involving 6732 individuals with metabolic syndrome and different grades of alcohol consumption showed that even low amounts of alcohol independently predict a severe phenotype of NAFLD [92]. Currently, less than moderate levels of alcohol, defined as 210 g/week for men and 140 g/week for women, are recommended most frequently, with absolute abstinence in patients with cirrhosis.

6. Alternative Dietary Approaches: Which Best Strategy?

The Mediterranean diet, despite its comprehensively beneficial action on NAFLD and metabolic syndrome, has several limitations, due to geographical and ethnical disparities throughout the globe. Therefore, different approaches have been evaluated in order to assess the benefit of a specific dietary pattern on the glycometabolic profile, regardless of the actions of single nutrients (Table 1).

Reducing overall daily calories leads to weight loss due to the negative energetic balance, strengthened by adding aerobic physical activity. In one prospective study conducted on obese individuals, a low-fat restricted-calorie diet, Mediterranean restricted-calorie diet, and low-carbohydrate non-restricted diet were randomly assigned for 2 years. The impact of a specific dietary pattern on weight loss varied according to gender, being higher for the low-carbohydrate group in males (mean reduction of 4.9 kg in men versus 2.4 kg in women) and for the Mediterranean diet group in women (mean reduction of 4.0 kg in males versus 6.2 kg in women). Low-fat diet resulted in smaller changes (3.4 kg in males and 0.1 kg in women). The low-carbohydrate diet showed the best rate of reduction in total cholesterol/HDL ratio (20% versus 12% in the low-fat arm), whereas the best impact on the glucose and insulin levels, in particular in the subgroup of diabetic individuals, was observed in the Mediterranean diet group [102]. These results suggest that a reduction in carbohydrates may be the preferable strategy in dietary patterns. In addition, the different impact on metabolic parameters emphasizes the importance of a tailored approach.

Table 1. Clinical studies evaluating the impact of different dietary pattern on both hepatic and metabolic outcomes.

Type of Intervention	Study	Year	Design	Population	Duration of the Study	Staging and Grading (Liver Phenotype)	Results	Other Outcomes
Diet and physical activity								
Low-calorie, low-fat	Vilar-Gomez et al. [9].	2015	Observational study	293 White American obese/overweight biopsied NASH	52 weeks	Histology	25% resolution of NASH ($p < 0.01$) and 19% regression of fibrosis ($p < 0.01$)	Improvement in NAS associated with weight loss > 5% ($p < 0.001$)
Low-fat, low glycemic index	Wong et al. [93].	2013	RCT	154 Asian MRS-proven NAFLD	48 weeks	Radiology (MRS)	Mean change in hepatic fat of 6.7% intervention versus 2.1% control ($p < 0.001$)	Median weight loss of 5.6 kg intervention versus 0.6 kg control ($p < 0.001$)
Low glycemic index Mediterranean diet	Franco et al. [46].	2020	RCT	144 Caucasian CAP-based NAFLD	12 weeks	Non-invasive assessment of steatosis (CAP)	Median reduction of 61 points in CAP ($p < 0.0001$)	Reduction in HOMA-IR (not statistically significant)
Low-calorie	Promrat et al. [10].	2010	RCT	31 White American obese/overweight biopsied NASH	48 weeks	Histology	72% improvement in NAS score ($p = 0.03$)	Improvement in NAS associated with weight loss ≥ 7% ($p < 0.001$)
Mediterranean diet								
	Katsagoni et al. [47].	2018	RCT	63 Caucasian obese/overweight U.S.-based NAFLD	24 weeks	Non-invasive assessment of fibrosis	Reduction in liver stiffness, not statistically significant	Median weight loss of 13.7 kg ($p < 0.05$)
	Properzi et al. [94].	2018	RCT	56 Australian MRS-proven NAFLD	12 weeks	Radiology (MRS)	Mean change in hepatic fat of 32.4% intervention versus 25% control (low-fat diet) ($p = 0.32$)	0.2% reduction in HbA1c in the intervention arm ($p < 0.045$)
	Gelli et al. [95].	2017	Observational study	46 Caucasian U.S.-based NAFLD	24 weeks	Radiology (US), liver function tests, non-invasive assessment of steatosis (FLI, NAFLD-LFS)	20% of steatosis regression at US, reduction in ALT ($p < 0.01$), mean reduction of 12.7 points in FLI ($p < 0.01$), reduction of 0.7 points in NAFLD-LFS ($p < 0.01$)	Reduction of 1.8 points of BMI ($p < 0.01$), improvement in HOMA-IR ($p < 0.01$)
	Ryan et al. [41].	2013	Randomized cross-over trial	12 Australian obese non-diabetic biopsied NAFLD	6 weeks	Radiology (MRS)	Mean change in hepatic fat of 39% intervention ($p = 0.012$)	Improvement in insulin sensitivity by euglycemic clamp ($p = 0.03$)

Table 1. *Cont.*

Type of Intervention	Study	Year	Design	Population	Duration of the Study	Staging and Grading (Liver Phenotype)	Results	Other Outcomes
Single nutrient evaluation								
Fiber supplementation	Stachowska et al. [89].	2020	Meta-analysis of 6 RCTs	242 Caucasian U.S.-based NAFLD	12 weeks	Liver function tests	Reduction in ALT ($p = 0.001$)	Reduction in 0.5 points of BMI ($p = 0.009$), improvement in HOMA-IR ($p = 0.003$)
Omega-3 supplementation (fish oil)	Qin et al. [84].	2015	RCT	80 Asian U.S.-based NAFLD	48 weeks	Liver function tests	Reduction in ALT ($p < 0.05$)	
Omega-3 supplementation (flaxseed, fish oil)	Nogueira et al. [81].	2016	RCT	60 Latin American biopsied NASH	24 weeks	Histology	Improvement in lobular inflammation ($p = 0.05$)	
Low-fructose, low glycemic index	Mager et al. [42].	2015	Observational study	26 White American children/adolescent biopsied NAFLD	24 weeks	Liver function tests	Reduction in ALT ($p = 0.004$)	Improvement in systolic blood pressure ($p = 0.01$), HOMA-IR ($p = 0.03$)
Omega-3 supplementation (olive oil)	Sofi et al. [82].	2010	Observational study	11 Caucasian U.S.-based NAFLD	48 weeks	Radiology (US), liver function tests	0.04-point reduction in Doppler Perfusion Index ($p < 0.05$), improvement in ALT ($p = 0.04$)	
Low-calorie diet								
Low-fat	Kabisch et al. [96].	2018	RCT (low-fat versus low-carbohydrate)	140 Caucasian pre-diabetic MRS-proven NAFLD	48 weeks	Radiology (MRS)	11% versus 10.2% mean change in hepatic fat ($p = 0.59$)	
DASH diet [1]	Razavi et al. [97].	2016	RCT	60 Caucasian obese/overweight U.S.-based NAFLD	8 weeks	Radiology (US), liver function tests	Decreasing of steatosis grade in 80% of patients ($p = 0.003$), reduction in ALT ($p = 0.02$)	Mean weight loss of 3.8 kg ($p = 0.006$), Improvement in HOMA-IR ($p = 0.01$) and CRP ($p = 0.03$)
Very low-calorie diet (800 kcal/day)	Scragg et al. [98].	2020	Observational study	45 Caucasian U.S.-based NAFLD	8 weeks	Non-invasive assessment of fibrosis	Mean reduction of 5.1 points at transient elastography ($p = 0.001$)	Mean weight loss of 10.3 kg (34% of patients reached > 10% weight loss), improvement in HOMA-IR

Table 1. *Cont.*

Type of Intervention	Study	Year	Design	Population	Duration of the Study	Staging and Grading (Liver Phenotype)	Results	Other Outcomes
Low-carbohydrate	Sevastianova et al. [22].	2011	Observational study	8 Caucasian homozygous rs738409 PNPLA3 G allele versus 10 homozygous rs738409 PNPLA3 C allele MRS-proven NAFLD carriers	6 days	Radiology (MRS)	Reduction in liver fat of 45% ($p < 0.001$) in G allele carriers versus 18% in C allele carriers ($p < 0.01$)	
Isocaloric diet								
Low-carbohydrate, high protein	Mardinoglu et al. [99].	2018	Observational study	10 Caucasian obese MRS-proven NAFLD	2 weeks	Radiology (MRS)	Mean reduction in liver fat of 43.8% ($p = 0.027$)	
High protein	Markova et al. [100].	2017	RCT	37 Caucasian diabetic MRS-proven NAFLD	6 weeks	Radiology (MRS)	Reduction in liver fat of 36–48% ($p = 0.0002$)	Improvement in insulin sensitivity by euglycemic clamp
Low-fat, low glycemic index	Utzschneider et al. [101].	2013	RCT	35 White American MRS-proven NAFLD	4 weeks	Radiology (MRS)	Mean reduction in liver fat of 2.2% ($p = 0.002$)	Improvement insulin sensitivity by Matsuda index ($p < 0.05$)

[1] Low-calorie, low-fat diet, restricted in refined grains and rich in fruits/vegetables. Abbreviations: ALT, alanine aminotransferase; BMI: Body Mass Index; CAP, controlled attenuation parameter; CRP, C-reactive protein; DASH, Dietary Approaches to Stop Hypertension; FLI, Fatty Liver Index; HOMA-IR, Homeostatic Model Assessment of Insulin Resistance; IL-6, interleukin-6; MRS, magnetic resonance spectroscopy; NAFLD, non-alcoholic fatty liver disease; NAFLD-LFS: NAFLD-liver fat score; NAS, NAFLD activity score; NASH, non-alcoholic steatohepatitis; PNPLA3, Patatin-like phospholipase domain-containing 3; RCT, randomized controlled trial; TNF-α, tumor necrosis factor-α; US, ultrasound.

Therefore, the common strategy, given the impact of weight loss in ameliorating liver histology, has been so far the hypocaloric diet (approximately 1000 kcal/day or less). In particular, the model of the very low calorie diet (800 kcal/day), of which benefits have been well-assessed in diabetic populations, was applied in 45 patients with NAFLD, and comprised 19.4% fat, 43.4% carbohydrate and 33.7% protein. Overall, 34% of participants achieved a 10% weight loss, while 51% of participants achieved a more than 7% weight reduction. Notably, this dietary pattern was well tolerated and provided long-term benefits after interruption of the study. Concomitant significant improvement of HOMA-IR, transaminases and liver stiffness were observed [98].

Reducing carbohydrates results in significant improvements in intrahepatic triglycerides, hepatic insulin sensitivity and glucose production, when compared to an approach based on reducing fats, despite long-term evaluations not showing differences between the two approaches [103]. It would seem that mere calorie restriction helps in achieving the endpoint, regardless of the nutrient composition. In fact, one randomized controlled trial conducted on 60 overweight or obese NAFLD patients showed the impact of a food pattern with low saturated fat and low dairy products (called DASH, Dietary Approaches to Stop Hypertension, with less than 1000 kcal/day). After 8 weeks of intervention, a greater reduction in weight loss was observed in the DASH group as compared to the control arm (3.8 kg versus 2.3 kg), and BMI (−1.3 points versus −0.8 points). A significant reduction in HOMA-IR was observed in the DASH group (−0.8 points versus −0.2 points in the control arm). Accordingly, serum triglycerides, total cholesterol/HDL ratio, transaminases and markers of oxidative stress improved significantly in the interventional arm. In this study, the proportion of nutrients was almost the same (55% carbohydrates, 15% proteins, 30% fats), with sole differences in fat compositions [97].

Alternatively, isocaloric approaches have been proposed, aiming at modifications in the proportion of nutrients, but keeping average daily calories. Increasing protein intake (30% of the total), either animal- or plant-derived, has been shown to reduce liver fat independently of body weight, and to improve insulin resistance [100]. The relative reduction in carbohydrate intake might also have contributed to the outcome. Moreover, a reduction in fat intake inside the isocaloric dietary pattern has been shown to positively impact on liver fat, as evaluated by proton magnetic resonance spectroscopy: limiting fat intake to less than 16% of the total daily calories led to a 20% reduction in liver fat [104], and similar results were achieved when combining low saturated fats with low glycemic index foods [101]. Furthermore, one 12 week interventional study compared the impacts of a low-fat diet or Mediterranean diet on hepatic steatosis, and showed no differences between the two strategies in terms of benefits, with stronger adherence among the latter [94]. However, one isocaloric low-carbohydrate approach with a relative increase in protein intake seemed to provide the same results, with further evidence of the improvement in beta-oxidation and de novo lipogenesis [99].

Overall, two assumptions could be made. Firstly, there seems to be no clear evidence of preferring one nutrient reduction in respect to others, and the heterogeneity of the studies does not allow for reliable conclusions to be drawn. Moreover, compound modifications in nutrient proportions show that the endpoint is rather achieved with a varied and balanced diet. One large, prospective, interventional study showed, with proton magnetic resonance spectroscopy, that a resolution of hepatic fat could be obtained after 12 months of a dietary approach based on a variety of foods, with emphasis on fruits and vegetables, and with moderate carbohydrate intake, low fats, low glycemic index foods and appropriate portions of meals [93].

Secondly, the short durations of the studies do not provide strong evidence for the impact on metabolic health and liver outcomes. This last point is of crucial relevance when approaching clinical management of the disease. Long-term adherence to a dietary approach may be difficult to achieve, and solid outcomes need to be addressed. One large multicenter study conducted in the United States (the Look Action for Health in Diabetes (AHEAD) Study) enrolled diabetic overweight or obese patients with T2DM and randomly assigned them to either an intensive lifestyle intervention aiming at weight loss (with both decreased caloric intake and increased physical activity), or to diabetes support and education. The primary endpoint was death from cardiovascular causes and incidence of cardiovascular events for a maximum follow-up of 13.5 years. The study was interrupted early at 10 years for futility: no difference between the two groups in the primary outcome occurrence was observed [105]. Again, this evidence highlights the still-unmet need for clear outcomes in dietary patterns, associated with specific, disease-related hard endpoints.

7. Intermittent Fasting to Improve Metabolic Health

Metabolic modifications with regard to lipid and glycemic profiles can also be achieved with other manipulations, in particular by acting on the mealtime, rather than on its composition. Protracted fasting, in particular, causes deep modifications in metabolic pathways that persist over the refeeding time and thus may be one valuable alternative for shaping energy intakes. With this approach, energy restriction is not required to obtain metabolic improvement, nor is weight loss regarded as a key endpoint [106].

This approach is defined as intermittent fasting (IF), and covers different strategies, including alternate-day fasting, the 5:2 diet, or the fast-mimicking diet. In particular, the model of time-restricted feeding (TRF) [107] looks attractive for the liver and metabolic health prospective. TRF allows food intake within a definite interval of hours, following hormone circadian rhythms. The concept of "chrononutrition" has progressively gained attention, because daily metabolic rhythms were found to be dictated by molecular "clocks" acting on specific subsets of genes [108]. This complex metabolic crosstalk is hierarchically overseen by the hypothalamus, in response to exogenous and endogenous stimuli (the sleep–wake cycle according to light–dark courses as a paradigm). Peripheral organs, such as the liver, pancreas, skeletal muscle and adipose tissue, the key drivers of the systemic metabolic homeostasis, allow the perfect coordination (a true "synchronization") between the environment and the biochemical processes. Glucose homeostasis and insulin sensitivity are examples of time-related regulation; mealtimes, in return, regulate adiposity and body weight [109,110].

Animal studies conducted in rodents experiencing TRF have been shown to attenuate the harm of obesogenic diets [111], proportional to the fasting duration, reversing or stabilizing preexisting obesity, insulin resistance [112], as well as improving hepatic steatosis [113] and inflammatory markers [114]. Furthermore, TRF seems to have a modulating effect on GM, contributing to cyclical changes and diversity that impact on host metabolism [115]. Moreover, a long-term protective impact of TRF, related to the underlying gene regulation that provides durable effects, was detected even when it was temporarily interrupted in favor of unlimited access to food [116].

In humans, biological clocks suggest that metabolism is optimized for food intake in the morning, with insulin sensitivity and beta-cell responsiveness being higher in the morning than in the afternoon or evening. Individuals with T2DM who follow this pattern show a significant reduction in postprandial hyperglycemia [117], whereas insulin-resistant individuals show significant improvements in insulin sensitivity and, interestingly, a reduction in hunger [117]. In addition, overweight or obese individuals are more prone to weight loss following this alimentary timing [118], and a better modulation of adipokines (increase in adiponectin and decrease in leptin) has been reported [119]. Sutton et al. performed one elegant proof-of-concept study to show the benefits of an early TRF (allowing food intake for 6 h before 3 p.m.). This was one 5-week, randomized, crossover, isocaloric end eucaloric controlled feeding study conducted on overweight and pre-diabetes individuals. TRF improved insulin levels, insulin sensitivity, beta-cell responsiveness, blood pressure and oxidative stress, even in the absence of weight loss [120]. In particular, the improvement in blood pressure, even though participants had mean values in the pre-hypertensive range, is of great interest in terms of endpoints, as well as the smaller importance given to weight loss in the amelioration of metabolic health. Given the central role of insulin resistance in the pathogenesis of NAFLD, the TRF approach might provide interesting results.

8. Outlook and Summary—Open Issues in the Clinical Setting

The wide heterogeneity of lifestyle intervention approaches, together with the diverse endpoints assessed by the clinical studies and the high variability in study populations and duration of treatment, have considerable impacts in clinical practice.

In addition, a uniform effect to specific lifestyle intervention can hardly be achieved. One large study evaluated changes in glucose levels in 800 overweight and obese non-diabetic individuals following identical meal intakes. Here, a high variability in post-prandial glucose levels was observed, highlighting that individual recommendations might be required [121].

More than one approach seems beneficial in NAFLD populations, but a lack of standardization prevents clinicians from making reliable decisions, in particular with regard to the short-term assessment of either compliance or success of the chosen intervention. Weight loss would still be the primary endpoints of "bedside" recommendations, but more appropriate, tailored outcomes are required. In particular, in the absence of long-term hard endpoints in NAFLD populations, validation of simple tools may help in handling successes or failures in clinical settings. Reductions in CAP after 12 weeks of treatment is one example explored in the above-mentioned studies [46], as was the remission of steatosis through proton magnetic resonance spectroscopy after 12 months of intervention [93]. Alternatively, surrogate measures of insulin resistance, such as short-term improvements in HOMA-IR, might be used to assess the benefit of the intervention, even though the size effect may vary according to the chosen approach [97,103].

In fact, the assessment of one endpoint seems critically associated with the interventional strategy. The fatty liver index (FLI) and the NAFLD-liver fat score (NAFLD-LFS) are non-invasive liver fat indices introduced in clinical practice aiming for the early detection of NAFLD and are best validated for purposes in cross-sectional studies. However, their longitudinal accuracy has been proven to be diet-specific: changes in liver fat, obtained by a low-fat diet intervention, moderately correlate to changes in FLI and NAFLD-LFS, but this does not happen when undertaking a low-carbohydrate diet approach [96]. Additionally, one 6-month clinical trial on NAFLD patients undertaking a Mediterranean diet has shown significant improvements in FLI and NAFLD-LFS [95]. The impact of lifestyle interventions on fibrosis is less well-established, although long-term histologic evaluation has provided evidence of fibrosis amelioration [11]. In one cross-sectional study conducted on biopsy-proven NAFLD patients with T2DM, liver fibrosis was inversely associated with adherence to the Mediterranean diet at multivariate analysis, including multiple putative factors (age, gender, BMI, glycated hemoglobin) [122]. In addition, one prospective observational Greek study reported the same inverse association when evaluating liver fibrosis with non-invasive scores (namely, fibrosis-4 score (FIB-4), AST-to-platelet ratio (APRI) and BARD (BMI, AST/ALT Ratio, Diabetes) index) through cross-sectional analysis [123]. An inverse association between the Mediterranean diet and liver fibrosis also emerged through elastography after 6 months of intervention, suggesting a possible role for liver stiffness, together with CAP, in the assessment of treatment progress [47].

In conclusion, the nutritional landscape of interventional approaches for NAFLD treatment is burdened by the heterogeneity in study populations, duration of treatments, and the diversity of selected endpoints. The lack of longitudinal studies in nutritional patterns on the natural history of NAFLD makes it hard to assess strong benefits of either nutrients or dietary approaches, of which evidence relies on surrogate markers or short-term evaluation. The complex picture of the metabolic syndrome requires a careful, multidisciplinary method, because multiple causal agents impact on liver disease, and systemic morbidities are in return worsened by liver damage. The relative strong evidence of beneficial effects of the Mediterranean diet is counterbalanced by a lack of studies in other dietary models, of which usefulness is crucial for populations where the Mediterranean diet would be undertaken with difficultly, due to geographical and cultural discrepancies in food accessibility.

In addition, evidence of the harmful effect on liver fat content of late eating, rushed eating and attitude to snacking has highlighted the importance of behavioral aspects in nutrition, which are often underreported and thus require proper investigation. It is likely than more than one approach would ameliorate liver and systemic metabolism, and good communication with patients with respect to harmful nutrients, time of meals, and proper behavior might produce better results per se, as well as increase compliance and understanding. Aiming for improvements in insulin sensitivity can be the fil rouge of different approaches, which would not potentially be limited to restrictions of food, but also to eating times. In fact, the positive results on insulin sensitivity coming from the novel approaches of intermittent feeding have corroborated the concept of quality and timing of eating, with important implications in clinical management. The need for associating one nutritional intervention with inexpensive, non-invasive scores of either steatosis or fibrosis is essential in clinical settings, in order to assess benefits or futility. These algorithms should be included in future studies, with particular regard to liver fibrosis, which is the hardest prognostic factor in the setting of chronic liver disease.

Author Contributions: Conceptualization, methodology and finalization of the manuscript: J.M.S. Formal analysis, data curation, and writing—original draft preparation A.A. All authors have read and agreed to the published version of the manuscript.

Funding: This research received no external funding. J.M.S. is partly funded by the European Union Innovative Medicines Initiative 2 (IMI2) Joint Undertaking under grant agreement 777377: LITMUS (Liver Investigation: Testing Biomarker Utility in Steatohepatitis).

Institutional Review Board Statement: Not applicable.

Informed Consent Statement: Not applicable.

Data Availability Statement: All data are publicly available.

Conflicts of Interest: J.M.S. declares Consultancy: BMS, Boehringer Ingelheim, Echosens, Genfit, Gilead Sciences, Intercept Pharmaceuticals, Madrigal, Nordic Bioscience, Novartis, Pfizer, Roche, Sanofi, Siemens Healthcare GmbH. Research funding: Gilead Sciences, Siemens Healthcare GmbH, Boehringer Ingelheim. Speakers bureau: Falk Foundation. A.A. declares no conflict of interest. The companies had no role in the design of the study; in the collection, analyses, or interpretation of data; in the writing of the manuscript, or in the decision to publish the results.

References

1. Younossi, Z.M.; Koenig, A.B.; Abdelatif, D.; Fazel, Y.; Henry, L.; Wymer, M. Global Epidemiology of Nonalcoholic Fatty Liver Disease—Meta-Analytic Assessment of Prevalence, Incidence, and Outcomes. *Hepatology* **2016**, *64*, 73–84. [CrossRef] [PubMed]
2. Estes, C.; Anstee, Q.M.; Arias-Loste, M.T.; Bantel, H.; Bellentani, S.; Caballeria, J.; Colombo, M.; Craxi, A.; Crespo, J.; Day, C.P.; et al. Modeling NAFLD Disease Burden in China, France, Germany, Italy, Japan, Spain, United Kingdom, and United States for the Period 2016–2030. *J. Hepatol.* **2018**, *69*, 896–904. [CrossRef] [PubMed]
3. Schattenberg, J.M.; Lazarus, J.V.; Newsome, P.N.; Serfaty, L.; Aghemo, A.; Augustin, S.; Tsochatzis, E.; de Ledinghen, V.; Bugianesi, E.; Romero-Gomez, M.; et al. Disease Burden and Economic Impact of Diagnosed Non-Alcoholic Steatohepatitis in Five European Countries in 2018: A Cost-of-Illness Analysis. *Liver Int.* **2021**, *41*, 1227–1242. [CrossRef] [PubMed]
4. Gehrke, N.; Schattenberg, J.M. Metabolic Inflammation—A Role for Hepatic Inflammatory Pathways as Drivers of Comorbidities in Nonalcoholic Fatty Liver Disease? *Gastroenterology* **2020**, *158*, 1929–1947.e6. [CrossRef] [PubMed]
5. Marengo, A.; Jouness, R.I.K.; Bugianesi, E. Progression and Natural History of Nonalcoholic Fatty Liver Disease in Adults. *Clin. Liver Dis.* **2016**, 313–324. [CrossRef] [PubMed]
6. Hagström, H.; Nasr, P.; Ekstedt, M.; Hammar, U.; Stål, P.; Hultcrantz, R.; Kechagias, S. Fibrosis Stage but Not NASH Predicts Mortality and Time to Development of Severe Liver Disease in Biopsy-Proven NAFLD. *J. Hepatol.* **2017**, *67*, 1265–1273. [CrossRef] [PubMed]
7. Labenz, C.; Huber, Y.; Kalliga, E.; Nagel, M.; Ruckes, C.; Straub, B.K.; Galle, P.R.; Wörns, M.A.; Anstee, Q.M.; Schuppan, D.; et al. Predictors of Advanced Fibrosis in Non-Cirrhotic Non-Alcoholic Fatty Liver Disease in Germany. *Aliment. Pharmacol. Ther.* **2018**, *48*, 1109–1116. [CrossRef] [PubMed]
8. Younes, R.; Govaere, O.; Petta, S.; Miele, L.; Tiniakos, D.; Burt, A.; David, E.; Vecchio, F.M.; Maggioni, M.; Cabibi, D.; et al. Caucasian lean subjects with non-alcoholic fatty liver disease share long-term prognosis of non-lean: Time for reappraisal of BMI-driven approach? *Gut* **2021**. [CrossRef]

9. Vilar-Gomez, E.; Martinez-Perez, Y.; Calzadilla-Bertot, L.; Torres-Gonzalez, A.; Gra-Oramas, B.; Gonzalez-Fabian, L.; Friedman, S.L.; Diago, M.; Romero-Gomez, M. Weight Loss through Lifestyle Modification Significantly Reduces Features of Nonalcoholic Steatohepatitis. *Gastroenterology* **2015**, *149*, 367–378.e5. [CrossRef] [PubMed]
10. Promrat, K.; Kleiner, D.E.; Niemeier, H.M.; Jackvony, E.; Kearns, M.; Wands, J.R.; Fava, J.L.; Wing, R.R. Randomized Controlled Trial Testing the Effects of Weight Loss on Nonalcoholic Steatohepatitis. *Hepatology* **2010**, *51*, 121–129. [CrossRef] [PubMed]
11. Lassailly, G.; Caiazzo, R.; Buob, D.; Pigeyre, M.; Verkindt, H.; Labreuche, J.; Raverdy, V.; Leteurtre, E.; Dharancy, S.; Louvet, A.; et al. Bariatric Surgery Reduces Features of Nonalcoholic Steatohepatitis in Morbidly Obese Patients. *Gastroenterology* **2015**, *149*, 379–388. [CrossRef]
12. Lassailly, G.; Caiazzo, R.; Ntandja-Wandji, L.C.; Gnemmi, V.; Baud, G.; Verkindt, H.; Ningarhari, M.; Louvet, A.; Leteurtre, E.; Raverdy, V.; et al. Bariatric Surgery Provides Long-Term Resolution of Nonalcoholic Steatohepatitis and Regression of Fibrosis. *Gastroenterology* **2020**, *159*, 1290–1301.e5. [CrossRef]
13. Armstrong, M.J.; Gaunt, P.; Aithal, G.P.; Barton, D.; Hull, D.; Parker, R.; Hazlehurst, J.M.; Guo, K.; Abouda, G.; Aldersley, M.A.; et al. Liraglutide safety and efficacy in patients with non-alcoholic steatohepatitis (LEAN): A multicentre, double-blind, randomised, placebo-controlled phase 2 study. *Lancet* **2016**, *387*, 679–690. [CrossRef]
14. Kahl, S.; Gancheva, S.; Straßburger, K.; Herder, C.; Machann, J.; Katsuyama, H.; Kabisch, S.; Henkel, E.; Kopf, S.; Lagerpusch, M.; et al. Empagliflozin Effectively Lowers Liver Fat Content in Well-Controlled Type 2 Diabetes: A Randomized, Double-Blind, Phase 4, Placebo-Controlled Trial. *Diabetes Care.* **2020**, *43*, 298–305. [CrossRef]
15. Sanyal, A.J.; Chalasani, N.; Kowdley, K.V.; McCullough, A.; Diehl, A.M.; Bass, N.M.; Neuschwander-Tetri, B.A.; Lavine, J.E.; Tonascia, J.; Unalp, A.; et al. Pioglitazone, vitamin E, or placebo for nonalcoholic steatohepatitis. *N. Engl. J. Med.* **2010**, *362*, 1675–1685. [CrossRef]
16. Glass, O.; Filozof, C.; Noureddin, M.; Berner-Hansen, M.; Schabel, E.; Omokaro, S.O.; Schattenberg, J.M.; Barradas, K.; Miller, V.; Francque, S.; et al. Standardisation of Diet and Exercise in Clinical Trials of NAFLD-NASH: Recommendations from the Liver Forum. *J. Hepatol.* **2020**, 680–693. [CrossRef]
17. Han, M.A.T.; Altayar, O.; Hamdeh, S.; Takyar, V.; Rotman, Y.; Etzion, O.; Lefebvre, E.; Safadi, R.; Ratziu, V.; Prokop, L.J.; et al. Rates of and Factors Associated with Placebo Response in Trials of Pharmacotherapies for Nonalcoholic Steatohepatitis: Systematic Review and Meta-Analysis. *Clin. Gastroenterol. Hepatol.* **2019**, *17*, 616–629.e26. [CrossRef]
18. Romeo, S.; Kozlitina, J.; Xing, C.; Pertsemlidis, A.; Cox, D.; Pennacchio, L.A.; Boerwinkle, E.; Cohen, J.C.; Hobbs, H.H. Genetic Variation in PNPLA3 Confers Susceptibility to Nonalcoholic Fatty Liver Disease. *Nat. Genet.* **2008**, *40*, 1461–1465. [CrossRef]
19. Mancina, R.M.; Dongiovanni, P.; Petta, S.; Pingitore, P.; Meroni, M.; Rametta, R.; Borén, J.; Montalcini, T.; Pujia, A.; Wiklund, O.; et al. The MBOAT7-TMC4 Variant Rs641738 Increases Risk of Nonalcoholic Fatty Liver Disease in Individuals of European Descent. *Gastroenterology* **2016**, *150*, 1219–1230.e6. [CrossRef]
20. Kozlitina, J.; Smagris, E.; Stender, S.; Nordestgaard, B.G.; Zhou, H.H.; Tybjærg-Hansen, A.; Vogt, T.F.; Hobbs, H.H.; Cohen, J.C. Exome-Wide Association Study Identifies a TM6SF2 Variant That Confers Susceptibility to Nonalcoholic Fatty Liver Disease. *Nat. Genet.* **2014**, *46*, 352–356. [CrossRef]
21. Shen, J.; Wong, G.L.H.; Chan, H.L.Y.; Chan, R.S.M.; Chan, H.Y.; Chu, W.C.W.; Cheung, B.H.K.; Yeung, D.K.W.; Li, L.S.; Sea, M.M.M.; et al. PNPLA3 Gene Polymorphism and Response to Lifestyle Modification in Patients with Nonalcoholic Fatty Liver Disease. *J. Gastroenterol. Hepatol.* **2015**, *30*, 139–146. [CrossRef] [PubMed]
22. Sevastianova, K.; Kotronen, A.; Gastaldelli, A.; Perttilä, J.; Hakkarainen, A.; Lundbom, J.; Suojanen, L.; Orho-Melander, M.; Lundbom, N.; Ferrannini, E.; et al. Genetic Variation in PNPLA3 (Adiponutrin) Confers Sensitivity to Weight Loss—Induced Decrease in Liver Fat in Humans. *Am. J. Clin. Nutr.* **2011**, *94*, 104–111. [CrossRef] [PubMed]
23. Degasperi, E.; Galmozzi, E.; Pelusi, S.; D'Ambrosio, R.; Soffredini, R.; Borghi, M.; Perbellini, R.; Facchetti, F.; Iavarone, M.; Sangiovanni, A.; et al. Hepatic Fat—Genetic Risk Score Predicts Hepatocellular Carcinoma in Patients with Cirrhotic HCV Treated with DAAs. *Hepatology* **2020**. [CrossRef] [PubMed]
24. Gil-Ramírez, A.; Caz, V.; Martin-Hernandez, R.; Marín, F.R.; Largo, C.; Rodríguez-Casado, A.; Tabernero, M.; Ruiz-Rodríguez, A.; Reglero, G.; Soler-Rivas, C. Modulation of cholesterol-related gene expression by ergosterol and ergosterol-enriched extracts obtained from Agaricus bisporus. *Eur. J. Nutr.* **2016**, *55*, 1041–1057. [CrossRef]
25. Mouries, J.; Brescia, P.; Silvestri, A.; Spadoni, I.; Sorribas, M.; Wiest, R.; Mileti, E.; Galbiati, M.; Invernizzi, P.; Adorini, L.; et al. Microbiota-Driven Gut Vascular Barrier Disruption Is a Prerequisite for Non-Alcoholic Steatohepatitis Development. *J. Hepatol.* **2019**, *71*, 1216–1228. [CrossRef]
26. Wu, G.D.; Chen, J.; Hoffmann, C.; Bittinger, K.; Chen, Y.Y.; Keilbaugh, S.A.; Bewtra, M.; Knights, D.; Walters, W.A.; Knight, R.; et al. Linking Long-Term Dietary Patterns with Gut Microbial Enterotypes. *Science* **2011**, *334*, 105–108. [CrossRef]
27. Cotillard, A.; Kennedy, S.P.; Kong, L.C.; Prifti, E.; Pons, N.; Le Chatelier, E.; Almeida, M.; Quinquis, B.; Levenez, F.; Galleron, N.; et al. Dietary Intervention Impact on Gut Microbial Gene Richness. *Nature* **2013**, *500*, 585–588. [CrossRef]
28. Saad, M.J.A.; Santos, A.; Prada, P.O. Linking Gut Microbiota and Inflammation to Obesity and Insulin Resistance. *Physiology* **2016**. [CrossRef]
29. Wang, D.D.; Nguyen, L.H.; Li, Y.; Yan, Y.; Ma, W.; Rinott, E.; Ivey, K.L.; Shai, I.; Willett, W.C.; Hu, F.B.; et al. The Gut Microbiome Modulates the Protective Association between a Mediterranean Diet and Cardiometabolic Disease Risk. *Nat. Med.* **2021**, *27*, 333–343. [CrossRef]

30. Zhou, F.; Zhou, J.; Wang, W.; Zhang, X.J.; Ji, Y.X.; Zhang, P.; She, Z.G.; Zhu, L.; Cai, J.; Li, H. Unexpected Rapid Increase in the Burden of NAFLD in China From 2008 to 2018: A Systematic Review and Meta-Analysis. *Hepatology* **2019**, *70*, 1119–1133. [CrossRef]
31. Katsagoni, C.N.; Georgoulis, M.; Papatheodoridis, G.V.; Panagiotakos, D.B.; Kontogianni, M.D. Effects of lifestyle interventions on clinical characteristics of patients with non-alcoholic fatty liver disease: A meta-analysis. *Metabolism* **2017**, *68*, 119–132. [CrossRef]
32. Bugianesi, E.; McCullough, A.J.; Marchesini, G. Insulin Resistance: A Metabolic Pathway to Chronic Liver Disease. *Hepatology* **2005**, 987–1000. [CrossRef]
33. Armandi, A.; Rosso, C.; Caviglia, G.P.; Bugianesi, E. Insulin Resistance across the Spectrum of Nonalcoholic Fatty Liver Disease. *Metabolites* **2021**, *11*, 155. [CrossRef]
34. Schattenberg, J.M.; Wang, Y.; Singh, R.; Rigoli, R.M.; Czaja, M.J. Hepatocyte CYP2E1 Overexpression and Steatohepatitis Lead to Impaired Hepatic Insulin Signaling. *J. Biol. Chem.* **2005**, *280*, 9887–9894. [CrossRef]
35. Schuster-Gaul, S.; Geisler, L.J.; McGeough, M.D.; Johnson, C.D.; Zagorska, A.; Li, L.; Wree, A.; Barry, V.; Mikaelian, I.; Jih, L.J.; et al. ASK1 Inhibition Reduces Cell Death and Hepatic Fibrosis in an Nlrp3 Mutant Liver Injury Model. *JCI Insight* **2020**, *5*. [CrossRef]
36. Solinas, G.; Becattini, B. JNK at the Crossroad of Obesity, Insulin Resistance, and Cell Stress Response. *Mol. Metab.* **2017**, 174–184. [CrossRef]
37. Rosso, C.; Kazankov, K.; Younes, R.; Esmaili, S.; Marietti, M.; Sacco, M.; Carli, F.; Gaggini, M.; Salomone, F.; Møller, H.J.; et al. Crosstalk between Adipose Tissue Insulin Resistance and Liver Macrophages in Non-Alcoholic Fatty Liver Disease. *J. Hepatol.* **2019**, *71*, 1012–1021. [CrossRef]
38. Musso, G.; Gambino, R.; De Michieli, F.; Cassader, M.; Rizzetto, M.; Durazzo, M.; Fagà, E.; Silli, B.; Pagano, G. Dietary Habits and Their Relations to Insulin Resistance and Postprandial Lipemia in Nonalcoholic Steatohepatitis. *Hepatology* **2003**, *37*, 909–916. [CrossRef]
39. Romero-Gómez, M.; Zelber-Sagi, S.; Trenell, M. Treatment of NAFLD with Diet, Physical Activity and Exercise. *J. Hepatol.* **2017**, 829–846. [CrossRef]
40. Ter Horst, K.W.; Serlie, M.J. Fructose Consumption, Lipogenesis, and Non-Alcoholic Fatty Liver Disease. *Nutrients* **2017**, *9*, 981. [CrossRef]
41. Ryan, M.C.; Itsiopoulos, C.; Thodis, T.; Ward, G.; Trost, N.; Hofferberth, S.; O'Dea, K.; Desmond, P.V.; Johnson, N.A.; Wilson, A.M. The Mediterranean Diet Improves Hepatic Steatosis and Insulin Sensitivity in Individuals with Non-Alcoholic Fatty Liver Disease. *J. Hepatol.* **2013**, *59*, 138–143. [CrossRef]
42. Mager, D.R.; Iñiguez, I.R.; Gilmour, S.; Yap, J. The Effect of a Low Fructose and Low Glycemic Index/Load (FRAGILE) Dietary Intervention on Indices of Liver Function, Cardiometabolic Risk Factors, and Body Composition in Children and Adolescents with Nonalcoholic Fatty Liver Disease (NAFLD). *J. Parenter. Enter. Nutr.* **2015**, *39*, 73–84. [CrossRef] [PubMed]
43. Simpson, S.J.; Le Couteur, D.G.; James, D.E.; George, J.; Gunton, J.E.; Solon-Biet, S.M.; Raubenheimer, D. The Geometric Framework for Nutrition as a Tool in Precision Medicine. *Nutr. Healthy Aging* **2017**, *4*, 217–226. [CrossRef] [PubMed]
44. Solon-Biet, S.M.; McMahon, A.C.; Ballard, J.W.O.; Ruohonen, K.; Wu, L.E.; Cogger, V.C.; Warren, A.; Huang, X.; Pichaud, N.; Melvin, R.G.; et al. The Ratio of Macronutrients, Not Caloric Intake, Dictates Cardiometabolic Health, Aging, and Longevity in Ad Libitum-Fed Mice. *Cell Metab.* **2014**, *19*, 418–430. [CrossRef] [PubMed]
45. Kastorini, C.M.; Milionis, H.J.; Esposito, K.; Giugliano, D.; Goudevenos, J.A.; Panagiotakos, D.B. The Effect of Mediterranean Diet on Metabolic Syndrome and Its Components: A Meta-Analysis of 50 Studies and 534,906 Individuals. *J. Am. Coll. Cardiol.* **2011**, *57*, 1299–1313. [CrossRef]
46. Franco, I.; Bianco, A.; Mirizzi, A.; Campanella, A.; Bonfiglio, C.; Sorino, P.; Notarnicola, M.; Tutino, V.; Cozzolongo, R.; Giannuzzi, V.; et al. Physical Activity and Low Glycemic Index Mediterranean Diet: Main and Modification Effects on NAFLD Score. Results from a Randomized Clinical Trial. *Nutrients* **2021**, *13*, 66. [CrossRef]
47. Katsagoni, C.N.; Papatheodoridis, G.V.; Ioannidou, P.; Deutsch, M.; Alexopoulou, A.; Papadopoulos, N.; Papageorgiou, M.V.; Fragopoulou, E.; Kontogianni, M.D. Improvements in Clinical Characteristics of Patients with Non-Alcoholic Fatty Liver Disease, after an Intervention Based on the Mediterranean Lifestyle: A Randomised Controlled Clinical Trial. *Br. J. Nutr.* **2018**, *120*, 164–175. [CrossRef]
48. Khalatbari-Soltani, S.; Imamura, F.; Brage, S.; De Lucia Rolfe, E.; Griffin, S.J.; Wareham, N.J.; Marques-Vidal, P.; Forouhi, N.G. The Association between Adherence to the Mediterranean Diet and Hepatic Steatosis: Cross-Sectional Analysis of Two Independent Studies, the UK Fenland Study and the Swiss CoLaus Study. *BMC Med.* **2019**, *17*, 1–14. [CrossRef]
49. Torres, M.C.P.; Aghemo, A.; Lleo, A.; Bodini, G.; Furnari, M.; Marabotto, E.; Miele, L.; Giannini, E.G. Mediterranean Diet and NAFLD: What We Know and Questions That Still Need to Be Answered. *Nutrients* **2019**, *11*, 2971. [CrossRef]
50. Rizza, W.; De Gara, L.; Antonelli Incalzi, R.; Pedone, C. Prototypical versus Contemporary Mediterranean Diet. *Clin. Nutr. ESPEN* **2016**, *15*, 44–48. [CrossRef]
51. Oddy, W.H.; Herbison, C.E.; Jacoby, P.; Ambrosini, G.L.; O'Sullivan, T.A.; Ayonrinde, O.T.; Olynyk, J.K.; Black, L.J.; Beilin, L.J.; Mori, T.A.; et al. The Western Dietary Pattern Is Prospectively Associated with Nonalcoholic Fatty Liver Disease in Adolescence. *Am. J. Gastroenterol.* **2013**, *108*, 778–785. [CrossRef]
52. Koopman, K.E.; Caan, M.W.A.; Nederveen, A.J.; Pels, A.; Ackermans, M.T.; Fliers, E.; la Fleur, S.E.; Serlie, M.J. Hypercaloric Diets with Increased Meal Frequency, but Not Meal Size, Increase Intrahepatic Triglycerides: A Randomized Controlled Trial. *Hepatology* **2014**, *60*, 545–553. [CrossRef]

53. Nishi, T.; Babazono, A.; Maeda, T.; Imatoh, T.; Une, H. Effects of Eating Fast and Eating before Bedtime on the Development of Nonalcoholic Fatty Liver Disease. *Popul. Health Manag.* **2016**, *19*, 279–283. [CrossRef]
54. Han, A.L. Association between Non-Alcoholic Fatty Liver Disease and Dietary Habits, Stress, and Health-Related Quality of Life in Korean Adults. *Nutrients* **2020**, *12*, 1555. [CrossRef]
55. Takahashi, F.; Hashimoto, Y.; Kawano, R.; Kaji, A.; Sakai, R.; Kawate, Y.; Okamura, T.; Ushigome, E.; Kitagawa, N.; Majima, S.; et al. Eating Fast Is Associated with Nonalcoholic Fatty Liver Disease in Men but Not in Women with Type 2 Diabetes: A Cross-Sectional Study. *Nutrients* **2020**, *12*, 2174. [CrossRef]
56. Lee, S.; Ko, B.J.; Gong, Y.; Han, K.; Lee, A.; Han, B.D.; Yoon, Y.J.; Park, S.; Kim, J.H.; Mantzoros, C.S. Self-Reported Eating Speed in Relation to Non-Alcoholic Fatty Liver Disease in Adults. *Eur. J. Nutr.* **2016**, *55*, 327–333. [CrossRef]
57. Bernsmeier, C.; Weisskopf, D.M.; Pflueger, M.O.; Mosimann, J.; Campana, B.; Terracciano, L.; Beglinger, C.; Heim, M.H.; Cajochen, C. Sleep Disruption and Daytime Sleepiness Correlating with Disease Severity and Insulin Resistance in Non-Alcoholic Fatty Liver Disease: A Comparison with Healthy Controls. *PLoS ONE* **2015**, *10*, e0143293. [CrossRef]
58. Canivet, C.M.; Perney, P.; Cherick, F.; Orlowski, M.; Patouraux, S.; Bailly-Maitre, B.; Tran, A.; Iannelli, A.; Gual, P.; Anty, R. No Association between Binge Eating Disorder and Severity of Non-Alcoholic Fatty Liver Disease in Severely Obese Patients. *JGH Open* **2020**, *4*, 525–531. [CrossRef]
59. Zhang, J.; Abbasi, O.; Malevanchik, L.; Mohan, N.; Denicola, R.; Tarangelo, N.; Halegoua-De Marzio, D. Pilot Study of the Prevalence of Binge Eating Disorder in Non-Alcoholic Fatty Liver Disease Patients. *Ann. Gastroenterol.* **2017**, *30*, 664–669. [CrossRef]
60. Lichtman, S.W.; Pisarska, K.; Berman, E.R.; Pestone, M.; Dowling, H.; Offenbacher, E.; Weisel, H.; Heshka, S.; Matthews, D.E.; Heymsfield, S.B. Discrepancy between Self-Reported and Actual Caloric Intake and Exercise in Obese Subjects. *N. Engl. J. Med.* **1992**, *327*, 1893–1898. [CrossRef]
61. Boushey, C.J.; Spoden, M.; Zhu, F.M.; Delp, E.J.; Kerr, D.A. New Mobile Methods for Dietary Assessment: Review of Image-Assisted and Image-Based Dietary Assessment Methods. In Proceedings of the Nutrition Society; Cambridge University Press: Cambridge, UK, 2017; Volume 76, pp. 283–294. [CrossRef]
62. Hassannejad, H.; Matrella, G.; Ciampolini, P.; De Munari, I.; Mordonini, M.; Cagnoni, S. Automatic Diet Monitoring: A Review of Computer Vision and Wearable Sensor-Based Methods. *Int. J. Food Sci. Nutr.* **2017**, *68*, 656–670. [CrossRef] [PubMed]
63. Gill, S.; Panda, S. A Smartphone App Reveals Erratic Diurnal Eating Patterns in Humans That Can Be Modulated for Health Benefits. *Cell Metab.* **2015**, *22*, 789–798. [CrossRef] [PubMed]
64. Zelber-Sagi, S.; Ivancovsky-Wajcman, D.; Fliss Isakov, N.; Webb, M.; Orenstein, D.; Shibolet, O.; Kariv, R. High Red and Processed Meat Consumption Is Associated with Non-Alcoholic Fatty Liver Disease and Insulin Resistance. *J. Hepatol.* **2018**, *68*, 1239–1246. [CrossRef] [PubMed]
65. Gerber, P.A.; Geidl, B.; Hochuli, M.; Tappy, L.; Spinas, G. Fructose-Sweetened Beverage Consumption Induces Hepatic de Novo Lipogenesis in Healthy Humans: A Double Blind Randomized Controlled Trial. *Clin. Nutr. ESPEN* **2020**, *40*, 622–623. [CrossRef]
66. Bawden, S.; Stephenson, M.; Falcone, Y.; Lingaya, M.; Ciampi, E.; Hunter, K.; Bligh, F.; Schirra, J.; Taylor, M.; Morris, P.; et al. Increased Liver Fat and Glycogen Stores after Consumption of High versus Low Glycaemic Index Food: A Randomized Crossover Study. *Diabetes Obes. Metab.* **2017**, *19*, 70–77. [CrossRef]
67. Parker, A.; Kim, Y. The Effect of Low Glycemic Index and Glycemic Load Diets on Hepatic Fat Mass, Insulin Resistance, and Blood Lipid Panels in Individuals with Nonalcoholic Fatty Liver Disease. *Metab. Syndr. aRelat. Disord.* **2019**, *17*, 389–396. [CrossRef]
68. Ojo, O.; Ojo, O.O.; Adebowale, F.; Wang, X.H. The Effect of Dietary Glycaemic Index on Glycaemia in Patients with Type 2 Diabetes: A Systematic Review and Meta-Analysis of Randomized Controlled Trials. *Nutrients* **2018**, *10*, 373. [CrossRef]
69. Polimeni, L.; Del Ben, M.; Baratta, F.; Perri, L.; Albanese, F.; Pastori, D.; Violi, F.; Angelico, F. Oxidative stress: New insights on the association of non-alcoholic fatty liver disease and atherosclerosis. *World J. Hepatol.* **2015**, *7*, 1325–1336. [CrossRef]
70. Mendes, I.K.S.; Matsuura, C.; Aguila, M.B.; Daleprane, J.B.; Martins, M.A.; Mury, W.V.; Brunini, T.M.C. Weight loss enhances hepatic antioxidant status in a NAFLD model induced by high-fat diet. *Appl. Physiol. Nutr. Metab.* **2018**, *43*, 23–29. [CrossRef]
71. Cabré, N.; Luciano-Mateo, F.; Fernández-Arroyo, S.; Baiges-Gayà, G.; Hernández-Aguilera, A.; Fibla, M.; Fernández-Julià, R.; París, M.; Sabench, F.; Castillo, D.D.; et al. Laparoscopic sleeve gastrectomy reverses non-alcoholic fatty liver disease modulating oxidative stress and inflammation. *Metabolism* **2019**, *99*, 81–89. [CrossRef]
72. Fernando, D.H.; Forbes, J.M.; Angus, P.W.; Herath, C.B. Development and Progression of Non-Alcoholic Fatty Liver Disease: The Role of Advanced Glycation End Products. *Int. J. Mol. Sci.* **2019**, *20*, 5037. [CrossRef]
73. Kim, Y.; Keogh, J.B.; Clifton, P.M. Effects of Two Different Dietary Patterns on Inflammatory Markers, Advanced Glycation End Products and Lipids in Subjects without Type 2 Diabetes: A Randomised Crossover Study. *Nutrients* **2017**, *9*, 336. [CrossRef]
74. Luevano-Contreras, C.; Chapman-Novakofski, K. Dietary Advanced Glycation End Products and Aging. *Nutrients* **2010**, *2*, 1247. [CrossRef]
75. Smedsrød, B.; Melkko, J.; Araki, N.; Sano, H.; Horiuchi, S. Advanced Glycation End Products Are Eliminated by Scavenger-Receptor-Mediated Endocytosis in Hepatic Sinusoidal Kupffer and Endothelial Cells. *Biochem. J.* **1997**, *322*, 567–573. [CrossRef]
76. Palma-Duran, S.A.; Kontogianni, M.D.; Vlassopoulos, A.; Zhao, S.; Margariti, A.; Georgoulis, M.; Papatheodoridis, G.; Combet, E. Serum Levels of Advanced Glycation End-Products (AGEs) and the Decoy Soluble Receptor for AGEs (SRAGE) Can Identify Non-Alcoholic Fatty Liver Disease in Age-, Sex- and BMI-Matched Normo-Glycemic Adults. *Metabolism* **2018**, *83*, 120–127. [CrossRef]

77. Świderska, M.; Maciejczyk, M.; Zalewska, A.; Pogorzelska, J.; Flisiak, R.; Chabowski, A. Oxidative Stress Biomarkers in the Serum and Plasma of Patients with Non-Alcoholic Fatty Liver Disease (NAFLD). Can Plasma AGE Be a Marker of NAFLD? Oxidative Stress Biomarkers in NAFLD Patients. *Free Radic. Res.* **2019**, *53*, 841–850. [CrossRef]
78. Vlassara, H.; Cai, W.; Tripp, E.; Pyzik, R.; Yee, K.; Goldberg, L.; Tansman, L.; Chen, X.; Mani, V.; Fayad, Z.A.; et al. Oral AGE Restriction Ameliorates Insulin Resistance in Obese Individuals with the Metabolic Syndrome: A Randomised Controlled Trial. *Diabetologia* **2016**, *59*, 2181–2192. [CrossRef]
79. Luukkonen, P.K.; Sädevirta, S.; Zhou, Y.; Kayser, B.; Ali, A.; Ahonen, L.; Lallukka, S.; Pelloux, V.; Gaggini, M.; Jian, C.; et al. Saturated Fat Is More Metabolically Harmful for the Human Liver than Unsaturated Fat or Simple Sugars. *Diabetes Care* **2018**, *41*, 1732–1739. [CrossRef]
80. Petersen, M.C.; Shulman, G.I. Roles of Diacylglycerols and Ceramides in Hepatic Insulin Resistance. *Trends Pharmacol. Sci.* **2017**, *38*, 649–665. [CrossRef]
81. Nogueira, M.A.; Oliveira, C.P.; Ferreira Alves, V.A.; Stefano, J.T.; dos Rodrigues, L.S.R.; Torrinhas, R.S.; Cogliati, B.; Barbeiro, H.; Carrilho, F.J.; Waitzberg, D.L. Omega-3 Polyunsaturated Fatty Acids in Treating Non-Alcoholic Steatohepatitis: A Randomized, Double-Blind, Placebo-Controlled Trial. *Clin. Nutr.* **2016**, *35*, 578–586. [CrossRef]
82. Sofi, F.; Giangrandi, I.; Cesari, F.; Corsani, I.; Abbate, R.; Gensini, G.F.; Casini, A. Effects of a 1-Year Dietary Intervention with n-3 Polyunsaturated Fatty Acid-Enriched Olive Oil on Non-Alcoholic Fatty Liver Disease Patients: A Preliminary Study. *Int. J. Food Sci. Nutr.* **2010**, *61*, 792–802. [CrossRef]
83. Nigam, P.; Bhatt, S.; Misra, A.; Chadha, D.S.; Vaidya, M.; Dasgupta, J.; Pasha, Q.M.A. Effect of a 6-Month Intervention with Cooking Oils Containing a High Concentration of Monounsaturated Fatty Acids (Olive and Canola Oils) Compared with Control Oil in Male Asian Indians with Nonalcoholic Fatty Liver Disease. *Diabetes Technol. Ther.* **2014**, *16*, 255–261. [CrossRef]
84. Qin, Y.; Zhou, Y.; Chen, S.H.; Zhao, X.L.; Ran, L.; Zeng, X.L.; Wu, Y.; Chen, J.L.; Kang, C.; Shu, F.R.; et al. Fish Oil Supplements Lower Serum Lipids and Glucose in Correlation with a Reduction in Plasma Fibroblast Growth Factor 21 and Prostaglandin E2 in Nonalcoholic Fatty Liver Disease Associated with Hyperlipidemia: A Randomized Clinical Trial. *PLoS ONE* **2015**, *10*, e0133496. [CrossRef]
85. Makki, K.; Deehan, E.C.; Walter, J.; Bäckhed, F. The Impact of Dietary Fiber on Gut Microbiota in Host Health and Disease. *Cell Host Microbe* **2018**, *23*, 705–715. [CrossRef]
86. Howarth, N.C.; Saltzman, E.; Roberts, S.B. Dietary Fiber and Weight Regulation. *Nutr. Rev.* **2001**, *59*, 129–139. [CrossRef]
87. Patterson, E.E.; Ryan, P.M.; Cryan, J.F.; Dinan, T.G.; Paul Ross, R.; Fitzgerald, G.F.; Stanton, C.E. Gut Microbiota, Obesity and Diabetes. *Postgrad. Med. J.* **2016**, *92*, 286–300. [CrossRef]
88. Tarantino, G.; Finelli, C. Systematic Review on Intervention with Prebiotics/Probiotics in Patients with Obesity-Related Nonalcoholic Fatty Liver Disease. *Future Microbiol.* **2015**, *10*, 889–902. [CrossRef]
89. Stachowska, E.; Portincasa, P.; Jamioł-Milc, D.; Maciejewska-Markiewicz, D.; Skonieczna-żydecka, K. The Relationship between Prebiotic Supplementation and Anthropometric and Biochemical Parameters in Patients with Nafld—A Systematic Review and Meta-Analysis of Randomized Controlled Trials. *Nutrients* **2020**, *12*, 3460. [CrossRef]
90. Dunn, W.; Sanyal, A.J.; Brunt, E.M.; Unalp-Arida, A.; Donohue, M.; McCullough, A.J.; Schwimmer, J.B. Modest Alcohol Consumption Is Associated with Decreased Prevalence of Steatohepatitis in Patients with Non-Alcoholic Fatty Liver Disease (NAFLD). *J. Hepatol.* **2012**, *57*, 384–391. [CrossRef]
91. Mitchell, T.; Jeffrey, G.P.; de Boer, B.; MacQuillan, G.; Garas, G.; Ching, H.; Hamdorf, J.; Adams, L.A. Type and Pattern of Alcohol Consumption Is Associated with Liver Fibrosis in Patients with Non-Alcoholic Fatty Liver Disease. *Am. J. Gastroenterol.* **2018**, *113*, 1484–1493. [CrossRef]
92. Åberg, F.; Helenius-Hietala, J.; Puukka, P.; Färkkilä, M.; Jula, A. Interaction between Alcohol Consumption and Metabolic Syndrome in Predicting Severe Liver Disease in the General Population. *Hepatology* **2018**, *67*, 2141–2149. [CrossRef] [PubMed]
93. Wong, V.W.S.; Chan, R.S.M.; Wong, G.L.H.; Cheung, B.H.K.; Chu, W.C.W.; Yeung, D.K.W.; Chim, A.M.L.; Lai, J.W.Y.; Li, L.S.; Sea, M.M.M.; et al. Community-Based Lifestyle Modification Programme for Non-Alcoholic Fatty Liver Disease: A Randomized Controlled Trial. *J. Hepatol.* **2013**, *59*, 536–542. [CrossRef] [PubMed]
94. Properzi, C.; O'Sullivan, T.A.; Sherriff, J.L.; Ching, H.L.; Jeffrey, G.P.; Buckley, R.F.; Tibballs, J.; MacQuillan, G.C.; Garas, G.; Adams, L.A. Ad Libitum Mediterranean and Low-Fat Diets Both Significantly Reduce Hepatic Steatosis: A Randomized Controlled Trial. *Hepatology* **2018**, *68*, 1741–1754. [CrossRef] [PubMed]
95. Gelli, C.; Tarocchi, M.; Abenavoli, L.; Di Renzo, L.; Galli, A.; De Lorenzo, A. Effect of a Counseling-Supported Treatment with the Mediterranean Diet and Physical Activity on the Severity of the Non-Alcoholic Fatty Liver Disease. *World J. Gastroenterol.* **2017**, *23*, 3150–3162. [CrossRef]
96. Kabisch, S.; Bäther, S.; Dambeck, U.; Kemper, M.; Gerbracht, C.; Honsek, C.; Sachno, A.; Pfeiffer, A.F.H. Liver Fat Scores Moderately Reflect Interventional Changes in Liver Fat Content by a Low-Fat Diet but Not by a Low-Carb Diet. *Nutrients* **2018**, *10*, 157. [CrossRef]
97. Razavi Zade, M.; Telkabadi, M.H.; Bahmani, F.; Salehi, B.; Farshbaf, S.; Asemi, Z. The Effects of DASH Diet on Weight Loss and Metabolic Status in Adults with Non-Alcoholic Fatty Liver Disease: A Randomized Clinical Trial. *Liver Int.* **2016**, *36*, 563–571. [CrossRef]

98. Scragg, J.; Avery, L.; Cassidy, S.; Taylor, G.; Haigh, L.; Boyle, M.; Trenell, M.I.; Anstee, Q.M.; McPherson, S.; Hallsworth, K. Feasibility of a Very Low Calorie Diet to Achieve a Sustainable 10% Weight Loss in Patients With Nonalcoholic Fatty Liver Disease. *Clin. Transl. Gastroenterol.* **2020**, *11*, e00231. [CrossRef]
99. Mardinoglu, A.; Wu, H.; Bjornson, E.; Zhang, C.; Hakkarainen, A.; Räsänen, S.M.; Lee, S.; Mancina, R.M.; Bergentall, M.; Pietiläinen, K.H.; et al. An Integrated Understanding of the Rapid Metabolic Benefits of a Carbohydrate-Restricted Diet on Hepatic Steatosis in Humans. *Cell Metab.* **2018**, *27*, 559–571.e5. [CrossRef]
100. Markova, M.; Pivovarova, O.; Hornemann, S.; Sucher, S.; Frahnow, T.; Wegner, K.; Machann, J.; Petzke, K.J.; Hierholzer, J.; Lichtinghagen, R.; et al. Isocaloric Diets High in Animal or Plant Protein Reduce Liver Fat and Inflammation in Individuals with Type 2 Diabetes. *Gastroenterology* **2017**, *152*, 571–585.e8. [CrossRef]
101. Utzschneider, K.M.; Bayer-Carter, J.L.; Arbuckle, M.D.; Tidwell, J.M.; Richards, T.L.; Craft, S. Beneficial Effect of a Weight-Stable, Low-Fat/Low-Saturated Fat/Low-Glycaemic Index Diet to Reduce Liver Fat in Older Subjects. *Br. J. Nutr.* **2013**, *109*, 1096–1104. [CrossRef]
102. Shai, I.; Schwarzfuchs, D.; Henkin, Y.; Shahar, D.R.; Witkow, S.; Greenberg, I.; Golan, R.; Fraser, D.; Bolotin, A.; Vardi, H.; et al. Weight Loss with a Low-Carbohydrate, Mediterranean, or Low-Fat Diet. *N. Engl. J. Med.* **2008**, *359*, 229–241. [CrossRef]
103. Kirk, E.; Reeds, D.N.; Finck, B.N.; Mayurranjan, M.S.; Patterson, B.W.; Klein, S. Dietary Fat and Carbohydrates Differentially Alter Insulin Sensitivity During Caloric Restriction. *Gastroenterology* **2009**, *136*, 1552–1560. [CrossRef]
104. Westerbacka, J.; Lammi, K.; Häkkinen, A.M.; Rissanen, A.; Salminen, I.; Aro, A.; Yki-Järvinen, H. Dietary Fat Content Modifies Liver Fat in Overweight Nondiabetic Subjects. *J. Clin. Endocrinol. Metab.* **2005**, *90*, 2804–2809. [CrossRef]
105. Look AHEAD Research Group; Wing, R.R.; Bolin, P.; Brancati, F.L.; Bray, G.A.; Clark, J.M.; Coday, M.; Crow, R.S.; Curtis, J.M.; Egan, C.M.; et al. Cardiovascular Effects of Intensive Lifestyle Intervention in Type 2 Diabetes. *N. Engl. J. Med.* **2013**, *369*, 145–154. [CrossRef]
106. Antoni, R.; Johnston, K.L.; Collins, A.L.; Robertson, M.D. Effects of Intermittent Fasting on Glucose and Lipid Metabolism. In *Proceedings of the Nutrition Society*; Cambridge University Press: Cambridge, UK, 2017; Volume 76, pp. 361–368. [CrossRef]
107. Chaix, A.; Manoogian, E.N.C.; Melkani, G.C.; Panda, S. Time-Restricted Eating to Prevent and Manage Chronic Metabolic Diseases. *Annu. Rev. Nutr.* **2019**, 291–315. [CrossRef]
108. Garaulet, M.; Ordovás, J.M.; Madrid, J.A. The Chronobiology, Etiology and Pathophysiology of Obesity. *Int. J. Obes.* **2010**, *34*, 1667–1683. [CrossRef]
109. Johnston, J.D. Physiological Responses to Food Intake throughout the Day. *Nutr. Res. Rev.* **2014**, *27*, 107–118. [CrossRef]
110. Poggiogalle, E.; Jamshed, H.; Peterson, C.M. Circadian Regulation of Glucose, Lipid, and Energy Metabolism in Humans. *Metabolism* **2018**, *84*, 11–27. [CrossRef]
111. Woodie, L.N.; Luo, Y.; Wayne, M.J.; Graff, E.C.; Ahmed, B.; O'Neill, A.M.; Greene, M.W. Restricted Feeding for 9 h in the Active Period Partially Abrogates the Detrimental Metabolic Effects of a Western Diet with Liquid Sugar Consumption in Mice. *Metabolism* **2018**, *82*, 1–13. [CrossRef]
112. Hatori, M.; Vollmers, C.; Zarrinpar, A.; DiTacchio, L.; Bushong, E.A.; Gill, S.; Leblanc, M.; Chaix, A.; Joens, M.; Fitzpatrick, J.A.J.; et al. Time-Restricted Feeding without Reducing Caloric Intake Prevents Metabolic Diseases in Mice Fed a High-Fat Diet. *Cell Metab.* **2012**, *15*, 848–860. [CrossRef]
113. Chung, H.; Chou, W.; Sears, D.D.; Patterson, R.E.; Webster, N.J.G.; Ellies, L.G. Time-Restricted Feeding Improves Insulin Resistance and Hepatic Steatosis in a Mouse Model of Postmenopausal Obesity. *Metabolism* **2016**, *65*, 1743–1754. [CrossRef] [PubMed]
114. Sherman, H.; Frumin, I.; Gutman, R.; Chapnik, N.; Lorentz, A.; Meylan, J.; le Coutre, J.; Froy, O. Long-Term Restricted Feeding Alters Circadian Expression and Reduces the Level of Inflammatory and Disease Markers. *J. Cell. Mol. Med.* **2011**, *15*, 2745–2759. [CrossRef] [PubMed]
115. Zarrinpar, A.; Chaix, A.; Yooseph, S.; Panda, S. Diet and Feeding Pattern Affect the Diurnal Dynamics of the Gut Microbiome. *Cell Metab.* **2014**, *20*, 1006–1017. [CrossRef] [PubMed]
116. Chaix, A.; Zarrinpar, A.; Miu, P.; Panda, S. Time-Restricted Feeding Is a Preventative and Therapeutic Intervention against Diverse Nutritional Challenges. *Cell Metab.* **2014**, *20*, 991–1005. [CrossRef]
117. Jakubowicz, D.; Wainstein, J.; Ahrén, B.; Bar-Dayan, Y.; Landau, Z.; Rabinovitz, H.R.; Froy, O. High-Energy Breakfast with Low-Energy Dinner Decreases Overall Daily Hyperglycaemia in Type 2 Diabetic Patients: A Randomised Clinical Trial. *Diabetologia* **2015**, *58*, 912–919. [CrossRef]
118. Garaulet, M.; Gómez-Abellán, P.; Alburquerque-Béjar, J.J.; Lee, Y.C.; Ordovás, J.M.; Scheer, F.A.J.L. Timing of Food Intake Predicts Weight Loss Effectiveness. *Int. J. Obes.* **2013**, *37*, 604–611. [CrossRef]
119. Moro, T.; Tinsley, G.; Bianco, A.; Marcolin, G.; Pacelli, Q.F.; Battaglia, G.; Palma, A.; Gentil, P.; Neri, M.; Paoli, A. Effects of Eight Weeks of Time-Restricted Feeding (16/8) on Basal Metabolism, Maximal Strength, Body Composition, Inflammation, and Cardiovascular Risk Factors in Resistance-Trained Males. *J. Transl. Med.* **2016**, *14*, s12967-016. [CrossRef]
120. Sutton, E.F.; Beyl, R.; Early, K.S.; Cefalu, W.T.; Ravussin, E.; Peterson, C.M. Early Time-Restricted Feeding Improves Insulin Sensitivity, Blood Pressure, and Oxidative Stress Even without Weight Loss in Men with Prediabetes. *Cell Metab.* **2018**, *27*, 1212–1221.e3. [CrossRef]
121. Zeevi, D.; Korem, T.; Zmora, N.; Israeli, D.; Rothschild, D.; Weinberger, A.; Ben-Yacov, O.; Lador, D.; Avnit-Sagi, T.; Lotan-Pompan, M.; et al. Personalized Nutrition by Prediction of Glycemic Responses. *Cell* **2015**, *163*, 1079–1094. [CrossRef]

122. Aller, R.; Sigüenza, R.; Pina, M.; Laserna, C.; Antolín, B.; Burgueño, B.; Durà, M.; Izaola, O.; Primo, D.; de Luis, D.A. Insulin Resistance Is Related with Liver Fibrosis in Type 2 Diabetic Patients with Non-Alcoholic Fatty Liver Disease Proven Biopsy and Mediterranean Diet Pattern as a Protective Factor. *Endocrine* **2020**, *68*, 557–563. [CrossRef]
123. Kouvari, M.; Boutari, C.; Chrysohoou, C.; Fragkopoulou, E.; Antonopoulou, S.; Tousoulis, D.; Pitsavos, C.; Panagiotakos, D.B.; Mantzoros, C.S. Mediterranean Diet Is Inversely Associated with Steatosis and Fibrosis and Decreases Ten-Year Diabetes and Cardiovascular Risk in NAFLD Subjects: Results from the ATTICA Prospective Cohort Study. *Clin. Nutr.* **2020**, *40*, 3314–3324. [CrossRef]

nutrients

MDPI

Review

Malnutrition in Patients with Liver Cirrhosis

Julia Traub [1], Lisa Reiss [1], Benard Aliwa [2] and Vanessa Stadlbauer [2,*]

[1] Department of Clinical Medical Nutrition, University Hospital Graz, 8036 Graz, Austria; Julia.Traub@klinikum-graz.at (J.T.); lisamaria.reiss@edu.fh-joanneum.at (L.R.)
[2] Department of Gastroenterology and Hepatology, Medical University of Graz, 8036 Graz, Austria; benard.aliwa@medunigraz.at
* Correspondence: vanessa.stadlbauer@medunigraz.at; Tel.: +43-316-385-82282

Abstract: Liver cirrhosis is an increasing public health threat worldwide. Malnutrition is a serious complication of cirrhosis and is associated with worse outcomes. With this review, we aim to describe the prevalence of malnutrition, pathophysiological mechanisms, diagnostic tools and therapeutic targets to treat malnutrition. Malnutrition is frequently underdiagnosed and occurs—depending on the screening methods used and patient populations studied—in 5–92% of patients. Decreased energy and protein intake, inflammation, malabsorption, altered nutrient metabolism, hypermetabolism, hormonal disturbances and gut microbiome dysbiosis can contribute to malnutrition. The stepwise diagnostic approach includes a rapid prescreen, the use of a specific screening tool, such as the Royal Free Hospital Nutritional Prioritizing Tool and a nutritional assessment by dieticians. General dietary measures—especially the timing of meals—oral nutritional supplements, micronutrient supplementation and the role of amino acids are discussed. In summary malnutrition in cirrhosis is common and needs more attention by health care professionals involved in the care of patients with cirrhosis. Screening and assessment for malnutrition should be carried out regularly in cirrhotic patients, ideally by a multidisciplinary team. Further research is needed to better clarify pathogenic mechanisms such as the role of the gut-liver-axis and to develop targeted therapeutic strategies.

Keywords: malnutrition; cirrhosis; nutritional screening; nutritional assessment; gut–liver axis; macronutrients; micronutrients; dysbiosis

Citation: Traub, J.; Reiss, L.; Aliwa, B.; Stadlbauer, V. Malnutrition in Patients with Liver Cirrhosis. *Nutrients* **2021**, *13*, 540. https://doi.org/10.3390/nu13020540

Academic Editors: Ina Bergheim and Roberto Iacone

Received: 4 December 2020
Accepted: 4 February 2021
Published: 7 February 2021

1. Introduction

The Hepahealth report from 2018 reported a prevalence of chronic liver disease and cirrhosis in Europe between 500 and 1100 cases per 100.000 inhabitants [1]. Data from the USA show a 65% increase in cirrhosis associated mortality between 1999 and 2016 [2]. Cirrhosis is a systemic disease and malnutrition is a key feature as well as an important complication of the disease. This implicates that malnutrition diagnosis is not only relevant as one of the clinical characteristics of cirrhosis, but also needs to be considered as an important complication, that warrants timely and appropriate therapy to improve prognosis. Our review should highlight the importance of early diagnosis, should help to understand the pathophysiology and define appropriate therapeutic measures. Additionally, knowledge gaps are identified. In this review the term malnutrition is used to describe undernutrition. The discussion of overnutrition in liver disease is beyond the scope of this review.

The reported prevalence of malnutrition in cirrhosis is highly variable, ranging from 5–92%, indicating either a knowledge gap or difficulties in diagnosing malnutrition or both. The knowledge gap is underpinned by a survey, in which only 20% of gastroenterologists gave correct answers regarding the prevalence of malnutrition in cirrhotic patients [3]. The problem of underdiagnosis can be derived from two very large studies from the USA using the national inpatient sample, that showed a much lower prevalence of 6–12%, whereas studies that used active screening for malnutrition exhibit higher rates of detection [4,5]. Supplementary Table S1 summarizes studies describing the prevalence of malnutrition in cirrhosis over the past five years, depicting the large differences depending on the study design and the tools used for diagnosis. Malnutrition should therefore be routinely screened for and assessed among this group of high-risk patients to avoid underdiagnosis. However, screening of malnutrition in liver cirrhosis patients is challenging because of the influence of fluid retention, ascites and peripheral edema [5]. Malnutrition prevalence, as a key feature of cirrhosis, increases with increasing disease severity [6,7]. However, frequently also patients with compensated liver disease are malnourished and malnutrition is a considerable risk factor for mortality even in patients with a model of end-stage liver disease (MELD) score < 15 [8]. Furthermore, patients with chronic liver disease but without cirrhosis are also frequently malnourished [9]. In this population, malnutrition may often be masked by obesity [9]. Malnutrition can also be seen as a complication of cirrhosis, since it has a negative impact on disease progression and outcome. Rate of hospitalization and mortality is doubled in malnourished compared to adequately nourished patients and malnutrition is an independent predictor of outcome [10–12]. Malnutrition is a predictor of other complications of cirrhosis [13]. Especially infections and hepatic encephalopathy are associated with malnutrition [14,15]. Furthermore, other malnutrition related diagnoses, namely sarcopenia, hepatic osteodystrophy and frailty are commonly found in liver cirrhosis. The presence of sarcopenia further impairs prognosis [16–25]. The interplay between malnutrition, sarcopenia and frailty has recently been reviewed [26].

This review summarizes the current knowledge on pathogenesis of malnutrition in cirrhosis and discusses the best clinically applicable strategies to diagnose malnutrition in order to raise awareness for this still often underappreciated complication of cirrhosis. Treating malnutrition by a multidisciplinary team improves survival rates and quality of life in patients with liver cirrhosis [27] and nutritional education leads to more nutrition consultations and a lower 90-day readmission rate in cirrhosis [28]. Therefore this review also discusses current and potential future therapeutic options to treat malnutrition in cirrhosis.

2. Pathogenesis

The etiology of malnutrition in liver cirrhosis is multifactorial (Figure 1). Decreased energy and protein intake, inflammation, malabsorption, altered nutrient metabolism, hormonal disturbances, hypermetabolism and gut microbiome dysbiosis can contribute to malnutrition. Additionally, fasting periods and external factors such as alcohol consumption have impact on malnutrition.

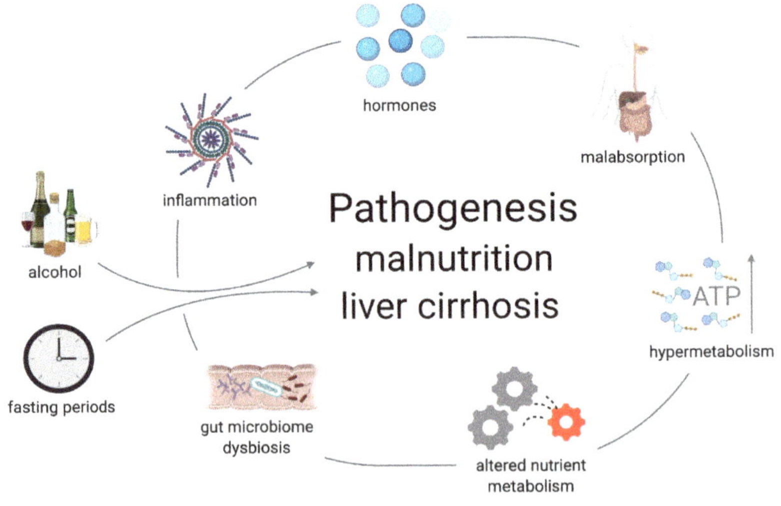

Figure 1. Factors contributing to malnutrition in cirrhosis. Created with BioRender.com.

2.1. Decreased Energy and Protein Intake

In patients with liver cirrhosis, decreased energy and protein intake are the commonest reason leading to malnutrition [29–35]. The percentage of patients with inadequate energy intake ranges from 9.2% to 100% in different studies, depending on the method of assessment and the patient population. Energy intake is reduced by 13–34%, indication a large variation in different studies. Supplementary Table S2 shows the design and results of different studies assessing nutritional intake. Several upstream mechanisms are known as reasons for decreased energy and protein intake (Figure 2). Impaired gastric motility and relaxation due to portal hypertension leads to reduced nutritional intake [36,37]. The presence of ascites can reduce food intake due to early feeling of fullness [38]. A decreased sense of smell and/or dysgeusia, which can be caused by micronutrient deficiencies can also be responsible for a decreased intake [39,40]. Additionally, recommended dietary restriction like a low-salt diet are discussed as possible factors for inadequate nutritional intake [41,42]. In cirrhosis, interestingly, high levels of ghrelin were observed [31,43,44]. Ghrelin is the only known peripherally-derived orexigenic hormone that normally increases appetite and food intake. However, despite high ghrelin levels, appetite is not increased in cirrhotic patients. High ghrelin levels can therefore be considered as an ineffective compensatory mechanism in cirrhosis [31]. To date, it is unknown, which of these factors plays the most important role. Therefore, all factors should be considered during the assessment of nutritional intake in each patient as a personalized approach to detect and adequately treat the most likely reasons for reduced nutritional intake.

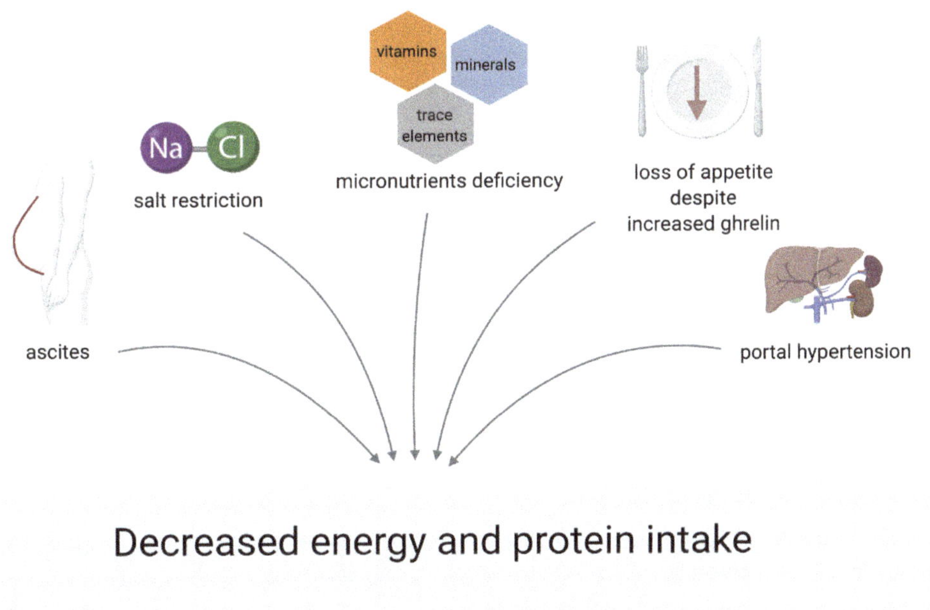

Figure 2. Reasons for decreased energy and protein intake in cirrhosis. Created with BioRender.com.

2.2. Malabsorption and Altered Metabolism of Macro- and Micronutrients

Fat malabsorption is commonly seen in cirrhosis [45,46]. Impaired bile acid metabolism which affects the formation of micelles that are necessary for fat digestion and absorption of fat-soluble vitamins, [47,48] small intestinal bacterial overgrowth, which is common in liver cirrhosis patients, [49] can lead to fat malabsorption via deconjugation of bile acids [50]. Chronic pancreatitis, secondary to alcohol abuse and common in liver cirrhosis patients, may contributes to fat malabsorption as well [51,52]. Protein loss due to portal hypertensive enteropathy has been described [53–55]. No data is available regarding impairment of carbohydrate absorption in cirrhosis. Malabsorption needs to be considered in the nutritional assessment and diagnosed, using biomarkers such as fecal elastase or fecal alpha-1-antitrypsin and tests for micronutrient deficiencies (see below). A useful stepwise diagnostic algorithm, starting with noninvasive routine blood tests and specific biomarkers has been proposed by Nikaki K. in 2016 [56].

In addition to altered absorption, also fat, protein and carbohydrate metabolism are altered in cirrhosis, with differing mechanisms depending on the etiology. Much research has been done to elucidate the mechanisms: chronic alcohol consumption alters lipid metabolism by stimulating lipogenesis, decreasing the export of very low-density lipoprotein, activate de novo lipogenesis and inhibiting fatty acid oxidation which also contributes to alcoholic fatty liver disease [57–61]. Alcohol consumption also impairs fatty acid catabolism predominantly through inhibition of mitochondrial ß-oxidation, which is the most significant contribution to alcohol-induced hepatic lipid accumulation and leads to triglyceride accumulation in the liver [62–64]. Also in non-alcoholic fatty liver disease, adipose tissue and hepatic triglyceride metabolism is altered [65]. Protein metabolism is impaired due to increased protein catabolism and decreased protein synthesis. Furthermore, decreased serum branched chained amino acids (BCAA) concentration and increased levels of aromatic amino acids are observed in cirrhosis, which play a role in the pathogenesis of hepatic encephalopathy and muscle wasting [66–70]. Glucose metabolism is severely altered as well: peripheral insulin resistance but normal or enhanced uptake into the liver as well as alterations in glycolytic enzymes and changes in glucose and insulin transporters have been described. This contributes to decreased hepatic glucose production and lower hepatic glycogen reserves, associated with increased gluconeogenesis from amino acids and secondary protein breakdown. The above described metabolic abnormalities in carbohydrate metabolism also lead to a state of accelerated starvation already after an overnight fast [71–76]. The role of portal hypertension and portosystemic shunting in protein energy metabolism is not fully elucidated yet: on the one hand, the placement of a transjugular intrahepatic portosystemic shunt can lead to improvement of fat free mass and thereby prognosis [77–79]. On the other hand, there is also evidence that portosystemic shunting may have deleterious nutritional effects due to a reduction in hepatic nutrient flow [80]. While the molecular principles of changes in macronutrient metabolism in cirrhosis are already well described, the direct therapeutic implications of these findings are not well defined yet. Timing and composition of meals as therapeutic measures to account for changes in macronutrient metabolism are described below. Further research is needed to understand the effect of portal hypertension and portosystemic shunting in humans on protein anabolism and catabolism including the role of the gut-liver axis.

But not only macronutrient metabolism is altered in liver cirrhosis patients. Deficiencies in trace elements, minerals and vitamins are common, due to fat malabsorption, diuretic use and inadequate intake. In addition, liver dysfunction itself can lead to alterations in trace element metabolism [81,82]. Zinc, selenium, iron and magnesium are commonly decreased in liver cirrhosis [83–88] whereas copper and manganese can be increased [81,89]. Fat-soluble vitamin deficiencies are common in liver cirrhosis [90] which can in turn impair absorption of other nutrients, such as protein and fat [91]. For the absorption of fat-soluble vitamins, bile acids are required to form micelles which are absorbed by enterocytes into the circulation. If there is inadequate delivery of bile acids, as it is common in liver cirrhosis patients, this can lead to a deficiency of fat-soluble vitamins, especially in jaundiced patients [92,93]. Trace element and vitamin deficiencies can in turn impact negatively on nutrition intake indicating a vicious cycle of malnutrition in cirrhosis: zinc and vitamin A deficiency can impair taste and olfaction and therefore impair food intake. [39,94] Vitamin D deficiency is of prognostic relevance, since it is associated with poor outcome, increased mortality and higher complications rate, however it is yet unclear whether vitamin D levels are a mere surrogate of advanced liver disease or if there exists a direct pathophysiological relation [95–98]. Water-soluble vitamins, especially vitamins C, B1, B2, B6 and folic acid [99–101] are decreased whereas vitamin B12 levels can be falsely increased in liver cirrhosis, possibly due to a flooding of vitamin B12 from damaged liver cells into the circulation [102,103]. For a summary of changes of micronutrients in cirrhosis see Supplementary Table S3.

Not only are absorption and metabolism altered, but also energy expenditure contributes to malnutrition: 15–30% of cirrhotic patients are hypermetabolic with a resting energy expenditure of >120%, which negatively effects nutrition status [104–107]. Hypermetabolism compromises overall transplant-free and early post-transplant survival [108–110]. The cause of hypermetabolism in liver cirrhosis is not yet clarified in full detail. From rheumatoid disease it is known that inflammation drives hypermetabolism [111]. Elevated levels of interleukin-1, interleukin-6 and transforming growth factor are also common in chronic alcoholic liver diseases [112–116]. Therefore, inflammation can be considered as a contributing factor to hypermetabolism and malnutrition [117], alongside with increased beta-adrenergic activity [72]. Additionally, elevated levels of proinflammatory cytokines may be directly responsible for decreased appetite [118,119]. Since inflammation in cirrhosis is tightly linked to changes in the gut–liver axis, [120–122] the relation to hypermetabolism needs further research to elucidate pathophysiology and define possible therapeutic interventions.

Hormones, as superordinate control of nutritional intake and metabolism also impact on nutrition in cirrhosis. The role of ghrelin in appetite regulation was already described above. In addition, ghrelin has a wide spectrum of other metabolic functions in glucose metabolism and weight control and posttranslational modification is essential to exert its metabolic function. The diverse roles of ghrelin in liver disease has recently been extensively reviewed [123]. Ghrelin as well as leptin are known to influence energy expenditure [124,125]. Leptin, which helps to regulate energy balance, circulates in free and bound form. The basal concentrations of leptin are higher in patients with liver cirrhosis and can lead to inadequate energy expenditure [126]. Hyperinsulinemia and insulin resistance are also common in liver cirrhosis; increased insulin levels induce satiety, leading to a reduction in energy intake [127]. Testosterone is reduced in about 90% of men with liver cirrhosis [128] and plays an important role in protein synthesis and protein breakdown [129].

2.3. Gut Microbiome Dysbiosis as Potential Contributor to Malnutrition

Altered nutritional status is associated with distinct gut microbiome dysbiosis in cirrhosis [130]. The gut microbiome is a nutrient signal transducer with the capacity to synthesize or modify nutrient signaling molecules such as short-chain fatty acid (SCFA) and branched chain amino acid (BCAA) [131,132]. Several bacterial genera are known to produce SCFA, such as *Bacteroides*, *Faecalibacterium*, *Succinivibrio* and *Butyricimonas* among others [133,134]. Undernourished children for example showed a lower abundance of different *Bacteroides* species, suggesting a loss in SCFA-producing species [135,136]. In cirrhosis, SCFA-producing bacterial species are reduced [137]. The observed alteration in the gut microbiome composition in cirrhosis is associated with increased protein catabolism mediated by inflammatory responses leading to muscle loss [138]. Gut microbiome dysbiosis is further associated with increased gut permeability and bacterial translocation, which is associated with inflammation [122] and complications of cirrhosis [122,139,140]. It is not known to date whether gut microbiome dysbiosis precedes the development of malnutrition in cirrhosis or if it is a consequence of the disease and the drug treatment of the disease. This question would be of high relevance to answer, in order to develop microbiome targeted therapeutic strategies to improve malnutrition in the clinical setting.

3. Diagnosis

Since malnutrition is a common key feature and a complication of liver cirrhosis and related to a poor prognosis, early diagnosis is important. Unfortunately, the common models to determine the prognosis in patients with liver cirrhosis, such as the model of end-stage liver disease (MELD) [141] and the Child–Pugh score [142], do not include nutritional screening or assessment. Of note, the original score developed by the surgeons Child and Turcotte, contained malnutrition as a variable, which was later substituted by prothrombin time [143]. All liver cirrhosis patients should be rapidly prescreened for the risk of malnutrition at each contact by assessing Child–Pugh score and BMI and when at high risk (Child–Pugh score C irrespective of the BMI or BMI < 18.5 kg/m^2 irrespective of the Child–Pugh score) a nutritional assessment, including assessment of sarcopenia as a complication of malnutrition, should be completed immediately to confirm the presence and determine the severity of malnutrition [144–146]. This prescreening can be done by skilled personnel from different disciplines, since it contains routine clinical data, which are normally collected at each outpatient visit or at hospitalization. In the future, automated prescreening combining routinely assessed data from electronic patient records, is thinkable. A specific screening tool, which is validated for patients with liver cirrhosis, is advised to account for special circumstances such as fluid retention in cirrhosis. The Royal Free Hospital–Nutritional Prioritizing Tool (RFH–NPT) fulfills this requirement [145,147]. Common malnutrition screening tools for the general hospital population do not perform well in cirrhosis [148]. If a general malnutrition tool is intended to be used in cirrhosis, it needs to be validated first. In patients with medium or high risk in the RFH–NPT (1 point or more), a detailed nutritional assessment using cirrhosis specific assessment tools such as the Subjective Global Assessment or the Royal Free Hospital Global Assessment as well as a detailed assessment of dietary intake are required [145,147]. In patients, who are at high risk for malnutrition, the assessment of sarcopenia in addition to a detailed nutritional assessment is recommended to confirm and characterize complications of malnutrition and identify modifiable variables for nutrition support. Methods for the assessment of sarcopenia in liver cirrhosis have recently been reviewed [149]. In obese patients with cirrhosis Child–Pugh A or B (BMI > 30 kg/m^2), a nutritional and lifestyle intervention targeting obesity is indicated [145,147]. As nutritional assessment is more comprehensive, time consuming and requires interpretation of multiple nutrition indicators, it should be performed by a dietitian [150]. In patients with low risk of malnutrition, rescreening should be performed every year, in all other patients the assessment should be repeated every one to six months in the outpatient setting and at admission and for inpatients periodically during the hospital stay [145]. Figure 3 shows a comprehensive algorithm to screen for and assess malnutrition in cirrhosis, adapted from the European Association for the Study of the Liver (EASL) clinical practices guidelines [145].

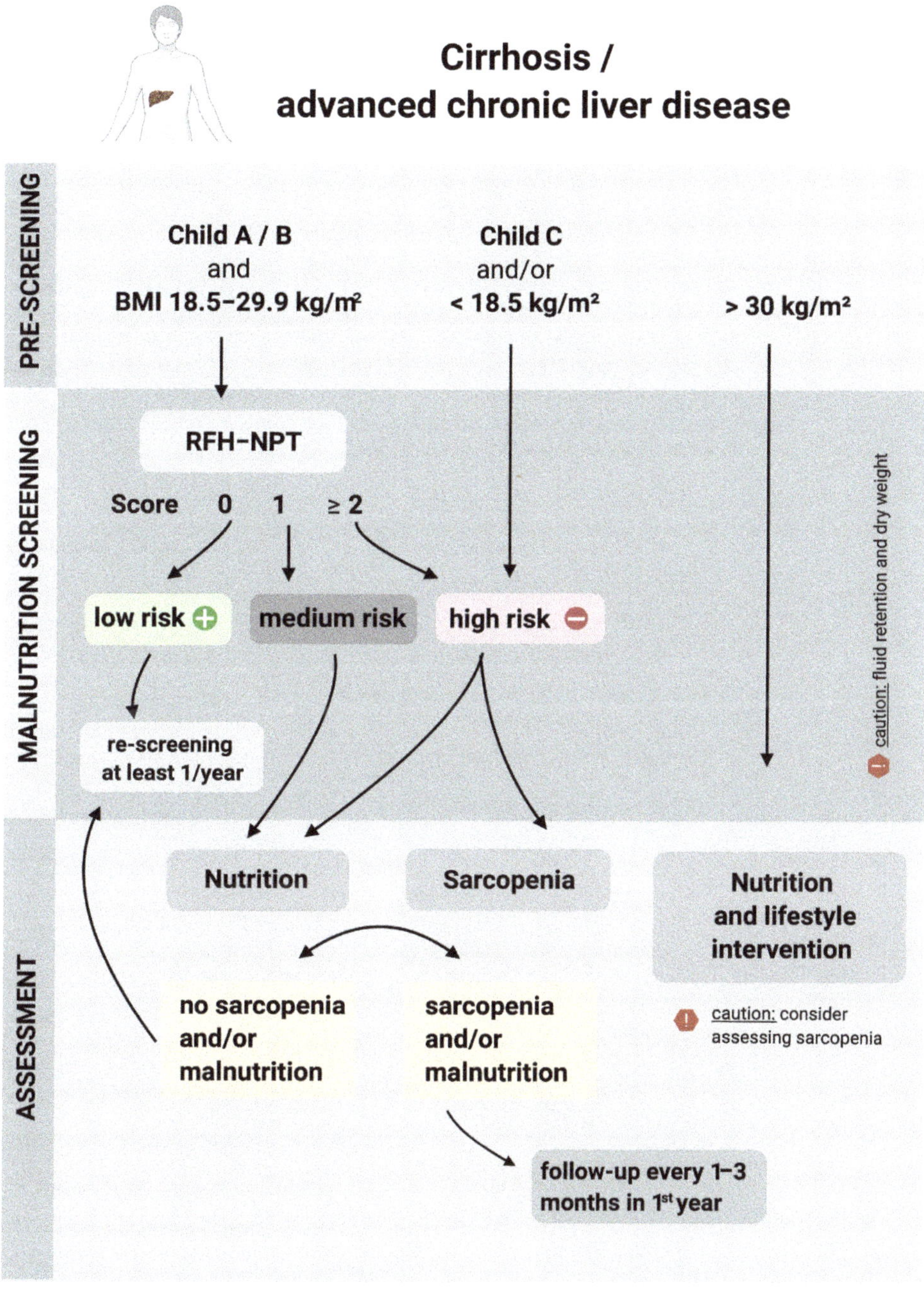

Figure 3. Algorithm to screen for and assess malnutrition in cirrhosis, adapted from the European Association for the Study of the Liver (EASL) clinical practices guidelines [145]. Created with BioRender.com.

4. Therapeutic Strategies

4.1. Diet

The positive effect of dietary interventions on prognosis in cirrhosis has been clearly shown: optimizing the nutritional status of liver cirrhosis patients improves morbidity and mortality [13,17,151–153] even in patients with acute on chronic liver failure [154]. Nutritional therapeutic interventions by a multidisciplinary team, especially through dietary counseling from dieticians, improves biomarkers of malnutrition, [155,156] quality of life [27,157] and survival rate [27].

The recommended macronutrient composition in cirrhosis mainly focusses on protein intake. Cirrhotic patients have an increased protein requirement based on the increased protein turnover and catabolism [158,159]. A high protein intake improves in nutritional status [155,160]. Even patients with hepatic encephalopathy, where in the past protein restriction was advocated, [161] benefit from normal to high protein intake [157,162–165]. The recommendations regarding protein intake differ slightly, but not relevantly, between different guidelines, depending on the nutritional status and range from 1.2–1.5 g protein/kg bodyweight (Table 1). There are no specific recommendations regarding carbohydrate and fat intake for patients with liver cirrhosis.

Table 1. Summary of dietary recommendations in cirrhosis from different populations.

Energy	General cirrhosis patients	>35 kcal/kg bodyweight [1]
	Malnourished cirrhosis patients	30–35 kcal/kg bodyweight [3]
	Cirrhosis patients with HE	35–40 kcal/kg bodyweight [1,2,4]
Protein	General cirrhosis patients	1.2–1.5 g/kg bodyweight [1]
	Replenish malnourished and/or sarcopenic cirrhosis patients	1.5 g/kg bodyweight [3]
	Nonmalnourished cirrhosis patients	1.2 g/kg bodyweight per day [3]
	Cirrhosis patients with HE	1.2–1.5 g/kg bodyweight [1,2,4] no protein restriction [3]
Fat		No specific recommendations
Carbohydrates		No specific recommendations
	Cirrhosis patients with HE	25–45 g of fiber [4] Late evening snack with 50 g complex carbohydrates [4]
Dietary pattern	General cirrhosis patients	3–5 meals a day [3] Late evening snack [3]
	Malnourished decompensated cirrhotic patients	Late evening ONS and breakfast [1]
	Cirrhosis patients with HE	Small meals evenly distributed throughout the day [2,4] Late- night snack [2,4]
Salt		"No added salt" diet with 5–6 g salt per day, care should be taken when salt reduction leads to reduced palatability [1,3]

HE, hepatic encephalopathy; ONS oral nutritional supplements. [1] European Association for the Study of the Liver (EASL). [2] American Association for the Study of Liver Diseases (AASLD). [3] European Society for Clinical Nutrition and Metabolism (ESPEN). [4] International Society for Hepatic Encephalopathy and Nitrogen Metabolism Consensus (ISHEN).

Current evidence suggests that timing and frequency of meals is of importance to improve malnutrition in cirrhosis. After overnight fasting, glycogen stores in cirrhotic livers are emptied [74]. A late evening snack with 50 g complex carbohydrates can improve nitrogen metabolism, increase lean body mass and reverse anabolic resistance and sarcopenia [166–169]. Eating breakfast improves cognitive function in cirrhosis, indicating that, depending on personal preferences in dietary habits, an individualized approach for timing of meals should be developed during dietary counselling [170]. A higher frequency of 5–6 meals/day also shortens episodes of catabolism during the day [171]. It is particularly important to pay attention to the timing and frequency of meals in hospitalized patients to avoid long and often unnecessary "nil per os" periods due to planned diagnostic tests.

The main emphasis in dietary counselling should be to ensure adequate oral intake. If oral intake including oral nutrition supplements (see below) is insufficient despite adequate nutritional advice, enteral tube feeding may be considered for cirrhotic patients to achieve their nutritional and energy goals [172]. Nonbleeding esophageal varices are no contraindication for the placement of a nasogastric tube [163,173]. An endoscopic gastrostomy, on the other hand, is associated with a higher risk of complications, especially bleeding, and is therefore not recommended for patients with advanced chronic liver disease [174]. In moderately or severely malnourished cirrhosis patients, who are unable to eat oral food or cannot be fed sufficiently enterally, parenteral nutrition should be started according to the ESPEN recommendations [175]. Additionally, parenteral nutrition should be given when fasting periods last longer than 72 h [175]. Since cirrhotic patients are more prone to sepsis or infections, care should be taken to avoid infections from central venous lines [176].

When ascites is diagnosed in patient with liver cirrhosis, a "no added salt" diet restricted to 90 mmol salt per day (5.2 g) is recommended by many guidelines. However, as we recently reviewed in detail, a low-sodium diet, although leading to a faster disappearance of ascites and less need for diuretics can lead to poor diet adherence because of impaired taste of the meals, reduced energy and protein intake and increased risk of malnutrition [42]. Therefore the risks and benefits of salt restriction have to be weighed carefully in each patient, again showing the necessity of a multidisciplinary team approach to treat malnutrition in cirrhosis.

4.2. Oral Nutritional Supplements and Micronutrients Supplementation

Oral nutritional supplements can help to achieve nutritional goals in cirrhosis. A significant improvement in anthropometric nutritional parameters such as lean muscle mass and body mass index as well as serum proteins can be achieved by oral nutritional supplements [160,177]. A meta-analysis concluded that oral nutritional supplements may also improve outcome [152]. Furthermore, an improvement in quality of life, functional status and rehabilitation of malnourished cirrhosis patients can be achieved [157,178]. In terms of administration time, nocturnal oral nutritional supplements have a better effect in improving the total body protein status than at daytime [166] (Table 2). Since oral nutritional supplements also contain micronutrients and vitamins, this may be of additional benefit in patients with cirrhosis, however, so far no clear benefit of micronutrient supplementation could be shown [179]. Moreover, no clinical data comparing different products are available, therefore choice can be based on personal preference and price.

Although some studies demonstrate the positive effect of micronutrients supplementation, due to the lack of robust data, no clear recommendations can be made. Additional supplementation is currently only recommended in cases of confirmed or clinically suspected deficiency. Data on zinc supplementation is conflicting: some studies report positive effects on zinc supplementation in hepatic encephalopathy [180–182], while others report no significant improvements [183,184]. Zinc supplementation in case of deficiency improves liver function and nutritional status [185,186] and may even impact positively on clinical outcome [187,188]. Normalizing zinc and vitamin A levels can also indirectly improve the nutritional status by a positive effect on sense of taste and thereby increased food intake [94,189]. A meta-analysis found no evidence to support or refute antioxidant supplements such as beta-carotene, vitamins A, C, E and selenium in liver disease [179]. Also, for vitamin D supplementation, evidence is not sufficient. Vitamin D deficiency is common in cirrhosis and associated with increased mortality [190] and deficiency can be corrected by oral supplementation in cirrhosis [191,192]. However, due to inadequate overall data quality, there is not sufficient evidence to prove or disprove an effect on morbidity and quality of life of vitamin D supplementation [193] (Table 2).

Table 2. Summary of recommendations for oral nutritional supplements, micronutrient supplementation and branched chain amino acid (BCAA) supplementation from different populations.

Oral nutritional supplements	General cirrhosis patients	Late evening oral nutritional supplement [1]
	Cirrhosis patients with HE	Liquid nutritional supplements evenly distributed throughout the day [2]
Micronutrient supplements	General cirrhosis patients	Administration of micronutrients to treat confirmed or clinically suspected deficiency [1,3] Assessment of vitamin D levels based on frequent deficiency and supplement at vitamin D levels < 20 ng/mL, to reach serum vitamin D (25(OH)D) > 30 ng/mL [1]
	Cirrhosis patients with HE	For patients with decompensated cirrhosis or those at risk for malnutrition a 2-week course of a multivitamin preparation could be justified [4] Specific treatment of clinically apparent vitamin deficiencies [4]
Amino acids	General cirrhosis patients	Decompensated cirrhosis: BCAA supplements and leucine enriched amino acid supplements when inadequate nitrogen intake by oral diet [1] Oral vegetable proteins or BCAA (0.25 g/kg bodyweight/day) for cirrhosis patients who are protein "intolerant" to facilitate adequate protein intake [3] Long-term oral BCAA supplements (0.25 g/kg bodyweight/day) in patients with advanced cirrhosis in order to improve event-free survival or quality of life [3]
	Cirrhosis patients with HE	BCAA supplementation for improving neuropsychiatric performance and to reach the recommended nitrogen intake [1] Oral BCAA supplementation for cirrhosis patients who are "intolerant" of dietary protein to achieve and maintain recommended nitrogen intake [2,4]

HE, hepatic encephalopathy; BCAA, branched chain amino acids. [1] European Association for the Study of the Liver (EASL). [2] American Association for the Study of Liver Diseases (AASLD). [3] European Society for Clinical Nutrition and Metabolism (ESPEN). [4] International Society for Hepatic Encephalopathy and Nitrogen Metabolism Consensus (ISHEN).

4.3. Amino Acids

BCAA serum levels are low in cirrhosis because they are preferentially used as energy substrates, but are also essential for protein synthesis and ammonia detoxification [194]. BCAA supplementation has been shown to prevent lipolysis and proteolysis and improves nitrogen balance, muscle mass, nutritional status, complication free survival, quality of live and hepatocellular carcinoma risk [194–202]. However, despite this quite clear evidence of positive effects, current guidelines recommend BCAA supplementation only in decompensated cirrhosis, when adequate protein intake cannot be achieved by oral diet or in case of complications. For hepatic encephalopathy, vegetarian protein, which is rich in BCAA, is considered to be the ideal protein source, not only because it is better tolerated than animal protein [164,203,204] but also because vegetarian protein may positively influence gut microbiome composition [205]. BCAA can also be used as late evening snack to improve nutritional status [73] especially in protein-intolerant patients [144–146,206] (Table 2).

5. Summary

Malnutrition is a common and dangerous key feature and complication of liver cirrhosis. Diagnosis is challenging and often overlooked. A repetitive diagnostic workup with a rapid screening for malnutrition using the Child–Pugh score and BMI, and in selected patients the RHF-NPT to stage patients in low, medium and high risk and conducting a detailed nutritional assessment including assessment of complications of malnutrition in patients at medium and high risk should be implemented in every center. Education on malnutrition in cirrhosis for all health care professionals who treat patients with liver cirrhosis and the availability of trained dieticians seem crucial in the future management process of malnutrition in cirrhosis, to account for the complexity of the disease and the need for individualized management. From a pathophysiological point of view, more work is needed to identify the drivers of malnutrition, especially considering the complex interplay between the gut microbiome and nutrient metabolism. Therapeutic efforts should consider alternative pathophysiological mechanisms in the development of malnutrition such as the role of inflammation and dysbiosis to identify potential therapeutic targets beyond the pure increase of nutritional intake. Further clinical studies, ideally as multicenter, multidisciplinary initiatives are needed to diagnose and adequately treat malnutrition in cirrhosis.

Supplementary Materials: The following are available online at https://www.mdpi.com/2072-6643/13//540/s1.

Author Contributions: Conceptualization, V.S.; writing—original draft preparation, V.S., J.T., L.R., B.A.; writing—review and editing, V.S., J.T., L.R., B.A.; visualization, J.T., L.R.; supervision, V.S. All authors have read and agreed to the published version of the manuscript.

Funding: This research was funded by a grant from the Austrian Science Fund (FWF) KLI 741 to VS.

Institutional Review Board Statement: Not applicable.

Informed Consent Statement: Not applicable.

Data Availability Statement: Not applicable.

Conflicts of Interest: The authors declare no conflict of interest.

References

1. Pimpin, L.; On Behalf of the HEPAHEALTH Steering Committee. *HEPAHEALTH Project Report*; Europeal Association for the Study of the Liver: Geneva, Switzerland, 2018.
2. Tapper, E.B.; Parikh, N.D. Mortality due to cirrhosis and liver cancer in the United States, 1999–2016: Observational study. *BMJ* **2018**, *362*, k2817. [CrossRef]
3. Gundling, F.; Seidl, H.; Pehl, C.; Schmidt, T.; Schepp, W. How close do gastroenterologists follow specific guidelines for nutrition recommendations in liver cirrhosis? A survey of current practice. *Eur. J. Gastroenterol. Hepatol.* **2009**, *21*, 756–761. [CrossRef]
4. Patel, A.; Silverman, S.; Baghdadi, J.; Shah, O.; Sundaram, V. Osteoporotic Fracture Risk and Health Care Burden in Patients with Cirrhosis. *J. Clin. Gastroenterol.* **2019**, *53*, 543–548. [CrossRef]
5. Sam, J.; Nguyen, G.C. Protein-calorie malnutrition as a prognostic indicator of mortality among patients hospitalized with cirrhosis and portal hypertension. *Liver Int.* **2009**, *29*, 1396–1402. [CrossRef] [PubMed]
6. Chaudhry, A.; Toori, K.U.; Shaikh, J.I. To determine correlation between biochemical parameters of nutritional status with disease severity in HCV related liver cirrhosis. *Pak. J. Med. Sci.* **2018**, *34*, 154–158. [CrossRef] [PubMed]
7. Naqvi, I.H.; Mahmood, K.; Salekeen, S.; Akhter, S.T. Determining the frequency and severity of malnutrition and correlating it with the severity of liver cirrhosis. *Turk. J. Gastroenterol.* **2013**, *24*, 415–422. [CrossRef]
8. Atiemo, K.; Skaro, A.; Maddur, H.; Zhao, L.; Montag, S.; VanWagner, L.; Goel, S.; Kho, A.; Ho, B.; Kang, R.; et al. Mortality Risk Factors among Patients with Cirrhosis and a Low Model for End-Stage Liver Disease Sodium Score (≤15): An Analysis of Liver Transplant Allocation Policy Using Aggregated Electronic Health Record Data. *Am. J. Transplant.* **2017**, *17*, 2410–2419. [CrossRef]
9. Bruch, J.P.; Alvares, D.A.S.M.R.; Alves, B.C.; Dall'alba, V. Reduced Hand Grip Strength in Overweight and Obese Chronic Hepatitis C Patients. *Arq. Gastroenterol.* **2016**, *53*, 31–35. [CrossRef]
10. Bunchorntavakul, C.; Supanun, R.; Atsawarungruangkit, A. Nutritional Status and its Impact on Clinical Outcomes for Patients Admitted to Hospital with Cirrhosis. *J. Med. Assoc. Thai.* **2016**, *99* (Suppl. 2), S47–S55.
11. Maharshi, S.; Sharma, B.C.; Srivastava, S. Malnutrition in cirrhosis increases morbidity and mortality. *J. Gastroenterol. Hepatol.* **2015**, *30*, 1507–1513. [CrossRef]
12. Ribeiro, H.S.; Mauricio, S.F.; Antonio da Silva, T.; de Vasconcelos Generoso, S.; Lima, A.S.; Toulson Davisson Correia, M.I. Combined nutritional assessment methods to predict clinical outcomes in patients on the waiting list for liver transplantation. *Nutrition* **2018**, *47*, 21–26. [CrossRef]
13. Huisman, E.J.; Trip, E.J.; Siersema, P.D.; van Hoek, B.; van Erpecum, K.J. Protein energy malnutrition predicts complications in liver cirrhosis. *Eur. J. Gastroenterol. Hepatol.* **2011**, *23*, 982–989. [CrossRef] [PubMed]

14. Ruiz-Margain, A.; Macias-Rodriguez, R.U.; Ampuero, J.; Cubero, F.J.; Chi-Cervera, L.; Rios-Torres, S.L.; Duarte-Rojo, A.; Espinosa-Cuevas, A.; Romero-Gomez, M.; Torre, A. Low phase angle is associated with the development of hepatic encephalopathy in patients with cirrhosis. *World J. Gastroenterol.* **2016**, *22*, 10064–10070. [CrossRef]
15. Lindqvist, C.; Majeed, A.; Wahlin, S. Body composition assessed by dual-energy X-ray absorptiometry predicts early infectious complications after liver transplantation. *J. Hum. Nutr. Diet* **2017**, *30*, 284–291. [CrossRef]
16. Bering, T.; Diniz, K.G.D.; Coelho, M.P.P.; Vieira, D.A.; Soares, M.M.S.; Kakehasi, A.M.; Correia, M.; Teixeira, R.; Queiroz, D.M.M.; Rocha, G.A.; et al. Association between pre-sarcopenia, sarcopenia, and bone mineral density in patients with chronic hepatitis C. *J. Cachexia Sarcopenia Muscle* **2018**, *9*, 255–268. [CrossRef]
17. Carey, E.J.; Lai, J.C.; Wang, C.W.; Dasarathy, S.; Lobach, I.; Montano-Loza, A.J.; Dunn, M.A.; Fitness, L.E.; Exercise in Liver Transplantation, C. A multicenter study to define sarcopenia in patients with end-stage liver disease. *Liver Transplant.* **2017**, *23*, 625–633. [CrossRef]
18. Chang, K.V.; Chen, J.D.; Wu, W.T.; Huang, K.C.; Lin, H.Y.; Han, D.S. Is sarcopenia associated with hepatic encephalopathy in liver cirrhosis? A systematic review and meta-analysis. *J. Formos. Med. Assoc.* **2019**, *118*, 833–842. [CrossRef]
19. Chinnaratha, M.A.; Chaudhary, S.; Doogue, M.; McCormick, R.J.; Woodman, R.J.; Wigg, A.J. Prevalence of hepatic osteodystrophy and vitamin D deficiency in cirrhosis. *Intern. Med. J.* **2015**, *45*, 1230–1235. [CrossRef] [PubMed]
20. Hanai, T.; Shiraki, M.; Nishimura, K.; Ohnishi, S.; Imai, K.; Suetsugu, A.; Takai, K.; Shimizu, M.; Moriwaki, H. Sarcopenia impairs prognosis of patients with liver cirrhosis. *Nutrition* **2015**, *31*, 193–199. [CrossRef]
21. Hara, N.; Iwasa, M.; Sugimoto, R.; Mifuji-Moroka, R.; Yoshikawa, K.; Terasaka, E.; Hattori, A.; Ishidome, M.; Kobayashi, Y.; Hasegawa, H.; et al. Sarcopenia and Sarcopenic Obesity Are Prognostic Factors for Overall Survival in Patients with Cirrhosis. *Intern. Med.* **2016**, *55*, 863–870. [CrossRef]
22. Kalafateli, M.; Mantzoukis, K.; Choi Yau, Y.; Mohammad, A.O.; Arora, S.; Rodrigues, S.; de Vos, M.; Papadimitriou, K.; Thorburn, D.; O'Beirne, J.; et al. Malnutrition and sarcopenia predict post-liver transplantation outcomes independently of the Model for End-stage Liver Disease score. *J. Cachexia Sarcopenia Muscle* **2017**, *8*, 113–121. [CrossRef]
23. Lai, J.C.; Covinsky, K.E.; Dodge, J.L.; Boscardin, W.J.; Segev, D.L.; Roberts, J.P.; Feng, S. Development of a novel frailty index to predict mortality in patients with end-stage liver disease. *Hepatology* **2017**, *66*, 564–574. [CrossRef]
24. Lai, J.C.; Dodge, J.L.; McCulloch, C.E.; Covinsky, K.E.; Singer, J.P. Frailty and the Burden of Concurrent and Incident Disability in Patients With Cirrhosis: A Prospective Cohort Study. *Hepatol. Commun.* **2020**, *4*, 126–133. [CrossRef]
25. Lai, J.C.; Rahimi, R.S.; Verna, E.C.; Kappus, M.R.; Dunn, M.A.; McAdams-DeMarco, M.; Haugen, C.E.; Volk, M.L.; Duarte-Rojo, A.; Ganger, D.R.; et al. Frailty Associated With Waitlist Mortality Independent of Ascites and Hepatic Encephalopathy in a Multicenter Study. *Gastroenterology* **2019**, *156*, 1675–1682. [CrossRef]
26. Bunchorntavakul, C.; Reddy, K.R. Review article: Malnutrition/sarcopenia and frailty in patients with cirrhosis. *Aliment. Pharmacol. Ther.* **2020**, *51*, 64–77. [CrossRef]
27. Iwasa, M.; Iwata, K.; Hara, N.; Hattori, A.; Ishidome, M.; Sekoguchi-Fujikawa, N.; Mifuji-Moroka, R.; Sugimoto, R.; Fujita, N.; Kobayashi, Y.; et al. Nutrition therapy using a multidisciplinary team improves survival rates in patients with liver cirrhosis. *Nutrition* **2013**, *29*, 1418–1421. [CrossRef]
28. Reuter, B.; Shaw, J.; Hanson, J.; Tate, V.; Acharya, C.; Bajaj, J.S. Nutritional Assessment in Inpatients With Cirrhosis Can Be Improved After Training and Is Associated With Lower Readmissions. *Liver Transplant.* **2019**, *25*, 1790–1799. [CrossRef]
29. Ferreira, L.G.; Ferreira Martins, A.I.; Cunha, C.E.; Anastacio, L.R.; Lima, A.S.; Correia, M.I. Negative energy balance secondary to inadequate dietary intake of patients on the waiting list for liver transplantation. *Nutrition* **2013**, *29*, 1252–1258. [CrossRef]
30. Ferreira, L.G.; Anastacio, L.R.; Lima, A.S.; Correia, M.I. Malnutrition and inadequate food intake of patients in the waiting list for liver transplant. *Rev. Assoc. Med. Bras.* **2009**, *55*, 389–393. [CrossRef]
31. Marchesini, G.; Bianchi, G.; Lucidi, P.; Villanova, N.; Zoli, M.; De Feo, P. Plasma ghrelin concentrations, food intake, and anorexia in liver failure. *J. Clin. Endocrinol. Metab.* **2004**, *89*, 2136–2141. [CrossRef]
32. Nielsen, K.; Kondrup, J.; Martinsen, L.; Stilling, B.; Wikman, B. Nutritional assessment and adequacy of dietary intake in hospitalized patients with alcoholic liver cirrhosis. *Br. J. Nutr.* **1993**, *69*, 665–679. [CrossRef] [PubMed]
33. Ney, M.; Abraldes, J.G.; Ma, M.; Belland, D.; Harvey, A.; Robbins, S.; Den Heyer, V.; Tandon, P. Insufficient Protein Intake Is Associated With Increased Mortality in 630 Patients With Cirrhosis Awaiting Liver Transplantation. *Nutr. Clin. Pract.* **2015**, *30*, 530–536. [CrossRef]
34. Ney, M.; Gramlich, L.; Mathiesen, V.; Bailey, R.J.; Haykowsky, M.; Ma, M.; Abraldes, J.G.; Tandon, P. Patient-perceived barriers to lifestyle interventions in cirrhosis. *Saudi J. Gastroenterol.* **2017**, *23*, 97–104. [CrossRef]
35. Pashayee-Khamene, F.; Saber-Firoozi, M.; Hatami, B.; Rashidkhani, B.; Aghamohammadi, V.; Mohammadi, E.; Hekmatdoost, A. Food groups intake of cirrhotic patients, comparison with the nutritional status and disease stage. *Gastroenterol. Hepatol. Bed Bench* **2019**, *12*, 226–232.
36. Izbeki, F.; Kiss, I.; Wittmann, T.; Varkonyi, T.T.; Legrady, P.; Lonovics, J. Impaired accommodation of proximal stomach in patients with alcoholic liver cirrhosis. *Scand. J. Gastroenterol.* **2002**, *37*, 1403–1410. [CrossRef]
37. Aprile, L.R.; Meneghelli, U.G.; Martinelli, A.L.; Monteiro, C.R. Gastric motility in patients with presinusoidal portal hypertension. *Am. J. Gastroenterol.* **2002**, *97*, 3038–3044. [CrossRef]
38. Aqel, B.A.; Scolapio, J.S.; Dickson, R.C.; Burton, D.D.; Bouras, E.P. Contribution of ascites to impaired gastric function and nutritional intake in patients with cirrhosis and ascites. *Clin. Gastroenterol. Hepatol.* **2005**, *3*, 1095–1100. [CrossRef]

39. Grungreiff, K.; Reinhold, D.; Wedemeyer, H. The role of zinc in liver cirrhosis. *Ann. Hepatol.* **2016**, *15*, 7–16. [CrossRef] [PubMed]
40. Madden, A.M.; Bradbury, W.; Morgan, M.Y. Taste perception in cirrhosis: Its relationship to circulating micronutrients and food preferences. *Hepatology* **1997**, *26*, 40–48. [CrossRef]
41. Gu, X.B.; Yang, X.J.; Zhu, H.Y.; Xu, B.Y. Effect of a diet with unrestricted sodium on ascites in patients with hepatic cirrhosis. *Gut Liver* **2012**, *6*, 355–361. [CrossRef] [PubMed]
42. Haberl, J.; Zollner, G.; Fickert, P.; Stadlbauer, V. To salt or not to salt?-That is the question in cirrhosis. *Liver Int.* **2018**, *38*, 1148–1159. [CrossRef]
43. Howick, K.; Griffin, B.T.; Cryan, J.F.; Schellekens, H. From Belly to Brain: Targeting the Ghrelin Receptor in Appetite and Food Intake Regulation. *Int. J. Mol. Sci.* **2017**, *18*, 273. [CrossRef]
44. El-Shehaby, A.M.; Obaia, E.M.; Alwakil, S.S.; Hiekal, A.A. Total and acylated ghrelin in liver cirrhosis: Correlation with clinical and nutritional status. *Scand. J. Clin. Lab. Investig.* **2010**, *70*, 252–258. [CrossRef]
45. Romiti, A.; Merli, M.; Martorano, M.; Parrilli, G.; Martino, F.; Riggio, O.; Truscelli, A.; Capocaccia, L.; Budillon, G. Malabsorption and nutritional abnormalities in patients with liver cirrhosis. *Ital. J. Gastroenterol.* **1990**, *22*, 118–123. [PubMed]
46. Linscheer, W.G.; Patterson, J.F.; Moore, E.W.; Clermont, R.J.; Robins, S.J.; Chalmers, T.C. Medium and long chain fat absorption in patients with cirrhosis. *J. Clin. Investig.* **1966**, *45*, 1317–1325. [CrossRef]
47. Lanspa, S.J.; Chan, A.T.; Bell, J.S., 3rd; Go, V.L.; Dickson, E.R.; DiMagno, E.P. Pathogenesis of steatorrhea in primary biliary cirrhosis. *Hepatology* **1985**, *5*, 837–842. [CrossRef] [PubMed]
48. Hofmann, A.F. The continuing importance of bile acids in liver and intestinal disease. *Arch. Intern. Med.* **1999**, *159*, 2647–2658. [CrossRef] [PubMed]
49. Maslennikov, R.; Pavlov, C.; Ivashkin, V. Small intestinal bacterial overgrowth in cirrhosis: Systematic review and meta-analysis. *Hepatol. Int.* **2018**, *12*, 567–576. [CrossRef]
50. Shindo, K.; Machida, M.; Miyakawa, K.; Fukumura, M. A syndrome of cirrhosis, achlorhydria, small intestinal bacterial overgrowth, and fat malabsorption. *Am. J. Gastroenterol.* **1993**, *88*, 2084–2091. [PubMed]
51. Li, B.R.; Pan, J.; Du, T.T.; Liao, Z.; Ye, B.; Zou, W.B.; Chen, H.; Ji, J.T.; Zheng, Z.H.; Wang, D.; et al. Risk Factors for Steatorrhea in Chronic Pancreatitis: A Cohort of 2,153 Patients. *Sci. Rep.* **2016**, *6*, 21381. [CrossRef]
52. Dumasy, V.; Delhaye, M.; Cotton, F.; Deviere, J. Fat malabsorption screening in chronic pancreatitis. *Am. J. Gastroenterol.* **2004**, *99*, 1350–1354. [CrossRef] [PubMed]
53. Davcev, P.; Vanovski, B.; Sestakov, D.; Tadzer, I. Protein-losing enteropathy in patients with liver cirrhosis. *Digestion* **1969**, *2*, 17–22. [CrossRef]
54. Dahlqvist, G.E.; Jamar, F.; Zech, F.; Geubel, A.P. In-111 transferrin scintigraphy in cirrhosis with hypoalbuminemia: Evidence for protein-losing enteropathy in a small group of selected cases. *Scand. J. Gastroenterol.* **2012**, *47*, 1247–1252. [CrossRef] [PubMed]
55. Takada, A.; Kobayashi, K.; Takeuchi, J. Gastroenteric clearance of albumin in liver icrrhosis; relative protein-losing gastroenteropathy. *Digestion* **1970**, *3*, 154–164. [CrossRef]
56. Nikaki, K.; Gupte, G.L. Assessment of intestinal malabsorption. *Best Pract. Res. Clin. Gastroenterol.* **2016**, *30*, 225–235. [CrossRef]
57. Suter, P.M.; Schutz, Y.; Jequier, E. The effect of ethanol on fat storage in healthy subjects. *N. Engl. J. Med.* **1992**, *326*, 983–987. [CrossRef] [PubMed]
58. Sonko, B.J.; Prentice, A.M.; Murgatroyd, P.R.; Goldberg, G.R.; van de Ven, M.L.; Coward, W.A. Effect of alcohol on postmeal fat storage. *Am. J. Clin. Nutr.* **1994**, *59*, 619–625. [CrossRef] [PubMed]
59. Ontko, J.A. Effects of ethanol on the metabolism of free fatty acids in isolated liver cells. *J. Lipid Res.* **1973**, *14*, 78–86. [CrossRef]
60. Siler, S.Q.; Neese, R.A.; Hellerstein, M.K. De novo lipogenesis, lipid kinetics, and whole-body lipid balances in humans after acute alcohol consumption. *Am. J. Clin. Nutr.* **1999**, *70*, 928–936. [CrossRef]
61. You, M.; Fischer, M.; Deeg, M.A.; Crabb, D.W. Ethanol induces fatty acid synthesis pathways by activation of sterol regulatory element-binding protein (SREBP). *J. Biol. Chem.* **2002**, *277*, 29342–29347. [CrossRef] [PubMed]
62. Correnti, J.M.; Gottshall, L.; Lin, A.; Williams, B.; Oranu, A.; Beck, J.; Chen, J.; Bennett, M.J.; Carr, R.M. Ethanol and C2 ceramide activate fatty acid oxidation in human hepatoma cells. *Sci. Rep.* **2018**, *8*, 12923. [CrossRef]
63. Cederbaum, A.I.; Lieber, C.S.; Beattie, D.S.; Rubin, E. Effect of chronic ethanol ingestion on fatty acid oxidation by hepatic mitochondria. *J. Biol. Chem.* **1975**, *250*, 5122–5129. [CrossRef]
64. Blomstrand, R.; Kager, L.; Lantto, O. Studies on the ethanol-induced decrease of fatty acid oxidation in rat and human liver slices. *Life Sci.* **1973**, *13*, 1131–1141. [CrossRef]
65. Fabbrini, E.; Mohammed, B.S.; Magkos, F.; Korenblat, K.M.; Patterson, B.W.; Klein, S. Alterations in adipose tissue and hepatic lipid kinetics in obese men and women with nonalcoholic fatty liver disease. *Gastroenterology* **2008**, *134*, 424–431. [CrossRef]
66. Changani, K.K.; Jalan, R.; Cox, I.J.; Ala-Korpela, M.; Bhakoo, K.; Taylor-Robinson, S.D.; Bell, J.D. Evidence for altered hepatic gluconeogenesis in patients with cirrhosis using in vivo 31-phosphorus magnetic resonance spectroscopy. *Gut* **2001**, *49*, 557–564. [CrossRef] [PubMed]
67. Kinny-Koster, B.; Bartels, M.; Becker, S.; Scholz, M.; Thiery, J.; Ceglarek, U.; Kaiser, T. Plasma Amino Acid Concentrations Predict Mortality in Patients with End-Stage Liver Disease. *PLoS ONE* **2016**, *11*, e0159205. [CrossRef]
68. McCullough, A.J.; Mullen, K.D.; Kalhan, S.C. Body cell mass and leucine metabolism in cirrhosis. *Gastroenterology* **1992**, *102*, 1325–1333. [CrossRef]

69. Morgan, M.Y.; Marshall, A.W.; Milsom, J.P.; Sherlock, S. Plasma amino-acid patterns in liver disease. *Gut* **1982**, *23*, 362–370. [CrossRef]
70. Record, C.O.; Buxton, B.; Chase, R.A.; Curzon, G.; Murray-Lyon, I.M.; Williams, R. Plasma and brain amino acids in fulminant hepatic failure and their relationship to hepatic encephalopathy. *Eur. J. Clin. Investig.* **1976**, *6*, 387–394. [CrossRef] [PubMed]
71. Collins, J.R.; Lacy, W.W.; Stiel, J.N.; Crofford, O.B. Glucose intolerance and insulin resistance in patients with liver disease. II. A study of etiologic factors and evaluation of insulin actions. *Arch. Intern. Med.* **1970**, *126*, 608–614. [CrossRef] [PubMed]
72. Muller, M.J.; Bottcher, J.; Selberg, O.; Weselmann, S.; Boker, K.H.; Schwarze, M.; von zur Muhlen, A.; Manns, M.P. Hypermetabolism in clinically stable patients with liver cirrhosis. *Am. J. Clin. Nutr.* **1999**, *69*, 1194–1201. [CrossRef] [PubMed]
73. Nakaya, Y.; Harada, N.; Kakui, S.; Okada, K.; Takahashi, A.; Inoi, J.; Ito, S. Severe catabolic state after prolonged fasting in cirrhotic patients: Effect of oral branched-chain amino-acid-enriched nutrient mixture. *J. Gastroenterol.* **2002**, *37*, 531–536. [CrossRef] [PubMed]
74. Owen, O.E.; Reichle, F.A.; Mozzoli, M.A.; Kreulen, T.; Patel, M.S.; Elfenbein, I.B.; Golsorkhi, M.; Chang, K.H.; Rao, N.S.; Sue, H.S.; et al. Hepatic, gut, and renal substrate flux rates in patients with hepatic cirrhosis. *J. Clin. Investig.* **1981**, *68*, 240–252. [CrossRef]
75. Petersen, K.F.; Krssak, M.; Navarro, V.; Chandramouli, V.; Hundal, R.; Schumann, W.C.; Landau, B.R.; Shulman, G.I. Contributions of net hepatic glycogenolysis and gluconeogenesis to glucose production in cirrhosis. *Am. J. Physiol.* **1999**, *276*, E529–E535. [CrossRef]
76. Yeung, R.T.; Wang, C.C. A study of carbohydrate metabolism in postnecrotic cirrhosis liver. *Gut* **1974**, *15*, 907–912. [CrossRef]
77. Dasarathy, J.; Alkhouri, N.; Dasarathy, S. Changes in body composition after transjugular intrahepatic portosystemic stent in cirrhosis: A critical review of literature. *Liver Int.* **2011**, *31*, 1250–1258. [CrossRef]
78. Holland-Fischer, P.; Vilstrup, H.; Frystyk, J.; Nielsen, D.T.; Flyvbjerg, A.; Gronbaek, H. The IGF system after insertion of a transjugular intrahepatic porto-systemic shunt in patients with liver cirrhosis. *Eur. J. Endocrinol.* **2009**, *160*, 957–963. [CrossRef] [PubMed]
79. Plauth, M.; Schutz, T.; Buckendahl, D.P.; Kreymann, G.; Pirlich, M.; Grungreiff, S.; Romaniuk, P.; Ertl, S.; Weiss, M.L.; Lochs, H. Weight gain after transjugular intrahepatic portosystemic shunt is associated with improvement in body composition in malnourished patients with cirrhosis and hypermetabolism. *J. Hepatol.* **2004**, *40*, 228–233. [CrossRef]
80. Rosemurgy, A.S.; Zervos, E.E.; Goode, S.E.; Black, T.J.; Zwiebel, B.R. Differential effects on portal and effective hepatic blood flow. A comparison between transjugular intrahepatic portasystemic shunt and small-diameter H-graft portacaval shunt. *Ann. Surg.* **1997**, *225*, 601–607; discussion 607–608. [CrossRef] [PubMed]
81. Agarwal, A.; Avarebeel, S.; Choudhary, N.S.; Goudar, M.; Tejaswini, C.J. Correlation of Trace Elements in Patients of Chronic Liver Disease with Respect to Child- Turcotte- Pugh Scoring System. *J. Clin. Diagn. Res.* **2017**, *11*, OC25–OC28. [CrossRef]
82. Loguercio, C.; De Girolamo, V.; Federico, A.; Feng, S.L.; Crafa, E.; Cataldi, V.; Gialanella, G.; Moro, R.; Del Vecchio Blanco, C. Relationship of blood trace elements to liver damage, nutritional status, and oxidative stress in chronic nonalcoholic liver disease. *Biol. Trace Elem. Res.* **2001**, *81*, 245–254. [CrossRef]
83. Yoshida, Y.; Higashi, T.; Nouso, K.; Nakatsukasa, H.; Nakamura, S.I.; Watanabe, A.; Tsuji, T. Effects of zinc deficiency/zinc supplementation on ammonia metabolism in patients with decompensated liver cirrhosis. *Acta Med. Okayama* **2001**, *55*, 349–355. [CrossRef]
84. Solis-Herruzo, J.; De Cuenca, B.; Munoz-Rivero, M.C. Intestinal zinc absorption in cirrhotic patients. *Z. Gastroenterol.* **1989**, *27*, 335–338. [PubMed]
85. Pasqualetti, P.; Casale, R.; Colantonio, D.; Di Lauro, G.; Festuccia, V.; Natali, L.; Natali, G. Serum levels of magnesium in hepatic cirrhosis. *Quad. Sclavo Diagn.* **1987**, *23*, 12–17. [PubMed]
86. Nangliya, V.; Sharma, A.; Yadav, D.; Sunder, S.; Nijhawan, S.; Mishra, S. Study of trace elements in liver cirrhosis patients and their role in prognosis of disease. *Biol. Trace Elem. Res.* **2015**, *165*, 35–40. [CrossRef]
87. Buyukasik, N.S.; Nadir, I.; Akin, F.E.; Cakal, B.; Kav, T.; Ersoy, O.; Buyukasik, Y. Serum iron parameters in cirrhosis and chronic hepatitis: Detailed description. *Turk. J. Gastroenterol.* **2011**, *22*, 606–611. [CrossRef] [PubMed]
88. Jurczyk, K.; Wawrzynowicz-Syczewska, M.; Boron-Kaczmarska, A.; Sych, Z. Serum iron parameters in patients with alcoholic and chronic cirrhosis and hepatitis. *Med. Sci. Monit.* **2001**, *7*, 962–965. [PubMed]
89. Rahelic, D.; Kujundzic, M.; Romic, Z.; Brkic, K.; Petrovecki, M. Serum concentration of zinc, copper, manganese and magnesium in patients with liver cirrhosis. *Coll. Antropol.* **2006**, *30*, 523–528.
90. Teriaky, A.; Mosli, M.; Chandok, N.; Al-Judaibi, B.; Marotta, P.; Qumosani, K. Prevalence of fat-soluble vitamin (A, D, and E) and zinc deficiency in patients with cirrhosis being assessed for liver transplantation. *Acta Gastroenterol. Belg.* **2017**, *80*, 237–241.
91. Chen, W.; Chen, G. The Roles of Vitamin A in the Regulation of Carbohydrate, Lipid, and Protein Metabolism. *J. Clin. Med.* **2014**, *3*, 453–479. [CrossRef] [PubMed]
92. Saeed, A.; Hoekstra, M.; Hoeke, M.O.; Heegsma, J.; Faber, K.N. The interrelationship between bile acid and vitamin A homeostasis. *Biochim. Biophys. Acta Mol. Cell Biol. Lipids* **2017**, *1862*, 496–512. [CrossRef]
93. Kaplan, M.M.; Elta, G.H.; Furie, B.; Sadowski, J.A.; Russell, R.M. Fat-soluble vitamin nutriture in primary biliary cirrhosis. *Gastroenterology* **1988**, *95*, 787–792. [CrossRef]
94. Garrett-Laster, M.; Russell, R.M.; Jacques, P.F. Impairment of taste and olfaction in patients with cirrhosis: The role of vitamin A. *Hum. Nutr. Clin. Nutr.* **1984**, *38*, 203–214.

95. Paternostro, R.; Wagner, D.; Reiberger, T.; Mandorfer, M.; Schwarzer, R.; Ferlitsch, M.; Trauner, M.; Peck-Radosavljevic, M.; Ferlitsch, A. Low 25-OH-vitamin D levels reflect hepatic dysfunction and are associated with mortality in patients with liver cirrhosis. *Wien. Klin. Wochenschr.* **2017**, *129*, 8–15. [CrossRef] [PubMed]
96. Stokes, C.S.; Krawczyk, M.; Reichel, C.; Lammert, F.; Grunhage, F. Vitamin D deficiency is associated with mortality in patients with advanced liver cirrhosis. *Eur. J. Clin. Investig.* **2014**, *44*, 176–183. [CrossRef]
97. Finkelmeier, F.; Kronenberger, B.; Koberle, V.; Bojunga, J.; Zeuzem, S.; Trojan, J.; Piiper, A.; Waidmann, O. Severe 25-hydroxyvitamin D deficiency identifies a poor prognosis in patients with hepatocellular carcinoma—A prospective cohort study. *Aliment. Pharmacol. Ther.* **2014**, *39*, 1204–1212. [CrossRef]
98. Peres, W.A.; Chaves, G.V.; Goncalves, J.C.; Ramalho, A.; Coelho, H.S. Vitamin A deficiency in patients with hepatitis C virus-related chronic liver disease. *Br. J. Nutr.* **2011**, *106*, 1724–1731. [CrossRef]
99. Zhou, Y.J.; Liang, M.Y.; Zhang, X.Q. Changes in serum folic acid and vitamin B12 levels in liver cirrhosis and its clinical significance. *Zhonghua Nei Ke Za Zhi* **1992**, *30*, 625–627, 658.
100. Majumdar, S.K.; Shaw, G.K.; O'Gorman, P.; Aps, E.J.; Offerman, E.L.; Thomson, A.D. Blood vitamin status (B1, B2, B6, folic acid and B12) in patients with alcoholic liver disease. *Int. J. Vitam. Nutr. Res.* **1982**, *52*, 266–271. [PubMed]
101. Beattie, A.D.; Sherlock, S. Ascorbic acid deficiency in liver disease. *Gut* **1976**, *17*, 571–575. [CrossRef]
102. Sugihara, T.; Koda, M.; Okamoto, T.; Miyoshi, K.; Matono, T.; Oyama, K.; Hosho, K.; Okano, J.I.; Isomoto, H.; Murawaki, Y. Falsely Elevated Serum Vitamin B12 Levels Were Associated with the Severity and Prognosis of Chronic Viral Liver Disease. *Yonago Acta Med.* **2017**, *60*, 31–39.
103. Areekul, S.; Panatampon, P.; Doungbarn, J. Vitamin B12 and vitamin B12 binding proteins in liver diseases. *Southeast. Asian J. Trop. Med. Public Health* **1977**, *8*, 322–328. [PubMed]
104. Prieto-Frias, C.; Conchillo, M.; Payeras, M.; Inarrairaegui, M.; Davola, D.; Fruhbeck, G.; Salvador, J.; Rodriguez, M.; Richter, J.A.; Mugueta, C.; et al. Factors related to increased resting energy expenditure in men with liver cirrhosis. *Eur. J. Gastroenterol. Hepatol.* **2016**, *28*, 139–145. [CrossRef] [PubMed]
105. Peng, S.; Plank, L.D.; McCall, J.L.; Gillanders, L.K.; McIlroy, K.; Gane, E.J. Body composition, muscle function, and energy expenditure in patients with liver cirrhosis: A comprehensive study. *Am. J. Clin. Nutr.* **2007**, *85*, 1257–1266. [CrossRef]
106. Guglielmi, F.W.; Panella, C.; Buda, A.; Budillon, G.; Caregaro, L.; Clerici, C.; Conte, D.; Federico, A.; Gasbarrini, G.; Guglielmi, A.; et al. Nutritional state and energy balance in cirrhotic patients with or without hypermetabolism. Multicentre prospective study by the 'Nutritional Problems in Gastroenterology' Section of the Italian Society of Gastroenterology (SIGE). *Dig. Liver Dis.* **2005**, *37*, 681–688. [CrossRef]
107. Greco, A.V.; Mingrone, G.; Benedetti, G.; Capristo, E.; Tataranni, P.A.; Gasbarrini, G. Daily energy and substrate metabolism in patients with cirrhosis. *Hepatology* **1998**, *27*, 346–350. [CrossRef] [PubMed]
108. Mathur, S.; Peng, S.; Gane, E.J.; McCall, J.L.; Plank, L.D. Hypermetabolism predicts reduced transplant-free survival independent of MELD and Child-Pugh scores in liver cirrhosis. *Nutrition* **2007**, *23*, 398–403. [CrossRef]
109. Tajika, M.; Kato, M.; Mohri, H.; Miwa, Y.; Kato, T.; Ohnishi, H.; Moriwaki, H. Prognostic value of energy metabolism in patients with viral liver cirrhosis. *Nutrition* **2002**, *18*, 229–234. [CrossRef]
110. Selberg, O.; Bottcher, J.; Tusch, G.; Pichlmayr, R.; Henkel, E.; Muller, M.J. Identification of high- and low-risk patients before liver transplantation: A prospective cohort study of nutritional and metabolic parameters in 150 patients. *Hepatology* **1997**, *25*, 652–657. [CrossRef]
111. Roubenoff, R.; Roubenoff, R.A.; Cannon, J.G.; Kehayias, J.J.; Zhuang, H.; Dawson-Hughes, B.; Dinarello, C.A.; Rosenberg, I.H. Rheumatoid cachexia: Cytokine-driven hypermetabolism accompanying reduced body cell mass in chronic inflammation. *J. Clin. Investig.* **1994**, *93*, 2379–2386. [CrossRef]
112. Tilg, H.; Wilmer, A.; Vogel, W.; Herold, M.; Nolchen, B.; Judmaier, G.; Huber, C. Serum levels of cytokines in chronic liver diseases. *Gastroenterology* **1992**, *103*, 264–274. [CrossRef]
113. Huang, Y.S.; Hwang, S.J.; Chan, C.Y.; Wu, J.C.; Chao, Y.; Chang, F.Y.; Lee, S.D. Serum levels of cytokines in hepatitis C-related liver disease: A longitudinal study. *Zhonghua Yi Xue Za Zhi (Taipei)* **1999**, *62*, 327–333.
114. Laso, F.J.; Lapena, P.; Madruga, J.I.; San Miguel, J.F.; Orfao, A.; Iglesias, M.C.; Alvarez-Mon, M. Alterations in tumor necrosis factor-alpha, interferon-gamma, and interleukin-6 production by natural killer cell-enriched peripheral blood mononuclear cells in chronic alcoholism: Relationship with liver disease and ethanol intake. *Alcohol. Clin. Exp. Res.* **1997**, *21*, 1226–1231. [CrossRef] [PubMed]
115. Muller, C.; Zielinski, C.C. Interleukin-6 production by peripheral blood monocytes in patients with chronic liver disease and acute viral hepatitis. *J. Hepatol.* **1992**, *15*, 372–377. [CrossRef]
116. Khoruts, A.; Stahnke, L.; McClain, C.J.; Logan, G.; Allen, J.I. Circulating tumor necrosis factor, interleukin-1 and interleukin-6 concentrations in chronic alcoholic patients. *Hepatology* **1991**, *13*, 267–276. [CrossRef]
117. Lin, S.Y.; Wang, Y.Y.; Sheu, W.H. Increased serum leptin concentrations correlate with soluble tumour necrosis factor receptor levels in patients with cirrhosis. *Clin. Endocrinol.* **2002**, *57*, 805–811. [CrossRef]
118. Shinsyu, A.; Bamba, S.; Kurihara, M.; Matsumoto, H.; Sonoda, A.; Inatomi, O.; Andoh, A.; Takebayashi, K.; Kojima, M.; Iida, H.; et al. Inflammatory cytokines, appetite-regulating hormones, and energy metabolism in patients with gastrointestinal cancer. *Oncol. Lett.* **2020**, *20*, 1469–1479. [CrossRef]

119. Paulsen, O.; Laird, B.; Aass, N.; Lea, T.; Fayers, P.; Kaasa, S.; Klepstad, P. The relationship between pro-inflammatory cytokines and pain, appetite and fatigue in patients with advanced cancer. *PLoS ONE* **2017**, *12*, e0177620. [CrossRef] [PubMed]
120. Bajaj, J.S.; Kakiyama, G.; Zhao, D.; Takei, H.; Fagan, A.; Hylemon, P.; Zhou, H.; Pandak, W.M.; Nittono, H.; Fiehn, O.; et al. Continued Alcohol Misuse in Human Cirrhosis is Associated with an Impaired Gut-Liver Axis. *Alcohol. Clin. Exp. Res.* **2017**, *41*, 1857–1865. [CrossRef]
121. Chen, P.; Starkel, P.; Turner, J.R.; Ho, S.B.; Schnabl, B. Dysbiosis-induced intestinal inflammation activates tumor necrosis factor receptor I and mediates alcoholic liver disease in mice. *Hepatology* **2015**, *61*, 883–894. [CrossRef]
122. Horvath, A.; Rainer, F.; Bashir, M.; Leber, B.; Schmerboeck, B.; Klymiuk, I.; Groselj-Strele, A.; Durdevic, M.; Freedberg, D.E.; Abrams, J.A.; et al. Biomarkers for oralization during long-term proton pump inhibitor therapy predict survival in cirrhosis. *Sci. Rep.* **2019**, *9*, 12000. [CrossRef]
123. Quinones, M.; Ferno, J.; Al-Massadi, O. Ghrelin and liver disease. *Rev. Endocr. Metab. Disord.* **2020**, *21*, 45–56. [CrossRef] [PubMed]
124. Tschop, M.; Smiley, D.L.; Heiman, M.L. Ghrelin induces adiposity in rodents. *Nature* **2000**, *407*, 908–913. [CrossRef]
125. Ahima, R.S.; Prabakaran, D.; Mantzoros, C.; Qu, D.; Lowell, B.; Maratos-Flier, E.; Flier, J.S. Role of leptin in the neuroendocrine response to fasting. *Nature* **1996**, *382*, 250–252. [CrossRef]
126. Ockenga, J.; Bischoff, S.C.; Tillmann, H.L.; Rifai, K.; Widjaja, A.; Boker, K.H.; Manns, M.P.; Brabant, G. Elevated bound leptin correlates with energy expenditure in cirrhotics. *Gastroenterology* **2000**, *119*, 1656–1662. [CrossRef] [PubMed]
127. Richardson, R.A.; Davidson, H.I.; Hinds, A.; Cowan, S.; Rae, P.; Garden, O.J. Influence of the metabolic sequelae of liver cirrhosis on nutritional intake. *Am. J. Clin. Nutr.* **1999**, *69*, 331–337. [CrossRef]
128. Grossmann, M.; Hoermann, R.; Gani, L.; Chan, I.; Cheung, A.; Gow, P.J.; Li, A.; Zajac, J.D.; Angus, P. Low testosterone levels as an independent predictor of mortality in men with chronic liver disease. *Clin. Endocrinol.* **2012**, *77*, 323–328. [CrossRef] [PubMed]
129. Griggs, R.C.; Kingston, W.; Jozefowicz, R.F.; Herr, B.E.; Forbes, G.; Halliday, D. Effect of testosterone on muscle mass and muscle protein synthesis. *J. Appl. Physiol.* **1989**, *66*, 498–503. [CrossRef]
130. Stadlbauer, V.; Komarova, I.; Klymiuk, I.; Durdevic, M.; Reisinger, A.; Blesl, A.; Rainer, F.; Horvath, A. Disease severity and proton pump inhibitor use impact strongest on faecal microbiome composition in liver cirrhosis. *Liver Int.* **2020**, *40*, 866–877. [CrossRef]
131. Ticinesi, A.; Lauretani, F.; Milani, C.; Nouvenne, A.; Tana, C.; Del Rio, D.; Maggio, M.; Ventura, M.; Meschi, T. Aging Gut Microbiota at the Cross-Road between Nutrition, Physical Frailty, and Sarcopenia: Is There a Gut-Muscle Axis? *Nutrients* **2017**, *9*, 1303. [CrossRef] [PubMed]
132. de Sire, R.; Rizzatti, G.; Ingravalle, F.; Pizzoferrato, M.; Petito, V.; Lopetuso, L.; Graziani, C.; de Sire, A.; Mentella, M.C.; Mele, M.C.; et al. Skeletal muscle-gut axis: Emerging mechanisms of sarcopenia for intestinal and extra intestinal diseases. *Minerva Gastroenterol. Dietol.* **2018**, *64*, 351–362. [CrossRef]
133. den Besten, G.; Lange, K.; Havinga, R.; van Dijk, T.H.; Gerding, A.; van Eunen, K.; Müller, M.; Groen, A.K.; Hooiveld, G.J.; Bakker, B.M.; et al. Gut-derived short-chain fatty acids are vividly assimilated into host carbohydrates and lipids. *Am. J. Physiol. Gastrointest. Liver. Physiol.* **2013**, *305*, G900–G910. [CrossRef] [PubMed]
134. Grosicki, G.J.; Fielding, R.A.; Lustgarten, M.S. Gut Microbiota Contribute to Age-Related Changes in Skeletal Muscle Size, Composition, and Function: Biological Basis for a Gut-Muscle Axis. *Calcif. Tissue Int.* **2018**, *102*, 433–442. [CrossRef] [PubMed]
135. Li, D.; Li, Y.; Dai, W.; Wang, H.; Qiu, C.; Feng, S.; Zhou, Q.; Wang, W.; Feng, X.; Yao, K.; et al. Intestinal Bacteroides sp. Imbalance Associated With the Occurrence of Childhood Undernutrition in China. *Front. Microbiol.* **2019**, *10*, 2635. [CrossRef] [PubMed]
136. Monira, S.; Nakamura, S.; Gotoh, K.; Izutsu, K.; Watanabe, H.; Alam, N.H.; Endtz, H.P.; Cravioto, A.; Ali, S.I.; Nakaya, T.; et al. Gut microbiota of healthy and malnourished children in bangladesh. *Front. Microbiol.* **2011**, *2*, 228. [CrossRef] [PubMed]
137. Qin, N.; Yang, F.; Li, A.; Prifti, E.; Chen, Y.; Shao, L.; Guo, J.; Le Chatelier, E.; Yao, J.; Wu, L.; et al. Alterations of the human gut microbiome in liver cirrhosis. *Nature* **2014**, *513*, 59–64. [CrossRef]
138. Meyer, F.; Bannert, K.; Wiese, M.; Esau, S.; Sautter, L.F.; Ehlers, L.; Aghdassi, A.A.; Metges, C.C.; Garbe, L.-A.; Jaster, R.; et al. Molecular Mechanism Contributing to Malnutrition and Sarcopenia in Patients with Liver Cirrhosis. *Int. J. Mol. Sci.* **2020**, *21*, 5357. [CrossRef]
139. Bajaj, J.S.; Heuman, D.M.; Hylemon, P.B.; Sanyal, A.J.; White, M.B.; Monteith, P.; Noble, N.A.; Unser, A.B.; Daita, K.; Fisher, A.R.; et al. Altered profile of human gut microbiome is associated with cirrhosis and its complications. *J. Hepatol.* **2014**, *60*, 940–947. [CrossRef]
140. Bajaj, J.S.; Hylemon, P.B.; Ridlon, J.M.; Heuman, D.M.; Daita, K.; White, M.B.; Monteith, P.; Noble, N.A.; Sikaroodi, M.; Gillevet, P.M. Colonic mucosal microbiome differs from stool microbiome in cirrhosis and hepatic encephalopathy and is linked to cognition and inflammation. *Am. J. Physiol. Gastrointest. Liver Physiol.* **2012**, *303*, G675–G685. [CrossRef]
141. Malinchoc, M.; Kamath, P.S.; Gordon, F.D.; Peine, C.J.; Rank, J.; ter Borg, P.C. A model to predict poor survival in patients undergoing transjugular intrahepatic portosystemic shunts. *Hepatology* **2000**, *31*, 864–871. [CrossRef]
142. Pugh, R.N.; Murray-Lyon, I.M.; Dawson, J.L.; Pietroni, M.C.; Williams, R. Transection of the oesophagus for bleeding oesophageal varices. *Br. J. Surg.* **1973**, *60*, 646–649. [CrossRef]
143. Garcia-Tsao, G. The Child-Turcotte Classification: From Gestalt to Sophisticated Statistics and Back. *Dig. Dis. Sci.* **2016**, *61*, 3102–3104. [CrossRef]
144. Plauth, M.; Bernal, W.; Dasarathy, S.; Merli, M.; Plank, L.D.; Schutz, T.; Bischoff, S.C. ESPEN guideline on clinical nutrition in liver disease. *Clin. Nutr.* **2019**, *38*, 485–521. [CrossRef]

145. European Association for the Study of the Liver. EASL Clinical Practice Guidelines on nutrition in chronic liver disease. *J. Hepatol.* **2019**, *70*, 172–193. [CrossRef] [PubMed]
146. European Association for the Study of the Liver. EASL Clinical Practice Guidelines for the management of patients with decompensated cirrhosis. *J. Hepatol.* **2018**, *69*, 406–460. [CrossRef] [PubMed]
147. Borhofen, S.M.; Gerner, C.; Lehmann, J.; Fimmers, R.; Gortzen, J.; Hey, B.; Geiser, F.; Strassburg, C.P.; Trebicka, J. The Royal Free Hospital-Nutritional Prioritizing Tool Is an Independent Predictor of Deterioration of Liver Function and Survival in Cirrhosis. *Dig. Dis. Sci.* **2016**, *61*, 1735–1743. [CrossRef]
148. Traub, J.; Bergheim, I.; Horvath, A.; Stadlbauer, V. Validation of Malnutrition Screening Tools in Liver Cirrhosis. *Nutrients* **2020**, *12*, 1306. [CrossRef]
149. Buchard, B.; Boirie, Y.; Cassagnes, L.; Lamblin, G.; Coilly, A.; Abergel, A. Assessment of Malnutrition, Sarcopenia and Frailty in Patients with Cirrhosis: Which Tools Should We Use in Clinical Practice? *Nutrients* **2020**, *12*, 186. [CrossRef]
150. Tandon, P.; Raman, M.; Mourtzakis, M.; Merli, M. A practical approach to nutritional screening and assessment in cirrhosis. *Hepatology* **2017**, *65*, 1044–1057. [CrossRef]
151. Fialla, A.D.; Israelsen, M.; Hamberg, O.; Krag, A.; Gluud, L.L. Nutritional therapy in cirrhosis or alcoholic hepatitis: A systematic review and meta-analysis. *Liver Int.* **2015**, *35*, 2072–2078. [CrossRef]
152. Ney, M.; Vandermeer, B.; van Zanten, S.J.; Ma, M.M.; Gramlich, L.; Tandon, P. Meta-analysis: Oral or enteral nutritional supplementation in cirrhosis. *Aliment. Pharmacol. Ther.* **2013**, *37*, 672–679. [CrossRef] [PubMed]
153. Montano-Loza, A.J.; Meza-Junco, J.; Prado, C.M.; Lieffers, J.R.; Baracos, V.E.; Bain, V.G.; Sawyer, M.B. Muscle wasting is associated with mortality in patients with cirrhosis. *Clin. Gastroenterol. Hepatol.* **2012**, *10*, 166–173, 173.e1. [CrossRef]
154. Chang, Y.; Liu, Q.Y.; Zhang, Q.; Rong, Y.M.; Lu, C.Z.; Li, H. Role of nutritional status and nutritional support in outcome of hepatitis B virus-associated acute-on-chronic liver failure. *World J. Gastroenterol.* **2020**, *26*, 4288–4301. [CrossRef] [PubMed]
155. Bories, P.N.; Campillo, B. One-month regular oral nutrition in alcoholic cirrhotic patients. Changes of nutritional status, hepatic function and serum lipid pattern. *Br. J. Nutr.* **1994**, *72*, 937–946. [CrossRef] [PubMed]
156. Manguso, F.; D'Ambra, G.; Menchise, A.; Sollazzo, R.; D'Agostino, L. Effects of an appropriate oral diet on the nutritional status of patients with HCV-related liver cirrhosis: A prospective study. *Clin. Nutr.* **2005**, *24*, 751–759. [CrossRef]
157. Maharshi, S.; Sharma, B.C.; Sachdeva, S.; Srivastava, S.; Sharma, P. Efficacy of Nutritional Therapy for Patients With Cirrhosis and Minimal Hepatic Encephalopathy in a Randomized Trial. *Clin. Gastroenterol. Hepatol.* **2016**, *14*, 454–460.e3; quiz e433. [CrossRef]
158. Dichi, J.B.; Dichi, I.; Maio, R.; Correa, C.R.; Angeleli, A.Y.; Bicudo, M.H.; Rezende, T.A.; Burini, R.C. Whole-body protein turnover in malnourished patients with child class B and C cirrhosis on diets low to high in protein energy. *Nutrition* **2001**, *17*, 239–242. [CrossRef]
159. Swart, G.R.; van den Berg, J.W.; van Vuure, J.K.; Rietveld, T.; Wattimena, D.L.; Frenkel, M. Minimum protein requirements in liver cirrhosis determined by nitrogen balance measurements at three levels of protein intake. *Clin. Nutr.* **1989**, *8*, 329–336. [CrossRef]
160. Putadechakum, S.; Klangjareonchai, T.; Soponsaritsuk, A.; Roongpisuthipong, C. Nutritional status assessment in cirrhotic patients after protein supplementation. *ISRN Gastroenterol.* **2012**, *2012*, 690402. [CrossRef]
161. Dawson, A.M.; Sherlock, S.; Summerskill, W.H. The treatment and prognosis of hepatic coma. *Lancet* **1956**, *271*, 689–694. [CrossRef]
162. Cordoba, J.; Lopez-Hellin, J.; Planas, M.; Sabin, P.; Sanpedro, F.; Castro, F.; Esteban, R.; Guardia, J. Normal protein diet for episodic hepatic encephalopathy: Results of a randomized study. *J. Hepatol.* **2004**, *41*, 38–43. [CrossRef]
163. Kearns, P.J.; Young, H.; Garcia, G.; Blaschke, T.; O'Hanlon, G.; Rinki, M.; Sucher, K.; Gregory, P. Accelerated improvement of alcoholic liver disease with enteral nutrition. *Gastroenterology* **1992**, *102*, 200–205. [CrossRef]
164. Gheorghe, L.; Iacob, R.; Vadan, R.; Iacob, S.; Gheorghe, C. Improvement of hepatic encephalopathy using a modified high-calorie high-protein diet. *Rom. J. Gastroenterol.* **2005**, *14*, 231–238.
165. Morgan, T.R.; Moritz, T.E.; Mendenhall, C.L.; Haas, R. Protein consumption and hepatic encephalopathy in alcoholic hepatitis. VA Cooperative Study Group #275. *J. Am. Coll. Nutr.* **1995**, *14*, 152–158. [CrossRef]
166. Plank, L.D.; Gane, E.J.; Peng, S.; Muthu, C.; Mathur, S.; Gillanders, L.; McIlroy, K.; Donaghy, A.J.; McCall, J.L. Nocturnal nutritional supplementation improves total body protein status of patients with liver cirrhosis: A randomized 12-month trial. *Hepatology* **2008**, *48*, 557–566. [CrossRef]
167. Swart, G.R.; Zillikens, M.C.; van Vuure, J.K.; van den Berg, J.W. Effect of a late evening meal on nitrogen balance in patients with cirrhosis of the liver. *BMJ* **1989**, *299*, 1202–1203. [CrossRef] [PubMed]
168. Tsien, C.D.; McCullough, A.J.; Dasarathy, S. Late evening snack: Exploiting a period of anabolic opportunity in cirrhosis. *J. Gastroenterol. Hepatol.* **2012**, *27*, 430–441. [CrossRef] [PubMed]
169. Guo, Y.J.; Tian, Z.B.; Jiang, N.; Ding, X.L.; Mao, T.; Jing, X. Effects of Late Evening Snack on Cirrhotic Patients: A Systematic Review and Meta-Analysis. *Gastroenterol. Res. Pract.* **2018**, *2018*, 9189062. [CrossRef] [PubMed]
170. Vaisman, N.; Katzman, H.; Carmiel-Haggai, M.; Lusthaus, M.; Niv, E. Breakfast improves cognitive function in cirrhotic patients with cognitive impairment. *Am. J. Clin. Nutr.* **2010**, *92*, 137–140. [CrossRef] [PubMed]
171. Verboeket-van de Venne, W.P.; Westerterp, K.R.; van Hoek, B.; Swart, G.R. Energy expenditure and substrate metabolism in patients with cirrhosis of the liver: Effects of the pattern of food intake. *Gut* **1995**, *36*, 110–116. [CrossRef]
172. Plauth, M.; Cabre, E.; Riggio, O.; Assis-Camilo, M.; Pirlich, M.; Kondrup, J.; DGEM; Ferenci, P.; Holm, E.; Vom Dahl, S.; et al. ESPEN Guidelines on Enteral Nutrition: Liver disease. *Clin. Nutr.* **2006**, *25*, 285–294. [CrossRef] [PubMed]

173. Cabre, E.; Rodriguez-Iglesias, P.; Caballeria, J.; Quer, J.C.; Sanchez-Lombrana, J.L.; Pares, A.; Papo, M.; Planas, R.; Gassull, M.A. Short- and long-term outcome of severe alcohol-induced hepatitis treated with steroids or enteral nutrition: A multicenter randomized trial. *Hepatology* **2000**, *32*, 36–42. [CrossRef] [PubMed]
174. Baltz, J.G.; Argo, C.K.; Al-Osaimi, A.M.; Northup, P.G. Mortality after percutaneous endoscopic gastrostomy in patients with cirrhosis: A case series. *Gastrointest. Endosc.* **2010**, *72*, 1072–1075. [CrossRef]
175. Plauth, M.; Cabre, E.; Campillo, B.; Kondrup, J.; Marchesini, G.; Schutz, T.; Shenkin, A.; Wendon, J. ESPEN Guidelines on Parenteral Nutrition: Hepatology. *Clin. Nutr.* **2009**, *28*, 436–444. [CrossRef]
176. Jalan, R.; Fernandez, J.; Wiest, R.; Schnabl, B.; Moreau, R.; Angeli, P.; Stadlbauer, V.; Gustot, T.; Bernardi, M.; Canton, R.; et al. Bacterial infections in cirrhosis: A position statement based on the EASL Special Conference 2013. *J. Hepatol.* **2014**, *60*, 1310–1324. [CrossRef]
177. Cunha, L.; Happi Nono, M.; Guibert, A.L.; Nidegger, D.; Beau, P.; Beauchant, M. Effects of prolonged oral nutritional support in malnourished cirrhotic patients: Results of a pilot study. *Gastroenterol. Clin. Biol.* **2004**, *28*, 36–39. [CrossRef]
178. Norman, K.; Kirchner, H.; Freudenreich, M.; Ockenga, J.; Lochs, H.; Pirlich, M. Three month intervention with protein and energy rich supplements improve muscle function and quality of life in malnourished patients with non-neoplastic gastrointestinal disease–a randomized controlled trial. *Clin. Nutr.* **2008**, *27*, 48–56. [CrossRef] [PubMed]
179. Bjelakovic, G.; Gluud, L.L.; Nikolova, D.; Bjelakovic, M.; Nagorni, A.; Gluud, C. Meta-analysis: Antioxidant supplements for liver diseases—The Cochrane Hepato-Biliary Group. *Aliment. Pharmacol. Ther.* **2010**, *32*, 356–367. [CrossRef]
180. Shen, Y.C.; Chang, Y.H.; Fang, C.J.; Lin, Y.S. Zinc supplementation in patients with cirrhosis and hepatic encephalopathy: A systematic review and meta-analysis. *Nutr. J.* **2019**, *18*, 34. [CrossRef]
181. Katayama, K.; Saito, M.; Kawaguchi, T.; Endo, R.; Sawara, K.; Nishiguchi, S.; Kato, A.; Kohgo, H.; Suzuki, K.; Sakaida, I.; et al. Effect of zinc on liver cirrhosis with hyperammonemia: A preliminary randomized, placebo-controlled double-blind trial. *Nutrition* **2014**, *30*, 1409–1414. [CrossRef]
182. Takuma, Y.; Nouso, K.; Makino, Y.; Hayashi, M.; Takahashi, H. Clinical trial: Oral zinc in hepatic encephalopathy. *Aliment. Pharmacol. Ther.* **2010**, *32*, 1080–1090. [CrossRef]
183. Bresci, G.; Parisi, G.; Banti, S. Management of hepatic encephalopathy with oral zinc supplementation: A long-term treatment. *Eur. J. Med.* **1993**, *2*, 414–416.
184. Riggio, O.; Ariosto, F.; Merli, M.; Caschera, M.; Zullo, A.; Balducci, G.; Ziparo, V.; Pedretti, G.; Fiaccadori, F.; Bottari, E.; et al. Short-term oral zinc supplementation does not improve chronic hepatic encephalopathy. Results of a double-blind crossover trial. *Dig. Dis. Sci.* **1991**, *36*, 1204–1208. [CrossRef] [PubMed]
185. Rossi, R.E.; Conte, D.; Massironi, S. Diagnosis and treatment of nutritional deficiencies in alcoholic liver disease: Overview of available evidence and open issues. *Dig. Liver Dis.* **2015**, *47*, 819–825. [CrossRef] [PubMed]
186. Somi, M.H.; Rezaeifar, P.; Ostad Rahimi, A.; Moshrefi, B. Effects of low dose zinc supplementation on biochemical markers in non-alcoholic cirrhosis: A randomized clinical trial. *Arch. Iran. Med.* **2012**, *15*, 472–476.
187. Hosui, A.; Kimura, E.; Abe, S.; Tanimoto, T.; Onishi, K.; Kusumoto, Y.; Sueyoshi, Y.; Matsumoto, K.; Hirao, M.; Yamada, T.; et al. Long-Term Zinc Supplementation Improves Liver Function and Decreases the Risk of Developing Hepatocellular Carcinoma. *Nutrients* **2018**, *10*, 1955. [CrossRef] [PubMed]
188. Matsuoka, S.; Matsumura, H.; Nakamura, H.; Oshiro, S.; Arakawa, Y.; Hayashi, J.; Sekine, N.; Nirei, K.; Yamagami, H.; Ogawa, M.; et al. Zinc supplementation improves the outcome of chronic hepatitis C and liver cirrhosis. *J. Clin. Biochem. Nutr.* **2009**, *45*, 292–303. [CrossRef]
189. Weismann, K.; Christensen, E.; Dreyer, V. Zinc supplementation in alcoholic cirrhosis. A double-blind clinical trial. *Acta. Med. Scand.* **1979**, *205*, 361–366. [CrossRef]
190. Yang, F.; Ren, H.; Gao, Y.; Zhu, Y.; Huang, W. The value of severe vitamin D deficiency in predicting the mortality risk of patients with liver cirrhosis: A meta-analysis. *Clin. Res. Hepatol. Gastroenterol.* **2019**, *43*, 722–729. [CrossRef]
191. Pilz, S.; Putz-Bankuti, C.; Gaksch, M.; Spindelboeck, W.; Haselberger, M.; Rainer, F.; Posch, A.; Kreuzer, P.; Stojakovic, T.; Stadlbauer, V.; et al. Effects of Vitamin D Supplementation on Serum 25-Hydroxyvitamin D Concentrations in Cirrhotic Patients: A Randomized Controlled Trial. *Nutrients* **2016**, *8*, 278. [CrossRef]
192. Rode, A.; Fourlanos, S.; Nicoll, A. Oral vitamin D replacement is effective in chronic liver disease. *Gastroenterol. Clin. Biol.* **2010**, *34*, 618–620. [CrossRef]
193. Bjelakovic, G.; Nikolova, D.; Bjelakovic, M.; Gluud, C. Vitamin D supplementation for chronic liver diseases in adults. *Cochrane Database Syst. Rev.* **2017**, *11*, CD011564. [CrossRef] [PubMed]
194. Kato, M.; Miwa, Y.; Tajika, M.; Hiraoka, T.; Muto, Y.; Moriwaki, H. Preferential use of branched-chain amino acids as an energy substrate in patients with liver cirrhosis. *Intern. Med.* **1998**, *37*, 429–434. [CrossRef] [PubMed]
195. Yamauchi, M.; Takeda, K.; Sakamoto, K.; Ohata, M.; Toda, G. Effect of oral branched chain amino acid supplementation in the late evening on the nutritional state of patients with liver cirrhosis. *Hepatol. Res.* **2001**, *21*, 199–204. [CrossRef]
196. Nakaya, Y.; Okita, K.; Suzuki, K.; Moriwaki, H.; Kato, A.; Miwa, Y.; Shiraishi, K.; Okuda, H.; Onji, M.; Kanazawa, H.; et al. BCAA-enriched snack improves nutritional state of cirrhosis. *Nutrition* **2007**, *23*, 113–120. [CrossRef]

197. Muto, Y.; Sato, S.; Watanabe, A.; Moriwaki, H.; Suzuki, K.; Kato, A.; Kato, M.; Nakamura, T.; Higuchi, K.; Nishiguchi, S.; et al. Overweight and obesity increase the risk for liver cancer in patients with liver cirrhosis and long-term oral supplementation with branched-chain amino acid granules inhibits liver carcinogenesis in heavier patients with liver cirrhosis. *Hepatol. Res.* **2006**, *35*, 204–214. [CrossRef]
198. Muto, Y.; Sato, S.; Watanabe, A.; Moriwaki, H.; Suzuki, K.; Kato, A.; Kato, M.; Nakamura, T.; Higuchi, K.; Nishiguchi, S.; et al. Effects of oral branched-chain amino acid granules on event-free survival in patients with liver cirrhosis. *Clin. Gastroenterol. Hepatol.* **2005**, *3*, 705–713. [CrossRef]
199. Marchesini, G.; Bianchi, G.; Merli, M.; Amodio, P.; Panella, C.; Loguercio, C.; Rossi Fanelli, F.; Abbiati, R.; Italian, B.S.G. Nutritional supplementation with branched-chain amino acids in advanced cirrhosis: A double-blind, randomized trial. *Gastroenterology* **2003**, *124*, 1792–1801. [CrossRef]
200. Horst, D.; Grace, N.D.; Conn, H.O.; Schiff, E.; Schenker, S.; Viteri, A.; Law, D.; Atterbury, C.E. Comparison of dietary protein with an oral, branched chain-enriched amino acid supplement in chronic portal-systemic encephalopathy: A randomized controlled trial. *Hepatology* **1984**, *4*, 279–287. [CrossRef]
201. Dam, G.; Aamann, L.; Vistrup, H.; Gluud, L.L. The role of Branched Chain Amino Acids in the treatment of hepatic Encephalopathy. *J. Clin. Exp. Hepatol.* **2018**, *8*, 448–451. [CrossRef] [PubMed]
202. Les, I.; Doval, E.; Garcia-Martinez, R.; Planas, M.; Cardenas, G.; Gomez, P.; Flavia, M.; Jacas, C.; Minguez, B.; Vergara, M.; et al. Effects of branched-chain amino acids supplementation in patients with cirrhosis and a previous episode of hepatic encephalopathy: A randomized study. *Am. J. Gastroenterol.* **2011**, *106*, 1081–1088. [CrossRef] [PubMed]
203. Gluud, L.L.; Dam, G.; Borre, M.; Les, I.; Cordoba, J.; Marchesini, G.; Aagaard, N.K.; Risum, N.; Vilstrup, H. Oral branched-chain amino acids have a beneficial effect on manifestations of hepatic encephalopathy in a systematic review with meta-analyses of randomized controlled trials. *J. Nutr.* **2013**, *143*, 1263–1268. [CrossRef]
204. Bianchi, G.P.; Marchesini, G.; Fabbri, A.; Rondelli, A.; Bugianesi, E.; Zoli, M.; Pisi, E. Vegetable versus animal protein diet in cirrhotic patients with chronic encephalopathy. A randomized cross-over comparison. *J. Intern. Med.* **1993**, *233*, 385–392. [CrossRef]
205. Merli, M.; Iebba, V.; Giusto, M. What is new about diet in hepatic encephalopathy. *Metab. Brain Dis.* **2016**, *31*, 1289–1294. [CrossRef]
206. Amodio, P.; Bemeur, C.; Butterworth, R.; Cordoba, J.; Kato, A.; Montagnese, S.; Uribe, M.; Vilstrup, H.; Morgan, M.Y. The nutritional management of hepatic encephalopathy in patients with cirrhosis: International Society for Hepatic Encephalopathy and Nitrogen Metabolism Consensus. *Hepatology* **2013**, *58*, 325–336. [CrossRef]

nutrients

MDPI

Review

Malnutrition in Pediatric Chronic Cholestatic Disease: An Up-to-Date Overview

Maria Tessitore [1,†], Eduardo Sorrentino [1,†], Giuseppe Schiano Di Cola [1], Angelo Colucci [1], Pietro Vajro [1] and Claudia Mandato [2,*]

[1] Department of Medicine, Surgery and Dentistry "Scuola Medica Salernitana", Chair of Pediatrics and Residency Program of Pediatrics, Via S. Allende, University of Salerno, 84081 Baronissi, SA, Italy; mariatessitore20@gmail.com (M.T.); eduardosorrentino30@gmail.com (E.S.); schianodicolag@gmail.com (G.S.D.C.); angelocolucci2@gmail.com (A.C.); pvajro@unisa.it (P.V.)

[2] Department of Pediatrics, Santobono-Pausilipon Children's Hospital Via M. Fiore, 80129 Naples, Italy

* Correspondence: cla.mandato@gmail.com; Tel.: +39-081-2205424; Fax: +39-081-2205428

† These authors contributed equally to this work.

Abstract: Despite recent advances, the causes of and effective therapies for pediatric chronic cholestatic diseases remain elusive, and many patients progress to liver failure and need liver transplantation. Malnutrition is a common complication in these patients and is a well-recognized, tremendous challenge for the clinician. We undertook a narrative review of both recent and relevant older literature, published during the last 20 years, for studies linking nutrition to pediatric chronic cholestasis. The collected data confirm that malnutrition and failure to thrive are associated with increased risks of morbidity and mortality, and they also affect the outcomes of liver transplantation, including long-term survival. Malnutrition in children with chronic liver disease is multifactorial and with multiple potential nutritional deficiencies. To improve life expectancy and the quality of life, patients require careful assessments and appropriate management of their nutritional statuses by multidisciplinary teams, which can identify and/or prevent specific deficiencies and initiate appropriate interventions. Solutions available for the clinical management of these children in general, as well as those directed to specific etiologies, are summarized. We particularly focus on fat-soluble vitamin deficiency and malnutrition due to fat malabsorption. Supplemental feeding, including medium-chain triglycerides, essential fatty acids, branched-chain amino acids, and the extra calories needed to overcome the consequences of anorexia and high energy requirements, is reviewed. Future studies should address the need for further improving commercially available and nutritionally complete infant milk formulae for the dietary management of this fragile category of patients. The aid of a specialist dietitian, educational training regarding nutritional guidelines for stakeholders, and improving family nutritional health literacy appear essential.

Keywords: cholestasis; chronic liver diseases; malnutrition; nutritional needs; pediatrics

Citation: Tessitore, M.; Sorrentino, E.; Schiano Di Cola, G.; Colucci, A.; Vajro, P.; Mandato, C. Malnutrition in Pediatric Chronic Cholestatic Disease: An Up-to-Date Overview. *Nutrients* **2021**, *13*, 2785. https://doi.org/10.3390/nu13082785

Academic Editor: Antonio Colecchia

Received: 13 July 2021
Accepted: 11 August 2021
Published: 13 August 2021

1. Introduction

Cholestasis is regarded as reduced bile formation or flow, leading to a decreased concentration of bile acids in the intestine, and the retention within the blood and the liver of biliary substances which are normally excreted into bile [1]. In the pediatric population, cholestasis predominantly affects neonates and infants, around 1 in every 2500 term infants [2]. A substantial proportion have chronic courses of disease (chronic cholestatic liver diseases (CCLD)). Despite recent advances, effective therapies remain elusive, so that many conditions progress to liver failure and may necessitate early liver transplantation. The most common causes of prevalently extrahepatic (surgical) cholestatic jaundice are biliary atresia (BA) (25–40%) and choledochal malformations. As shown in Figure 1, the etiologies of intrahepatic (medical) disorders are more numerous and include a mounting group of genetic disorders related to bile acid transport or synthesis, inborn errors of protein and carbohydrate metabolism, syndromes, mitochondrial and endocrine diseases, hepatic infections, and parenteral nutrition-associated cholestasis (PNAC)) [3–5].

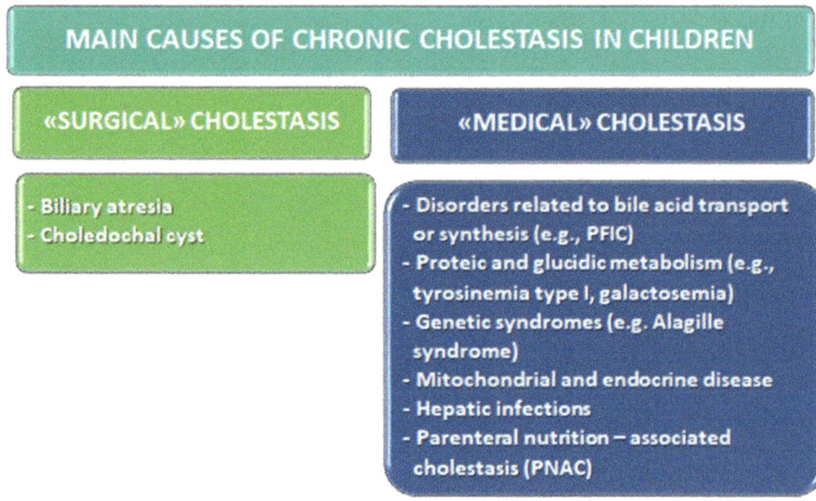

Figure 1. Main etiologies of prevalently extrahepatic ("surgical") or intrahepatic ("medical") pediatric chronic cholestatic liver disease. HFI, hereditary fructose intolerance; PFIC, progressive familial intrahepatic cholestasis.

Malnutrition is a common complication of CCLD, which may increase morbidity and mortality at all ages, particularly in pediatric patients due to their specific developmental aspects. The proportion of malnutrition due to intensely reduced bile acid-dependent absorption of fats and fat-soluble nutrients is, however, difficult to quantify compared to that caused by fibrosis/cirrhosis per se.

Chronic enteropathy secondary to associated portal hypertension, poor nutrient intake, increased energy needs, and endocrine dysfunction are further mechanisms [6,7]. Pediatricians and dietitians should be specifically experienced in pediatric CCLD. Data show that improved nutritional consultations after stakeholders' educational training on nutritional guidelines were associated with lower readmissions of adult patients [8]. Similarly, parental health illiteracy necessitates appropriate interventions targeting social and cultural family circumstances [9].

In this context, we performed a comprehensive review of the literature in the PubMed and Google Scholar databases for studies published during the past 20 years linking nutrition to pediatric chronic cholestatic liver disease. This search was completed by personal knowledge, a snowball strategy by searching for any relevant previous studies within the list of references of analyzed articles and/or citation tracking by a computer aided manual search (Figure 2). Here, we report and put together the multiple pieces of the malnutrition puzzle in children and summarize solutions that are available in everyday clinical practice.

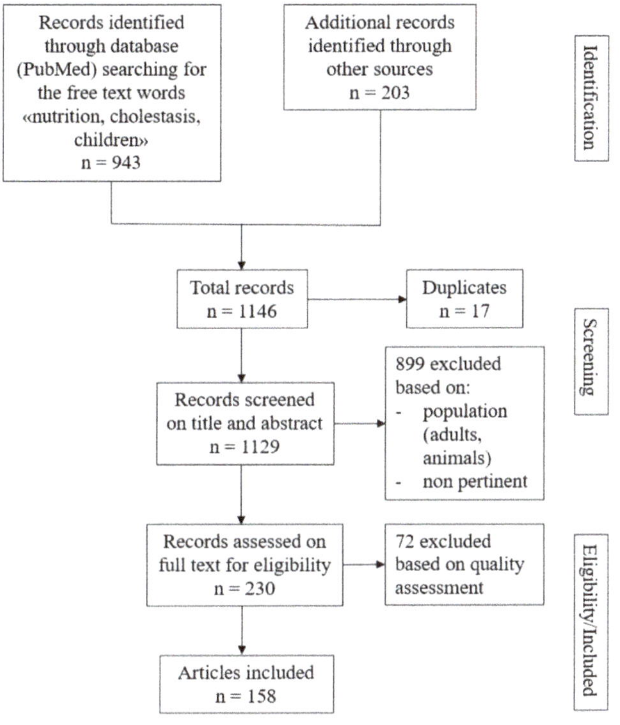

Figure 2. Flow chart showing the selection process to identify studies included in the article.

2. Causes of Malnutrition in Cholestatic Children

As outlined earlier, malnutrition in CCLD depends on several cooperating factors (Figure 3).

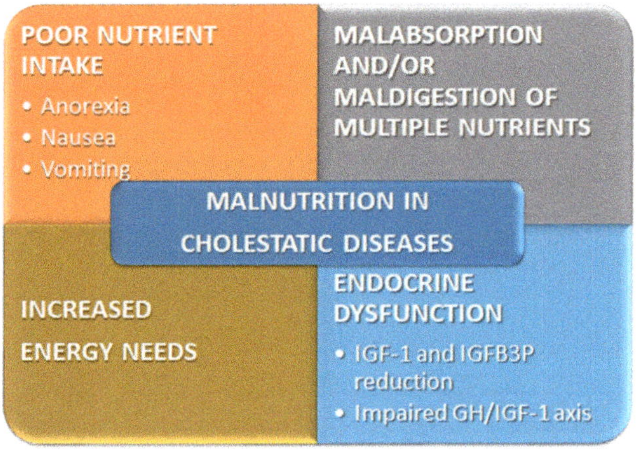

Figure 3. Main factors that determine malnutrition in pediatric chronic cholestatic liver disease. GH, growth hormone; IGFB3P, insulin-like growth factor binding protein 3; IGF-1, insulin-like growth factor-1.

2.1. Metabolic Changes

The liver plays an important role in the regulation of nutrient metabolism. Therefore, when the liver is damaged by any type of injury, nutrient digestion, absorption, storage, and utilization are affected. The decrease in hepatic and muscular glycogen reserves results in the activation of alternative metabolic pathways with an increased release of amino acids (and hyperammonemia), increased fat oxidation, and rapid use of fat, resulting in hypercholesterolemia and low triglyceride levels. The associated cholestatic component results in further specific problems.

2.2. Poor Nutrient Intake

The reduced dietary intake in cholestatic children may result from anorexia, nausea, vomiting, changes in taste perception, and early satiety [10].

Anorexia may presumably be the result of a change in amino acid metabolism with increased tryptophan levels and ensuing increases in brain serotonergic activity, which are reportedly involved in the regulation of eating behavior [11]. Nausea and vomiting are triggered by increased pro-inflammatory cytokines [12], organomegaly, and ascites, resulting in reduced gastric capacity and consequent early satiety [13].

Chronic pruritus, due to accumulation of bile acid (BA) and other pruritogenic substances, may be particularly severe and disturbing especially in PFIC and Alagille syndrome. It may dramatically reduce patients' quality of life and therefore warrants serious and prompt attention with adequate treatment [1,14–18].

Finally, a deficiency in zinc or magnesium contributes to a change in taste perception, which may be further aggravated by the use of poorly palatable special formulas [8] and/or dietary modifications with sodium, fluid, or protein restrictions.

2.3. Increased Requirements or Malabsorption/Maldigestion of Multiple Nutrients

2.3.1. Increased Energy Needs

Children with cholestatic disease have often (but not unanimously) [19] been reported to suffer from a *hypermetabolic state,* with increased energy expenditure, probably due to the intracellular activation of thyroid hormone by bile acids [20,21]. In biliary atresia, a 29% increase in energy requirements, compared to healthy controls, has been calculated [20]. In children with end-stage liver disease, the energy demand may further increase up to 150% of that of predicted normal for a given height and weight, especially in cases of complications such as episodes of sepsis from peritonitis, cholangitis, or variceal bleeding [10,22].

Because the equations commonly used to predict resting energy expenditure (cREE), i.e., the Food and Agriculture Organization/World Health Organization/United Nations University Schofield (weight, and weight and height) equations [23] (Table 1) perform inadequately, especially in the case of end-stage liver disease cholestasis, indirect calorimetry should be used when available to guide energy provisions, particularly in children who are already malnourished [24].

Table 1. Weight-based equation for calculating energy requirements (kcal/day) according to the FAO/WHO/UNO [23].

Age	Males	Females
3–10 years	22.7 (Weight) + 495	22.5 (Weight) + 499
10–18 years	17.5 (Weight) + 651	12.2 (Weight) + 746

2.3.2. Water and Electrolytes

The fluid and electrolytes requirements are normal for maintaining weight, unless restriction is needed because of ascites or edema. In infancy, a sodium intake of at least 1 mmol/kg/day and potassium intake of about 2 mmol/kg/day are usually appropriate [6].

Sodium should not be added to correct hyponatremia due to the systemic vasodilation and arterial underfilling in patients with advanced liver disease who have developed ascites.

2.3.3. Carbohydrates

For CCLD patients, carbohydrates are a major source of energy (about two-thirds of non-protein energy) as they are better accepted than lipids or proteins due to a more pleasant taste [25]. They can be given as monomers, polymers, and starch. Short-chain polymers (maltodextrin) are generally used because of their low osmotic load, which prevents the onset of diarrhea. Starch is a possible alternative, but it can cause abdominal bloating and diarrhea because it has poor digestibility due to the enzymatic immaturity of amylases at an early age.

2.3.4. Proteins

In patients with biliary atresia, increased oxidation of both exogenous and endogenous proteins has been shown, which results in zero nitrogen balance oxidation leading to muscle proteolysis [20,25,26]. Children with advanced CCLD require a higher protein intake for catch-up growth (a protein/energy ratio of 10%). Currently, the specific needs in cholestasis are not known; infants with severe cholestatic liver disease have a daily protein intake need of approximately 2–3 g/kg/day to ensure a positive nitrogen balance and proper growth [27]. In childhood, the requirement becomes progressively lower, except when complications (e.g., cholangitis) increase proteolysis and lead to negative protein balance [28].

In the absence of encephalopathy, hyperammonemia up to 120 mmol is well tolerated, without evident side-effects, and does not require protein restriction [29]. Protein restriction is rarely needed in children with cholestatic liver disease, unless there is a urea cycle disorder (UCD) related-cholestasis. Protein restriction can be considered if encephalopathy occurs as a result of liver failure and of portal hypertension-related portosystemic shunts. In this case, protein intake should be limited to 0.5–1.0 g/kg/day although restriction to <2 g/kg/day should not be continued in the long term, as it can induce endogenous muscle protein consumption. If unresponsive, patients may benefit from branched-chain amino acid (BCAA) supplementation [27].

Children with cholestatic liver disease have low levels of serum BCAAs and elevated ratios of aromatic amino acids to BCAAs, due to an alteration in amino acid kinetics and increased expenditure of BCAAs in the muscle [30]. In other words, low levels of BCAAs reflect increased BCAA utilization in muscle. Therefore, diets rich in BCAAs may have significant advantages. In experimental biliary atresia, a formula enriched with BCAAs improved weight gain, protein mass, muscle mass, nitrogen balance, body composition, and bone mineral density [31]. In addition, a diet enriched in BCAAs in children with end-stage liver disease resulted in an improvement in nutritional status and total body potassium [32]. To improve palatability without reducing energy intake, an ideally formulated product should preferably include whey protein (3 g/kg/day), at 2.6 g/100 mL in reconstituted formula, enriched in BCAAs to 10% [32–35].

2.3.5. Lipids and Bile–Acid-Dependent Absorption of Fats and Fat-Soluble Nutrients

Lipids are the main source of energy supplied through breast milk. During the first years of life, lipids are necessary for growth, development, and providing essential polyunsaturated fatty acids (PUFAs) and fat-soluble vitamins [36]. Although lipids are less palatable than other macronutrients, they are an important supplement because of their high energy, low osmolarity, and content of essential PUFAs [23,37]. In cholestasis, the considerable decrease in long-chain triglyceride absorption due to impaired micellization requires balancing energy losses with extra energy supply. Despite the possible development of steatorrhea, animal studies suggest that overall nutritional status benefits from a high rather than a restricted fat intake [38]. Oral bile salt substitution therapy would be cumbersome and impractical. However, some cholestatic diseases are treated with ursodeoxycholic acid (UDCA) (20–30 mg/kg/day), as it increases bile formation and counteracts the hydrophobic effect of retained bile acids on cell membranes. However, it has no effect on micelle formation and lipid absorption.

2.3.6. Medium Chain Triglycerides and Long-Chain Triglycerides

Although medium-chain triglycerides (MCTs; C-8 to C-12 fatty acids) have a lower energy content (about 16% lower than long-chain triglycerides LCTs), they are used as a lipid supplement (MCT-enriched formula) because their shorter chains allow them to spread passively (bound to albumin) through the gastrointestinal tract and be directly absorbed into the portal circulation. In fact, unlike LCTs, MCTs do not require micellar solubilization and re-esterification [39] because they completely bypass the lymphatic system [40], with approximately 95% bioavailability, even in very cholestatic children [41]. Unless the levels of MCT exceed the metabolic capacity of the liver, they undergo liver metabolism, with energy release. This happens independently of carnitine, which is required for the transport of long-chain fatty acids through the mitochondrial membrane. However, MCTs may reduce appetite, probably due to interaction with the peptides YY and cholecystokinin, with possible interference with the metabolism of adipose tissue [42].

In CCLD children, the ideal fat content and ratio of MCT to LCT are difficult to determine. The optimal proportion of total lipids as MCTs for nutritional management is between 30 and 50% [10,43]. Much higher MCT content in the diet (i.e., >80%) without adequate supplementation of PUFA should be avoided since it can lead to a deficiency in essential fatty acids [44]. Limited data have shown that infants fed with 30 or 70% MCT instead of a 50/50% mixture of MCT/LCT have better fat solubilization and growth [45,46]. For these reasons, most infant formulae with MCTs have an MCT/LCT ratio of about 1/1 and are supplemented with essential fatty acids. Lipid intake and the MCT ratio should be tailored for optimal weight gain and growth. [46].

2.3.7. Essential Fatty Acids

Essential fatty acids (EFAs) are macronutrients that must necessarily be included in the diet because humans cannot synthesize their precursors. Linoleic acid (C18:2n–6) and linolenic acid (C18:3n–3) are the two main EFAs. They undergo hepatic and cerebral elongation into long-chain polyunsaturated fatty acids (LCPs, PUFAs, i.e., arachidonic acid (AA) and docosahexaenoic acid (DHA)). The latter are important for the growth and development of membrane-rich tissues such as the brain and retina [44]. PUFAs are also precursors of eicosanoids, which improve immune function, reduce systemic inflammation [47], and participate in platelet aggregation. These important biological roles make PUFA deficiency in cholestatic liver disease a concern. PUFA and EFA deficiency may derive from low intake, fat maldigestion/malabsorption, and inefficient elongation of EFA precursors secondary to dysfunctional hepatocytes and enhanced peroxidation of lipids [6]. It can also be iatrogenic, when diets are exceedingly high in MCT and low in LCT [48]. Clinically, PUFA and EFA deficiencies present with dry and rough skin, poor growth, numbness, paresthesia, and vision impairment. These clinical signs may go unrecognized or be misdiagnosed as vitamin deficiencies [49]. Testing for PUFA and EFA deficiency, which requires the total fatty acid profile in red blood cells, is costly and not commonly available. It is recommended every 3–6 months especially in the case of severe maldigestion/malabsorption or if the diet comprises exclusive MCT lipids (>80%), or lower in severe cholestasis [6]. Attention is required when linoleic acid, α-linolenic acid, eicosapentanoic acid, and/or docosahexaenoic acid levels are low and either clinical signs or severe deficiencies in fat-soluble vitamins are resistant to supplementation. Importantly, the classic marker of a ratio of triene/tetraene >0.2 is not a sufficient marker when testing for PUFA and EFA deficiency [50]. Unlike in breast milk, formulae contain little LCP [51], and this can lead to possible malnutrition unless egg yolks or fish oil are administered at weaning [22,51].

Estimating appropriate PUFA and LCP intakes for healthy infants and children is not easy and is even more difficult for children with cholestasis. Supplementation should exceed 10% of total energy, but even this may not be sufficient. In fact, when liver damage increases, hepatic PUFA conversion to LCPs is impaired, and LCP deficiency occurs even with an adequate EFA supply. In advanced CCLD, LCP deficiency is difficult to correct; it has been documented that even 1 year after liver transplantation, the LCP status may still not be entirely reversed [52].

That said, there are currently no studies showing the functional effects of LCP supplementation in cholestatic children based on which exact recommendations can be provided. In infants, breast milk can be supplemented with breast milk fortifiers. If breastfeeding is not possible, an MCT-enriched formula should be used. To better meet the energy needs of infants, another option is to increase the caloric density of the formula to 0.8–1 kcal/mL [53]. Older children can improve their EFA intake by adding canola oil, sunflower oil, soybean oil, fish oil, and egg yolks to their diet [22,54].

2.3.8. Fat-Soluble Vitamins

In cholestasis, the reduced secretion of bile acids into the intestinal lumen also induces the malabsorption of fat-soluble vitamins (FSVs) (vitamins A, D, E, and K). This is more frequent when direct-reacting (i.e., conjugated) bilirubin serum levels are greater than 2 mg/dL [55]. In infants with biliary atresia, total serum bilirubin appears to be a better, although still imperfect, predictor of FSV deficiency [56]. The role of serum bile acid as a surrogate marker for guiding the monitoring of FSV deficiency in chronic intrahepatic cholestasis is still undefined.

When cholestasis begins in infancy, vitamin stores present at birth are rapidly depleted; this results in biochemical and clinical signs of fat-soluble vitamin deficiency as early as 4–12 months of age if supplementation has not begun [27]. The serum levels of vitamins and prothrombin time should be monitored to allow the proper adjustment of dosages to the patient's specific needs and the prompt treatment of possible side effects. In particular, vitamin A levels must be closely monitored during supplementation because high levels may cause neurologic and hepatic damage [1].

Vitamin D insufficiency, defined as 25-OH vitamin D less than 30 ng/mL, is present in more than half of cholestatic patients, and this is positively correlated with serum calcium [57]. Decreased bone mineral density (BMD) was present in more than half of the studied cholestatic patients and was correlated with low serum calcium rather than vitamin D levels. Decreased BMD and dental disorders in cholestatic children are related to the level of hyperbilirubinemia [57]. Table 2 synoptically summarizes the effects of individual fat-soluble vitamin deficiencies and toxicities as reported in pediatric practice [25,27,33,57,58].

Table 2. Synopsis of fat-soluble vitamin deficiencies and toxicities.

Vit	Signs and Symptoms of Deficiency	How to Monitor	Supplementation	Toxicity
A	Dry skin Xerophthalmia Night blindness.	Plasma retinol/retinol binding protein molar ratio > 0.8 relative dose response When serum retinol levels < 20 μg/dL, RDR test is indicative of Vit A deficiency when the plasma retinol concentration increases after exogenous administration of Vit A dose.	3000–10,000 IU/day <10 kg: 5000 UI/day, Oral >10 kg: 10,000 UI/day, Oral 50,000 UI/1–3 monthly, IM	Hepatic and neurologic toxicity Development of long bone fractures

Table 2. *Cont.*

Vit	Signs and Symptoms of Deficiency	How to Monitor	Supplementation	Toxicity
D	Hypocalcemia/hypophosphatemia/ tetany Osteomalacia and rickets History of reduced intake, decreased cutaneous synthesis, altered absorption/impaired metabolism in the liver (i.e., where Vit D2 and D3 undergo 25-hydroxylation). Phenobarbital treatment. Breastfeeding	Serum 25-OH-D (Vit D deficiency < 20 ng/mL; insufficiency < 30 ng/mL) Ca, P, AP, PTH, Bone radiography/Bone densitometry used to identify osteomalacia, osteopenia or rickets	Cholecalciferol: 800–5000 IU/day, Oral 1.25-OH cholecalciferol: 0.05–0.2 μg/kg/day, Oral	Hypercalcemia leading to depression of the central nervous system and ectopic calcification. Hypercalciuria leading to nephrocalcinosis
E	Hypo- or a-reflexia Ataxia Impaired vibratory sensation Proximal muscle weakness Ophthalmoplegia Degenerative lesions of the retina Irreversible neurological lesions if Vit E deficiency remains untreated	Vit E/total lipids ratio (increased lipoprotein levels in cholestasis may falsely elevate serum Vit E levels in a patient with Vit E deficiency) Vit E deficiency: <0.6 mg/g (age <1 year) <0.8 mg/g (age >1 year)	Alpha-tocopherol acetate: 15–25 to 25–200 UI/kg/day, Oral TPGS (tocopheryl polyethylene glycol-1000 succinate): 15–25 UI/Kg, Oral	Potentiation of Vit K deficiency coagulopathy Diarrhea Hyperosmolality (TPGS)
K	Hemorrhagic disease (other risks of bleeding: portal hypertension gastrointestinal bleeds, thrombocytopenia, platelet dysfunction, reduced hepatic synthesis of other coagulation factors	Prothrombin time International normalized ratio Protein induced in Vit K absence II (PIVKA II) <3 ng/mL Deficiency can be diagnosed if these values improve after a dose of parenteral vitamin K	2.5–5.0 mg/day from twice a week to every day 5–10 kg: 5 mg, oral >10 kg: 10 mg, oral 5–10 mg/day every two weeks, IM	Hemolytic anemia in glucose 6-phosphate dehydrogenase-deficient infants

AP, alkaline phosphatase; IM, intramuscular; PTH, parathyroid hormone; Vit, vitamin.

2.3.9. Water-Soluble Vitamins and Minerals

Cholestatic children may also present with malabsorption of water-soluble vitamins and minerals. Therefore, it is necessary to supplement vitamins normally present in the diet, using standard pediatric multivitamins. The doses recommended are twice the RDAs [59].

The liver and gut microbiota play important roles in regulating most trace elements; therefore, the impairment of liver function and/or dysbiosis can negatively affect metabolism. In addition, the administration or depletion of these trace elements may also cause liver dysfunction [60] Zinc, selenium, calcium, phosphate, magnesium, and iron may be deficient as well, and they should be supplemented according to the plasma levels. Zinc and selenium are important antioxidants. Reduced zinc concentrations lead to poor linear growth, hypogeusia, anorexia, impaired immune function [42], skin rashes, and diarrhea [61,62]. Identifying zinc deficiency may be difficult because plasma zinc levels do not correlate with tissue zinc content; thus, clinicians should suspect zinc deficiency in patients on the basis of gastrointestinal and dermatological manifestations [63].

Calcium and phosphate deficiency may also develop during cholestasis, leading to bone abnormalities that are unresponsive to the normalization of vitamin D status [27]. Magnesium deficiency may occur as well, contributing to the metabolic bone disease of CCLD. Liver transplantation has favorable effects on osteopenia and vitamin D deficiency. It has been shown that in infants and children <2 years of age, the bone mineral content normalizes approximately 11 months after transplantation, provided there is a sufficient period of normal serum 25-OH-D levels [64].

Iron deficiency is uncommon in chronic liver disease. It may occur, however, in cases of reduced intake and chronic blood loss from gastrointestinal bleeds (esophageal varices or portal hypertensive gastropathy) [22]. As iron can promote oxidative stress, carcinogenesis, and fibrogenesis, it should not be supplemented on a regular basis [65]. On the contrary, chronic cholestasis may be accompanied by the excessive accumulation of copper in the liver [66] and possibly contribute to its further damage.

2.4. Endocrine Dysfunction

The liver is the main source of insulin-like growth factor (IGF-1) and its major circulating binding protein, IGF binding protein 3 (IGF-BP3). IGF-1 stimulates the anabolic actions of growth hormone (GH). In cholestatic liver diseases, IGF-1 and IGF-BP3 formation is reduced, resulting in an impaired GH/IGF-1 axis. IGF-1 levels further decrease due to GH resistance, caused by the downregulation of GH receptors [22]. This condition contributes to aggravating growth failure in these children, who may already have an underlying predisposition to short stature (e.g., Alagille Syndrome) [67].

3. Issues in the Nutritional Management of Children with Cholestasis

According to the above considerations, it is evident that CCLD negatively affects the nutritional status in infancy (i.e., when growth rates are the highest), thus compromising clinical outcomes for cholestatic children who have end-stage liver disease [68], and is present in about 80% of cases [69]. CCLD increases the mortality and morbidity associated with underlying diseases and significantly influences the outcomes of liver transplantation in children [70]. It is therefore necessary to ensure proper nutritional support to prevent further liver damage and improve the likelihood of successful liver transplantation [36]. All children with cholestatic liver disease should have a clinical nutritional evaluation with an intervention and follow-up plan reported in their care records, with a frequency appropriate for the patient's clinical course. The evaluation of the nutritional status of these patients is based on anthropometric, biochemical, and instrumental indicators [71,72], as summarized below and in Figure 4.

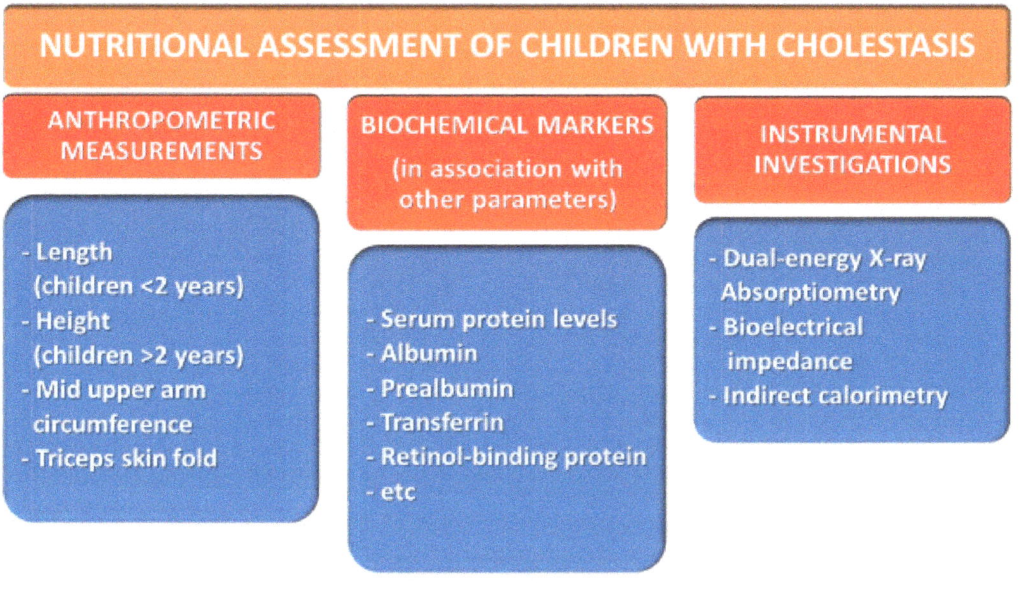

Figure 4. Nutritional assessment of pediatric chronic cholestatic liver diseases.

3.1. Anthropometric Measurements

In children with cholestatic liver disease, the use of weight-for-age and weight-for-height measurements can be inaccurate for nutritional evaluation, owing to visceromegaly, subclinical edema, and/or ascites as a result of excessive tissue sequestration of water, which results from an abnormal intravascular colloid osmotic status, renal retention of salt and water, and hyperaldosteronism [1,73].

Stunting is determined by serial measurements of the height–age index (height-for-age ≤ -2 SD of the WHO child growth standards median). Length (for children < 2 years) or height (for children > 2 years) may be a good indicator of chronic malnutrition, while short-term changes in nutritional status need to be tracked using other parameters such as the mid upper arm circumference (MUAC) and triceps skin fold (TSF). These measures provide information regarding the patient's body composition and are less affected by fluid overload or other complications of end-stage liver disease [10], and they are reported to decline before changes in weight or height become apparent [74].

The MUAC captures both muscle and adipose tissue mass, and it is relatively stable in the first years of life. An absolute value < 12.5–13.0 cm [75] or a Z score < -2 [76] indicate moderate to severe malnutrition. Therefore, by using the MUAC as a screening tool, it is possible to quickly identify children with moderate to severe malnutrition, who are at increased risk for death and need nutrient supplementation and treatment for their underlying disease [36]. The TSF reflects adiposity and is a good indicator of energy reserve depletion during cholestatic liver disease [77]. It has been shown to be more sensitive for acute malnutrition and wasting than weight-for-height Z scores [78].

Globally, children with cholestatic liver disease have decreased MUAC, TSF, and mid-upper arm muscle area Z scores than healthy controls [79]. The frequency of the measurement is variable, depending on the grade of nutritional status, and can range from every 2 weeks to every 3 months [6].

3.2. Biochemical Markers

Serum protein levels (albumin, prealbumin, transferrin, and retinol-binding protein) alone cannot be used as an indicator of malnutrition, as their production is influenced by hepatic disease, sepsis, inflammation, and hydration status [80]. Albumin is not sensitive to acute changes in the nutritional statuses of children with chronic liver disease, as it has a long half-life (18–20 days). In addition, it may be depressed due to inflammation or acute physiologic stress [81]. Prealbumin is a more sensitive marker for the severity of malnutrition and/or adequacy of nutritional support [74]; it has a half-life of 2 days [15], low body reserves, responds quickly to nutritional status [80], and its production in the liver is maintained until late in liver disease [82]. Retinol-binding protein has a half-life of 12 h, which makes it the best indicator of recent dietary changes [80].

3.3. Other Investigations

The evaluation of nutritional status in cholestatic children also includes instrumental investigations that assess body composition: dual-energy X-ray absorptiometry (DXA), bioelectrical impedance (BIA), and indirect calorimetry. DXA and BIA provide a measure of fat and free-fat mass, which are helpful when designing a nutritional rehabilitation approach, although the accuracy of these tools can decrease due to fluid overload [83].

Indirect calorimetry measures oxygen consumption and can be used to estimate resting energy expenditure (REE), which increases in children with chronic liver disease. Although non-invasive, it is a technically difficult procedure and may not be easily performed in uncooperative children [80].

The above predictive equations, however, appear to perform poorly in infants and young children with ESLD. Efforts should be made to guide energy provisions with the aid of indirect calorimetry, especially when malnutrition is already present [24].

Accurately monitoring the nutritional status of these patients is essential in early interventions to correct deficiencies, thus improving growth and reducing both morbidity and mortality.

Oral nutrition should always be encouraged whenever possible because it is physiologic, maintains gastrointestinal tract immunity and gut barrier integrity, and reduces bacterial overgrowth [41]. As discussed in detail above, ready-to-use commercial preparations for infants with chronic liver disease should contain higher calorie amounts, MCT fat contents of about 50%, and branched-chain amino acids. Powder formulations allow one to customize the nutritional contributions in relation to the patient's needs. More concentrated formulae containing maltodextrin may be useful for improving carbohydrate intake [25,53].

A commercially available and nutritionally complete powdered infant milk for the dietary management of acute and chronic liver disease is composed of dried glucose syrup, 49% MCT, soya and canola oil, whey and casein, vitamins, minerals and trace elements. The formula contains a higher-than-average content of branched-chain amino acids (30% of total protein) and lower-than-average sodium content (0.56 mmol/100 mL) than any other infant formula. The low lactose content, avoiding the saturation of the intestinal lactase capacity, is also an advantage of this formula when it is necessary to administer high amounts to increase the caloric input (e.g., by enteral feeding). There are no available published studies on this product, other than one report on its safety and tolerability, and there was no evidence of increased growth parameters when cholestatic children were given the product as their primary energy source [84].

Oral feeding is preferable only if sufficient energy and nutrient supply can be secured. In infants, except in some cases such as those due to galactosemia, breastfeeding should be encouraged for its numerous advantages, provided that supplementation with a breast milk fortifier containing MCT and proteins assures adequate intake of calories and nutrients. Later in life, children with CCLD require a complete nutritional assessment and nutritional therapy. Again, besides some rare inborn errors of metabolism (IEM) that may need special treatment, a standardized, hypercaloric diet with MCT should be used for preventing and/or correcting malnutrition. Small and frequent feeding may be useful to reduce anorexia and early satiety, prevent hypoglycemia, and avoid muscle catabolism.

When oral nutrition does not guarantee an adequate nutritional intake and causes poor growth because of intolerance to a large intake via the oral route, thus necessitating moderate calorie and high protein intake, patients should start nutritional supplementation. In this case, constant attention to palatability, osmolarity, or exceedingly high lipid levels is necessary When rapid growth is needed (e.g., before liver transplantation), enteral feeding with a nasogastric tube may be the best option. Bolus feeding is preferred in the first instance because it is physiological. When bolus feeding is not tolerated, it is appropriate to use continuous infusion administered by a peristaltic pump for up to 20 h/day [22]. One study, comparing an MCT-fortified formula administered orally versus by enteral nutrition (EN), showed that EN prevented malnutrition and growth impairment in infants with BA waiting for a liver transplant [46].

Night-time nasogastric tube feeding is another option; the child is allowed to eat ad libitum during the day and receives extra energy when sleeping, with the chance therefore of maintaining some normal feeding behavior [38]. Importantly, regular non-nutritive sucking and oral stimulation should be implemented to reduce the development of oral aversion. Feeding aversion may be prevented, in part, by promoting daytime oral intake as well as encouraging children to experiment with various age- flavors and textures [41]. Last, but not least, the oral route has trophic action on the intestine and liver, and it protects against infections.

When nasogastric tube feeding is not feasible or inadequate for meeting caloric needs, jejunal tube feeding [85], or even an endoscopic gastrostomy if portal hypertension is mild [86–88] may be proposed if a multidisciplinary team can provide active follow-up and care for the child and the device is available. Gastrostomy feeding, however, should be avoided in children with CCLD because there is a risk of peritoneal infection and variceal bleeding (if portal hypertension is present) [36,89]. Newer approaches, such as low-profile gastrostomy, designed to avoid worsening existing portal hypertension and avoid peristomal variceal complications, remain scarcely documented in this category of pediatric patients [90].

Some children with severe liver disease may require parenteral nutrition (PN). These include children who cannot tolerate oral nutrition due to osmotic diarrhea or vomiting, and those with recurrent variceal bleeding. PN is also indicated when enteral feeding fails to realize growth targets. The use of PN in children with compensated liver disease follows standard principles, and if it is a short-term indication, it is not associated with hepatobiliary dysfunction or worsened cholestasis [91]. On the other hand, in children with decompensated liver disease (usually also awaiting liver transplantation), prolonged PN might lead to worsening liver disease (parenteral nutrition-associated liver disease (PNALD) or cholestasis (PNAC) [92]. The risk is generally higher in infants, especially if premature, and in those with associated extreme short gut [93]. To prevent the onset of PNALD or PNAC, it is advisable to monitor bilirubin and to administer, when possible, simultaneous enteral calories, even in small volumes (so-called "minimal enteral intake"). Enteral feeding, at least intermittently, is protective for liver function and may restore mucosal integrity, minimizing small intestinal bacterial overgrowth and promoting bile flow [92]. In addition, the administration of UDCA has also been shown to have a beneficial effect in PNAC in adults and children, and it may therefore be indicated in these patients [94,95]. The prophylactic use of UDCA for parenteral nutrition cholestasis is of unproven efficacy but acceptable. PN requires strict monitoring, as it may result in fluid and sodium overload and worsening ascites, and central line-associated bloodstream infections have also been reported [96]. PN with a partially or completely fish-oil-based lipid content may be advisable for halting and reversing liver disease [97].

Altogether, PN has pros and cons; it may significantly improve the nutritional status prior to LT, with a beneficial effect on the outcome [96,98], but in some cases, it may aggravate jaundice [92] and introduce a risk of fatal vascular access-related complications in an already fragile subject [99]. The home management of enteral/parenteral nutrition is possible after appropriate family education [100,101].

4. Special Diets in Some Common or Special IEM Causing Cholestasis

4.1. *Tyrosinemia Type 1*

Tyrosinemia type 1 (OMIM # 276700) is a rare autosomal recessive disorder of the tyrosine degradation pathway, resulting in the accumulation of toxic metabolites (succinyl acetone) that are substrates for fumarylacetoacetate hydrolase, i.e., the enzyme that is deficient. The clinical manifestations vary and include neonatal cholestasis with acute liver failure, hepatocellular carcinoma, growth retardation, renal dysfunction, and porphyria-like syndrome with neuropathy [102]. The current specific treatment consists of 2-(2 nitro-4-3 trifluoro-methylbenzoyl)-1,3-cyclohexanedione (NTBC) paired with a tyrosine- and phenylalanine-restricted diet Many commercial products containing tyrosine- and phenylalanine-free amino acid mixtures and supplemented with vitamins and minerals are available [103]. The aim of combined treatment (NTBC and diet) is to provide adequate nutrition that allows normal growth and development, keeping the tyrosine levels in the blood and tissues under control [104].

4.2. *Galactosemia*

Classic galactosemia (OMIM # 23040) is a rare inborn disorder of carbohydrate metabolism caused by a defect in the GALT gene encoding the galactose-1-phosphate uridyltransferase, resulting in an inability to metabolize galactose to glucose. If not treated with a lactose- and galactose-free diet, this condition can cause neonatal cholestasis, liver failure, kidney damage, sepsis, mental retardation, and death. A galactose-restricted diet should be promptly initiated using soy formulas, not only in symptomatic infants but also in those with a highly suspicious newborn screening result [105]. With the introduction of solid foods, nutritional issues become more complicated, as trace amounts of hidden galactose are naturally found in fruits, vegetables, bread, and legumes [106]. The treatment guidelines for galactosemia, published in 2017, recommend eliminating sources of lactose and galactose from dairy products but allowing any amount and type of fruit, legume, vegetable, or mature cheese (with galactose contents <25 mg/100 g) in the diet, as they only have small amounts of galactose [107]. The restriction of dairy products is associated with inadequate calcium intake and with a risk of diminished bone mineral density, unless proper supplementation is initiated [108,109].

4.3. Hereditary Fructose Intolerance

Hereditary fructose intolerance (OMIM# 229600) is a disorder that arises in infancy, at the start of weaning, when fructose and sucrose are introduced into the diet, or even before if an infant is given medications containing sucrose. Generally, the child presents with recurrent vomiting, abdominal pain, (sometimes fatal) hypoglycemia, growth retardation, liver failure, and renal tubulopathy (in the case of long-term fructose exposure). Older patients who survive infancy develop a natural avoidance of sweets and fruits [110].

4.4. Citrin Deficiency

Citrin deficiency, or neonatal-onset type II citrullinemia (OMIM # 605814), is a genetic disorder caused by a homozygous or compound heterozygous mutation in the SLC25A13 gene (603859) causing various metabolic abnormalities. It may present different clinical manifestations during the course of infancy to adulthood, ranging from cholestasis, fatty liver, and growth retardation in infancy, to liver dysfunction and neuropsychiatric symptoms in childhood and adulthood. Although in almost all infants it is a self-limiting condition, with symptoms and biochemical markers improving with age, the occurrence of severe hepatic dysfunction requiring liver transplantation has also been described [111,112]. It is recommended to maintain a low carbohydrate intake and a protein- and fat-rich diet with a protein–fat–carbohydrate ratio of 15–25%:40–50%:30–40% [113]. In fact, due to the impairment of NADH shuttling and glucose metabolism, patients have an aversion to carbohydrates and a peculiar preference for high-protein and high-fat foods, in contrast to patients with other urea cycle defects [114]. Excessive carbohydrate intake causes different symptoms such as fatigue, anorexia, weight loss, psychiatric symptoms, and liver failure, which is the most significant risk factor for adult-onset type II citrullinemia [113,115]. In patients with hypergalactosemia, the use of lactose-free milk containing MCT is associated with improved energy metabolism and liver function [116]. The administration of sodium pyruvate (0.1–0.3 g/kg/day) [83] improves clinical symptoms and helps in weight recovery [117,118].

5. Pre- and Post-Transplant Nutritional Status of Children with End-Stage Cholestatic Liver Disease

Although some causes of neonatal cholestasis have no specific treatment, affected children may benefit from appropriate nutritional support to prevent malnutrition and to correct macro/micronutrient deficiencies. This is paramount because a better pre-transplant nutritional status is associated with better post-transplant outcomes, and lower mortality and morbidity [18,68,119–121]. The nutritional needs of children with liver disease are outlined schematically in Figure 5.

A comparison of nutritional needs of children with chronic cholestatic liver diseases before and after liver transplantation [6,41,54,120] is shown in Table 3.

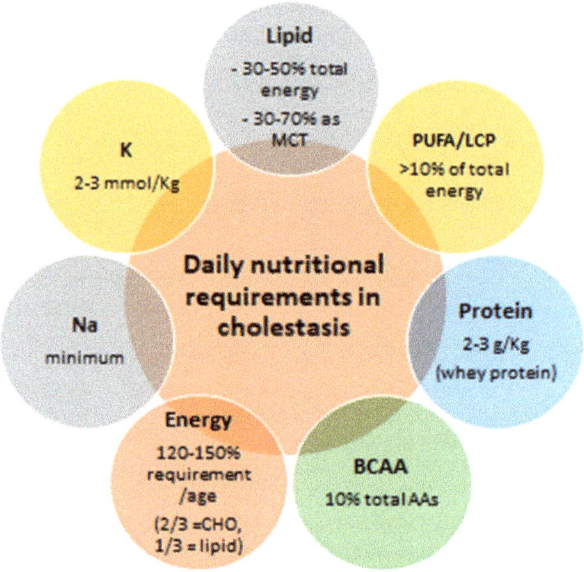

Figure 5. Nutritional needs of children with liver disease. BCAA, branched-chain amino acids; CHO, carbohydrates; LCP, long chain polyunsaturated fatty acids; MCT, medium-chain triglycerides; PUFA, polyunsaturated fatty acids.

Table 3. Nutritional needs of children with chronic cholestatic liver diseases before and after liver transplantation.

Before Liver Transplant	After Liver Transplant
Energy intake 130–150% EAR	**Energy intake** 120% EAR
Carbohydrates (40% to 60% of total energy) 15–20 g/kg/day as monomers, polymers, and starch Balancing the hypoglycemia from end-stage liver disease and hyperglycemia from insulin resistance	**Carbohydrates** 6–8 g/kg/day as monomers, polymers, and starch Warning: Consider the diabetogenic potential of tacrolimus when it is used for immunosuppression
Proteins (9% of total energy) 3–4 g/kg/day BCAA-enriched formula can be used (10% of total amino acid) Low protein-diet is needed only when severe encephalopathy is present. Once encephalopathy is resolved, the patient should resume a diet with appropriate protein supply because long-term restriction <2 g/kg/day can induce endogenous muscle protein consumption	**Proteins** 2.5–3 g/kg/day
Fats (40% of total energy; 10% of which as LCPUFA) 8 g/kg/day with 30–50% as MCTs. Warning: MCT contents >80% without adequate supplementation of PUFA can lead to a deficiency of essential fatty acids	**Fats** 5–6 g/kg/day After liver transplantation, when bile flow is established and malabsorption is resolved, children fed with high MCT-containing supplementation pre-transplant can transition to standard formula
Fluids and electrolytes Fluid requirement is normal for actual weight, unless restriction is needed because of ascites or edema. Sodium intake is 1 mmol/kg/day and potassium about 2 mmol/kg/day	**Fluids and electrolytes** A "no added salt" diet (3 g sodium/day) is recommended to prevent water retention associated with steroid therapy

BCAA, branched-chain amino acids; EAR, estimated average requirement; LCPUFA, long-chain polyunsaturated fatty acids; MCT, medium-chain triglycerides; PUFA, polyunsaturated fatty acids. Bold characters indicate the different nutrients' groups. Modified from Yang et al. (2017) [41].

Useful literature on the nutritional management of children undergoing liver transplantation is limited. There are no pediatric studies showing which approach to feeding is superior (e.g., nasogastric, oral, or nasojejunal) and what impact it has on the post-liver transplantation outcomes. However, if the child eats orally up until the transplant, enteral feeding should be encouraged with a healthy, age-appropriate diet and also trying foods that they disliked before the transplant, as taste changes are common. Post-transplant feeding should start as soon as possible, ideally during the first 72 h if the child is stable [54]. After liver transplantation, weight gain seems to recover completely, despite previous malnutrition. Even mid-arm muscle mass and mid-arm fat start to rapidly improve within 3–6 months [41,120]. Although marked catch-up growth is observed in those who are more stunted before transplantation, transplanted children generally achieve final heights below their genetic targets [120,122]. Rejection, re-transplantation, and metabolic disease are independent risk factors associated with shorter stature. The use of steroids post-transplant for autoimmune (e.g., de novo autoimmune hepatitis) or allergic complications also contributes [123,124]. However, underlying genetic conditions (e.g., Alagille syndrome) may contribute to permanent low stature.

6. Post-Transplant Obesity with Fatty Liver and MetS Risk: The Malnutrition in Excess paradox

Most children catch up to their peers in weight by one-year post-transplant [125–127], but it may take up to five years to catch up in height [93]. This imbalance may lead to overweight and obesity within two years post-transplant. Careful nutritional counselling and close follow-up for the identification of children at risk for persistent overweight/obesity may help with targeted interventions to prevent not only obesity [128] but also long-term obesity-related comorbidities, such as metabolic syndrome [129,130] and non-alcoholic fatty liver disease (NAFLD), which has now been more correctly renamed metabolic-associated fatty liver disease (MAFLD) [131].

Post-transplant metabolic syndrome (PTMS) is increasingly recognized as a significant contributor to long-term morbidity and even mortality after solid-organ transplantation [132,133]. However, data on the prevalence of PTMS in children are scarce. Hypertension and glucose intolerance seem to be most common in the early post-transplant period. However, even five to ten years post-transplant, these comorbidities are much more common in children after liver transplantation than in the general population [134]. Liver transplant recipients are predisposed to these comorbidities predominantly due to the side effects of the immunosuppressive therapy used to prevent rejection. The side effects of glucocorticoids (GCs) are widely documented. GCs cause glucose intolerance by reducing peripheral insulin sensitivity (particularly in the muscles, the liver, and adipose tissue) and decreasing glucose utilization, and inducing dyslipidemia, hypertension, and hepatic changes with steatosis [135–137]. The high prevalence of impaired glucose tolerance (IGT) in liver transplant recipients is also due to the prolonged use of calcineurin inhibitors (CNIs). Tacrolimus is more diabetogenic than cyclosporine [138], as it impairs beta cell function [139] and/or stimulates insulin resistance [140]. CNI exposure is also associated with systolic hypertension owing to its direct nephrotoxicity, enhancement of sodium reabsorption, and induction of systemic vasoconstriction [141,142], and with increased circulating LDL cholesterol [143]. PTMS, influenced by weight gain and drug-induced hyperlipidemia and diabetes, is a crucial factor associated with the development of de novo NAFLD in liver transplant recipients [144]. Guidelines from the American Association for the Study of Liver Disease and the American Society of Transplantation recommend annual screening for obesity, hypertension, dyslipidemia, and diabetes mellitus with physical examination and fasting blood tests [145,146]. For this reason, it is essential to identify and test interventions to improve the screening and management of PTMS in order to improve long-term outcomes.

Sarcopenia

Sarcopenia is a reduction in muscle mass/muscle wasting. Few studies have evaluated its impact in this special pediatric population. Reduced muscle mass may be studied using several methods requiring cooperation to stay motionless during measurements, which is challenging for infants and young children. The tests include dual-energy X-ray absorptiometry (DEXA), CT, MRI, and bioelectric impedance analysis (BIA). The risks from radiation exposure restrict the use of CT, if not included in clinical staging (e.g., before OLT), while radiation-free MRI is expensive and may require narcosis in pediatric-age subjects. DEXA is a useful substitute, but whether all-body skeletal muscle mass and appendicular lean mass performs equally well, as markers of sarcopenia are controversial. The poor availability of age- and gender-matched normative data is a relevant obstacle.

CCLD pediatric patients may develop liver insufficiency and need early liver transplantation. As in adults, both malnutrition and sarcopenia compromise pediatric post-transplantation outcomes [147,148]. In liver-transplanted children, sarcopenia is associated with relevant clinical outcomes (growth retardation, length of hospitalization, and rate of readmission) [149]. The use of an internal comparison for the diagnosis of sarcopenia may cause different series to have different cutoffs, which hampers the generalization of results [150,151].

Myopenia, caused by decreased protein synthesis and increased protein degradation, is another clinical manifestation of end-stage liver disease (ESLD). It is associated with adverse clinical outcomes, confirming the importance of rehabilitation strategies [152].

Although several valid malnutrition screening tools with which to identify nutritional risk are available, an increased risk of protein malnutrition may be masked by edema, and further work regarding tools with which to distinguish between protein–energy malnutrition (PEM), sarcopenia, and cachexia are required for the pediatric age group. Functional assessment of muscle sarcopenia in pediatric CCLD is not generally performed because appropriate muscle function tests have not been developed for early childhood, and a standardized assessment of muscle function for the diagnosis of sarcopenia in young pediatric patients is currently lacking [150,151].

CT-derived body metrics (e.g., the skeletal muscle index (SMI), psoas muscle index (PMI), psoas muscle surface area (PMSA), and subcutaneous fat index (ScFI)) are measurable components of sarcopenia, frailty, and nutrition. The relationship between preoperative metrics and post-LT outcomes in pediatric recipients < 1 year-old with cirrhotic liver disease is unknown. Fragmentary data suggest that the preoperative values are significantly reduced compared with controls, and the values are correlated with moderate to severe postoperative infections and with longer hospital and ICU stays [153]. A higher re-operation rate and longer hospital stays following transplantation, but not waitlist mortality, were associated with lower PMI [154] or psoas muscle surface area (PMSA) [155]. Sarcopenia, evaluated by total psoas muscle area (tPMA) measurements from computed tomography (CT) imaging, was prevalent in patients with pediatric ESLD awaiting LT, highlighting the need for nutritional support before LT and/or after LT in the PICU [156]. Only one study [150] examined both muscle mass and muscle function in children with CLD, as recommended by the recent EWGSO consensus [157].

7. Conclusions

The causes of malnutrition in CCLD are varied. Although some causes of cholestasis have no specific treatment, all children with CCLD should have a periodic evaluation of clinical, anthropometric, biochemical, and instrumental nutritional indicators. In particular, it is important to identify specific nutritional needs and any deficiencies that require early preventive or corrective treatment. Supplemental nutrition, including MCT, essential fatty acids, branched-chain amino acids, and vitamins, is crucial for overcoming anorexia and preventing growth retardation.

Malnutrition is associated with increased risks of morbidity and mortality, and it also affects the outcomes of liver transplantation and long-term survival, so that failure of nutritional care is an indication for reviewing liver transplant timing. To improve life expectancy and quality of life, patients with CCLD need careful assessment of their nutritional statuses by a multidisciplinary team capable of initiating appropriate interventions. The complex dietary management for this type of malnourished patient requires new studies to improve commercially available and nutritionally complete infant milk formulae.

Author Contributions: Conceptualization C.M. and P.V.; literature search and data curation, M.T., E.S., A.C. and G.S.D.C.; writing—original draft preparation M.T., E.S., C.M., G.S.D.C. and A.C.; writing—review and editing C.M. and P.V. All authors were involved in the writing of the draft. All authors have read and agreed to the published version of the manuscript.

Funding: This research received no external funding.

Institutional Review Board Statement: Not applicable.

Informed Consent Statement: Not applicable.

Conflicts of Interest: The authors declare no conflict of interest.

References

1. Catzola, A.; Vajro, P. Management options for cholestatic liver disease in children. *Expert Rev. Gastroenterol. Hepatol.* **2017**, *11*, 1019–1030. [CrossRef] [PubMed]
2. Venigalla, S.; Gourley, G.R. Neonatal cholestasis. *Semin Perinatol.* **2004**, *28*, 348–355. [CrossRef] [PubMed]
3. Fawaz, R.; Baumann, U.; Ekong, U.; Fischler, B.; Hadzic, N.; Mack, C.L.; McLin, V.A.; Molleston, J.P.; Neimark, E.; Ng, V.L.; et al. Guideline for the Evaluation of Cholestatic Jaundice in Infants: Joint Recommendations of the North American Society for Pediatric Gastroenterology, Hepatology, and Nutrition and the European Society for Pediatric Gastroenterology, Hepatology, and Nutrition. *J. Pediatr. Gastroenterol. Nutr.* **2017**, *64*, 154–168. [CrossRef] [PubMed]
4. Mandato, C.; Siano, M.A.; Nazzaro, L.; Gelzo, M.; Francalanci, P.; Rizzo, F.; D'Agostino, Y.; Morleo, M.; Brillante, S.; Weisz, A.; et al. A ZFYVE19 gene mutation associated with neonatal cholestasis and cilia dysfunction: Case report with a novel pathogenic variant. *Orphanet J. Rare Dis.* **2021**, *16*, 1–9. [CrossRef]
5. Mandato, C.; Zollo, G.; Vajro, P. Cholestatic jaundice in infancy: Struggling with many old and new phenotypes. *Ital. J. Pediatr.* **2019**, *45*, 1–5. [CrossRef]
6. Mouzaki, M.; Bronsky, J.; Gupte, G.; Hojsak, I.; Jahnel, J.; Pai, N.; Quiros-Tejeira, R.E.; Wieman, R.; Sundaram, S. Nutrition Support of Children with Chronic Liver Diseases: A Joint Position Paper of the North American Society for Pediatric Gastroenterology, Hepatology, and Nutrition and the European Society for Pediatric Gastroenterology, Hepatology, and Nutrition. *J. Pediatr. Gastroenterol. Nutr.* **2019**, *69*, 498–511. [CrossRef]
7. Mandato, C.; Di Nuzzi, A.; Vajro, P. Nutrition and Liver Disease. *Nutrients* **2017**, *10*, 9. [CrossRef]
8. Reuter, B.; Shaw, J.; Hanson, J.; Tate, V.; Acharya, C.; Bajaj, J.S. Nutritional Assessment in Inpatients with Cirrhosis Can Be Improved After Training and Is Associated with Lower Readmissions. *Liver Transplant.* **2019**, *25*, 1790–1799. [CrossRef] [PubMed]
9. Chen, X.; Wang, J.; Lu, Y.; Xie, X.; Gu, Y.; Latour, J.M.; Zhang, Y. Feeding practices in 6–24-month-old children with chronic cholestatic liver diseases: A mixed-method study. *BMC Pediatr.* **2020**, *20*, 395. [CrossRef]
10. Nightingale, S.; Ng, V.L. Optimizing Nutritional Management in Children with Chronic Liver Disease. *Pediatr. Clin. N. Am.* **2009**, *56*, 1161–1183. [CrossRef]
11. Laviano, A.; Cangiano, C.; Preziosa, I. Plasma tryptophan levels and anorexia in liver cirrhosis. *Int. J. Eat. Disord.* **1997**, *21*, 181–186. [CrossRef]
12. Aranda-Michel, J. Nutrition in hepatic failure and liver transplantation. *Curr. Gastroenterol. Rep.* **2001**, *3*, 362–370. [CrossRef] [PubMed]
13. Aqel, B.A.; Scolapio, J.; Dickson, R.C.; Burton, D.D.; Bouras, E.P. Contribution of Ascites to Impaired Gastric Function and Nutritional Intake in Patients with Cirrhosis and Ascites. *Clin. Gastroenterol. Hepatol.* **2005**, *3*, 1095–1100. [CrossRef]
14. Portincasa, P.; Di Ciaula, A.; Garruti, G.; Vacca, M.; De Angelis, M.; Wang, D.Q.-H. Bile Acids and GPBAR-1: Dynamic Interaction Involving Genes, Environment and Gut Microbiome. *Nutrients* **2020**, *12*, 3709. [CrossRef] [PubMed]
15. Kriegermeier, A.; Green, R. Pediatric Cholestatic Liver Disease: Review of Bile Acid Metabolism and Discussion of Current and Emerging Therapies. *Front. Med.* **2020**, *7*. [CrossRef]
16. Shneider, B.L. Intestinal Bile Acid Transport: Biology, Physiology, and Pathophysiology. *J. Pediatr. Gastroenterol. Nutr.* **2001**, *32*, 407–417. [CrossRef]
17. Lemoine, C.; Superina, R. Surgical diversion of enterohepatic circulation in pediatric cholestasis. *Semin. Pediatr. Surg.* **2020**, *29*, 150946. [CrossRef]
18. Utterson, E.C.; Shepherd, R.W.; Sokol, R.J.; Bucuvalas, J.; Magee, J.C.; McDiarmid, S.V.; Anand, R.; the SPLIT Research Group. Biliary Atresia: Clinical Profiles, Risk Factors, and Outcomes of 755 Patients Listed for Liver Transplantation. *J. Pediatr.* **2005**, *147*, 180–185. [CrossRef]

19. Kyrana, E.; Williams, J.E.; Wells, J.C.; Dhawan, A. Resting Energy Expenditure of Children with End-stage Chronic Liver Disease Before and After Liver Transplantation. *J. Pediatr. Gastroenterol. Nutr.* **2019**, *69*, 102–107. [CrossRef] [PubMed]
20. Watanabe, M.; Houten, S.; Mataki, C.; Christoffolete, M.; Kim, B.W.; Sato, H.; Messaddeq, N.; Harney, J.W.; Ezaki, O.; Kodama, T.; et al. Bile acids induce energy expenditure by promoting intracellular thyroid hormone activation. *Nat. Cell Biol.* **2006**, *439*, 484–489. [CrossRef]
21. DeRusso, P.A.; Ye, W.; Shepherd, R.; Haber, B.A.; Shneider, B.L.; Whitington, P.F.; Schwarz, K.B.; Bezerra, J.A.; Rosenthal, P.; Karpen, S.; et al. Growth failure and outcomes in infants with biliary atresia: A report from the Biliary Atresia Research Consortium. *Hepatology* **2007**, *46*, 1632–1638. [CrossRef]
22. Kelly, D.A.; Proteroe, S.; Clarke, S. Acute and chronic liver disease. In *Nutrition in Pediatris*, 5th ed.; Duggan, C., Watkins, J.B., Koletzko, B., Walker, W.A., Eds.; People's Medical Publishing House-USA: Shelton, CT, USA, 2016; pp. 851–863. [CrossRef]
23. Schofield, W.N. Predicting basal metabolic rate, new standards and review of previous work. Hum. *Nutr. Clin. Nutr.* **1985**, *39* (Suppl. 1), 5–41.
24. Carpenter, A.; Ng, V.L.; Chapman, K.; Ling, S.C.; Mouzaki, M. Predictive equations are inaccurate in the estimation of the resting energy expenditure of children with end-stage liver disease. *J. Parenter. Enter Nutr.* **2017**, *41*, 507–511. [CrossRef]
25. Socha, P. Nutritional Management of Cholestatic Syndromes in Childhood. *Ann. Nestlé* **2008**, *66*, 137–147. [CrossRef]
26. Tavill, A.S. The synthesis and degradation of liver-produced proteins. *Gut* **1972**, *13*, 225–241. [CrossRef] [PubMed]
27. Feranchak, A.P.; Suchy, F.J.; Sokol, R.J. Medical management of cholestasis in infants and children. In *Liver Disease in Children*, 4th ed.; Suchy, F.J., Sokol, R.J., Balistreri, W., Eds.; Published by Cambridge University Press: Cambridge, UK, 2014; pp. 111–139. [CrossRef]
28. Tomkins, A.M.; Garlick, P.J.; Schofield, W.N.; Waterlow, J.C. The Combined Effects of Infection and Malnutrition on Protein Metabolism in Children. *Clin. Sci.* **1983**, *65*, 313–324. [CrossRef]
29. Charlton, C.P.; Buchanan, E.; Holden, C.E.; Preece, M.A.; Green, A.; Booth, I.W.; Tarlow, M.J. Intensive enteral feeding in advanced cirrhosis: Reversal of malnutrition without precipitation of hepatic encephalopathy. *Arch. Dis. Child.* **1992**, *67*, 603–607. [CrossRef]
30. Weisdorf, S.A.; Freese, D.K.; Fath, J.J.; Tsai, M.Y.; Cerra, F.B. Amino Acid Abnormalities in Infants with Extrahepatic Biliary Atresia and Cirrhosis. *J. Pediatr. Gastroenterol. Nutr.* **1987**, *6*, 860–864. [CrossRef]
31. Sokal, E.M.; Baudoux, M.C.; Collette, E.; Hausleithner, V.; Lambotte, L.; Buts, J.P. Branched Chain Amino Acids Improve Body Composition and Nitrogen Balance in a Rat Model of Extra Hepatic Biliary Atresia. *Pediatr. Res.* **1996**, *40*, 66–71. [CrossRef]
32. Chin, S.E.; Shepherd, R.; Thomas, B.J.; Cleghorn, G.; Patrick, M.K.; Wilcox, J.A.; Ong, T.H.; Lynch, S.V.; Strong, R. Nutritional support in children with end-stage liver disease: A randomized crossover trial of a branched-chain amino acid supplement. *Am. J. Clin. Nutr.* **1992**, *56*, 158–163. [CrossRef] [PubMed]
33. Baker, A.; Stevenson, R.; Dhawan, A.; Goncalves, I.; Socha, P.; Sokal, E. Guidelines for nutritional care for infants with cholestatic liver disease before liver transplantation. *Pediatr. Transplant.* **2007**, *11*, 825–834. [CrossRef] [PubMed]
34. Mager, D.R.; Wykes, L.J.; Roberts, E.A.; Ball, R.O.; Pencharz, P.B. Branched-Chain Amino Acid Needs in Children with Mild-to-Moderate Chronic Cholestatic Liver Disease. *J. Nutr.* **2006**, *136*, 133–139. [CrossRef]
35. Plauth, M.; Bernal, W.; Dasarathy, S.; Merli, M.; Plank, L.D.; Schütz, T.; Bischoff, S.C. ESPEN guideline on clinical nutrition in liver disease. *Clin. Nutr.* **2019**, *38*, 485–521. [CrossRef]
36. Young, S.; Kwarta, E.; Azzam, R.; Sentongo, T. Nutrition Assessment and Support in Children with End-Stage Liver Disease. *Nutr. Clin. Pract.* **2013**, *28*, 317–329. [CrossRef] [PubMed]
37. Koletzko, B.; Demmelmair, H.; Socha, P. 3 Nutritional support of infants and children: Supply and metabolism of lipids. *Baillieres Clin Gastroenterol.* **1998**, *12*, 671–696. [CrossRef]
38. Kalivianakis, M.; Minich, D.M.; Havinga, R.; Kuipers, F.; Stellaard, F.; Vonk, R.J.; Verkade, H.J. Detection of impaired intestinal absorption of long-chain fatty acids: Validation studies of a novel test in a rat model of fat malabsorption. *Am. J. Clin. Nutr.* **2000**, *72*, 174–180. [CrossRef] [PubMed]
39. Carey, M.C.; Small, D.M.; Bliss, C.M. Lipid Digestion and Absorption. *Annu. Rev. Physiol.* **1983**, *45*, 651–677. [CrossRef]
40. Shah, N.; Limketkai, B. The use of medium chain triglycerides in gastrointestinal disorders. *Pract. Gastroenterol.* **2017**, *XLI*, 20–28. Available online: https://med.virginia.edu/ginutrition/wp-content/uploads/sites/199/2014/06/Parrish-February-17.pdf (accessed on 12 August 2021).
41. Beath, S.V.; Booth, I.W.; Kelly, D.A. Nutritional support in liver disease. *Arch. Dis. Child.* **1993**, *69*, 545–547. [CrossRef]
42. Medium-Chain Triglycerides. Available online: https://www.drugbank.ca/drugs/DB13959 (accessed on 12 August 2021).
43. Kelly, D.A.; Davenport, M. Current management of biliary atresia. *Arch. Dis. Child.* **2007**, *92*, 1132–1135. [CrossRef]
44. Koletzko, B.; Agostoni, C.; Carlson, S.; Clandinin, T.; Hornstra, G.; Neuringer, M.; Uauy, R.; Yamashiro, Y.; Willatts, P. Long chain polyunsaturated fatty acids (LC-PUFA) and perinatal development. *Acta Paediatr.* **2007**, *90*, 460–464. [CrossRef]
45. Kaufman, S.S.; Scrivner, D.J.; Murray, N.D.; Vanderhoof, J.A.; Hart, M.H.; Antonson, D.L. Influence of portagen and pregestimil on essential fatty acid status in infantile liver disease. *Pediatrics* **1992**, *89*, 151–154.
46. Macías-Rosales, R.; Larrosa-Haro, A.; Ortíz-Gabriel, G.; Trujillo-Hernandez, B. Effectiveness of Enteral Versus Oral Nutrition with a Medium-Chain Triglyceride Formula to Prevent Malnutrition and Growth Impairment in Infants with Biliary Atresia. *J. Pediatr. Gastroenterol. Nutr.* **2016**, *62*, 101–109. [CrossRef]

47. Kumar, N.G.; Contaifer, D.; Madurantakam, P.; Carbone, S.; Price, E.T.; Van Tassell, B.; Brophy, D.F.; Wijesinghe, D.S. Dietary Bioactive Fatty Acids as Modulators of Immune Function: Implications on Human Health. *Nutrients* **2019**, *11*, 2974. [CrossRef] [PubMed]
48. Pettei, M.J.; Daftary, S.; Levine, J.J. Essential fatty acid deficiency associated with the use of a medium-chain-triglyceride infant formula in pediatric hepatobiliary disease. *Am. J. Clin. Nutr.* **1991**, *53*, 1217–1221. [CrossRef] [PubMed]
49. Holman, R.T. The Slow Discovery of the Importance of ω3 Essential Fatty Acids in Human Health. *J. Nutr.* **1998**, *128*, 427S–433S. [CrossRef]
50. Pavlovski, C.J. Screening for essential fatty acid deficiency in at risk infants. *Med. Hypotheses* **2009**, *73*, 910–916. [CrossRef] [PubMed]
51. Koletzko, B. Fats for brains. *Eur. J. Clin. Nutr.* **1992**, *46*, S51–S62.
52. Yang, C.H.; Perumpail, B.J.; Yoo, E.R.; Ahmed, A.; Kerner, J.A., Jr. Nutritional Needs and Support for Children with Chronic Liver Disease. *Nutrients* **2017**, *9*, 1127. [CrossRef]
53. Sundaram, S.S.; Mack, C.L.; Feldman, A.G.; Sokol, R.J. Biliary atresia: Indications and timing of liver transplantation and optimization of pretransplant care. *Liver Transplant.* **2016**, *23*, 96–109. [CrossRef]
54. Smart, K.M.; Alex, G.; Hardikar, W. Feeding the child with liver disease: A review and practical clinical guide. *J. Gastroenterol. Hepatol.* **2011**, *26*, 810–815. [CrossRef]
55. Shneider, B.L.; Magee, J.C.; Karpen, S.J.; Rand, E.B.; Narkewicz, M.R.; Bass, L.M.; Schwarz, K.; Whitington, P.F.; Bezerra, J.A.; Kerkar, N.; et al. Total Serum Bilirubin within 3 Months of Hepatoportoenterostomy Predicts Short-Term Outcomes in Biliary Atresia. *J. Pediatr.* **2015**, *170*, 211–217. [CrossRef]
56. Venkat, V.L.; Shneider, B.L.; Magee, J.C.; Turmelle, Y.; Arnon, R.; Bezerra, J.A.; Hertel, P.M.; Karpen, S.J.; Kerkar, N.; Loomes, K.M.; et al. Total Serum Bilirubin Predicts Fat-Soluble Vitamin Deficiency Better Than Serum Bile Acids in Infants with Biliary Atresia. *J. Pediatr. Gastroenterol. Nutr.* **2014**, *59*, 702–707. [CrossRef] [PubMed]
57. Samra, N.M.; El Abrak, S.E.; El Dash, H.H.; Raziky, M.E.S.E.; El Sheikh, M.A. Evaluation of vitamin D status bone mineral density and dental health in children with cholestasis. *Clin. Res. Hepatol. Gastroenterol.* **2018**, *42*, 368–377. [CrossRef] [PubMed]
58. Sokol, R.J. Frequency and Clinical Progression of the Vitamin E Deficiency Neurologic Disorder in Children with Prolonged Neonatal Cholestasis. *Arch. Pediatr. Adolesc. Med.* **1985**, *139*, 1211–1215. [CrossRef] [PubMed]
59. Ramaccioni, V.; Soriano, H.E.; Arumugam, R.; Klish, W.J. Nutritional Aspects of Chronic Liver Disease and Liver Transplantation in Children. *J. Pediatr. Gastroenterol. Nutr.* **2000**, *30*, 361–367. [CrossRef] [PubMed]
60. Himoto, T.; Masaki, T. Current Trends of Essential Trace Elements in Patients with Chronic Liver Diseases. *Nutrients* **2020**, *12*, 2084. [CrossRef]
61. Karakochuk, C.D.; Barr, S.I.; Boy, E.; Bahizire, E.; Tugirimana, P.L.; Akilimali, P.Z.; Houghton, L.A.; Green, T.J. The effect of inflammation on serum zinc concentrations and the prevalence estimates of population-level zinc status among Congolese children aged 6–59 months. *Eur. J. Clin. Nutr.* **2017**, *71*, 1467–1470. [CrossRef]
62. Gibson, R.S.; Hess, S.Y.; Hotz, C.; Brown, K.H. Indicators of zinc status at the population level: A review of the evidence. *Br. J. Nutr.* **2008**, *99* (Suppl. 3), S14–S23. [CrossRef]
63. Göksu, N.; Özsoylu, S. Hepatic and Serum Levels of Zinc, Copper, and Magnesium in Childhood Cirrhosis. *J. Pediatr. Gastroenterol. Nutr.* **1986**, *5*, 459–462. [CrossRef]
64. Argao, E.A.; Balistreri, W.F.; Hollis, B.W.; Ryckman, F.C.; Heubi, J.E. Effect of Orthotopic Liver Transplantation on Bone Mineral Content and Serum Vitamin D Metabolites in Infants and Children with Chronic Cholestasis. *Hepatology* **1994**, *20*, 598–603. [CrossRef]
65. Bonkovsky, H.L.; Banner, B.F.; Lambrecht, R.W.; Rubin, R.B. Iron in Liver Diseases Other than Hemochromatosis. *Semin. Liver Dis.* **1996**, *16*, 65–82. [CrossRef]
66. Burt, A.; Ferrell, L.; Hubscher, S.; Portmann, B. *Mac Sween's Pathology of the Liver*, 6th ed.; Churchill Livingstone: Edinburgh, UK, 2012.
67. Kohut, T.J.; Gilbert, M.A.; Loomes, K.M. Alagille Syndrome: A Focused Review on Clinical Features, Genetics, and Treatment. *Semin. Liver Dis.* **2021**. [CrossRef]
68. Los, E.L.; Lukovac, S.; Werner, A.; Dijkstra, T.; Verkade, H.J.; Rings, E.H. Nutrition for Children with Cholestatic Liver Disease. *Nestle Nutr. Workshop Ser. Pediatr. Program.* **2007**, *59*, 147–159. [CrossRef] [PubMed]
69. Nastasio, S.; Maggiore, G. Malattie croniche epatobiliari. In *Manuale Sigenp di Nutrizione Pediatrica*; Catassi, C., Diamanti, A., Agostoni, C., Eds.; Pensiero Scientifico Editore: Roma, Italy, 2016; pp. 251–258. ISBN 9788849005592.
70. Barshes, N.R.; Chang, I.-F.; Karpen, S.J.; Carter, B.A.; Goss, J.A. Impact of Pretransplant Growth Retardation in Pediatric Liver Transplantation. *J. Pediatr. Gastroenterol. Nutr.* **2006**, *43*, 89–94. [CrossRef]
71. Da Silva, F.V.; Ferri, P.M.; Queiroz, T.C.N.; Barbosa, P.D.S.H.; De Oliveira, M.C.C.; Pereira, L.J.D.M.; e Silva, A.C.S.; Penna, F.J.; Fagundes, E.D.T.; Ferreira, A.R. Nutritional evaluation of children with chronic cholestatic disease. *J. Pediatr.* **2016**, *92*, 197–205. [CrossRef] [PubMed]
72. Diamanti, A.; Capriati, T.; Elia, D. Malnutrizione in difetto. In *Manuale Sigenp di Gastroenterologia ed Epatologia Pediatrica*; Il Pensiero Scientifico Editore: Rome, Italy, 2014; pp. 3–6.
73. Goulet, O.J.; de Ville de Goyet, J.; Otte, J.B.; Ricour, C. Preoperative Nutritional Evaluation and Support for Liver Transplantation in Children. *Transpl. Proc.* **1987**, *19*, 3249–3255.

74. Yuksekkaya, H.A.; Cakir, M.; Tumgor, G.; Baran, M.; Arikan, C.; Yagci, R.V.; Aydogdu, S. Nutritional status of infants with neonatal cholestasis. *Dig. Dis. Sci.* **2007**, *53*, 803–808. [CrossRef]
75. Trowbridge, F.L.; Sommer, A. Nutritional anthropometry and mortality risk. *Am. J. Clin. Nutr.* **1981**, *34*, 2591–2592. [CrossRef] [PubMed]
76. Hall, G.; Chowdhury, S.; Bloem, M. Use of mid-upper-arm circumference Z scores in nutritional assessment. *Lancet* **1993**, *341*, 1481. [CrossRef]
77. Zemel, B.S.; Riley, E.M.; Stallings, V.A. EVALUATION OF METHODOLOGY FOR NUTRITIONAL ASSESSMENT IN CHILDREN: Anthropometry, Body Composition, and Energy Expenditure. *Annu. Rev. Nutr.* **1997**, *17*, 211–235. [CrossRef]
78. Sokol, R.J.; Stall, C. Anthropometric evaluation of children with chronic liver disease. *Am. J. Clin. Nutr.* **1990**, *52*, 203–208. [CrossRef] [PubMed]
79. Wasserman, D.; Zemel, B.S.; Mulberg, A.E.; John, H.A.; Emerick, K.M.; Barden, E.M.; Piccoli, D.A.; Stallings, V.A. Growth, nutritional status, body composition, and energy expenditure in prepubertal children with Alagille syndrome. *J. Pediatr.* **1999**, *134*, 172–177. [CrossRef]
80. Taylor, R.M.; Dhawan, A. Assessing nutritional status in children with chronic liver disease. *J. Gastroenterol. Hepatol.* **2005**, *20*, 1817–1824. [CrossRef]
81. Klein, S. The myth of serum albumin as a measure of nutritional status. *Gastroenterology* **1990**, *99*, 1845–1846. [CrossRef]
82. Beck, F.K.; Rosenthal, T.C. Prealbumin: A marker for nutritional evaluation. *Am. Fam. Physician* **2002**, *65*, 1575–1578.
83. Mouzaki, M.; Ng, V.; Kamath, B.M.; Selzner, N.; Pencharz, P.; Ling, S. Enteral Energy and Macronutrients in End-Stage Liver Disease. *J. Parenter. Enter. Nutr.* **2014**, *38*, 673–681. [CrossRef] [PubMed]
84. Houchin, M.; Kamat, D.; Leibbrandt, C.; Hind, J.; Dhawan, A. Safety and efficacy of Heparon Junior in infants with cholestatic liver disease. *J. Pediatr. Gastroenterol. Nutr.* **2010**, *50*, E53. [CrossRef]
85. Broekaert, I.J.; Falconer, J.; Bronsky, J.; Gottrand, F.; Dall'Oglio, L.; Goto, E.; Hojsak, I.; Hulst, J.; Kochavi, B.; Papadopoulou, A.; et al. The Use of Jejunal Tube Feeding in Children: A Position Paper by the Gastroenterology and Nutrition Committees of the European Society for Paediatric Gastroenterology, Hepatology, and Nutrition 2019. *J. Pediatr. Gastroenterol. Nutr.* **2019**, *69*, 239–258. [CrossRef]
86. Duché, M.; Habès, D.; Lababidi, A.; Chardot, C.; Wenz, J.; Bernard, O. Percutaneous Endoscopic Gastrostomy for Continuous Feeding in Children with Chronic Cholestasis. *J. Pediatr. Gastroenterol. Nutr.* **1999**, *29*, 42–45. [CrossRef]
87. Heuschkel, R.; Gottrand, F.; Devarajan, K.; Poole, H.; Callan, J.; Dias, J.; Karkelis, S.; Papadopoulou, A.; Husby, S.; Ruemmele, F.; et al. ESPGHAN Position Paper on Management of Percutaneous Endoscopic Gastrostomy in Children and Adolescents. *J. Pediatr. Gastroenterol. Nutr.* **2015**, *60*, 131–141. [CrossRef]
88. Burgmaier, K.; Brandt, J.; Shroff, R.; Witters, P.; Weber, L.T.; Dötsch, J.; Schaefer, F.; Mekahli, D.; Liebau, M.C. Gastrostomy Tube Insertion in Pediatric Patients with Autosomal Recessive Polycystic Kidney Disease (ARPKD): Current Practice. *Front. Pediatr.* **2018**, *6*, 164. [CrossRef] [PubMed]
89. Protheroe, S.M. Feeding the child with chronic liver disease. *Nutrition* **1998**, *14*, 796–800. [CrossRef]
90. Cortez, A.R.; Warren, P.W.; Goddard, G.R.; Jenkins, T.M.; Sauser, J.A.; Gerrein, B.T.; Rymeski, B.A. Primary Placement of a Low-Profile Gastrostomy Button Is Safe and Associated with Improved Outcomes in Children. *J. Surg. Res.* **2020**, *249*, 156–162. [CrossRef] [PubMed]
91. Kelly, D.A.; Tong, C. Neonatal and paediatric infection. In *Viral Hepatitis*, 4th ed.; Thomas, H.C., Lok, A.S., Locarnini, S.A., Arie, J., Zuckerman, A.J., Eds.; Wiley Blackwell: Hoboken, NJ, USA, 2012; pp. 529–543.
92. Orso, G.; Mandato, C.; Veropalumbo, C.; Cecchi, N.; Garzi, A.; Vajro, P. Pediatric parenteral nutrition-associated liver disease and cholestasis: Novel advances in pathomechanisms-based prevention and treatment. *Dig. Liver Dis.* **2015**, *48*, 215–222. [CrossRef] [PubMed]
93. Cahova, M.; Bratova, M.; Wohl, P. Parenteral Nutrition-Associated Liver Disease: The Role of the Gut Microbiota. *Nutrients* **2017**, *9*, 987. [CrossRef]
94. Lindor, K.D.; Burnes, J. Ursodeoxycholic acid for the treatment of home parenteral nutrition-associated cholestasis. *Gastroenterology* **1991**, *101*, 250–253. [CrossRef]
95. Spagnuolo, M.I.; Iorio, R.; Vegnente, A.; Guarino, A. Ursodeoxycholic acid for treatment of cholestasis in children on long- term total parenteral nutrition: A pilot study. *Gastroenterology* **1996**, *111*, 716–719. [CrossRef]
96. Sullivan, J.S.; Sundaram, S.S.; Pan, Z.; Sokol, R.J. Parenteral nutrition supplementation in biliary atresia patients listed for liver transplantation. *Liver Transplant.* **2011**, *18*, 120–128. [CrossRef]
97. Rochling, F. Intravenous Lipid Emulsions in the Prevention and Treatment of Liver Disease in Intestinal Failure. *Nutrients* **2021**, *13*, 895. [CrossRef]
98. Wendel, D.; Mortensen, M.; Harmeson, A.; Shaffer, M.L.; Hsu, E.; Horslen, S. Resolving Malnutrition With Parenteral Nutrition Before Liver Transplant in Biliary Atresia. *J. Pediatr. Gastroenterol. Nutr.* **2018**, *66*, 212–217. [CrossRef] [PubMed]
99. Holt, R.I.; Broide, E.; Buchanan, C.R.; Miell, J.P.; Baker, A.J.; Mowat, A.P.; Mieli-Vergani, G. Orthotopic liver transplantation reverses the adverse nutritional changes of end-stage liver disease in children. *Am. J. Clin. Nutr.* **1997**, *65*, 534–542. [CrossRef]
100. Colomb, V.; Goulet, O.; Ricour, C. 12 Home enteral and parenteral nutrition in children. *Baillieres Clin. Gastroenterol.* **1998**, *12*, 877–894. [CrossRef]

101. Diamanti, A.; Calvitti, G.; Martinelli, D.; Santariga, E.; Capriati, T.; Bolasco, G.; Iughetti, L.; Pujia, A.; Knafelz, D.; Maggiore, G. Etiology and Management of Pediatric Intestinal Failure: Focus on the Non-Digestive Causes. *Nutrients* **2021**, *13*, 786. [CrossRef]
102. Van Ginkel, W.G.; Van Reemst, H.E.; Kienstra, N.S.; Daly, A.; Rodenburg, I.L.; Macdonald, A.; Burgerhof, J.G.; De Blaauw, P.; Van De Krogt, J.; Santra, S.; et al. The Effect of Various Doses of Phenylalanine Supplementation on Blood Phenylalanine and Tyrosine Concentrations in Tyrosinemia Type 1 Patients. *Nutrients* **2019**, *11*, 2816. [CrossRef]
103. Van Vliet, K.; Rodenburg, I.L.; Van Ginkel, W.G.; Lubout, C.M.; Wolffenbuttel, B.H.; Van Der Klauw, M.M.; Heiner-Fokkema, M.R.; Van Spronsen, F.J. Biomarkers of Micronutrients in Regular Follow-Up for Tyrosinemia Type 1 and Phenylketonuria Patients. *Nutrients* **2019**, *11*, 2011. [CrossRef]
104. Van Spronsen, F.J.; Van Rijn, M.; Meyer, U.; Das, A.M. Dietary Considerations in Tyrosinemia Type I. *Adv. Exp. Med. Biol.* **2017**, *959*, 197–204. [CrossRef]
105. Berry, G.T.; Walter, J.H. Disorders of galactose metabolism. In *Inborn Metabolic Diseases: Diagnosis and Treatment*; Saudubray, J.M., van den Berghe, G., Walter, J.H., Eds.; Springer: Berlin/Heidelberg, Germany, 2012. [CrossRef]
106. Bosch, A.M. Classic galactosemia: Dietary dilemmas. *J. Inherit. Metab. Dis.* **2010**, *34*, 257–260. [CrossRef] [PubMed]
107. Welling, L.; Bernstein, L.E.; Berry, G.T.; Burlina, A.B.; Eyskens, F.; Gautschi, M.; Grünewald, S.; Gubbels, C.S.; Knerr, I.; Labrune, P.; et al. International clinical guideline for the management of classical galactosemia: Diagnosis, treatment, and follow-up. *J. Inherit. Metab. Dis.* **2016**, *40*, 171–176. [CrossRef]
108. Coelho, A.I.; Rubio-Gozalbo, M.E.; Vicente, J.; Rivera, I. Sweet and sour: An update on classic galactosemia. *J. Inherit. Metab. Dis.* **2017**, *40*, 325–342. [CrossRef] [PubMed]
109. Batey, L.A.; Welt, C.K.; Rohr, F.; Wessel, A.; Anastasoaie, V.; Feldman, H.A.; Guo, C.-Y.; Rubio-Gozalbo, E.; Berry, G.; Gordon, C.M. Skeletal health in adult patients with classic galactosemia. *Osteoporos. Int.* **2012**, *24*, 501–509. [CrossRef] [PubMed]
110. Ali, M.; Rellos, P.; Cox, T.M. Hereditary fructose intolerance. *J. Med Genet.* **1998**, *35*, 353–365. [CrossRef] [PubMed]
111. Vajro, P.; Veropalumbo, C. Citrin Deficiency: Learn More, and Don't Forget to Add It to the List of Neonatal Cholestasis and the NASH Trash Bin. *J. Pediatr. Gastroenterol. Nutr.* **2010**, *50*, 578–579. [CrossRef]
112. Pinto, A.; Ashmore, C.; Batzios, S.; Daly, A.; Dawson, C.; Dixon, M.; Evans, S.; Green, D.; Gribben, J.; Hunjan, I.; et al. Dietary Management, Clinical Status and Outcome of Patients with Citrin Deficiency in the UK. *Nutrients* **2020**, *12*, 3313. [CrossRef] [PubMed]
113. Okano, Y.; Ohura, T.; Sakamoto, O.; Inui, A. Current treatment for citrin deficiency during NICCD and adaptation/compensation stages: Strategy to prevent CTLN2. *Mol. Genet. Metab.* **2019**, *127*, 175–183. [CrossRef]
114. Hayasaka, K.; Numakura, C. Adult-onset type II citrullinemia: Current insights and therapy. *Appl. Clin. Genet.* **2018**, *11*, 163–170. [CrossRef]
115. Okano, Y.; Okamoto, M.; Yazaki, M.; Inui, A.; Ohura, T.; Murayama, K.; Watanabe, Y.; Tokuhara, D.; Takeshima, Y. Analysis of daily energy, protein, fat, and carbohydrate intake in citrin-deficient patients: Towards prevention of adult-onset type II citrullinemia. *Mol. Genet. Metab.* **2021**, *133*, 63–70. [CrossRef]
116. Hayasaka, K.; Numakura, C.; Toyota, K.; Kimura, T. Treatment with Lactose (Galactose)-Restricted and Medium-Chain Triglyceride-Supplemented Formula for Neonatal Intrahepatic Cholestasis Caused by Citrin Deficiency. *JIMD Rep.* **2011**, *2*, 37–44. [CrossRef]
117. Inui, A.; Hashimoto, T.; Sogo, T.; Komatsu, H.; Saheki, T.; Fujisawa, T. Chronic hepatitis without hepatic steatosis caused by citrin deficiency in a child. *Hepatol. Res.* **2015**, *46*, 357–362. [CrossRef]
118. Mutoh, K.; Kurokawa, K.; Kobayashi, K.; Saheki, T. Treatment of a citrin-deficient patient at the early stage of adult-onset type II citrullinaemia with arginine and sodium pyruvate. *J. Inherit. Metab. Dis.* **2008**, *31* (Suppl. 2), 343–347. [CrossRef]
119. Figueiredo, F.; Dickson, E.R.; Pasha, T.; Kasparova, P.; Therneau, T.; Malinchoc, M.; DiCecco, S.; Francisco-Ziller, N.; Charlton, M. Impact of nutritional status on outcomes after liver transplantation1. *Transplantation* **2000**, *70*, 1347–1352. [CrossRef]
120. Hammad, A.; Kaido, T.; Aliyev, V.; Mandato, C.; Uemoto, S. Nutritional Therapy in Liver Transplantation. *Nutrients* **2017**, *9*, 1126. [CrossRef]
121. McDiarmid, S.V.; Anand, R.; Lindblad, A.S. Development of a pediatric end-stage liver disease score to predict poor outcome in children awaiting liver transplantation1. *Transplantation* **2002**, *74*, 173–181. [CrossRef] [PubMed]
122. Mohammad, S.; Grimberg, A.; Rand, E.; Anand, R.; Yin, W.; Alonso, E.M. Long-Term Linear Growth and Puberty in Pediatric Liver Transplant Recipients. *J. Pediatr.* **2013**, *163*, 1354–1360. [CrossRef] [PubMed]
123. Almaas, R.; Haflidadottir, S.; Kaldestad, R.H.; Matthews, I.L. Asthma, Eczema, and Food Allergy in Children Following Liver Transplantation. *J. Pediatr.* **2019**, *204*, 263–269. [CrossRef] [PubMed]
124. Valta, H.; Jalanko, H.; Holmberg, C.; Helenius, I.; Mäkitie, O. Impaired Bone Health in Adolescents After Liver Transplantation. *Arab. Archaeol. Epigr.* **2007**, *8*, 150–157. [CrossRef]
125. Ng, V.L.; Alonso, E.M.; Bucuvalas, J.C.; Cohen, G.; Limbers, C.A.; Varni, J.W.; Mazariegos, G.; Magee, J.; McDiarmid, S.V.; Anand, R. Health Status of Children Alive 10 Years after Pediatric Liver Transplantation Performed in the US and Canada: Report of the Studies of Pediatric Liver Transplantation Experience. *J. Pediatr.* **2011**, *160*, 820–826. [CrossRef]
126. Saito, T.; Mizuta, K.; Hishikawa, S.; Kawano, Y.; Sanada, Y.; Fujiwara, T.; Yasuda, Y.; Sugimoto, K.; Sakamoto, K.; Kawarasaki, H. Growth curves of pediatric patients with biliary atresia following living donor liver transplantation: Factors that influence post-transplantation growth. *Pediatr. Transplant.* **2007**, *11*, 764–770. [CrossRef]

127. McDiarmid, S.V.; Gornbein, J.A.; DeSilva, P.J.; Goss, J.A.; Vargas, J.H.; Martín, M.G.; Ament, M.E.; Busuttil, R.W. Factors affecting growth after pediatric liver transplantation. *Transplantation* **1999**, *67*, 404–411. [CrossRef]
128. Swenson, S.M.; Perito, E.R. Weight Gain Trajectory Predicts Long-term Overweight and Obesity After Pediatric Liver Transplant. *J. Pediatr. Gastroenterol. Nutr.* **2019**, *68*, 89–95. [CrossRef]
129. Nair, S.; Verma, S.; Thuluvath, P.J. Obesity and its effect on survival in patients undergoing orthotopic liver transplantation in the United States. *Hepatology* **2002**, *35*, 105–109. [CrossRef]
130. Conzen, K.D.; Vachharajani, N.; Collins, K.M.; Anderson, C.D.; Lin, Y.; Wellen, J.R.; Shenoy, S.; Lowell, J.A.; Doyle, M.B.M.; Chapman, W.C. Morbid obesity in liver transplant recipients adversely affects longterm graft and patient survival in a single-institution analysis. *HPB* **2015**, *17*, 251–257. [CrossRef] [PubMed]
131. Fouad, Y.; Waked, I.; Bollipo, S.; Gomaa, A.; Ajlouni, Y.; Attia, D. What's in a name? Renaming 'NAFLD' to 'MAFLD'. *Liver Int.* **2020**, *40*, 1254–1261. [CrossRef]
132. Prasad, G.V.R.; Huang, M.; Silver, S.; Al-Lawati, A.I.; Rapi, L.; Nash, M.M.; Zaltzman, J.S. Metabolic syndrome definitions and components in predicting major adverse cardiovascular events after kidney transplantation. *Transpl. Int.* **2014**, *28*, 79–88. [CrossRef]
133. Thoefner, L.B.; Rostved, A.A.; Pommergaard, H.-C.; Rasmussen, A. Risk factors for metabolic syndrome after liver transplantation: A systematic review and meta-analysis. *Transplant. Rev.* **2018**, *32*, 69–77. [CrossRef] [PubMed]
134. Perito, E.R.; Glidden, D.; Roberts, J.P.; Rosenthal, P. Overweight and obesity in pediatric liver transplant recipients: Prevalence and predictors before and after transplant, United Network for Organ Sharing Data, 1987–2010. *Pediatr. Transplant.* **2011**, *16*, 41–49. [CrossRef]
135. Nobili, V.; Goyet, J.D.V.D. Pediatric post-transplant metabolic syndrome: New clouds on the horizon. *Pediatr. Transplant.* **2013**, *17*, 216–223. [CrossRef] [PubMed]
136. Högler, W.; Baumann, U.; Kelly, D. Endocrine and Bone Metabolic Complications in Chronic Liver Disease and After Liver Transplantation in Children. *J. Pediatr. Gastroenterol. Nutr.* **2012**, *54*, 313–321. [CrossRef]
137. Seo, S.; Maganti, K.; Khehra, M.; Ramsamooj, R.; Tsodikov, A.; Bowlus, C.; McVicar, J.; Zern, M.; Torok, N. De novo nonalcoholic fatty liver disease after liver transplantation. *Liver Transplant.* **2007**, *13*, 844–847. [CrossRef]
138. Heisel, O.; Heisel, R.; Balshaw, R.; Keown, P. New Onset Diabetes Mellitus in Patients Receiving Calcineurin Inhibitors: A Systematic Review and Meta-Analysis. *Arab. Archaeol. Epigr.* **2004**, *4*, 583–595. [CrossRef]
139. Perito, E.R.; Lustig, R.H.; Rosenthal, P. Metabolic Syndrome Components After Pediatric Liver Transplantation: Prevalence and the Impact of Obesity and Immunosuppression. *Arab. Archaeol. Epigr.* **2016**, *16*, 1909–1916. [CrossRef]
140. Di Cosmo, N.; Vajro, P.; Debray, D.; Valerio, G.; Giugliano, M.; Buono, P.; Franzese, A. Normal -Cell Function in Post-Liver Transplantation Diabetes Treated with Tacrolimus. *Diabetes Care* **2004**, *27*, 1837–1838. [CrossRef]
141. Watt, K.D. Metabolic syndrome. Is immunosuppression to blame? *Liver Transpl.* **2011**, *17* (Suppl. 3), S38–S42. [CrossRef] [PubMed]
142. Vajro, P.; Pecoraro, C.; De Vincenzo, A.; Genovese, E.; Silvestre, C.; Brancato, T.; Migliaro, F. Monitoring of renal function in children before and after liver transplantation. *Transplant. Proc.* **1998**, *30*, 1991–1992. [CrossRef]
143. Munoz, S.J. Hyperlipidemia and other coronary risk factors after orthotopic liver transplantation: Pathogenesis, diagnosis and management. *Liver Transpl. Surg.* **1995**, *1* (Suppl. 1), 29–38. [PubMed]
144. Galvin, Z.; Rajakumar, R.; Chen, E.; Adeyi, O.; Selzner, M.; Grant, D.; Sapisochin, G.; Greig, P.; Cattral, M.; McGilvray, I.; et al. Predictors of De Novo Nonalcoholic Fatty Liver Disease After Liver Transplantation and Associated Fibrosis. *Liver Transplant.* **2019**, *25*, 56–67. [CrossRef] [PubMed]
145. Kelly, D.A.; Bucuvalas, J.C.; Alonso, E.M.; Karpen, S.J.; Allen, U.; Green, M.; Farmer, D.; Shemesh, E.; McDonald, R.A. Long-term medical management of the pediatric patient after liver transplantation: 2013 practice guideline by the American Association for the Study of Liver Diseases and the American Society of Transplantation. *Liver Transplant.* **2013**, *19*, 798–825. [CrossRef]
146. Hilk, K.; Zerofsky, M.; Rhee, S.; Rosenthal, P.; Perito, E.R. Center variation in screening for and management of metabolic syndrome in pediatric liver transplant recipients: A survey of SPLIT centers. *Pediatr. Transplant.* **2019**, *23*, e13347. [CrossRef] [PubMed]
147. Mangus, R.; Bush, W.J.; Miller, C.; Kubal, C. Severe Sarcopenia and Increased Fat Stores in Pediatric Patients with Liver, Kidney, or Intestine Failure. *J. Pediatr. Gastroenterol. Nutr.* **2017**, *65*, 579–583. [CrossRef] [PubMed]
148. Lurz, E.; Patel, H.; Frimpong, R.G.; Ricciuto, A.; Kehar, M.; Wales, P.W.; Towbin, A.J.; Chavhan, G.B.; Kamath, B.M.; Ng, V.L. Sarcopenia in patients with end-stage liver disease. *J. Pediatr. Gastroenterol. Nutr.* **2018**, *66*, 222–226. [CrossRef]
149. Mager, D.R.; Hager, A.; Ooi, P.H.; Siminoski, K.; Gilmour, S.M.; Yap, J.Y. Persistence of Sarcopenia After Pediatric Liver Transplantation Is Associated with Poorer Growth and Recurrent Hospital Admissions. *J. Parenter. Enter. Nutr.* **2018**, *43*, 271–280. [CrossRef] [PubMed]
150. Rezende, I.F.B.; Conceição-Machado, M.E.P.; Souza, V.S.; Dos Santos, E.M.; Silva, L.R. Sarcopenia in children and adolescents with chronic liver disease. *J. Pediatr.* **2019**, *96*, 439–446. [CrossRef] [PubMed]
151. Merli, M. Pediatric sarcopenia: Exploring a new concept in children with chronic liver disease. *J. Pediatr.* **2019**, *96*, 406–408. [CrossRef] [PubMed]

152. Meyer, F.; Bannert, K.; Wiese, M.; Esau, S.; Sautter, L.F.; Ehlers, L.; Aghdassi, A.A.; Metges, C.C.; Garbe, L.-A.; Jaster, R.; et al. Molecular Mechanism Contributing to Malnutrition and Sarcopenia in Patients with Liver Cirrhosis. *Int. J. Mol. Sci.* **2020**, *21*, 5357. [CrossRef]
153. Verhagen, M.V.; Levolger, S.; Hulshoff, J.B.; Werner, M.J.; van der Doef, H.P.; Viddeleer, A.R.; de Kleine, R.H.; de Haas, R.J. Utility of preoperative CT-based body metrics in relation to postoperative complications in pediatric liver transplant recipients. *Liver Transplant.* **2021**. [CrossRef]
154. Jitwongwai, S.; Lertudomphonwanit, C.; Junhasavasdikul, T.; Fuangfa, P.; Tanpowpong, P.; Gesprasert, G.; Treepongkaruna, S. Low psoas muscle index as an unfavorable factor in children with end-stage liver disease undergoing liver transplantation. *Pediatr. Transplant.* **2021**, *25*, e13996. [CrossRef]
155. Boster, J.M.; Browne, L.P.; Pan, Z.; Zhou, W.; Ehrlich, P.F.; Sundaram, S.S. Higher Mortality in Pediatric Liver Transplant Candidates with Sarcopenia. *Liver Transplant.* **2021**, *27*, 808–817. [CrossRef]
156. Woolfson, J.P.; Perez, M.; Chavhan, G.B.; Johara, F.T.; Lurz, E.; Kamath, B.M.; Ng, V.L. Sarcopenia in Children with End-Stage Liver Disease on the Transplant Waiting List. *Liver Transplant.* **2021**, *27*, 641–651. [CrossRef]
157. Cruz-Jentoft, A.J.; Baeyens, J.P.; Bauer, J.M.; Boirie, Y.; Cederholm, T.; Landi, F.; Martin, F.C.; Michel, J.-P.; Rolland, Y.; Schneider, S.; et al. Sarcopenia: European consensus on definition and diagnosis: Report of the European Working Group on Sarcopenia in Older People. *Age Ageing* **2010**, *39*, 412–423. [CrossRef]

nutrients

MDPI

Review

Malnutrition in Older Adults—Recent Advances and Remaining Challenges

Kristina Norman [1,2,3,4,*], Ulrike Haß [2,3] and Matthias Pirlich [5]

1 Department of Geriatrics and Medical Gerontology, Charité–Universitätsmedizin Berlin, Corporate Member of Freie Universität Berlin and Humboldt-Universität zu Berlin, 13347 Berlin, Germany
2 Department of Nutrition and Gerontology, German Institute of Human Nutrition Potsdam-Rehbrücke, 14458 Nuthetal, Germany; ulrike.hass@dife.de
3 Institute of Nutritional Science, University of Potsdam, 14458 Nuthetal, Germany
4 German Center for Cardiovascular Research (DZHK), Partner Site Berlin, 10115 Berlin, Germany
5 Imperial Oak Outpatient Clinic, Endocrinology, Gastroenterology & Clinical Nutrition, 12159 Berlin, Germany; pirlich@kaisereiche.de
* Correspondence: kristina.norman@charite.de; Tel.: +49-030450565139

Abstract: Malnutrition in older adults has been recognised as a challenging health concern associated with not only increased mortality and morbidity, but also with physical decline, which has wide ranging acute implications for activities of daily living and quality of life in general. Malnutrition is common and may also contribute to the development of the geriatric syndromes in older adults. Malnutrition in the old is reflected by either involuntary weight loss or low body mass index, but hidden deficiencies such as micronutrient deficiencies are more difficult to assess and therefore frequently overlooked in the community-dwelling old. In developed countries, the most cited cause of malnutrition is disease, as both acute and chronic disorders have the potential to result in or aggravate malnutrition. Therefore, as higher age is one risk factor for developing disease, older adults have the highest risk of being at nutritional risk or becoming malnourished. However, the aetiology of malnutrition is complex and multifactorial, and the development of malnutrition in the old is most likely also facilitated by ageing processes. This comprehensive narrative review summarizes current evidence on the prevalence and determinants of malnutrition in old adults spanning from age-related changes to disease-associated risk factors, and outlines remaining challenges in the understanding, identification as well as treatment of malnutrition, which in some cases may include targeted supplementation of macro- and/or micronutrients, when diet alone is not sufficient to meet age-specific requirements.

Keywords: malnutrition; ageing; inflammaging; sarcopenia; anorexia of aging; micronutrients; DoMAP; GLIM criteria

Citation: Norman, K.; Haß, U.; Pirlich, M. Malnutrition in Older Adults—Recent Advances and Remaining Challenges. *Nutrients* **2021**, *13*, 2764. https://doi.org/10.3390/nu13082764

Academic Editor: Ina Bergheim

Received: 29 June 2021
Accepted: 10 August 2021
Published: 12 August 2021

1. Introduction

The World Health Organization (WHO) has declared Healthy Ageing a priority of its work on ageing between 2016 and 2030 and developed a policy framework which emphasizes the need for action across multiple sectors [1]. The aim of the program is to enable older persons to develop and maintain functional ability which permits wellbeing and allows them to partake in society [1]. Older adults (individuals aged 65 or over) are the fastest growing age group and projections from the United Nations predict that by 2050, the number of adults aged 65 or over will be twice as big as the amount of children under the age of five and also surpass the number of adolescents aged between 15 and 24 years. In 2050, improvements in survival are expected to add approximately 5 years to the life expectancy at birth for the world's population which was 72.6 years in 2019 [2].

Biology of ageing is understood as the time related decline of physiological functions, leading to changes in functional performance of different organ systems as well as with reduced resilience to physical, cognitive and mental stressors, however, there are great individual differences in these changes. Advanced age is associated with reduced adaptive and regenerative capacity which is leading to higher rates of morbidity [3]. On the other hand, the presence of age-associated diseases in middle-aged individuals has been interpreted as a sign of accelerated ageing [4]. Maintaining an adequate nutritional status as well as a sufficient nutrient intake is key to health and quality of life and as such is one prerequisite for wellbeing in higher age and modulator of healthy ageing as defined by the WHO. However, older adults are susceptible to nutritional problems and ultimately also to malnutrition through a variety of mechanisms [5]. As age is one main risk factor for the development of chronic disease, older persons are particularly susceptible to disease-related weight loss, loss of muscle mass and strength (i.e., sarcopenia) and ultimately, the frailty syndrome, all of which can fundamentally impact recovery from disease and clinical outcome in general [6–8]. Weight loss, a marker of macronutrient deficiency and/or catabolism, is a common key initial phenomenon in old patients, which sets off a catabolic cascade of unfavourable events resulting in higher morbidity and mortality. The causes for weight loss in higher age are multifactorial, but can in part be attributed to both disease processes such as catabolic events, disease or age-related anorexia ("anorexia of aging") and subsequent insufficient dietary intake, but also to increased inflammatory status, depressive or cognitive disorders [9] as well as a decreased socio-economic status [10]. But even outside the context of disease or manifest disorders, both ageing processes and age-associated changes can slowly impact physiology and metabolism and thus effect a gradual change in the nutritional status of older adults [11].

The treatment of malnutrition requires early identification and multimodal intervention, in hospitalized patients as well as community dwelling older adults. However, treatment modality still poses a challenge for nutritional therapy with yet open questions [12,13]. This review summarizes the current state of evidence on the complex aetiology of malnutrition in old adults, considering both effects of ageing processes and disease-related factors. Also remaining challenges in the identification and treatment of malnutrition in the old are outlined.

2. Impact of Malnutrition in the Old

Although clinical malnutrition predominantly occurs in patients in hospitals, care situations or nursing homes, malnutrition, nutritional risk and specific nutrient deficiencies in particular, are a common albeit frequently overlooked occurrence in community-dwelling old people [14,15].

Consequences of malnutrition are deleterious and far reaching and have been described in detail [6]. While disease-related malnutrition is not limited to older adults, it is more frequent in higher age, and the consequences appear to be more severe in older persons due to their impaired regenerative capacity.

Malnutrition in general has serious implications for clinical outcome, for recovery from disease, trauma and surgery and is associated with increased morbidity and mortality both in acute and chronic disease [6] and has thus been acknowledged as serious burden for the health care system [16].

Depending on the type of malnutrition, protein catabolism can be pronounced. Disease-related malnutrition therefore leads to rapid skeletal muscle wasting, whereas age-related malnutrition is associated with a slower but progressive loss of muscle mass. The effects of protein catabolism are prominently reflected by lower muscle mass, muscle strength and function with severe implications for physical performance [17]. At the same time, malnutrition [18] but also reduced dietary protein intake per se [19] are associated with a decrease in bone mineral mass in higher age. Together with poor physical performance and coordination resulting in higher fall risk, these factors accelerate the age-related risk of osteoporosis and osteoporotic fractures [20]. Taken together, the risk of falling and the subsequent loss of independence and disability are greatly increased.

Malnutrition-related protein catabolism and micronutrient deficiency have been prominently linked to the impairment of immune function [21], which is already affected in higher age. In older malnourished adults, this manifests as the loss of cell-mediated immunity in particular [22], increasing the risk for infection and delaying recovery from disease [23]. Several studies have thus shown a close relationship between malnutrition and risk for infection, such as healthcare-associated infections [24], infectious complications and subsequent longer stays in intensive care units (ICU) and increased ICU mortality in older malnourished patients [25].

Wound healing as well as tissue recovery are also impaired in malnutrition, which in part can be attributed to micronutrient deficiency [26]. This clearly predisposes older adults to wound healing disorders and chronic wounds which are a great burden to patients as well as associated with decreased quality of life and furthermore with higher expenditure in the health care setting [27].

One particularity of malnutrition in older adults is that the impact appears to be more severe than in younger adults. Studies have not only shown that changes in body composition in malnutrition occur to a greater extent in older compared to younger adults [28], but also that recovery of low body cell mass or muscle mass is impaired in higher age following weight loss [17]. This predisposes malnourished older adults to the risk of developing the so-called geriatric syndromes which have been described as multifactorial syndromes typical in higher age [29]. Geriatric syndromes greatly compromise health status, cognitive functioning, functional ability, and compensatory capacity [29] and result in higher mortality, albeit predominantly in the younger old [30].

2.1. Role of Malnutrition in the Geriatric Syndromes Frailty, Fatigue, Sarcopenia

Malnutrition plays an important role in the development of certain geriatric syndromes. Geriatric syndromes are complex multifactorial conditions occurring in higher age with serious implications for health [29] and have been described as "phonotypical presentations of accumulated and underlying ageing-related dysfunctions spanning over different organ systems" [31]. They include (but are not limited to) dementia and delirium, depression, incontinence, fall risk, visual as well as hearing impairment, wound healing disorders, frailty, and sarcopenia [32].

Involuntary weight loss, a hallmark of malnutrition, is inevitably associated with loss of skeletal muscle mass, which appears to occur at a greater extent in higher age. This increases the risk of developing sarcopenia, a phenomenon which is characterized by the loss of both muscle mass as well as muscle strength and function. As these two entities frequently occur together, this has led to the new term "sarcopenia malnutrition syndrome" [33] and a need for new screening tools which reliably identify both conditions has been voiced [34].

A phenomenon, which is often overseen in this content but nonetheless is important in an increasingly obese society, is sarcopenic obesity [35], which describes a condition with reduced lean mass and increased fat mass, resulting in a high-risk body composition phenotype. This underlines the importance of evaluating muscle mass independently from weight loss, since age-related changes in body composition such as increases in (visceral) fat mass may well mask low lean mass. Furthermore, ectopic fat infiltration in the muscle lowers the quality of skeletal muscle and thereby impairs muscle functionality [35].

As sarcopenia has been called the biological substrate of frailty, the close relationship between malnutrition and sarcopenia suggests a link between malnutrition and frailty as well. Not surprisingly, involuntary weight loss, which is an indicator of catabolism, is a major risk factor for developing physical frailty [36]. Weight loss is therefore not only one of five factors such as fatigue, weakness, slow gait speed and low physical activity which constitute frailty as defined by the frailty phenotype [37], but is also linked causally to the other four factors [36]. As such, there is a close relationship between malnutrition and frailty which has been well documented [38], although they are considered conceptually distinct conditions. Numerous studies have shown a significant overlap of both entities in hospital patients [39] and community dwelling old [40,41]. There is even evidence that the consequences frequently attributed to malnutrition might in reality be due to the effects of frailty. An evaluation in 2804 community-dwelling older adults from the Singapore Longitudinal Aging Study II revealed that functional decline measured by decreased (instrumental) activities of daily living, disability, impaired quality of life and long-term mortality in particular was more apparent in older adults with physical frailty with or without malnutrition [42].

Moreover, fatigue is also one of the core elements of frailty. Fatigue has been described as a relentless exhaustion affecting the ability to carry out physical and mental activities [43] and has been linked to age-related mitochondrial dysfunction [44,45]. Old patients with severe involuntary weight loss at discharge from hospital had a significantly higher risk for severe fatigue which in turn compromises post hospital recovery [46]. Nutritional status including both malnutrition and obesity, has moreover been identified as an important modulator of fatigue [47], and more evidence is warranted on the role of dietary approaches, including anti-inflammatory diets, in the treatment of fatigue [48].

Due to the proposed relationship between long-term dietary patterns and cognitive function, malnutrition has also been linked to cognitive impairment [49,50], although the relationship is complex and difficult to disentangle and more studies are needed on this subject [50]. Similarly, there is a close interaction between malnutrition and depression [51–53], but causality is difficult to establish, as the relation is most likely mutual.

2.2. Prognostic Impact of Malnutrition in the Old on Mortality

It is well established that malnutrition is associated with increased mortality in both acute and chronic disease, and this effect is still observed in older adults. The risk for both short-term mortality in acute conditions [54] as well as for long-term mortality in chronic disease is significantly increased [55]. One large cohort study included 1767 older individuals with a variety of disease ranging from cancer to diseases of the circulatory or respiratory system and revealed that the increased risk of mortality due to malnutrition existed irrespective of the cause of death [56].

3. Malnutrition: Definition and Types and How to Screen for Them

Despite an ongoing debate, there is still no universally accepted definition of malnutrition [57]. Therefore, in 2016, the world's four leading Clinical Nutrition Societies (ESPEN, ASPEN, FELANPE, and PENSA) representing more than 70 national scientific societies started a consensus process in order to develop criteria for malnutrition which could be used in all clinical settings on a global scale. The resulting concept of the Global Leadership Initiative on Malnutrition (GLIM) [58,59] was published in 2019 and considers three phenotypic criteria for the diagnosis of malnutrition: weight loss (>5% within past 6 months, or >10% beyond 6 months), low body mass index (BMI) (<20 kg/m^2 if <70 years, or <22 kg/m^2 if >70 years), reduced muscle mass (according to validated body composition techniques), and two etiologic criteria: reduced food intake or assimilation ($\leq$50% of energy requirements >1 week, or any reduction for >2 weeks, or any chronic gastro-intestinal condition that adversely impacts food assimilation or absorption), and inflammation (acute disease/injury or chronic disease-related). It is proposed that the diagnosis of malnutrition is based upon the presence of at least one phenotypic and one etiologic criterion, and in a second step different thresholds of the criteria can be used for severity grading of malnutrition. The GLIM criteria are subject of numerous ongoing validation studies. A recently published study on community-dwelling older adults participating in a long-term osteoporosis trial in Hongkong demonstrated that the GLIM criteria were associated with a higher risk for sarcopenia, frailty, and mortality during a 14-year follow-up period [60]. While the GLIM criteria are not age-specific, they do include age as a risk factor among the components. For geriatric patients, the guidelines on Enteral Nutrition in Geriatrics by the European Society of Clinical Nutrition and Metabolism (ESPEN), have defined clinical malnutrition as the presence of either weight loss which reflects a catabolic state (>5%

in six months) and/or low BMI (i.e., BMI below 20 kg/m^2) which represents depleted physiological stores [61].

Nutritional risk is less well defined but commonly understood to be a condition in which the present nutritional status is at imminent risk of impairment due to a range of factors such as medical history, comorbidities or drugs which might increase dietary requirements or interfere with nutrient absorption or metabolism. Further factors may include physical, mental or cognitive status which might prevent the older person to properly care for themselves as well as socio-economic factors which hinder access to a varied high-quality diet [61].

Screening tools which are suitable for the use in older adults need to focus on the most important risk factors for malnutrition in high age, in order not only to diagnose manifest malnutrition but also to capture the risk to develop malnutrition as described above. That way, screening tools can be used to identify patients early in order to initiate nutritional treatment. Several screening tools have been established for the specific use in older adults, but the most widely used and most studied instrument is the Mini Nutritional Assessment (MNA) in its long or short form. However, due to its broad range of covered topics, the specificity of the MNA has been questioned, as it is associated with a high risk of "over-diagnosing" malnutrition in the old [62]. This problem might be countered by complementing the MNA with the GLIM criteria as suggested [59].

Screening for and treating malnutrition has also been acknowledged as one of the first necessary steps in the identification and treatment of sarcopenia [63]. Recent studies have compared the ability of screening tools for malnutrition such as the MNA short or long form to predict the development of sarcopenia to newer diagnostic tools such as the GLIM criteria [58,59]. In the SarcoPhAge cohort, the GLIM criteria did predict incident sarcopenia, whereas neither of the MNA forms did [17].

3.1. Macronutrient Deficiencies

Clinical malnutrition results from an imbalance between macronutrient intake and requirement [64], which causes a measurable reduction in tissue and ultimately weight. Due to the frequently insufficient protein and energy intake, it is therefore also commonly referred to as protein-energy malnutrition (PEM) or protein-energy undernutrition (PEU). Consequently, there is concomitant need for both adequate energy intake (25–30 kcal/kg body weight, depending on individual situation) and higher protein intake [65]. There is overwhelming evidence that protein requirements are generally higher in older age (1.0–1.2 g/kg body weight) and adequate protein intake [65,66] is crucial in order to prevent malnutrition and sarcopenia. Recommendations regarding protein intake are mainly based on the concept of anabolic resistance, which describes an impaired capacity of the muscle to respond to anabolic stimuli in higher age. However, protein intake in community-dwelling older adults has been reported to be frequently well below recommended intake [67,68] which is associated with a higher risk for the development of malnutrition.

Also, depending on the type of malnutrition and underlying disease, protein catabolism can be pronounced. Protein requirements are therefore further increased in older adults with malnutrition or disease (1.2–1.5 g/kg body weight) [65,66].

3.2. Micronutrient Deficiencies

One specific form of malnutrition frequent in older adults are micronutrient deficiencies [10,69]. In contrast to quantitative malnutrition, which is reflected by weight loss, micronutrient deficiencies are much harder to screen for and identify, in part also due to methodological issues such as missing suitable markers of stored or available micronutrients. The majority of dietary intake surveys have, however, identified an inadequate intake of a broad range of micronutrients in older adults [70]. Since micronutrient deficiency such as iron (Fe), vitamins C and D, vitamins B6 and B12, as well as folic acid and the trace element zinc (Zn), have been prominently linked to the impairment of immune function [71], the following chapter focusses on recent findings regarding these nutrients in particular.

The reasons for insufficient micronutrient intake vary including not only a low amount of food, but typically also the choice of food and the lack of variety. Prices, availability of foods rich in vitamins, trace elements and minerals can reduce the intake of micronutrients. On the other hand, ageing is also associated with changes that facilitate deficiencies in calcium, vitamin D, vitamin B12, Fe, magnesium, and Zn among other important nutrients. Large studies on the micronutrient serum status in older adults are rare due to the costs and efforts. The Population Based KORA-Age Study, a representative cohort study, recently showed subclinical micronutrient deficiencies in community-dwelling old [69]. 52.0% of the 1079 older study participants had a vitamin D deficiency (<50 nmol/L), 27.3% had low vitamin B12 levels (<221 pmol/L), 11.0% had insufficient Fe levels (men <11.6 μmol/L, women <9.0 μmol/L), and 8.7% had low folate levels (<13.6 nmol/L). Among the risk factors for the subclinical deficiencies were advanced age, frailty, lack of physical activity and no or irregular use of dietary supplements. Dietary supplement intake is an easily modifiable risk factor, and recent studies have confirmed that regular use of dietary supplements is associated with lower rates of subclinical deficiencies of nearly every micronutrient investigated in the study [72]. Age-specific requirements are not really clear, but higher intakes have been recommended due to the frequently impaired intestinal absorption or distribution [73]. Also, it has recently been postulated that diet alone might not be enough to meet the age-specific requirements of older adults, but tailored supplementation of micronutrients might well be necessary [74], in order to e.g., support the immune function. The intake of drugs may also interfere with the metabolism of micronutrients [75,76] which further compounds the problem as poly-pharmacy is common in higher age.

Moreover, it is known that inflammation, which is common in older adults, affects trace element status and their biomarkers, resulting in depleted plasma stores of Fe, Zn, and manganese, a phenomenon which is partly known as nutritional immunity and serves as a defence against invading pathogens [77].

Due to the multifactorial role of micronutrients in various metabolic processes as well as immune functioning, cell proliferation and growth, signalling processes and genomic stability, micronutrient deficiencies are in turn involved in the pathogenesis of a variety of conditions and age-related diseases [78]. One of the most well studied trace elements in higher age is Zn, and dietary Zn intake has also been reported to be frequently insufficient in older adults [78]. Age-associated alterations of intestinal absorption, problems regarding chewing and swallowing, drug-interactions, and impaired subcellular processes in Zn metabolism can further contribute to low Zn absorption and availability. In turn, low dietary Zn intake with subsequent Zn deficiency has been linked to depressive disorders [79], loss of appetite and cachexia in age-advanced old [80], and to increased muscle catabolism via inflammatory cytokine activation [81] as well as to immunosenescence [82] and frailty [83]. Moreover, in older adults, an inadequate intake of antioxidant micronutrients such as vitamin E, carotenoids, and vitamin C has also been linked to the development of impaired muscle strength and physical performance [84–87].

4. Prevalence of Malnutrition

Despite the body of evidence describing the personal and clinical consequences of malnutrition and its economic impact on the health care system, malnutrition in the old remains a considerable problem with reported high frequencies, especially in situations of dependency [88]. This has been attributed to poor awareness and lack of time or education in medical as well as nursing staff, but recognition and treatment of malnutrition in older adults is undeniably a challenge even when identified early. All in all, it is estimated that roughly a quarter of European adults over the age of 65 are at high risk of malnutrition across various settings [89,90].

Prevalence of malnutrition, however, strongly depends on the setting, on underlying or accompanying diseases as well as on screening and assessment methods. Numerous studies have investigated prevalence of malnutrition in hospital and nursing home settings; but recently, several meta-analyses have been published which also use meta-regression to further explore determinants. Although the specific number of malnutrition prevalence differs between meta-analyses, the main findings are comparable, showing the lowest percentage of malnourished individuals in the community setting and the highest percentage in acute and subacute care settings, higher prevalence rates in higher age as well as sex-specific differences, as women had the highest risk. Also, region and instrument used to identify malnutrition had a significant impact on prevalence. One systematic review and meta-analysis of prevalence data of malnutrition and nutritional risk in older adults across different healthcare settings showed a wide range of malnutrition from 3% in the community setting to approximately 30% in rehabilitation and subacute care, even though the review only included studies using the MNA [91]. This study was recently complemented by two systematic reviews and meta-analyses [89,90]. Leji-Halfwerk et al. included further studies using 22 malnutrition screening tools validated for use in adults aged 65 years or more. They reported pooled prevalence rates of high malnutrition risk across all countries and screening tools which ranged from 8.5% in the community setting to 28.0% for the hospital [89,90]. The meta-analysis and systematic review by Crichton and colleagues focussed on studies carried out in community-dwelling older adults who had been assessed with Subjective Global Assessment (SGA), Patient-generated (PG)-SGA or MNA and found clear differences across countries with low prevalence in Northern Europe and highest in Southeast Asia. Using meta-regression, both meta-analyses reported higher prevalence rates in adults aged above the age of 80 [89,90], in women, in patients with one or more comorbidities and a higher prevalence of malnutrition in rural rather than metropolitan regions [90].

Recent advances in prevalence research has also included re-evaluation of large pool datasets using harmonized definitions of malnutrition indicators such as low BMI, weight loss, low food intake and combinations of these. Wolters et al. were able to show varying prevalence of malnutrition according to the different criteria used and concluded that it might be more useful to consider the criteria separately as each may reflect a distinct nutritional problem [92].

5. Determinants of Malnutrition

Malnutrition in older adults is of complex and multifactorial origin. A variety of factors such as life-style factors, disease and ageing processes may be involved and interaction between these factors is common. Understanding risk factors is crucial in order to address malnutrition effectively, but the complex aetiology of malnutrition is still not perfectly understood.

In industrial countries, disease is one of the most common reasons for developing malnutrition and the onset of malnutrition can be both acute and slow. Age in itself is an established non-modifiable risk factor for malnutrition. Higher age is associated with physiological changes which can potentially slowly result in or further malnutrition such as impaired taste and smell, decreased gastric flexibility, reduced appetite, etc. As higher age clearly increases the risk for disease, there is considerable risk for potentiation of nutritional problems.

5.1. Modifiable Determinants of Malnutrition

Recent research has particularly focussed on the evidence of potentially modifiable risk factors of clinical malnutrition, as targeting these risk factors is a fundamental element in the prevention and treatment of malnutrition. A recent meta-analysis investigated which factors are associated with incident malnutrition and identified marital situation, hospitalisation and physical limitations as the most important predictors [93]. Another systematic review categorized the evidence grade of thirty potentially modifiable factors which have been associated with malnutrition; however, stating that robust evidence was lacking for most of the studied determinants and highlighting the need for high quality prospective studies [94].

In a modified Delphi process, an international expert group therefore developed a model for the theoretical framework on the aetiology of malnutrition termed DoMAP (Determinants of Malnutrition in Aged Persons) and specifically, potential causative mechanisms [95]. The model construct consists of three triangle-shaped levels illustrates the suggested direction of causality of the various risk factors for developing malnutrition. Malnutrition is at the inner core of the DoMAP model surrounded by different layers of risk factors. The immediate layer consists of the three principal conditions which result in malnutrition (low intake, increased requirements, and impaired nutrient bioavailability). The next layer consists of the factors which are believed to directly cause one of these conditions and the outermost level presents factors which impact the direct factors and indirectly cause one of the three main conditions. By illustrating the sequence of events which can cause malnutrition the DoMAP model offers a better understanding of the aetiology which can be used both in research as well as in clinical routine. By addressing factors, which can potentially impact the three principal components and thus cause malnutrition, the model also offers the possibility of early identification of patients at risk for malnutrition.

5.2. Age-Associated Changes as Risk Factors

Physiological factors which may precipitate malnutrition in higher age include sensory impairment such as diminished taste or olfactory dysfunction, delayed gastric emptying, and disturbed motility leading to a functional decline of the ageing gut [96]. Ageing is therefore also associated with an increase in colonic transit time, increased intestinal permeability, and, ultimately, altered intestinal microbiota [97], which includes loss of biodiversity, enrichment in opportunistic pathogens, and concomitant reduction of health-associated species, such as short chain fatty acid producing species [98]. The changes in the microbiome have recently been implicated in the development of loss of appetite and frailty and as such are also potentially promoting malnutrition [99,100], but more studies are needed in this context.

Moreover, a decrease in gastrointestinal hormones (e.g., ghrelin), with concomitant adverse changes in anorectic signalling (e.g., neuropeptide Y, peptide YY (PYY), orexin A, leptin, cholecystokinin (CCK)) [101–103] leading to an altered appetite regulation in higher age have been described. While ghrelin is the only appetite-enhancing peptide, other hormones such as CCK and PYY for instance are recognized as relevant mediators in satiety. Those gut-derived peripheral hormones as well as systemic insulin, glucose and fatty acid levels regulate central appetite and hunger in the hypothalamic region in a feedback loop [104–106]. As has been shown in a postprandial setting with 14 frail and 20 non-frail old (>70 years) as well as 19 young adults (20–65 years), older persons have higher levels of fasting insulin and glucose as well as increased and prolonged postprandial insulin, glucose and CCK levels compared to younger adults [107]. Furthermore, frail older persons exhibited significantly less hunger in the fasting condition and impaired gastric emptying and gallbladder contraction in the postprandial period [107]. Similarly, in a small study, the suppressed postprandial hunger in the old was paradoxically accompanied by sustained ghrelin levels [108].

Last but not least, in the case of reduced smell and taste, the reward system in the central nervous system (CNS) is hampered in the perception of pleasure associated with palatable foods [109] which may in turn further decrease dietary intake.

Some of the pathway dysregulations can be attributed to the chronically elevated inflammation observed in higher age since enhanced cytokine levels (e.g., tumour necrosis factor alpha (TNF-α), interleukin (IL)-1β) are able to affect appetite regions in the CNS [103,110]. The resulting loss of appetite and early satiety seen in older adults has been termed "anorexia of aging" and leads to an insufficient food intake with a higher risk for both quantitative and qualitative malnutrition [111,112]. One risk factor for anorexia in healthy older adults is, not surprisingly, ageing itself [113]. Furthermore, body weight dissatisfaction, which might lead to avid dieting, weight loss and eating disorders, is frequent in older adults [114] and has been studied as possible precursor of appetite loss. Ultimately, anorexia of ageing has even been proposed to be a geriatric syndrome [115] as it significantly and independently affects nutritional und functional status in the old [116,117].

Postprandial regulation in older adults has gained increasing attention in the last decade, especially with regard to muscle protein synthesis. In the old, a blunted postprandial muscle protein synthesis in response to dietary protein has been described [118–120], whereas the basal muscle protein synthesis does not appear to be affected [121]. Moreover, the lower protein balance response to hyperinsulinemia in older adults indicates insulin resistance of protein metabolism [122]. The so-called anabolic resistance, i.e., an impaired capacity of the muscle to respond to anabolic stimuli (such as dietary protein and resistance exercise) seen in higher age serves as one explanation leading to a slow onset of sarcopenia [123].

Recent research has also focussed on postprandial regulation of metabolic parameters. Age-related changes in postprandial glucose and insulin have been well established [107,124], but more recently, age-specific changes in the postprandial dynamics of fibroblast growth factor 21 (FGF21) have also been shown, resulting in considerably higher values in older compared to younger adults [125]. FGF21 is an important metabolic parameter which regulates glucose and lipid metabolism, but higher levels in older adults are paradoxically associated with higher mortality [126] and have been implicated in the loss of muscle [127,128] and bone mass [129] as well as in the cachexia anorexia syndrome in old hospitalized patients [130]. Altered postprandial regulation of FGF21 might contribute to explain the higher values. Similarly, concentrations of adiponectin, a mediator of FGF21 functions increase with age [131] and are also associated with all-cause mortality in the old [132]. Again, alterations in postprandial adiponectin response have been shown in higher age which were positively associated with the FGF21 response [133].

Taken together, in contrast to disease-related mechanisms such as catabolism, which can result in acute weight loss and malnutrition, altered appetite regulation in higher age and changes in postprandial metabolism are associated with gradual changes that may not necessarily result in manifest malnutrition but contribute to or exacerbate existent nutritional problems and facilitate nutritional deficiencies.

5.3. *Inflammaging as a Risk Factor for Malnutrition*

Under physiological conditions, inflammatory processes represent a desired, strictly controlled and usually self-limiting reaction. However, chronic inflammation has long-term negative effects on the entire system. Increasing knowledge indicates that ageing is accompanied by slightly but chronically elevated inflammation levels. This sterile and silent inflammation in old adults has been termed "inflammaging" [134]. Inflammaging has since been investigate a whole range of age-related diseases within the so-called "network theory of aging". Recently, in the pandemic of coronavirus disease 2019 (COVID-19), inflammaging gained additional attention in view of the observed cytokine storm and autoimmunity during infection among older patients, which can result in multiple organ failure and is in general associated with a worse disease outcome [135,136] and of course potentially increases the risk for malnutrition, though data are lacking [137]. This severe systemic inundation of cytokines might be further triggered by the dysregulated immune system in old age (immunosenescence), which is expressed in an imbalanced homeostasis of pro- and anti-inflammatory mediators [135].

Although underlying mechanisms are still not fully elucidated, inflammaging has been characterized as a chronic low-grade inflammatory status, which not only is reflected by higher pro-inflammatory levels, but by a remarkable imbalanced cytokine network [138,139]. Research in centenarians has shown that they have less burden of the typical age-related (co-) morbidities and that those "long-livers" have better immunological coping strategies which are attributed to a balanced and more anti-inflammatory system [4,139]. The senescence-associated secretory phenotype (SASP) is considered to support key cellular processes in inflammaging, since the secretion of those pro-inflammatory mediators do not only disrupt local tissue structures and functions, but primarily perpetuate the vicious inflammatory circle. Further, accumulation of waste products or inadequate elimination of cellular damage, as well as age-associated mitochondrial damages that result in a higher release of reactive oxygen species and extracellular release of mitochondrial DNA in particular, belong to the damage-associated molecular pattern which activates the innate immune system [140,141].

Meanwhile, it is widely recognized that inflammaging is partly responsible for triggering many age-related diseases, e.g., Alzheimer's Disease ("neuroinflammation") [142], atherosclerosis and cardiovascular events [143,144], type 2 diabetes mellitus ("metaflammation") [145], (osteo-) sarcopenia and frailty [146,147] as well as cancer [148–150]. Therefore, there is considerable potential that inflammaging also contributes to an impaired nutritional status [151], but more studies are needed to explore the relationship. In the old, increased levels of pro-inflammatory cytokines have early on been associated with cachexia ("geriatric cachexia") [152]. Cachexia is a well described complex wasting syndrome which is driven by inflammation and reflected by a striking loss of skeletal muscle mass and function [153]. It is the most pronounced condition within the disease-related PEM spectrum and occurs in catabolic disease such as cancer, chronic kidney disease, HIV/AIDS, chronic obstructive pulmonary disease. Pro-inflammatory cytokines, such as TNF-α, initially called "cachectin" [154] and IL-6, also known as the "cytokine for gerontologists" [155,156], play a dominant role in this phenomenon. The underlying inflammatory processes actively fuel the muscle protein breakdown via pro-inflammatory cytokines (TNF-α, IL-6, IL-1β) which on the one hand stimulate the ubiquitin-proteasome signalling pathway, while on the other hand inhibiting anabolic pathways [157]. Although cachexia is generally associated with severe disease, there is definite overlap in the mechanisms that have been used to explain malnutrition in the old caused by inflammaging [158] and in geriatric cachexia [152]. Inflammatory and related immunological processes in older persons is clearly associated with a higher risk for malnutrition (as indicated by the MNA) [158].

Taken together, the aetiology of malnutrition is complex, and a multitude of risk factors can contribute to or aggravate the development of malnutrition as illustrated in Figure 1. While it is important to further understand age-associated changes which impact nutritional intake, nutrient absorption, digestion and metabolism, many of these factors are not easily identified or modified. They, however, need to be accounted for when assessing the risk for malnutrition and in the concept of treatment.

6. Treatment of Malnutrition

Acknowledging the different and complex risk factors, which can result in or aggravate malnutrition or contribute to the risk of developing it, it becomes clear that treatment of malnutrition is as complicated and challenging. Dependent on both setting and the situation of the older adult, different therapy approaches are warranted.

Regarding the overwhelming evidence that protein requirements are higher in older age [65,66], adequate protein with appropriate energy intake is crucial in order to prevent malnutrition and sarcopenia. Further, due to the frequent micronutrient deficiencies in higher age, targeted supplementation of micronutrients might be useful, when diet alone is not sufficient to meet the age-specific requirements. However, there are still considerable gaps concerning the evidence of non-pharmacological treatment of malnutrition.

Available studies so far have delivered conflicting evidence on the benefits of nutritional theory. Recently, an international panel of nutrition experts identified the shortcomings of these studies which include heterogeneous study populations and too short or varying treatment duration as well as issues relating study methodology such as inadequately described control group and no placebo, no blinding of study personnel, frequently no intention-to-treat-analyses as well as selective outcome presentation [13]. Therefore, it remains unresolved which interventions are most effective in which patient groups and whether specific approaches are needed depending on the aetiology of malnutrition [13].

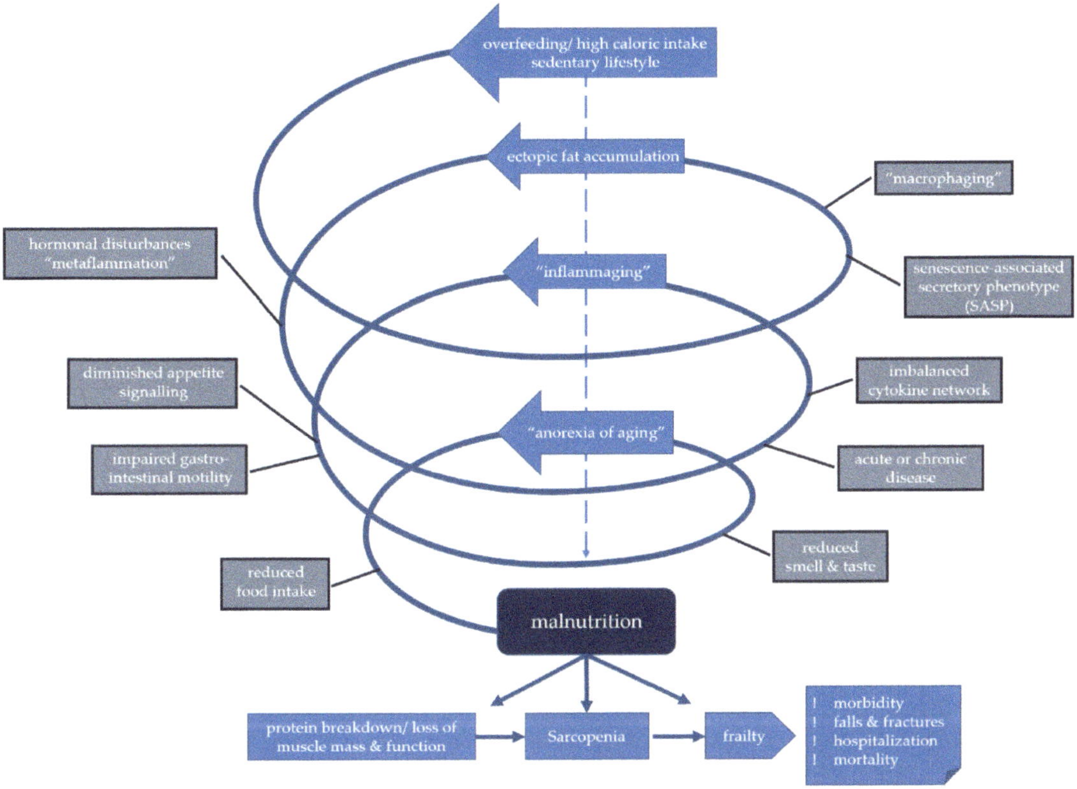

Figure 1. Model of downward spiral in the development of malnutrition in higher age.

A recent meta-analysis, which did not find convincing evidence of the use of oral nutritional supplements in the treatment of malnutrition in older patients, also highlighted the need for more and larger trials to identify suitable and effective interventions for treating malnutrition [159]. Pooling data from 9 randomized controlled trials from different care settings [160], Reinders and colleagues found evidence that dietary counselling with or without oral nutritional supplements was more effective across the settings than oral nutritional supplements alone, but the outcome parameters included merely energy intake and body weight, and no other relevant outcome parameters such as improved muscle strength or function were considered.

7. Remaining Challenges

Despite the overwhelming evidence regarding the negative outcome of malnourished older adults, there are still many remaining challenges in the understanding, identification and treatment of malnutrition in older adults. They concern both the prevention of slow onset age-associated malnutrition as well as the treatment of disease-related malnutrition. Figure 2 summarizes open questions and remaining challenges in this field.

Figure 2. Remaining challenges in the effective prevention and sustainable treatment of malnutrition in old adults. ↑↑↑ indicates a high prevalence, whereas ↑ indicates a lower prevalence.

One particular concern regarding the prevention of malnutrition is that some nutritional requirements are not well established in higher age. Micronutrient and trace element concentrations have been shown to change with age [74] and have even been implicated to play a role in the ageing processes [161]. Changes in body composition and physical activity may lower energy requirements, but not the requirements of nutrients in general. In order to ensure a sufficient supply of essential nutrients, this requires a high nutrient density in the diet. Therefore, requirements of micronutrients which are frequently deficient in older adults, clearly need elucidation given their multifactorial role in promoting health. Also, while the need for some nutrients may change in higher age, ageing processes themselves may impact the requirements of certain nutrients, in particular in the presence of inflammation. As inflammation alters the trace element profile (increasing copper, decreasing selenium and Zn among other trace elements), an inflammation adapted intake of these trace elements may be necessary. However, more research is needed to elucidate inflammation related needs of the older population and the possible benefit of a tailored supplementation. Furthermore, as higher age is frequently associated with disease, disease-specific nutrient requirements may further complicate the picture.

Taken together, a need to critically appraise current nutrient recommendations for the older age group has been identified, and the WHO has been challenged to issue guidelines [162]. There is also an increasing demand worldwide for WHO guidelines which competent national authorities can use to address the nutritional needs of their growing older populations [70].

Moreover, although recent research on determinants of malnutrition has added considerable new evidence, the impact of ageing-related changes on the long-term development of malnutrition, still needs further elucidation. One example is the ageing gut and its altered microbiome, which have gained attention in the last years. Emerging technologies, such as e.g., high-throughput culturing, will further the research on ageing microbiome, so its role in the development of malnutrition might soon be elucidated. Novel dietary approaches which modulate the ageing microbiome such as the Mediterranean diet [163] are of interest in the prevention of malnutrition, and more studies are warranted. Considering the importance of muscle mass maintenance [164], nutritional intervention needs to focus on muscle mass and also, needs to be combined with exercise. Therefore, more studies on nutritional therapy together with different kinds of exercise regimen are clearly needed.

Lastly, not only the prevention of malnutrition, but also the topic of treatment of malnutrition needs more research, in terms of well-designed and adequately powered clinical trials in order to ensure sufficient statistical power to identify true treatment effects. Moreover, relevant outcome parameters need to be included, and a careful selection of target populations with well-defined malnutrition is necessary. Also, more research is needed for treatment approaches which specifically target the underlying causes of malnutrition themselves [165] i.e., causative treatment. This, however, not only requires a longer time frame and a broader inter-disciplinary approach, but more research regarding the most relevant causes and their common pathophysiology which would allow a "causation-oriented" multimodal treatment [13].

8. Conclusions

Taken together, more research is needed to understand which ageing-related changes are early predictors/precursors of malnutrition that in turn can be addressed in order to prevent the development of nutritional deficiencies. For clinically manifest malnutrition, more studies need to be performed in older adults in order to identify the suitable treatment for the various settings.

Author Contributions: Conceptualization, K.N.; writing—original draft preparation, K.N. and U.H.; writing—review and editing, K.N. and M.P.; visualization, U.H. All authors have read and agreed to the published version of the manuscript.

Funding: This research received no external funding.

Conflicts of Interest: The authors declare no conflict of interest.

Abbreviations

ASPEN	American Society for Parenteral and Enteral Nutrition
BMI	body mass index
CCK	cholecystokinin
CNS	central nervous system
COVID-19	coronavirus disease 2019
DoMAP	Determinants of Malnutrition in Aged Persons
DNA	deoxyribonucleic acid
ESPEN	European Society of Clinical Nutrition and Metabolism
Fe	iron
FELANPE	Federación Latinoamericana de Terapia Nutricional, Nutrición Clínica y Metabolismo
FGF21	fibroblast growth factor 21
GLIM	Global Leadership Initiative on Malnutrition
HIV/AIDS	human immunodeficiency virus/acquired immunodeficiency syndrome
ICU	intensive care unit
IL	interleukin
MNA	Mini Nutritional Assessment
PENSA	The Parenteral and Enteral Nutrition Society of Asia
PEM	protein-energy malnutrition
PEU	protein-energy undernutrition
(PG-) SGA	(Patient-Generated) Subjective Global Assessment
PYY	peptide YY

SASP	senescence-associated secretory phenotype
TNF-α	tumour necrosis factor alpha
WHO	World Health Organization
Zn	zinc

References

1. Rudnicka, E.; Napierała, P.; Podfigurna, A.; Męczekalski, B.; Smolarczyk, R.; Grymowicz, M. The World Health Organization (WHO) approach to healthy ageing. *Maturitas* **2020**, *139*, 6–11. [CrossRef] [PubMed]
2. United Nations. *World Population Prospects 2019: Highlights*; Department of Economic and Social Affairs, Population Division: New York, NY, USA, 2019.
3. Niccoli, T.; Partridge, L. Ageing as a risk factor for disease. *Curr. Biol.* **2012**, *22*, R741–R752. [CrossRef] [PubMed]
4. Biagi, E.; Nylund, L.; Candela, M.; Ostan, R.; Bucci, L.; Pini, E.; Nikkila, J.; Monti, D.; Satokari, R.; Franceschi, C.; et al. Through ageing, and beyond: Gut microbiota and inflammatory status in seniors and centenarians. *PLoS ONE* **2010**, *5*, e10667. [CrossRef]
5. Hickson, M. Malnutrition and ageing. *Postgrad. Med. J.* **2006**, *82*, 2–8. [CrossRef]
6. Norman, K.; Pichard, C.; Lochs, H.; Pirlich, M. Prognostic impact of disease-related malnutrition. *Clin. Nutr.* **2008**, *27*, 5–15. [CrossRef]
7. Cunha, A.I.L.; Veronese, N.; de Melo Borges, S.; Ricci, N.A. Frailty as a predictor of adverse outcomes in hospitalized older adults: A systematic review and meta-analysis. *Ageing Res. Rev.* **2019**, *56*, 100960. [CrossRef]
8. Otten, L.; Stobäus, N.; Franz, K.; Genton, L.; Müller-Werdan, U.; Wirth, R.; Norman, K. Impact of sarcopenia on 1-year mortality in older patients with cancer. *Age Ageing* **2019**, *48*, 413–418. [CrossRef]
9. Bardon, L.A.; Streicher, M.; Corish, C.A.; Clarke, M.; Power, L.C.; Kenny, R.A.; O'Connor, D.M.; Laird, E.; O'Connor, E.M.; Visser, M.; et al. Predictors of Incident Malnutrition in Older Irish Adults from the Irish Longitudinal Study on Ageing Cohort-A MaNuEL study. *J. Gerontol. Ser. A* **2020**, *75*, 249–256. [CrossRef]
10. Dean, M.; Raats, M.M.; Grunert, K.G.; Lumbers, M. Factors influencing eating a varied diet in old age. *Public Health Nutr.* **2009**, *12*, 2421–2427. [CrossRef]
11. Amarya, S.; Singh, K.; Sabharwal, M. Changes during aging and their association with malnutrition. *J. Clin. Gerontol. Geriatr.* **2015**, *6*, 78–84. [CrossRef]
12. Gomes, F.; Schuetz, P.; Bounoure, L.; Austin, P.; Ballesteros-Pomar, M.; Cederholm, T.; Fletcher, J.; Laviano, A.; Norman, K.; Poulia, K.A.; et al. ESPEN guidelines on nutritional support for polymorbid internal medicine patients. *Clin. Nutr.* **2018**, *37*, 336–353. [CrossRef]
13. Volkert, D.; Beck, A.M.; Cederholm, T.; Cereda, E.; Cruz-Jentoft, A.; Goisser, S.; de Groot, L.; Großhauser, F.; Kiesswetter, E.; Norman, K.; et al. Management of Malnutrition in Older Patients-Current Approaches, Evidence and Open Questions. *J. Clin. Med.* **2019**, *8*, 974. [CrossRef] [PubMed]
14. Van Den Broeke, C.; De Burghgraeve, T.; Ummels, M.; Gescher, N.; Deckx, L.; Tjan-Heijnen, V.; Buntinx, F.; van den Akker, M. Occurrence of Malnutrition and Associated Factors in Community-Dwelling Older Adults: Those with a Recent Diagnosis of Cancer Are at Higher Risk. *J. Nutr. Health Aging* **2018**, *22*, 191–198. [CrossRef]
15. Eckert, C.; Gell, N.M.; Wingood, M.; Schollmeyer, J.; Tarleton, E.K. Malnutrition Risk, Rurality, and Falls among Community-Dwelling Older Adults. *J. Nutr. Health Aging* **2021**, *25*, 624–627. [CrossRef] [PubMed]
16. Khalatbari-Soltani, S.; Marques-Vidal, P. The economic cost of hospital malnutrition in Europe; a narrative review. *Clin. Nutr. ESPEN* **2015**, *10*, e89–e94. [CrossRef]
17. Lengelé, L.; Bruyère, O.; Beaudart, C.; Reginster, J.Y.; Locquet, M. Impact of Malnutrition Status on Muscle Parameter Changes over a 5-Year Follow-Up of Community-Dwelling Older Adults from the SarcoPhAge Cohort. *Nutrients* **2021**, *13*, 407. [CrossRef] [PubMed]
18. Coin, A.; Sergi, G.; Benincà, P.; Lupoli, L.; Cinti, G.; Ferrara, L.; Benedetti, G.; Tomasi, G.; Pisent, C.; Enzi, G. Bone mineral density and body composition in underweight and normal elderly subjects. *Osteoporos. Int.* **2000**, *11*, 1043–1050. [CrossRef]
19. Geinoz, G.; Rapin, C.H.; Rizzoli, R.; Kraemer, R.; Buchs, B.; Slosman, D.; Michel, J.P.; Bonjour, J.P. Relationship between bone mineral density and dietary intakes in the elderly. *Osteoporos. Int.* **1993**, *3*, 242–248. [CrossRef]
20. Bonjour, J.P.; Schurch, M.A.; Rizzoli, R. Nutritional aspects of hip fractures. *Bone* **1996**, *18*, 139s–144s. [CrossRef]
21. Schaible, U.E.; Kaufmann, S.H. Malnutrition and infection: Complex mechanisms and global impacts. *PLoS Med.* **2007**, *4*, e115. [CrossRef] [PubMed]
22. Kawakami, K.; Kadota, J.; Iida, K.; Shirai, R.; Abe, K.; Kohno, S. Reduced immune function and malnutrition in the elderly. *Tohoku J. Exp. Med.* **1999**, *187*, 157–171. [CrossRef] [PubMed]
23. Alam, I.; Almajwal, A.; Alam, W.; Alam, I.; Ullah, N.; Abulmeaaty, M.; Razak, S.; Khan, S.; Pawelec, G.; Paracha, P. The immune-nutrition interplay in aging facts and controversies. *Nutr. Healthy Aging* **2019**, *5*, 73–95. [CrossRef]
24. Fitzpatrick, F.; Skally, M.; O'Hanlon, C.; Foley, M.; Houlihan, J.; Gaughan, L.; Smith, O.; Moore, B.; Cunneen, S.; Sweeney, E.; et al. Food for thought. Malnutrition risk associated with increased risk of healthcare-associated infection. *J. Hosp. Infect.* **2019**, *101*, 300–304. [CrossRef]
25. Shpata, V.; Ohri, I.; Nurka, T.; Prendushi, X. The prevalence and consequences of malnutrition risk in elderly Albanian intensive care unit patients. *Clin. Interv. Aging* **2015**, *10*, 481–486. [CrossRef] [PubMed]

26. Barchitta, M.; Maugeri, A.; Favara, G.; Magnano San Lio, R.; Evola, G.; Agodi, A.; Basile, G. Nutrition and Wound Healing: An Overview Focusing on the Beneficial Effects of Curcumin. *Int. J. Mol. Sci.* **2019**, *20*, 1119. [CrossRef] [PubMed]
27. Olsson, M.; Järbrink, K.; Divakar, U.; Bajpai, R.; Upton, Z.; Schmidtchen, A.; Car, J. The humanistic and economic burden of chronic wounds: A systematic review. *Wound Repair Regen.* **2019**, *27*, 114–125. [CrossRef]
28. Hébuterne, X.; Bermon, S.; Schneider, S.M. Ageing and muscle: The effects of malnutrition, re-nutrition, and physical exercise. *Curr. Opin. Clin. Nutr. Metab. Care* **2001**, *4*, 295–300. [CrossRef]
29. Inouye, S.K.; Studenski, S.; Tinetti, M.E.; Kuchel, G.A. Geriatric syndromes: Clinical, research, and policy implications of a core geriatric concept. *J. Am. Geriatr. Soc.* **2007**, *55*, 780–791. [CrossRef]
30. Kane, R.L.; Shamliyan, T.; Talley, K.; Pacala, J. The association between geriatric syndromes and survival. *J. Am. Geriatr. Soc.* **2012**, *60*, 896–904. [CrossRef] [PubMed]
31. Rausch, C.; van Zon, S.K.R.; Liang, Y.; Laflamme, L.; Möller, J.; de Rooij, S.E.; Bültmann, U. Geriatric Syndromes and Incident Chronic Health Conditions Among 9094 Older Community-Dwellers: Findings from the Lifelines Cohort Study. *J. Am. Med. Dir. Assoc.* **2021**. [CrossRef] [PubMed]
32. Won, C.W.; Yoo, H.J.; Yu, S.H.; Kim, C.O.; Dumlao, L.C.I.; Dewiasty, E.; Rowland, J.; Chang, H.H.; Wang, J.; Akishita, M.; et al. Lists of geriatric syndromes in the Asian-Pacific geriatric societies. *Eur. Geriatr. Med.* **2013**, *4*, 335–338. [CrossRef]
33. Tan, V.M.H.; Pang, B.W.J.; Lau, L.K.; Jabbar, K.A.; Seah, W.T.; Chen, K.K.; Ng, T.P.; Wee, S.L. Malnutrition and Sarcopenia in Community-Dwelling Adults in Singapore: Yishun Health Study. *J. Nutr. Health Aging* **2021**, *25*, 374–381. [CrossRef] [PubMed]
34. Juby, A.G.; Mager, D.R. A review of nutrition screening tools used to assess the malnutrition-sarcopenia syndrome (MSS) in the older adult. *Clin. Nutr. ESPEN* **2019**, *32*, 8–15. [CrossRef]
35. Stenholm, S.; Harris, T.B.; Rantanen, T.; Visser, M.; Kritchevsky, S.B.; Ferrucci, L. Sarcopenic obesity: Definition, cause and consequences. *Curr. Opin. Clin. Nutr. Metab. Care* **2008**, *11*, 693–700. [CrossRef] [PubMed]
36. Fougère, B.; Morley, J.E. Editorial: Weight Loss is a Major Cause of Frailty. *J. Nutr. Health Aging* **2017**, *21*, 933–935. [CrossRef] [PubMed]
37. Boulos, C.; Salameh, P.; Barberger-Gateau, P. Malnutrition and frailty in community dwelling older adults living in a rural setting. *Clin. Nutr.* **2016**, *35*, 138–143. [CrossRef] [PubMed]
38. Jeejeebhoy, K.N. Malnutrition, fatigue, frailty, vulnerability, sarcopenia and cachexia: Overlap of clinical features. *Curr. Opin. Clin. Nutr. Metab. Care* **2012**, *15*, 213–219. [CrossRef]
39. Gingrich, A.; Volkert, D.; Kiesswetter, E.; Thomanek, M.; Bach, S.; Sieber, C.C.; Zopf, Y. Prevalence and overlap of sarcopenia, frailty, cachexia and malnutrition in older medical inpatients. *BMC Geriatr.* **2019**, *19*, 120. [CrossRef]
40. Norazman, C.W.; Adznam, S.N.; Jamaluddin, R. Malnutrition as Key Predictor of Physical Frailty among Malaysian Older Adults. *Nutrients* **2020**, *12*, 1713. [CrossRef]
41. Norazman, C.W.; Adznam, S.N.; Jamaluddin, R. Physical Frailty among Urban-Living Community-Dwelling Older Adults in Malaysia. *Int. J. Environ. Res. Public Health* **2020**, *17*, 6549. [CrossRef]
42. Wei, K.; Nyunt, M.S.; Gao, Q.; Wee, S.L.; Yap, K.B.; Ng, T.P. Association of Frailty and Malnutrition With Long-term Functional and Mortality Outcomes Among Community-Dwelling Older Adults: Results From the Singapore Longitudinal Aging Study 1. *JAMA Netw. Open* **2018**, *1*, e180650. [CrossRef]
43. Zengarini, E.; Ruggiero, C.; Pérez-Zepeda, M.U.; Hoogendijk, E.O.; Vellas, B.; Mecocci, P.; Cesari, M. Fatigue: Relevance and implications in the aging population. *Exp. Gerontol.* **2015**, *70*, 78–83. [CrossRef]
44. Filler, K.; Lyon, D.; Bennett, J.; McCain, N.; Elswick, R.; Lukkahatai, N.; Saligan, L.N. Association of Mitochondrial Dysfunction and Fatigue: A Review of the Literature. *BBA Clin.* **2014**, *1*, 12–23. [CrossRef]
45. Herpich, C.; Franz, K.; Klaus, S.; Müller-Werdan, U.; Ost, M.; Norman, K. Age-related fatigue is associated with reduced mitochondrial function in peripheral blood mononuclear cells. *Exp. Gerontol.* **2021**, *144*, 111177. [CrossRef]
46. Franz, K.; Otten, L.; Müller-Werdan, U.; Doehner, W.; Norman, K. Severe Weight Loss and Its Association with Fatigue in Old Patients at Discharge from a Geriatric Hospital. *Nutrients* **2019**, *11*, 2415. [CrossRef]
47. Azzolino, D.; Arosio, B.; Marzetti, E.; Calvani, R.; Cesari, M. Nutritional Status as a Mediator of Fatigue and Its Underlying Mechanisms in Older People. *Nutrients* **2020**, *12*, 444. [CrossRef]
48. Haß, U.; Herpich, C.; Norman, K. Anti-Inflammatory Diets and Fatigue. *Nutrients* **2019**, *11*, 2315. [CrossRef] [PubMed]
49. Ogawa, S. Nutritional management of older adults with cognitive decline and dementia. *Geriatr. Gerontol. Int.* **2014**, *14* (Suppl. 2), 17–22. [CrossRef] [PubMed]
50. Yu, W.; Yu, W.; Liu, X.; Wan, T.; Chen, C.; Xiong, L.; Zhang, W.; Lü, Y. Associations between malnutrition and cognitive impairment in an elderly Chinese population: An analysis based on a 7-year database. *Psychogeriatrics* **2021**, *21*, 80–88. [CrossRef] [PubMed]
51. Cabrera, M.A.; Mesas, A.E.; Garcia, A.R.; de Andrade, S.M. Malnutrition and depression among community-dwelling elderly people. *J. Am. Med. Dir. Assoc.* **2007**, *8*, 582–584. [CrossRef] [PubMed]
52. Smoliner, C.; Norman, K.; Wagner, K.H.; Hartig, W.; Lochs, H.; Pirlich, M. Malnutrition and depression in the institutionalised elderly. *Br. J. Nutr.* **2009**, *102*, 1663–1667. [CrossRef]
53. Yoshimura, K.; Yamada, M.; Kajiwara, Y.; Nishiguchi, S.; Aoyama, T. Relationship between depression and risk of malnutrition among community-dwelling young-old and old-old elderly people. *Aging Ment. Health* **2013**, *17*, 456–460. [CrossRef]

54. Gentile, S.; Lacroix, O.; Durand, A.C.; Cretel, E.; Alazia, M.; Sambuc, R.; Bonin-Guillaume, S. Malnutrition: A highly predictive risk factor of short-term mortality in elderly presenting to the emergency department. *J. Nutr. Health Aging* **2013**, *17*, 290–294. [CrossRef]
55. Komici, K.; Vitale, D.F.; Mancini, A.; Bencivenga, L.; Conte, M.; Provenzano, S.; Grieco, F.V.; Visaggi, L.; Ronga, I.; Cittadini, A.; et al. Impact of Malnutrition on Long-Term Mortality in Elderly Patients with Acute Myocardial Infarction. *Nutrients* **2019**, *11*, 224. [CrossRef]
56. Söderström, L.; Rosenblad, A.; Thors Adolfsson, E.; Bergkvist, L. Malnutrition is associated with increased mortality in older adults regardless of the cause of death. *Br. J. Nutr.* **2017**, *117*, 532–540. [CrossRef] [PubMed]
57. Teigen, L.M.; Kuchnia, A.J.; Nagel, E.M.; Price, K.L.; Hurt, R.T.; Earthman, C.P. Diagnosing clinical malnutrition: Perspectives from the past and implications for the future. *Clin. Nutr. ESPEN* **2018**, *26*, 13–20. [CrossRef]
58. Cederholm, T.; Jensen, G.L.; Correia, M.; Gonzalez, M.C.; Fukushima, R.; Higashiguchi, T.; Baptista, G.; Barazzoni, R.; Blaauw, R.; Coats, A.; et al. GLIM criteria for the diagnosis of malnutrition–A consensus report from the global clinical nutrition community. *Clin. Nutr.* **2019**, *38*, 1–9. [CrossRef] [PubMed]
59. De van der Schueren, M.A.E.; Keller, H.; Cederholm, T.; Barazzoni, R.; Compher, C.; Correia, M.; Gonzalez, M.C.; Jager-Wittenaar, H.; Pirlich, M.; Steiber, A.; et al. Global Leadership Initiative on Malnutrition (GLIM): Guidance on validation of the operational criteria for the diagnosis of protein-energy malnutrition in adults. *Clin. Nutr.* **2020**, *39*, 2872–2880. [CrossRef] [PubMed]
60. Yeung, S.S.Y.; Chan, R.S.M.; Kwok, T.; Lee, J.S.W.; Woo, J. Malnutrition According to GLIM Criteria and Adverse Outcomes in Community-Dwelling Chinese Older Adults: A Prospective Analysis. *J. Am. Med. Dir. Assoc.* **2020**. [CrossRef]
61. Cederholm, T.; Barazzoni, R.; Austin, P.; Ballmer, P.; Biolo, G.; Bischoff, S.C.; Compher, C.; Correia, I.; Higashiguchi, T.; Holst, M.; et al. ESPEN guidelines on definitions and terminology of clinical nutrition. *Clin. Nutr.* **2017**, *36*, 49–64. [CrossRef] [PubMed]
62. Cereda, E. Mini nutritional assessment. *Curr. Opin. Clin. Nutr. Metab. Care* **2012**, *15*, 29–41. [CrossRef]
63. Vandewoude, M.F.; Alish, C.J.; Sauer, A.C.; Hegazi, R.A. Malnutrition-sarcopenia syndrome: Is this the future of nutrition screening and assessment for older adults? *J. Aging Res.* **2012**, *2012*, 651570. [CrossRef] [PubMed]
64. Kinosian, B.; Jeejeebhoy, K.N. What is malnutrition? Does it matter? *Nutrition* **1995**, *11*, 196–197.
65. Bauer, J.; Biolo, G.; Cederholm, T.; Cesari, M.; Cruz-Jentoft, A.J.; Morley, J.E.; Phillips, S.; Sieber, C.; Stehle, P.; Teta, D.; et al. Evidence-based recommendations for optimal dietary protein intake in older people: A position paper from the PROT-AGE Study Group. *J. Am. Med. Dir. Assoc.* **2013**, *14*, 542–559. [CrossRef] [PubMed]
66. Deutz, N.E.; Bauer, J.M.; Barazzoni, R.; Biolo, G.; Boirie, Y.; Bosy-Westphal, A.; Cederholm, T.; Cruz-Jentoft, A.; Krznaric, Z.; Nair, K.S.; et al. Protein intake and exercise for optimal muscle function with aging: Recommendations from the ESPEN Expert Group. *Clin. Nutr.* **2014**, *33*, 929–936. [CrossRef]
67. Hengeveld, L.M.; Wijnhoven, H.A.H.; Olthof, M.R.; Brouwer, I.A.; Harris, T.B.; Kritchevsky, S.B.; Newman, A.B.; Visser, M. Prospective associations of poor diet quality with long-term incidence of protein-energy malnutrition in community-dwelling older adults: The Health, Aging, and Body Composition (Health ABC) Study. *Am. J. Clin. Nutr.* **2018**, *107*, 155–164. [CrossRef] [PubMed]
68. Hengeveld, L.M.; Boer, J.M.A.; Gaudreau, P.; Heymans, M.W.; Jagger, C.; Mendonça, N.; Ocké, M.C.; Presse, N.; Sette, S.; Simonsick, E.M.; et al. Prevalence of protein intake below recommended in community-dwelling older adults: A meta-analysis across cohorts from the PROMISS consortium. *J. Cachexia Sarcopenia Muscle* **2020**, *11*, 1212–1222. [CrossRef]
69. Conzade, R.; Koenig, W.; Heier, M.; Schneider, A.; Grill, E.; Peters, A.; Thorand, B. Prevalence and Predictors of Subclinical Micronutrient Deficiency in German Older Adults: Results from the Population-Based KORA-Age Study. *Nutrients* **2017**, *9*, 1276. [CrossRef]
70. IOM (Institute of Medicine). *Providing Healthy and Safe Foods as We Age: Workshop Summary*; The National Academies Press: Washington, DC, USA, 2010.
71. Hamer, D.H.; Sempértegui, F.; Estrella, B.; Tucker, K.L.; Rodríguez, A.; Egas, J.; Dallal, G.E.; Selhub, J.; Griffiths, J.K.; Meydani, S.N. Micronutrient deficiencies are associated with impaired immune response and higher burden of respiratory infections in elderly Ecuadorians. *J. Nutr.* **2009**, *139*, 113–119. [CrossRef]
72. Wallace, T.C.; Frankenfeld, C.L.; Frei, B.; Shah, A.V.; Yu, C.R.; van Klinken, B.J.; Adeleke, M. Multivitamin/Multimineral Supplement Use is Associated with Increased Micronutrient Intakes and Biomarkers and Decreased Prevalence of Inadequacies and Deficiencies in Middle-Aged and Older Adults in the United States. *J. Nutr. Gerontol. Geriatr.* **2019**, *38*, 307–328. [CrossRef]
73. Elmadfa, I.; Meyer, A.L. *Present Knowledge in Nutrition. Chapter 5–Nutrition, Aging, and Requirements in the Elderly*; Marriott, B.P., Birt, D.F., Stallings, V.A., Yates, A.A., Eds.; Academic Press: Cambridge, MA, USA, 2020; Volume 11.
74. Maggini, S.; Pierre, A.; Calder, P.C. Immune Function and Micronutrient Requirements Change over the Life Course. *Nutrients* **2018**, *10*, 1531. [CrossRef]
75. Karadima, V.; Kraniotou, C.; Bellos, G.; Tsangaris, G.T. Drug-micronutrient interactions: Food for thought and thought for action. *EPMA J.* **2016**, *7*, 10. [CrossRef] [PubMed]
76. Gröber, U.; Schmidt, J.; Kisters, K. Important drug-micronutrient interactions: A selection for clinical practice. *Crit. Rev. Food Sci. Nutr.* **2020**, *60*, 257–275. [CrossRef] [PubMed]
77. Kehl-Fie, T.E.; Skaar, E.P. Nutritional immunity beyond iron: A role for manganese and zinc. *Curr. Opin. Chem. Biol.* **2010**, *14*, 218–224. [CrossRef]

78. Mocchegiani, E.; Romeo, J.; Malavolta, M.; Costarelli, L.; Giacconi, R.; Diaz, L.E.; Marcos, A. Zinc: Dietary intake and impact of supplementation on immune function in elderly. *Age* **2013**, *35*, 839–860. [CrossRef] [PubMed]
79. Jung, A.; Spira, D.; Steinhagen-Thiessen, E.; Demuth, I.; Norman, K. Zinc Deficiency Is associated With Depressive Symptoms-Results From the Berlin Aging Study II. *J. Gerontol. Ser. A Biol. Sci. Med. Sci.* **2017**, *72*, 1149–1154. [CrossRef] [PubMed]
80. Suzuki, H.; Asakawa, A.; Li, J.B.; Tsai, M.; Amitani, H.; Ohinata, K.; Komai, M.; Inui, A. Zinc as an appetite stimulator–the possible role of zinc in the progression of diseases such as cachexia and sarcopenia. *Recent Pat. Food Nutr. Agric.* **2011**, *3*, 226–231. [CrossRef]
81. Klotz, L.O.; Kröncke, K.D.; Buchczyk, D.P.; Sies, H. Role of copper, zinc, selenium and tellurium in the cellular defense against oxidative and nitrosative stress. *J. Nutr.* **2003**, *133*, 1448s–1451s. [CrossRef]
82. Mocchegiani, E.; Malavolta, M. Zinc-gene interaction related to inflammatory/immune response in ageing. *Genes Nutr.* **2008**, *3*, 61–75. [CrossRef]
83. Mocchegiani, E.; Basso, A.; Giacconi, R.; Piacenza, F.; Costarelli, L.; Pierpaoli, S.; Malavolta, M. Diet (zinc)-gene interaction related to inflammatory/immune response in ageing: Possible link with frailty syndrome? *Biogerontology* **2010**, *11*, 589–595. [CrossRef]
84. Lauretani, F.; Semba, R.D.; Bandinelli, S.; Dayhoff-Brannigan, M.; Giacomini, V.; Corsi, A.M.; Guralnik, J.M.; Ferrucci, L. Low plasma carotenoids and skeletal muscle strength decline over 6 years. *J. Gerontol. Ser. A Biol. Sci. Med. Sci.* **2008**, *63*, 376–383. [CrossRef]
85. Lauretani, F.; Semba, R.D.; Bandinelli, S.; Dayhoff-Brannigan, M.; Lauretani, F.; Corsi, A.M.; Guralnik, J.M.; Ferrucci, L. Carotenoids as protection against disability in older persons. *Rejuvenation Res.* **2008**, *11*, 557–563. [CrossRef]
86. Bartali, B.; Frongillo, E.A.; Guralnik, J.M.; Stipanuk, M.H.; Allore, H.G.; Cherubini, A.; Bandinelli, S.; Ferrucci, L.; Gill, T.M. Serum micronutrient concentrations and decline in physical function among older persons. *JAMA* **2008**, *299*, 308–315. [CrossRef] [PubMed]
87. Cesari, M.; Pahor, M.; Bartali, B.; Cherubini, A.; Penninx, B.W.; Williams, G.R.; Atkinson, H.; Martin, A.; Guralnik, J.M.; Ferrucci, L. Antioxidants and physical performance in elderly persons: The Invecchiare in Chianti (InCHIANTI) study. *Am. J. Clin. Nutr.* **2004**, *79*, 289–294. [CrossRef] [PubMed]
88. Sánchez-Rodríguez, D.; Annweiler, C.; Ronquillo-Moreno, N.; Tortosa-Rodríguez, A.; Guillén-Solà, A.; Vázquez-Ibar, O.; Escalada, F.; Muniesa, J.M.; Marco, E. Clinical application of the basic definition of malnutrition proposed by the European Society for Clinical Nutrition and Metabolism (ESPEN): Comparison with classical tools in geriatric care. *Arch. Gerontol. Geriatr.* **2018**, *76*, 210–214. [CrossRef]
89. Leij-Halfwerk, S.; Verwijs, M.H.; van Houdt, S.; Borkent, J.W.; Guaitoli, P.R.; Pelgrim, T.; Heymans, M.W.; Power, L.; Visser, M.; Corish, C.A.; et al. Prevalence of protein-energy malnutrition risk in European older adults in community, residential and hospital settings, according to 22 malnutrition screening tools validated for use in adults >/=65 years: A systematic review and meta-analysis. *Maturitas* **2019**, *126*, 80–89. [CrossRef] [PubMed]
90. Crichton, M.; Craven, D.; Mackay, H.; Marx, W.; de van der Schueren, M.; Marshall, S. A systematic review, meta-analysis and meta-regression of the prevalence of protein-energy malnutrition: Associations with geographical region and sex. *Age Ageing* **2019**, *48*, 38–48. [CrossRef]
91. Cereda, E.; Pedrolli, C.; Klersy, C.; Bonardi, C.; Quarleri, L.; Cappello, S.; Turri, A.; Rondanelli, M.; Caccialanza, R. Nutritional status in older persons according to healthcare setting: A systematic review and meta-analysis of prevalence data using MNA®. *Clin. Nutr.* **2016**, *35*, 1282–1290. [CrossRef] [PubMed]
92. Wolters, M.; Volkert, D.; Streicher, M.; Kiesswetter, E.; Torbahn, G.; O'Connor, E.M.; O'Keeffe, M.; Kelly, M.; O'Herlihy, E.; O'Toole, P.W.; et al. Prevalence of malnutrition using harmonized definitions in older adults from different settings–A MaNuEL study. *Clin. Nutr.* **2019**, *38*, 2389–2398. [CrossRef]
93. Streicher, M.; van Zwienen-Pot, J.; Bardon, L.; Nagel, G.; Teh, R.; Meisinger, C.; Colombo, M.; Torbahn, G.; Kiesswetter, E.; Flechtner-Mors, M.; et al. Determinants of Incident Malnutrition in Community-Dwelling Older Adults: A MaNuEL Multicohort Meta-Analysis. *J. Am. Geriatr. Soc.* **2018**, *66*, 2335–2343. [CrossRef]
94. O'Keeffe, M.; Kelly, M.; O'Herlihy, E.; O'Toole, P.W.; Kearney, P.M.; Timmons, S.; O'Shea, E.; Stanton, C.; Hickson, M.; Rolland, Y.; et al. Potentially modifiable determinants of malnutrition in older adults: A systematic review. *Clin. Nutr.* **2019**, *38*, 2477–2498. [CrossRef] [PubMed]
95. Volkert, D.; Kiesswetter, E.; Cederholm, T.; Donini, L.M.; Eglseer, D.; Norman, K.; Schneider, S.M.; Ströbele-Benschop, N.; Torbahn, G.; Wirth, R.; et al. Development of a Model on Determinants of Malnutrition in Aged Persons: A MaNuEL Project. *Gerontol. Geriatr. Med.* **2019**, *5*, 2333721419858438. [CrossRef] [PubMed]
96. Rémond, D.; Shahar, D.R.; Gille, D.; Pinto, P.; Kachal, J.; Peyron, M.A.; Dos Santos, C.N.; Walther, B.; Bordoni, A.; Dupont, D.; et al. Understanding the gastrointestinal tract of the elderly to develop dietary solutions that prevent malnutrition. *Oncotarget* **2015**, *6*, 13858–13898. [CrossRef] [PubMed]
97. An, R.; Wilms, E.; Masclee, A.A.M.; Smidt, H.; Zoetendal, E.G.; Jonkers, D. Age-dependent changes in GI physiology and microbiota: Time to reconsider? *Gut* **2018**, *67*, 2213–2222. [CrossRef]
98. Nagpal, R.; Mainali, R.; Ahmadi, S.; Wang, S.; Singh, R.; Kavanagh, K.; Kitzman, D.W.; Kushugulova, A.; Marotta, F.; Yadav, H. Gut microbiome and aging: Physiological and mechanistic insights. *Nutr. Healthy Aging* **2018**, *4*, 267–285. [CrossRef] [PubMed]
99. Van de Wouw, M.; Schellekens, H.; Dinan, T.G.; Cryan, J.F. Microbiota-Gut-Brain Axis: Modulator of Host Metabolism and Appetite. *J. Nutr.* **2017**, *147*, 727–745. [CrossRef]

100. Ticinesi, A.; Lauretani, F.; Milani, C.; Nouvenne, A.; Tana, C.; Del Rio, D.; Maggio, M.; Ventura, M.; Meschi, T. Aging Gut Microbiota at the Cross-Road between Nutrition, Physical Frailty, and Sarcopenia: Is There a Gut-Muscle Axis? *Nutrients* **2017**, *9*, 1303. [CrossRef]
101. Moss, C.; Dhillo, W.S.; Frost, G.; Hickson, M. Gastrointestinal hormones: The regulation of appetite and the anorexia of ageing. *J. Hum. Nutr. Diet.* **2012**, *25*, 3–15. [CrossRef]
102. Atalayer, D.; Astbury, N.M. Anorexia of aging and gut hormones. *Aging Dis.* **2013**, *4*, 264–275. [CrossRef]
103. Morley, J.E.; Miller, D.K.; Perry, H.M., 3rd; Patrick, P.; Guigoz, Y.; Vellas, B. Anorexia of aging, leptin, and the Mini Nutritional Assessment. *Nestle Nutr. Workshop Ser. Clin. Perform. Programme* **1999**, *1*, 67–76. [CrossRef]
104. Ueno, H.; Nakazato, M. Mechanistic relationship between the vagal afferent pathway, central nervous system and peripheral organs in appetite regulation. *J. Diabetes Investig.* **2016**, *7*, 812–818. [CrossRef] [PubMed]
105. Kirsz, K.; Zieba, D.A. Ghrelin-mediated appetite regulation in the central nervous system. *Peptides* **2011**, *32*, 2256–2264. [CrossRef]
106. Pliquett, R.U.; Führer, D.; Falk, S.; Zysset, S.; von Cramon, D.Y.; Stumvoll, M. The effects of insulin on the central nervous system–focus on appetite regulation. *Horm. Metab. Res.* **2006**, *38*, 442–446. [CrossRef] [PubMed]
107. Serra-Prat, M.; Mans, E.; Palomera, E.; Clavé, P. Gastrointestinal peptides, gastrointestinal motility, and anorexia of aging in frail elderly persons. *Neurogastroenterol. Motil.* **2013**, *25*, 291-e245. [CrossRef]
108. Bauer, J.M.; Haack, A.; Winning, K.; Wirth, R.; Fischer, B.; Uter, W.; Erdmann, J.; Schusdziarra, V.; Sieber, C.C. Impaired postprandial response of active ghrelin and prolonged suppression of hunger sensation in the elderly. *J. Gerontol. Ser. A Biol. Sci. Med. Sci.* **2010**, *65*, 307–311. [CrossRef]
109. Morton, G.J.; Cummings, D.E.; Baskin, D.G.; Barsh, G.S.; Schwartz, M.W. Central nervous system control of food intake and body weight. *Nature* **2006**, *443*, 289–295. [CrossRef]
110. Brown, W.E.; Bradford, B.J. Invited review: Mechanisms of hypophagia during disease. *J. Dairy Sci.* **2021**. [CrossRef]
111. Wysokinski, A.; Sobow, T.; Kloszewska, I.; Kostka, T. Mechanisms of the anorexia of aging—A review. *Age* **2015**, *37*, 9821. [CrossRef]
112. Cox, N.J.; Ibrahim, K.; Sayer, A.A.; Robinson, S.M.; Roberts, H.C. Assessment and Treatment of the Anorexia of Aging: A Systematic Review. *Nutrients* **2019**, *11*, 144. [CrossRef] [PubMed]
113. Giezenaar, C.; Chapman, I.; Luscombe-Marsh, N.; Feinle-Bisset, C.; Horowitz, M.; Soenen, S. Ageing Is Associated with Decreases in Appetite and Energy Intake–A Meta-Analysis in Healthy Adults. *Nutrients* **2016**, *8*, 28. [CrossRef] [PubMed]
114. Roy, M.; Shatenstein, B.; Gaudreau, P.; Morais, J.A.; Payette, H. Seniors' body weight dissatisfaction and longitudinal associations with weight changes, anorexia of aging, and obesity: Results from the NuAge Study. *J. Aging Health* **2015**, *27*, 220–238. [CrossRef]
115. Morley, J.E. Anorexia of aging: A true geriatric syndrome. *J. Nutr. Health Aging* **2012**, *16*, 422–425. [CrossRef]
116. Landi, F.; Liperoti, R.; Russo, A.; Giovannini, S.; Tosato, M.; Barillaro, C.; Capoluongo, E.; Bernabei, R.; Onder, G. Association of anorexia with sarcopenia in a community-dwelling elderly population: Results from the ilSIRENTE study. *Eur. J. Nutr.* **2013**, *52*, 1261–1268. [CrossRef]
117. Tsutsumimoto, K.; Doi, T.; Makizako, H.; Hotta, R.; Nakakubo, S.; Makino, K.; Suzuki, T.; Shimada, H. The association between anorexia of aging and physical frailty: Results from the national center for geriatrics and gerontology's study of geriatric syndromes. *Maturitas* **2017**, *97*, 32–37. [CrossRef] [PubMed]
118. Wall, B.T.; Gorissen, S.H.; Pennings, B.; Koopman, R.; Groen, B.B.; Verdijk, L.B.; van Loon, L.J. Aging Is Accompanied by a Blunted Muscle Protein Synthetic Response to Protein Ingestion. *PLoS ONE* **2015**, *10*, e0140903. [CrossRef]
119. Dickinson, J.M.; Volpi, E.; Rasmussen, B.B. Exercise and nutrition to target protein synthesis impairments in aging skeletal muscle. *Exerc. Sport Sci. Rev.* **2013**, *41*, 216–223. [CrossRef]
120. Drummond, M.J.; Dreyer, H.C.; Pennings, B.; Fry, C.S.; Dhanani, S.; Dillon, E.L.; Sheffield-Moore, M.; Volpi, E.; Rasmussen, B.B. Skeletal muscle protein anabolic response to resistance exercise and essential amino acids is delayed with aging. *J. Appl. Physiol.* **2008**, *104*, 1452–1461. [CrossRef]
121. Markofski, M.M.; Dickinson, J.M.; Drummond, M.J.; Fry, C.S.; Fujita, S.; Gundermann, D.M.; Glynn, E.L.; Jennings, K.; Paddon-Jones, D.; Reidy, P.T.; et al. Effect of age on basal muscle protein synthesis and mTORC1 signaling in a large cohort of young and older men and women. *Exp. Gerontol.* **2015**, *65*, 1–7. [CrossRef]
122. Morais, J.A.; Jacob, K.W.; Chevalier, S. Effects of aging and insulin resistant states on protein anabolic responses in older adults. *Exp. Gerontol.* **2018**, *108*, 262–268. [CrossRef]
123. Barclay, R.D.; Burd, N.A.; Tyler, C.; Tillin, N.A.; Mackenzie, R.W. The Role of the IGF-1 Signaling Cascade in Muscle Protein Synthesis and Anabolic Resistance in Aging Skeletal Muscle. *Front. Nutr.* **2019**, *6*, 146. [CrossRef] [PubMed]
124. Basu, R.; Dalla Man, C.; Campioni, M.; Basu, A.; Klee, G.; Toffolo, G.; Cobelli, C.; Rizza, R.A. Effects of age and sex on postprandial glucose metabolism: Differences in glucose turnover, insulin secretion, insulin action, and hepatic insulin extraction. *Diabetes* **2006**, *55*, 2001–2014. [CrossRef]
125. Herpich, C.; Haß, U.; Kochlik, B.; Franz, K.; Laeger, T.; Klaus, S.; Bosy-Westphal, A.; Norman, K. Postprandial dynamics and response of fibroblast growth factor 21 in older adults. *Clin. Nutr.* **2021**, *40*, 3765–3771. [CrossRef]
126. Conte, M.; Ostan, R.; Fabbri, C.; Santoro, A.; Guidarelli, G.; Vitale, G.; Mari, D.; Sevini, F.; Capri, M.; Sandri, M.; et al. Human aging and longevity are characterized by high levels of mitokines. *J. Gerontol. Ser. A Biol. Sci. Med. Sci.* **2018**, *74*, 600–607. [CrossRef] [PubMed]

127. Oost, L.J.; Kustermann, M.; Armani, A.; Blaauw, B.; Romanello, V. Fibroblast growth factor 21 controls mitophagy and muscle mass. *J. Cachexia Sarcopenia Muscle* **2019**, *10*, 630–642. [CrossRef] [PubMed]
128. Tezze, C.; Romanello, V.; Desbats, M.A.; Fadini, G.P.; Albiero, M.; Favaro, G.; Ciciliot, S.; Soriano, M.E.; Morbidoni, V.; Cerqua, C.; et al. Age-Associated Loss of OPA1 in Muscle Impacts Muscle Mass, Metabolic Homeostasis, Systemic Inflammation, and Epithelial Senescence. *Cell Metab.* **2017**, *25*, 1374–1389.e1376. [CrossRef]
129. Sun, H.; Sherrier, M.; Li, H. Skeletal Muscle and Bone–Emerging Targets of Fibroblast Growth Factor-21. *Front. Physiol.* **2021**, *12*, 625287. [CrossRef]
130. Franz, K.; Ost, M.; Otten, L.; Herpich, C.; Coleman, V.; Endres, A.S.; Klaus, S.; Müller-Werdan, U.; Norman, K. Higher serum levels of fibroblast growth factor 21 in old patients with cachexia. *Nutrition* **2019**, *63–64*, 81–86. [CrossRef]
131. Lin, Z.; Tian, H.; Lam, K.S.; Lin, S.; Hoo, R.C.; Konishi, M.; Itoh, N.; Wang, Y.; Bornstein, S.R.; Xu, A.; et al. Adiponectin mediates the metabolic effects of FGF21 on glucose homeostasis and insulin sensitivity in mice. *Cell Metab.* **2013**, *17*, 779–789. [CrossRef]
132. Menzaghi, C.; Trischitta, V. The Adiponectin Paradox for All-Cause and Cardiovascular Mortality. *Diabetes* **2018**, *67*, 12–22. [CrossRef] [PubMed]
133. Herpich, C.; Kochlik, B.; Haß, U.; Weber, D.; Grune, T.; Norman, K. Altered Adiponectin Response in Older Women Following Dextrose and High-Fat Dietary Challenges. *Mol. Nutr. Food Res.* **2021**, e2100487. [CrossRef]
134. Franceschi, C.; Bonafè, M.; Valensin, S.; Olivieri, F.; De Luca, M.; Ottaviani, E.; De Benedictis, G. Inflamm-aging. An evolutionary perspective on immunosenescence. *Ann. N. Y. Acad. Sci.* **2000**, *908*, 244–254. [CrossRef] [PubMed]
135. Meftahi, G.H.; Jangravi, Z.; Sahraei, H.; Bahari, Z. The possible pathophysiology mechanism of cytokine storm in elderly adults with COVID-19 infection: The contribution of "inflame-aging". *Inflamm. Res.* **2020**, *69*, 825–839. [CrossRef]
136. Omarjee, L.; Perrot, F.; Meilhac, O.; Mahe, G.; Bousquet, G.; Janin, A. Immunometabolism at the cornerstone of inflammaging, immunosenescence, and autoimmunity in COVID-19. *Aging* **2020**, *12*, 26263–26278. [CrossRef]
137. Rothenberg, E. Coronavirus Disease 19 from the Perspective of Ageing with Focus on Nutritional Status and Nutrition Management-A Narrative Review. *Nutrients* **2021**, *13*, 1294. [CrossRef]
138. Morrisette-Thomas, V.; Cohen, A.A.; Fulop, T.; Riesco, E.; Legault, V.; Li, Q.; Milot, E.; Dusseault-Belanger, F.; Ferrucci, L. Inflamm-aging does not simply reflect increases in pro-inflammatory markers. *Mech. Ageing Dev.* **2014**, *139*, 49–57. [CrossRef] [PubMed]
139. Minciullo, P.L.; Catalano, A.; Mandraffino, G.; Casciaro, M.; Crucitti, A.; Maltese, G.; Morabito, N.; Lasco, A.; Gangemi, S.; Basile, G. Inflammaging and Anti-Inflammaging: The Role of Cytokines in Extreme Longevity. *Arch. Immunol. Ther. Exp.* **2016**, *64*, 111–126. [CrossRef]
140. Pinti, M.; Cevenini, E.; Nasi, M.; De Biasi, S.; Salvioli, S.; Monti, D.; Benatti, S.; Gibellini, L.; Cotichini, R.; Stazi, M.A.; et al. Circulating mitochondrial DNA increases with age and is a familiar trait: Implications for "inflamm-aging". *Eur. J. Immunol.* **2014**, *44*, 1552–1562. [CrossRef]
141. Conte, M.; Martucci, M.; Chiariello, A.; Franceschi, C.; Salvioli, S. Mitochondria, immunosenescence and inflammaging: A role for mitokines? In *Seminars in Immunopathology*; Springer: Berlin/Heidelberg, Germany, 2020. [CrossRef]
142. Onyango, I.G.; Jauregui, G.V.; Čarná, M.; Bennett, J.P., Jr.; Stokin, G.B. Neuroinflammation in Alzheimer's Disease. *Biomedicines* **2021**, *9*, 524. [CrossRef]
143. Liberale, L.; Montecucco, F.; Tardif, J.C.; Libby, P.; Camici, G.G. Inflamm-ageing: The role of inflammation in age-dependent cardiovascular disease. *Eur. Heart J.* **2020**, *41*, 2974–2982. [CrossRef]
144. Livshits, G.; Kalinkovich, A. Inflammaging as a common ground for the development and maintenance of sarcopenia, obesity, cardiomyopathy and dysbiosis. *Ageing Res. Rev.* **2019**, *56*, 100980. [CrossRef]
145. Prattichizzo, F.; De Nigris, V.; Spiga, R.; Mancuso, E.; La Sala, L.; Antonicelli, R.; Testa, R.; Procopio, A.D.; Olivieri, F.; Ceriello, A. Inflammageing and metaflammation: The yin and yang of type 2 diabetes. *Ageing Res. Rev.* **2018**, *41*, 1–17. [CrossRef] [PubMed]
146. Kirk, B.; Feehan, J.; Lombardi, G.; Duque, G. Muscle, Bone, and Fat Crosstalk: The Biological Role of Myokines, Osteokines, and Adipokines. *Curr. Osteoporos. Rep.* **2020**, *18*, 388–400. [CrossRef]
147. Marzetti, E.; Picca, A.; Marini, F.; Biancolillo, A.; Coelho-Junior, H.J.; Gervasoni, J.; Bossola, M.; Cesari, M.; Onder, G.; Landi, F.; et al. Inflammatory signatures in older persons with physical frailty and sarcopenia: The frailty "cytokinome" at its core. *Exp. Gerontol.* **2019**, *122*, 129–138. [CrossRef]
148. Ostan, R.; Lanzarini, C.; Pini, E.; Scurti, M.; Vianello, D.; Bertarelli, C.; Fabbri, C.; Izzi, M.; Palmas, G.; Biondi, F.; et al. Inflammaging and cancer: A challenge for the Mediterranean diet. *Nutrients* **2015**, *7*, 2589–2621. [CrossRef]
149. Olivieri, F.; Rippo, M.R.; Monsurrò, V.; Salvioli, S.; Capri, M.; Procopio, A.D.; Franceschi, C. MicroRNAs linking inflamm-aging, cellular senescence and cancer. *Ageing Res. Rev.* **2013**, *12*, 1056–1068. [CrossRef] [PubMed]
150. Bonafè, M.; Storci, G.; Franceschi, C. Inflamm-aging of the stem cell niche: Breast cancer as a paradigmatic example: Breakdown of the multi-shell cytokine network fuels cancer in aged people. *Bioessays* **2012**, *34*, 40–49. [CrossRef]
151. Calder, P.C.; Bosco, N.; Bourdet-Sicard, R.; Capuron, L.; Delzenne, N.; Dore, J.; Franceschi, C.; Lehtinen, M.J.; Recker, T.; Salvioli, S.; et al. Health relevance of the modification of low grade inflammation in ageing (inflammageing) and the role of nutrition. *Ageing Res. Rev.* **2017**, *40*, 95–119. [CrossRef]
152. Yeh, S.S.; Schuster, M.W. Geriatric cachexia: The role of cytokines. *Am. J. Clin. Nutr.* **1999**, *70*, 183–197. [CrossRef] [PubMed]
153. Evans, W.J.; Morley, J.E.; Argiles, J.; Bales, C.; Baracos, V.; Guttridge, D.; Jatoi, A.; Kalantar-Zadeh, K.; Lochs, H.; Mantovani, G.; et al. Cachexia: A new definition. *Clin. Nutr.* **2008**, *27*, 793–799. [CrossRef]

154. Tracey, K.J.; Wei, H.; Manogue, K.R.; Fong, Y.; Hesse, D.G.; Nguyen, H.T.; Kuo, G.C.; Beutler, B.; Cotran, R.S.; Cerami, A.; et al. Cachectin/tumor necrosis factor induces cachexia, anemia, and inflammation. *J. Exp. Med.* **1988**, *167*, 1211–1227. [CrossRef]
155. Oldenburg, H.S.; Rogy, M.A.; Lazarus, D.D.; Van Zee, K.J.; Keeler, B.P.; Chizzonite, R.A.; Lowry, S.F.; Moldawer, L.L. Cachexia and the acute-phase protein response in inflammation are regulated by interleukin-6. *Eur. J. Immunol.* **1993**, *23*, 1889–1894. [CrossRef]
156. Ershler, W.B. Interleukin-6: A cytokine for gerontologists. *J. Am. Geriatr. Soc.* **1993**, *41*, 176–181. [CrossRef]
157. Boirie, Y. Physiopathological mechanism of sarcopenia. *J. Nutr. Health Aging* **2009**, *13*, 717–723. [CrossRef]
158. Fatyga, P.; Pac, A.; Fedyk-Łukasik, M.; Grodzicki, T.; Skalska, A. The relationship between malnutrition risk and inflammatory biomarkers in outpatient geriatric population. *Eur. Geriatr. Med.* **2020**, *11*, 383–391. [CrossRef]
159. Correa-Pérez, A.; Abraha, I.; Cherubini, A.; Collinson, A.; Dardevet, D.; de Groot, L.C.P.G.M. Efficacy of non-pharmacological interventions on nutritional status and clinical outcomes in older people with malnutrition or at risk of malnutrition: A systematic review. The SENATOR project ONTOP series and MaNuEL knowledge hub project. *Aging Res. Rev.* **2019**, *49*, 27–48. [CrossRef]
160. Reinders, I.; Volkert, D.; de Groot, L.; Beck, A.M.; Feldblum, I.; Jobse, I.; Neelemaat, F.; de van der Schueren, M.A.E.; Shahar, D.R.; Smeets, E.; et al. Effectiveness of nutritional interventions in older adults at risk of malnutrition across different health care settings: Pooled analyses of individual participant data from nine randomized controlled trials. *Clin. Nutr.* **2019**, *38*, 1797–1806. [CrossRef]
161. Höhn, A.; Weber, D.; Jung, T.; Ott, C.; Hugo, M.; Kochlik, B.; Kehm, R.; König, J.; Grune, T.; Castro, J.P. Happily (n)ever after: Aging in the context of oxidative stress, proteostasis loss and cellular senescence. *Redox Biol.* **2017**, *11*, 482–501. [CrossRef]
162. Fisher, L. *Food Science and Nutrition*; ED-Tech Press: Waltham Abbey, UK, 2019.
163. Ghosh, T.S.; Rampelli, S.; Jeffery, I.B.; Santoro, A.; Neto, M.; Capri, M.; Giampieri, E.; Jennings, A.; Candela, M.; Turroni, S.; et al. Mediterranean diet intervention alters the gut microbiome in older people reducing frailty and improving health status: The NU-AGE 1-year dietary intervention across five European countries. *Gut* **2020**, *69*, 1218–1228. [CrossRef] [PubMed]
164. Deutz, N.E.P.; Ashurst, I.; Ballesteros, M.D.; Bear, D.E.; Cruz-Jentoft, A.J.; Genton, L.; Landi, F.; Laviano, A.; Norman, K.; Prado, C.M. The Underappreciated Role of Low Muscle Mass in the Management of Malnutrition. *J. Am. Med. Dir. Assoc.* **2019**, *20*, 22–27. [CrossRef]
165. Van der Pols-Vijlbrief, R.; Wijnhoven, H.A.H.; Bosmans, J.E.; Twisk, J.W.R.; Visser, M. Targeting the underlying causes of undernutrition. Cost-effectiveness of a multifactorial personalized intervention in community-dwelling older adults: A randomized controlled trial. *Clin. Nutr.* **2017**, *36*, 1498–1508. [CrossRef]

nutrients

MDPI

Communication

Enteral Nutrition by Nasogastric Tube in Adult Patients under Palliative Care: A Systematic Review

Eduardo Sánchez-Sánchez [1,2,*], María Araceli Ruano-Álvarez [3], Jara Díaz-Jiménez [4], Antonio Jesús Díaz [5] and Francisco Javier Ordonez [6]

1 Internal Medicine Department, Punta de Europa Hospital, Algeciras, 11207 Cádiz, Spain
2 Instituto de Investigación e Innovación Biomédica de Cádiz (INiBICA), Hospital Universitario Puerta del Mar, Universidad de Cádiz, 11009 Cádiz, Spain
3 Punta de Europa Hospital, Algeciras, 11207 Cádiz, Spain; malva757@gmail.com
4 Faculty of Education Sciences, University of Cádiz, 11519 Puerto Real, Spain; luna_nueva17@hotmail.com
5 Medicine Department, School of Nursing, University of Cadiz, Plaza Fragela s/n, 11003 Cadiz, Spain; antoniojesus.diaz@uca.es
6 Human Anatomy, School of Medicine, University of Cádiz, Plaza Fragela s/n, 11003 Cadiz, Spain; franciscojavier.ordonez@uca.es
* Correspondence: eduardo.sanchez.sanchez.sspa@juntadeandalucia.es; Tel.: +34-671-569-320

Abstract: Nutritional management of patients under palliative care can lead to ethical issues, especially when Enteral Nutrition (EN) is prescribed by nasogastric tube (NGT). The aim of this review is to know the current status in the management of EN by NG tube in patients under palliative care, and its effect in their wellbeing and quality of life. The following databases were used: PubMed, Web of Science (WOS), Scopus, Scielo, Embase and Medline. After inclusion and exclusion criteria were applied, as well as different qualities screening, a total of three entries were used, published between 2015 and 2020. In total, 403 articles were identified initially, from which three were selected for this review. The use of NGT caused fewer diarrhea episodes and more restrictions than the group that did not use NG tubes. Furthermore, the use of tubes increased attendances to the emergency department, although there was no contrast between NGT and PEG devices. No statistical difference was found between use of tubes (NGT and PEG) or no use, with respect to the treatment of symptoms, level of comfort, and satisfaction at the end of life. Nevertheless, it improved hospital survival compared with other procedures, and differences were found in hospital stays in relation to the use of other probes or devices. Finally, there are not enough quality studies to provide evidence on improving the health status and quality of life of the use of EN through NGT in patients receiving palliative care. For this reason, decision making in this field must be carried out individually, weighing the benefits and damages that they can cause in the quality of life of the patients.

Keywords: artificial nutrition; enteral nutrition; nasogastric feeding; nasogastric tube; palliative care

Citation: Sánchez-Sánchez, E.; Ruano-Álvarez, M.A.; Díaz-Jiménez, J.; Díaz, A.J.; Ordonez, F.J. Enteral Nutrition by Nasogastric Tube in Adult Patients under Palliative Care: A Systematic Review. *Nutrients* **2021**, *13*, 1562. https://doi.org/10.3390/nu13051562

Academic Editors: Ina Bergheim and Omorogieva Ojo

Received: 27 February 2021
Accepted: 29 April 2021
Published: 6 May 2021

1. Introduction

Initially, the aim of Palliative Care (PC) was to relieve suffering at the end of life. However, it is nowadays considered as a model to follow in patients in whom there is no curative treatment, and is therefore being implemented at earlier stages. Initially, PC was focused on cancer patients, but it currently covers other conditions such as advanced dementia, HIV/AIDS, heart disease, etc. [1].

Every year, 40 million people need palliative care, but only 3 million have access to such special attention [2]. Currently, the goal of PC is to promote comfort and to maintain an optimal quality of life for patients and their families under palliative care [3] through prevention and management of physical, psychosocial and spiritual issues in these patients [4]. It should not be forgotten that quality of life evaluates the subjective perception that each patient has around alterations or limitations that the disease undertakes in the physical, psychosocial and spiritual aspects of their lives [5].

Nutrition and hydration are basic elements for maintaining life, and they are considered signs of health in our society [6]. Occasionally, patients present failure at maintaining adequate oral intake for meeting their nutritional needs [7], and this can lead to physical and psychosocial issues such as anxiety and distress [8]. Therefore, it may be necessary to commence Artificial Nutrition (AN). In 2008, Cochrane published a review regarding the use of AN in adult patients receiving palliative care. The authors concluded that there was not enough evidence to guide the development of guidelines for practice [9]. Six years later, the update of this review presented the same results; therefore, there are no new quality studies regarding this subject [7].

If the patient takes less than 50% of their nutritional requirements and there are no contraindications or bronchoaspiration risks, and their life expectancy is less than 6 weeks [10], Enteral Nutrition (EN) must be prescribed through a nasogastric (NG) tube [11]. This is a widely used and easily accessible technique, although in the case of patients with advanced dementia who receive PC, evidence supporting the use of NG tube is limited, and this technique may have a negative impact on the quality of life of these patients [12]. The use of tubes in patients with advanced dementia does not improve survival, prevent aspiration [13], or improve their functional status. In addition, the use of tubes for artificial nutrition has been associated with agitation, increased physical restrictions, and complications related to the tubes [14,15].

The use of EN through NG tubes in patients under PC continues to be a controversial subject [16], since there is little evidence on the role of nutritional support and whether its implementation improves quality of life. In addition, it affects the psychological sphere of patients, because it can influence their social relationships and the way they interact with others. However, Mitchell et al. reported that more than a third of nursing home residents with dementia had been subjected to a feeding tube [17]. Decisions and/or choices may confront patients, family members, and health professionals. Therefore, having a good knowledge of the benefits and harms of the use of this technique is paramount in order to reduce ethical conflicts and to understand how the use of this technique can influence the physical, psychological and spiritual spheres, and therefore, the quality of life of patients receiving PC. Accordingly, the goal of the present study is to understand the current state of the management of EN using NG tubes in patients receiving palliative care, along with its effect on health status and quality of life.

2. Materials and Methods

A systematic review of the literature was made. The results were obtained by direct online access through the following database: PubMed, Web of Science (WOS), Scopus, y Scielo, Embase y Medline. The aim of this review was to address the next question: Is it appropriate the use of EN by NG tube in patients under palliative care?

To define the research, question the PICOS criteria (Table 1) was used.

Table 1. PICOS criteria (Population; Intervention; Comparison; Outcome; Study design).

P	Patients under Palliative Care
I	Enteral nutrition by nasogastric tube
C	The comparison could be made with any other type of dietary intervention or nutritional support.
O	Improving wellbeing and quality of life
S	Systematic review

The articles reviewed were published in any country, by any institution or individual investigator, and written in Spanish or English. The research was limited to those published in the last 5 years (between 2015 and 2020).

For the documentary retrieval, the following MeSH descriptors were used: "palliative care", "enteral nutrition", "terminal care" "terminally ill". Neither Subheadings nor Entry Term classifiers were used. The search strategy was: ("Palliative Care" OR "Terminal Care" OR "Terminally ill") AND "Enteral Nutrition". The final choice of articles was made following the inclusion criteria: (a) studies published in journals indexed in international databases subject to peer review, (b) published between 2015 and 2020, and (c) written in English or Spanish; and the exclusion criteria were: (a) studies based on pediatric age, (b) expert reports, editor's letters, books, monographs, clinical narratives or reviews. Due to the large number of articles found in the first search, and as a quality assessment, two screenings were carried out. The first was based on the title and summary, eliminating those articles that dealt with a topic other than the one proposed. In the second screening, review articles, editor's letters, etc., were eliminated.

To carry out the critical reading and evaluation of the articles found, the STROBE (Strengthening the Reporting of Observational studies in Epidemiology) statement was used for the observational studies [18] and the CONSORT guide (Consolidated Standards of Reporting Trials) for randomized clinical trials [19].

3. Results

A total of 403 articles were found: 32 (7.9%) PubMed, 30 (7.4%) WOS, 20 (4.9%) Scopus, 4 (1.0%) Scielo, 151 (37.4%) Embase, and 166 (41.2%) Medline. Of these recovered papers, 223 (55.3%) were redundant.

Once the first screening was applied based on the title and abstract, 168 articles were eliminated. After the second screening, nine articles were eliminated. The number of articles selected was three, all of which were observational studies, for which the STROBE statement was made. All of these articles fulfilled 90% of the points of the set declaration. The parameters of PRISMA (Preferred Reporting Items for Systematic Review and Meta-Analyses) were followed (Figure 1).

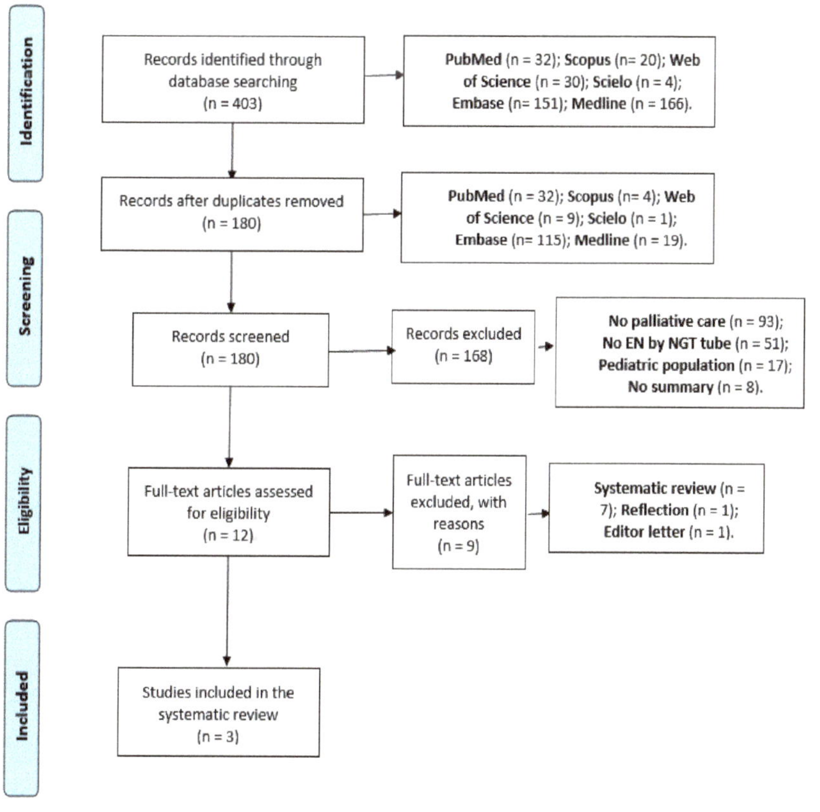

Figure 1. PRISMA (Preferred Reporting Items for Systematic Review and Meta-Analyses) diagram.

The results obtained showed different study parameters in the approach to the proposed topic (Table 2). No studies were found that addressed the use of NGT versus not using a feeding tube, but there was always a third group representing the use of either Percutaneous Endoscopic Gastrostomy (PEG) or esophageal stent. Therefore, the results obtained in relation to the use of NG tube and the other groups were taken.

In the study carried out by Bentur et al. 2015 [20], three groups were compared: subjects without feeding tubes, subjects with NG tubes and another group caring Percutaneous Endoscopic Gastrostomy (PEG). The results related to the use of NG tube versus the non-use of a catheter or the use of PEG were taken as a reference for this review. They concluded that the use of a feeding tube in people with advanced dementia in the community was associated with negative outcomes and increased caregiver burden. The use of an NG tube caused less diarrhea and more restrictions than the group that did not carry a catheter. The use of feeding tubes increased attendances to the emergency department, although they did not distinguish between NGT and PEG. No statistical difference was found between catheter use (NG tube and PEG) and non-use with respect to the treatment of symptoms at the end of life, comfort or satisfaction at the end of life.

Yang et al., in 2015, compared hospital stays and survival among patients with esophageal obstruction and a short life expectancy in subjects with EN by tube, with esophageal stent placement, and with nutritional support without oral intake. The results obtained showed that the patients with NGT and esophageal stent had a shorter hospital stay (19 and 12 days, respectively) and a longer median survival ($p < 0.01$) than the group with nutritional support. Concluding that enteral feeding by NG tube in palliative care was safe, inexpensive, and had a low complication rate [21].

The multicenter study carried out by Shinozaki et al. in 2017 in Japan, found that 74.6% of patients in the terminal phase required EN.

These authors suggest that the nutritional intake route may play a role in quality of life. No significant difference was found in quality of life between the different study groups. However, the mean hospitalization period was significantly shorter for gastrostomy-fed patients than for nasogastric tube-fed patients (21 vs. 64 days). Patients with PEG had a shorter period between study prescription and death than patients fed through an NG tube [22].

Table 2. Studies included in the systematic review.

Author, Year	Aim	Methodology	Results
Bentur N et al., 2015 [20].	To examine the prevalence of feeding tube use among older people with advanced dementia (OPAD) living in the community; to evaluate the characteristics, quality of care, and the burden on caregivers.	A cross-sectional survey of 117 caregivers of OPAD living in the community. They recorded clinical and care variables, treatment of symptoms at the end of life in dementia (SM-EOLD), evaluation of comfort in dementia at the end of life (CAD-EOLD) and satisfaction with the scales of care at the end of life with dementia (SWC-EOLD).	13% of patients carried NG tubes. The use of this type of device caused less diarrhea episodes than those subjects that did not use any feeding tube (6.6% vs. 32.5%) and more restrictions (60.0% vs. 9.9%, $p < 0.05$). Subjects with feeding tubes (NG tube or PEG) attended the emergency department at least once a day (40% vs. 34.2%, $p < 0.05$), and on more occasions (2.92 ± 1.68 vs. 1.6 ± 0.9 during the day and 2.9 ± 1.6 times compared to 1.4 ± 0.5 times during night time, $p < 0.05$). No statistical differences of significance were found between the use or non-use of feeding tubes in the scales of SM-EOLD, SWC-EOLD y CAD-EOLED, finding a difference in the wellbeing subscale of the last one, in subjects with or without feeding tubes, either NGT o PEG (6.9 ± 2.3 vs. 5.2 ± 2.0, $p < 0.05$, respectively), without finding differences between the types of feeding tubes.
Yang CW et al., 2015 [21].	Comparing clinical results of EN by tube and the placement of an esophageal stent in patients with malignant esophageal obstruction and a short life expectancy.	Retrospective observational study in 31 patients diagnosed with advanced-stage esophageal cancer, divided into three groups: patients with NG tube ($n = 12$), with esophageal stent, ($n = 10$) and patients with nutritional support but without oral intake ($n = 9$).	The average duration of hospital admissions was 19 days in the group of NGT, 12 days in the group of esophageal stents, and 39 days in the group without nutritional support ($p = 0.01$). The mean average survival after the diagnosis of malignant esophageal obstruction was 122 days in the group with NGT tube, 133 days in the group with esophageal stents and 51 days in those not receiving any nutritional support. The most common complication for the group using feeding tubes was pneumonia caused by aspiration (58%), although this was lower than in the group not receiving any nutritional support (100%).
Shinozaki et al., 2017 [22]	To examine the quality of life and functional state in terminal patients with brain and neck cancer.	Prospective and multicenter observational study with 11 oncology centers and hospitals in Japan. The survey EORTC QLQ-C15-PAL was used weekly formed by 15 items related with health wellbeing and quality of life. The sample was formed by 100 patients.	74.6% of patients required EN. Those with NGT showed longer hospital admissions than patients using PEG (64 compared to 21 days, $p < 0.05$). Patients using PEG presented shorter periods between the study prescription and death, compared to those fed by NGT. No significant difference was found in quality of life, between the starting point and week 3 of the study, among the different study groups.

4. Discussion

The results obtained show the limited bibliography in the field of EN through NG tube in patients receiving palliative care. There are studies on the use of tube feeding in these patients, but without distinction between the NG tube and PEG, so it was not possible to obtain individual and differentiated results between both routes of administration.

The articles in this research can be found to represent a low level of evidence, since they are observational studies, and no randomized clinical trials (RCTs) were performed. These results coincide with those reported by other studies, such as the systematic reviews carried out by Good et al. in 2008 and later in 2014 [7,9].

Malnutrition leads to increased comorbidities and decreased performance status and quality of life [10]. Therefore, nutritional support should be integrated into palliative care, and its implications with respect to quality of life and life expectancy should be assessed [23]. Within such nutritional support is included the use of nutrition through a tube, although its use remains controversial, especially in the case of the NG tube. The emergence of research and guidelines on the management of patients under palliative care has managed to reduce the use of tube feeding by 50% [24].

Some studies report that the use of enteral tube feeding is effective for improving the quality of life of patients [25], since it may improve physical, psychosocial and spiritual aspects. Although the quality of life of patients with NGT was not studied in the study carried out by Bentur et al. in 2015, they did find that these patients presented more diarrhea and restrictions, which can affect the physical and even psychosocial sphere, which could influence the quality of life of these patients.

Even though there was no distinction between patients with NG tube and PEG, it was concluded that these patients attended the emergency department more times than those who did not carry any type of feeding tube, which also negatively influences their quality of life, since they present more comorbidities, making it necessary for them to go to a health center more frequently, and causing changes in their daily life, as reflected in the well-being subscale of CAD-EOLED [20]. Another aspect that can negatively influence quality of life is the increase in the number of hospital stays and the decrease in survival. The use of NGT may decrease hospital stays and improve survival in patients receiving palliative care, and thus improve the quality of life perceived by these patients [21]. However, Shinozaki et al. concluded that subjects presenting NG tube had longer hospital admissions than those using PEG. Even though the survival period was longer, no significant differences in quality of life were found among the various groups [22]. This may be due to the choice of the measurement interval, since it was performed in patients with a short life expectancy. It should be noted that the perception of quality of life is related to reality and expectations. In patients receiving palliative care, the expectations for improvement are sometimes low, especially when their life expectancy is short [26]. Perhaps for this reason, no differences were found in quality of life in these investigations. The scant evidence on this topic has led to different interpretations and approaches in these patients.

The Ethics Work Group of the Spanish Society in Parenteral and Enteral Nutrition (SENPE) recently (2019) published a confirmation that the placement of tubes for nutrition in patients with advanced dementia was a futile treatment that only contributed to prolonged suffering and concluded that health care professionals should not make wide use of EN by tube [27]. Schwartz et al. considered that EN by tube could improve quality of life, but that the benefits in the last days phase were limited and did not exceed the loads [28].

Furthermore, there may be discrepancies between health professionals and patients when prescribing nutritional support through an NG tube. For example, Amano at al. found that 78.6% of subjects in their study did not wish to receive artificial nutrition by feeding tube, even though their intake was insufficient [29]. In the study undertaken by Pengo et al. in 2017, it was found that the numbers of doctors and nurses who agreed with the use of the AN declined when life expectancy decreased [30]. These decisions can create ethical dilemmas and are related to feelings, thoughts and beliefs [31].

Sometimes, it is the patients themselves who do not wish to receive EN by NG tube [28]. Therefore, it is necessary to make an individualized decision, even though no other contraindications may be found. This respects the principles of autonomy, beneficence and non-maleficence [32]. Furthermore, the team of health care professionals looking after such patients should establish what the aims and benefits of such treatment are, whether these are achievable, and any possible damage that may be encountered [33]. In addition, the principle of autonomy recognizes the right and the capacity of a person to make their own personal decisions. Self-determination includes the right to reject EN, although this refusal may be difficult to understand for family members and healthcare professionals [3]. Perhaps the means to avoid ethical conflicts and future dilemmas is the use of anticipated instructions, where patients can reflect their decisions regarding future treatments or techniques, although the prevalence of patients who make use of such mechanisms is very low [34].

Among the limitations in this review are the lack of studies with a large enough sample to be able to describe the results, and the subjectivity of the results.

Although it is a difficult field of research, conducting higher-quality research could result in the provision of recommendations or guidance to aid patients and healthcare professionals in decision making.

The results obtained lead us to consider the need to create a clinical practice guide on the nutritional management of these patients, which includes the use of EN by NGT. Progress must be continued in education so that these differences do not exist, and such clinical practice is common to all nurses. The benefits and risks of the use of EN by NGT in these patients should be investigated, in order to provide evidence-based care. Clear evidence would help to reduce variability in the management of these patients.

5. Conclusions

There are not enough quality studies to provide evidence regarding the benefits for wellbeing and quality of life in patients under palliative care receiving EN through an NG tube.

For this reason, decision making in this field must be carried out individually, weighing the benefits and damages that they can cause in the quality of life of the patients.

Author Contributions: Conceptualization and design, E.S.-S.; study selection, E.S.-S. and J.D.-J.; data extraction, E.S.-S. and J.D.-J.; data synthesis, E.S.-S. and J.D.-J.; writing—original draft preparation, E.S.-S. and M.A.R.-Á.; writing—review and editing, E.S.-S., J.D.-J., M.A.R.-Á., A.J.D. and F.J.O.; Translation of the manuscript: E.S.-S., M.A.R.-Á., A.J.D. and F.J.O. All authors have read and agreed to the published version of the manuscript.

Funding: This research received no external funding.

Informed Consent Statement: Not applicable.

Data Availability Statement: Data was collected from available literature. Data for doing the systematic review is available in the manuscript's tables.

Conflicts of Interest: The authors declare no conflict of interest.

References

1. Radbruch, L.; de Lima, L.; Knaul, F.; Wenk, R.; Ali, Z.; Bhatnaghar, S.; Blanchard, C.; Bruera, E.; Buitrago, R.; Burla, C.; et al. Redefining Palliative Care—A New Consensus-Based Definition. *J. Pain Symptom Manag.* **2020**, *60*, 754–764. [CrossRef] [PubMed]
2. Connor, S.R.; Gwyther, E. The Worldwide Hospice Palliative Care Alliance. *J. Pain Symptom Manag.* **2018**, *55*, S112–S116. [CrossRef]
3. Druml, C.; Ballmer, P.E.; Druml, W.; Oehmichen, F.; Shenkin, A.; Singer, P.; Soeters, P.; Weimann, A.; Bischoff, S.C. ESPEN guideline on ethical aspects of artificial nutrition and hydration. *Clin. Nutr.* **2016**, *35*, 545–556. [CrossRef] [PubMed]
4. Tziraki, C.; Grimes, C.; Ventura, F.; O'Caoimh, R.; Santana, S.; Zavagli, V.; Varani, S.; Tramontano, D.; Apóstolo, J.; Geurden, B.; et al. Rethinking palliative care in a public health context: Addressing the needs of persons with non-communicable chronic diseases. *Prim. Health Care Res. Dev.* **2020**, *21*, e60, Correction in *Prim Health Care Res Dev.* **2020**, *21*, e60. [CrossRef] [PubMed]

5. Sanchez Sanchez, E.; Lopez Aliaga, I.; Montes Romero, A.B.; Alferez, M.J.M. Correlation between loss of lean mass and quality of life in cancer patients. *Nutr. Clin. Diet. Hosp.* **2018**, *38*, 68–73. [CrossRef]
6. Henry, B. Evolving ethical and legal implications for feeding at the end of life. *Ann. Palliat. Med.* **2017**, *6*, 87–90. [CrossRef]
7. Good, P.; Richard, R.; Syrmis, W.; Jenkins-Marsh, S.; Stephens, J. Medically assisted nutrition for adult palliative care patients. *Cochrane Database Syst. Rev.* **2014**, CD006274. [CrossRef] [PubMed]
8. Reid, J.; McKenna, H.; Fitzsimons, D.; McCance, T. Fighting over food: Comprensión del paciente y la familia sobre la caquexia por cáncer. *Oncol. Nurs. Forum* **2009**, *36*, 439–445. [CrossRef] [PubMed]
9. Good, P.; Cavenagh, J.; Mather, M.; Ravenscroft, P. Medically assisted nutrition for palliative care in adult patients. *Cochrane Database Syst. Rev.* **2008**, CD006274. [CrossRef]
10. Cotogni, P.; Stragliotto, S.; Ossola, M.; Collo, A.; Riso, S.; on behalf of the Intersociety Italian Working Group for Nutritional Support in Cancer. The Role of Nutritional Support for Cancer Patients in Palliative Care. *Nutrients* **2021**, *13*, 306. [CrossRef] [PubMed]
11. Kokura, Y.; Suzuki, C.; Wakabayashi, H.; Maeda, K.; Sakai, K.; Momosaki, R. Semi-Solid Nutrients for Prevention of Enteral Tube Feeding-Related Complications in Japanese Population: A Systematic Review and Meta-Analysis. *Nutrients* **2020**, *12*, 1687. [CrossRef]
12. Anantapong, K.; Davies, N.; Chan, J.; McInnerney, D.; Sampson, E.L. Mapping and understanding the decision-making process for providing nutrition and hydration to people living with dementia: A systematic review. *BMC Geriatr.* **2020**, *20*, 520. [CrossRef]
13. Teno, J.M. Hospital Characteristics Associated with Feeding Tube Placement in Nursing Home Residents with Advanced Cognitive Impairment. *JAMA* **2010**, *303*, 544–550. [CrossRef]
14. American Geriatrics Society Ethics Committee and Clinical Practice and Models of Care Committee. American Geriatrics Society Feeding Tubes in Advanced Dementia Position Statement. *J. Am. Geriatr. Soc.* **2014**, *62*, 1590–1593. [CrossRef]
15. Ying, I. Artificial nutrition and hydration in advanced dementia. *Can. Fam. Physician* **2015**, *61*, 245–248. [PubMed]
16. Orrevall, Y.; Tishelman, C.; Permert, J.; Lundström, S. A National Observational Study of the Prevalence and Use of Enteral Tube Feeding, Parenteral Nutrition and Intravenous Glucose in Cancer Patients Enrolled in Specialized Palliative Care. *Nutrients* **2013**, *5*, 267–282. [CrossRef]
17. Mitchell, S.L. Clinical and Organizational Factors Associated with Feeding Tube Use among Nursing Home Residents with Advanced Cognitive Impairment. *JAMA* **2003**, *290*, 73–80. [CrossRef]
18. Vandenbroucke, J.; Von, E.; Altman, D. Improving communication in observational studies in epidemiology (STROBE): Explanation and elaboration. *GAC Sanit.* **2009**, *23*, 158e1–158e28. [CrossRef]
19. Schulz, K.F.; Altman, D.G.; Moher, D. CONSORT 2010 Statement: Updated Guidelines for Reporting Parallel Group Randomized Trials. *Ann. Intern. Med.* **2010**, *152*, 726–732. [CrossRef] [PubMed]
20. Bentur, N.; Sternberg, S.; Shuldiner, J.; Dwolatzky, T. Feeding Tubes for Older People with Advanced Dementia Living in the Community in Israel. *Am. J. Alzheimer's Dis. Other Demen.* **2014**, *30*, 165–172. [CrossRef] [PubMed]
21. Yang, C.; Lin, H.; Hsieh, T.; Chang, W. Palliative enteral feeding for patients with malignant esophageal obstruction: A retrospective study. *BMC Palliat. Care* **2015**, *14*, 58. [CrossRef] [PubMed]
22. Shinozaki, T.; Ebihara, M.; Iwase, S.; Yamaguchi, T.; Hirakawa, H.; Shimbashi, W.; Kamijo, T.; Okamoto, M.; Beppu, T.; Ohori, J.; et al. Quality of life and functional status of terminally ill head and neck cancer patients: A nation-wide, prospective observational study at tertiary cancer centers in Japan. *Jpn. J. Clin. Oncol.* **2016**, *47*, 47–53. [CrossRef] [PubMed]
23. Cotogni, P.; Pedrazzoli, P.; De Waele, E.; Aprile, G.; Farina, G.; Stragliotto, S.; De Lorenzo, F.; Caccialanza, R. Nutritional Therapy in Cancer Patients Receiving Chemoradiotherapy: Should We Need Stronger Recommendations to Act for Improving Outcomes? *J. Cancer* **2019**, *10*, 4318–4325. [CrossRef] [PubMed]
24. Mitchell, S.L.; Mor, V.; Gozalo, P.L.; Servadio, J.L.; Teno, J.M. Tube Feeding in US Nursing Home Residents with Advanced Dementia, 2000–2014. *JAMA* **2016**, *316*, 769–770. [CrossRef]
25. Ojo, O.; Keaveney, E.; Wang, X.-H.; Feng, P. The Effect of Enteral Tube Feeding on Patients' Health-Related Quality of Life: A Systematic Review. *Nutrients* **2019**, *11*, 1046. [CrossRef]
26. Tudela, L.L. La calidad de vida relacionada con la salud [Health-related quality of life]. *Atención Primaria* **2009**, *41*, 411–416. [CrossRef] [PubMed]
27. Blanco, A.C.; Fuster, F.M.L.; García, M.D.D.O.; Casas, M.N.V.; Lozano, C.W.-B.; Avilés, V.; Enguídanos, R.A.; López, I.F.; Soria, J.B.M.; González, J.C.M.; et al. Manejo nutricional de la demencia avanzada: Resumen de recomendaciones del Grupo de Trabajo de Ética de la SENPE. *Nutr. Hosp.* **2019**, *36*, 988–995. [CrossRef]
28. Schwartz, D.B.; Posthauer, M.E.; Maillet, J.O. Advancing Nutrition and Dietetics Practice: Dealing with Ethical Issues of Nutrition and Hydration. *J. Acad. Nutr. Diet.* **2020**, *121*, 823–831. [CrossRef] [PubMed]
29. Amano, K.; Morita, T.; Miyamoto, J.; Uno, T.; Katayama, H.; Tatara, R. Perception of need for nutritional support in advanced cancer patients with cachexia: A survey in palliative care settings. *Support. Care Cancer* **2018**, *26*, 2793–2799. [CrossRef]
30. Pengo, V.; Zurlo, A.; Voci, A.; Valentini, E.; De Zaiacomo, F.; Catarini, M.; Iasevoli, M.; Maggi, S.; Pegoraro, R.; Manzato, E.; et al. Advanced dementia: Opinions of physicians and nurses about antibiotic therapy, artificial hydration and nutrition in patients with different life expectancies. *Geriatr. Gerontol. Int.* **2017**, *17*, 487–493. [CrossRef]

31. Sánchez-Sánchez, E.; Ramírez-Vargas, G.; Peinado-Canas, A.; Martín-Estrada, F.; Díaz-Jimenez, J.; Ordonez, F.J. Nurse and Nursing Students' Opinions and Perceptions of Enteral Nutrition by Nasogastric Tube in Palliative Care. *Nutrients* **2021**, *13*, 402. [CrossRef]
32. Orlandoni, P.; Peladic, N.J.; Di Rosa, M.; Venturini, C.; Fagnani, D.; Sparvoli, D.; Giorgini, N.; Basile, R.; Cola, C. The outcomes of long term home enteral nutrition (HEN) in older patients with severe dementia. *Clin. Nutr.* **2018**, *38*, 1871–1876. [CrossRef] [PubMed]
33. Orrevall, Y. Nutritional support at the end of life. *Nutrients* **2015**, *31*, 615–616. [CrossRef] [PubMed]
34. Tamura, M.K.; Montez-Rath, M.E.; Hall, Y.N.; Katz, R.; O'Hare, A.M. Advance Directives and End-of-Life Care among Nursing Home Residents Receiving Maintenance Dialysis. *Clin. J. Am. Soc. Nephrol.* **2017**, *12*, 435–442. [CrossRef] [PubMed]

nutrients

MDPI

Review

Oral Nutritional Supplements and Enteral Nutrition in Patients with Gastrointestinal Surgery

Maria Wobith and Arved Weimann *

Department of General, Visceral, and Oncological Surgery, Klinikum St. Georg gGmbH, 04129 Leipzig, Germany; maria.wobith@sanktgeorg.de
* Correspondence: arved.weimann@sanktgeorg.de

Abstract: Nowadays, patients undergoing gastrointestinal surgery are following perioperative treatment in enhanced recovery after surgery (ERAS) protocols. Although oral feeding is supposed not to be stopped perioperatively with respect to ERAS, malnourished patients and inadequate calorie intake are common. Malnutrition, even in overweight or obese patients, is often underestimated. Patients at metabolic risk have to be identified early to confirm the indication for nutritional therapy. The monitoring of nutritional status postoperatively has to be considered in the hospital and after discharge, especially after surgery in the upper gastrointestinal tract, as normal oral food intake is decreased for several months. The article gives an overview of the current concepts of perioperative enteral nutrition in patients undergoing gastrointestinal surgery.

Keywords: enteral nutrition; oral nutritional supplements; perioperative nutrition; malnutrition; sarcopenia; gastrointestinal surgery

Citation: Wobith, M.; Weimann, A. Oral Nutritional Supplements and Enteral Nutrition in Patients with Gastrointestinal Surgery. *Nutrients* **2021**, *13*, 2655. https://doi.org/10.3390/nu13082655

Academic Editor: Ina Bergheim

Received: 29 June 2021
Accepted: 28 July 2021
Published: 30 July 2021

1. Introduction

In an overweight or even obese society, obvious malnutrition seems to be a rare phenomenon even in patients undergoing abdominal surgery. Furthermore, nutritional therapy, either the enteral or even the parenteral route, seems to be very traditional in the era of enhanced recovery after surgery (ERAS) protocols. Nevertheless, malnutrition can be overlooked easily and has been described as a "silent epidemic" [1]. This is why screening for malnutrition and, if necessary, nutritional therapy is one eminent part of ERAS protocols [2].

While the term malnutrition means loss of body mass by an inadequate supply of energy, the term sarcopenia describes the loss of muscle mass. Malnutrition is often associated with sarcopenia if, in addition to storage fat, a relevant proportion of muscle mass is lost. On the other hand, sarcopenia can—often in geriatrics—develop without weight loss and thus without the criterion of malnutrition [3].

The presence of sarcopenia is often masked, especially in overweight or obese patients. In clinical terms, sarcopenic obesity is usually considerably underestimated. Hospital mortality occurred in 7% of 750 obese patients who were also malnourished [4]. The terminological differentiation between sarcopenic obesity and malnutrition in obesity is a matter of discussion.

In 2019, the Global Leadership Initiative on Malnutrition (GLIM) developed a new definition of malnutrition that is endorsed by all major nutritional and medical societies worldwide [5]. A distinction is made between phenotypic (weight loss, low body mass index, reduced muscle mass) and etiological criteria (reduced food intake or assimilation, inflammation/disease burden). For diagnosing malnutrition, a two-step procedure is recommended: In the case of positive nutritional screening, at least one phenotypic and one etiological criterion must be met.

The prognostic influence of the nutritional status on the occurrence of postoperative complications and the length of hospital stay has been shown many times retrospectively and prospectively [6–8], resulting in evidence-based guidelines on nutritional therapy for the surgical patient: European Society for Clinical Nutrition and Metabolism (ESPEN) 2017 [7] and Short ESPEN Practice guideline 2021 [9].

The ESPEN surgical guidelines are in line with the aim of early oral feeding in ERAS [7]:

- Integration of nutrition and nutritional status in the overall management
- Avoiding prolonged periods of sobriety preoperatively
- Start nutritional therapy as soon as metabolic risk is apparent
- Metabolic monitoring of blood sugar levels
- Reduction of factors that trigger stress and catabolism or impair gastrointestinal motility and function
- Early mobilization to stimulate protein synthesis and to maintain muscle function

It is evident that in most patients, oral/enteral feeding can be started just a few hours after abdominal surgery. This also applies to anastomoses in the upper gastrointestinal tract [6,7]. Oral food intake as early as possible is a key intervention of the ERAS program [10]. In the case of inadequate coverage of the calorie and protein requirement, perioperative nutrition therapy strives for supplementing the primarily oral diet by oral nutritional supplements (ONS) and enteral nutrition.

Recent evidence comes from a meta-analysis of 56 randomized controlled studies with 6370 patients undergoing surgery for gastrointestinal cancer. Benefits of supplementation with a glucose drink, increased protein intake, and immunonutrition were shown to reduce postoperative complications (RR 0.74, 95% CI 0.69–0.80), postoperative infections (RR 0.71, 95% CI 0.64–0.79), and non-infectious complications (RR 0.79, 95% CI 0.71–0.87) with a reduction in length of hospital stay of 1.58 days (95% CI −1.83 to −1.32). Nevertheless, there is considerable heterogeneity of the included studies (I = 89%) [11].

2. Who Will Benefit from Perioperative Nutritional Supplementation?

Screening and Assessment of Nutritional Status

The functional status, which is pivotal for enhanced postoperative recovery, is largely determined by muscle mass. From a metabolic point of view, the combination of impaired functionality and nutritional status is crucial for postoperative mobilization and lung function. Sarcopenic patients have a significantly increased risk for severe ($p < 0.0001$) and the total number of postoperative complications ($p = 0.001$) [12]. This is why sarcopenia may be considered to be "forcing a vicious circle" [13]. The prevalence of sarcopenia in cancer patients with chemotherapy can be up to 83% [14].

Therefore, the identification of risk patients is essential preoperatively. In elderly people in particular, this should include functionality and nutritional status in a "complex geriatric assessment" [15,16]. As screening for malnutrition, the ESPEN nutritional risk score (NRS), according to Kondrup [17], has been well validated for surgical patients.

While in surgical patients, "severe nutritional risk" including NRS score ≥3, may be achieved in a 71-year-old patient without any weight loss or diminished food intake scheduled for surgery of colorectal cancer, it has been demonstrated impressively in medical patients that appropriate nutritional therapy in patients with an NRS score ≥3, treatment may already favor this outcome [18].

Current data from an observational study on the implementation of the ERAS program in hospitals in the Canadian province of Alberta revealed a positive NRS as a predictor of low patient compliance in the ERAS program (<70%: OR 2.77; 95% CI 2.11–3.64; $p < 0.001$) and a trend towards longer hospital stay >5 days (OR 1.40; 95% CI 1.00–1.96; $p = 0.052$). An ERAS compliance of <70% was associated with an increased rate of postoperative complications (OR 2.69; 95% CI 2.23–3.24; $p < 0.001$) [19].

When patients are screened at risk for malnutrition, body composition should be examined for establishing GLIM criteria [5]. In addition to more "traditional" tools such as bioelectrical impedance analysis (BIA) and dual X-ray absorptiometry (DEXA), the use of routine computer tomography (CT) in cross-section L3 has been introduced. From the section derived muscle and fat mass, conclusions can be drawn for the whole body with good correlation. The skeletal muscle index at level L3 can be used as a surrogate parameter for the presence of sarcopenia [13]. During the past few years, the prognostic impact of the CT-based measurement of the body composition with the determination of the muscle mass and the radiodensity of the muscle has been shown in numerous studies, especially for oncological patients with surgical therapy [20–22], as shown in Table 1.

Table 1. Overview of the literature that shows CT-derived sarcopenia to be predictive for outcomes in patients undergoing gastrointestinal surgery.

Year	Author	Cancer	N	Outcome
2017	Cespedes et al. [23]	Colorectal	2470	Higher mortality
2017	Palmela et al. [24]	Gastric	48	Higher mortality
2018	Kroenke et al. [25]	Colorectal	3262	Higher mortality
2018	Martin et al. [26]	Colorectal	1139	Longer LOS
2018	Pecorelli et al. [22]	Pancreas	120	Higher mortality
2019	Xiao et al. [27]	Colorectal	3051	Higher comorbidity
2019	Herrod et al. [28]	Colorectal	169	Higher comorbidity rate
2019	Lindner et al. [29]	Pancreas	139	Higher comorbidity rate
2019	Yassaie et al. [30]	Esophagus	53	Higher mortality
2020	Martin et al. [31]	Neck, Lung, GI	1157	Higher mortality
2020	Xie et al. [32]	Colorectal	132	Higher comorbidity rate, higher mortality
2020	Gruber et al. [33]	Pancreas	133	Higher comorbidity rate, higher mortality
2021	Ishida et al. [34]	Esophagus	333	Higher mortality
2021	Argillander et al. [35]	Colorectal	233	Higher mortality

3. Preoperative Nutritional Therapy

The indication for nutritional therapy should be based on nutritional screening and by the diagnosis of malnutrition. Depending on the degree of malnutrition, the underlying disease, and the expected period of inadequately low calorie take, the decision on the duration and route of nutritional therapy has to be made (Figure 1). In case of neoadjuvant treatment with the indication for major abdominal surgery, any weight loss and deterioration of the nutritional status should be prevented.

> The current ESPEN guideline [7] gives a "Good Clinical Practice (GCP)" recommendation: "Perioperative nutritional support therapy is indicated in patients with malnutrition and those at nutritional risk. Perioperative nutritional therapy should also be initiated, if it is anticipated that the patient will be unable to eat for more than five days perioperatively. It is also indicated in patients expected to have low oral intake and who cannot maintain above 50% of recommended intake for more than seven days".

This is in line with the updated ERAS guideline for patients undergoing elective colorectal surgery, which recommends preoperative screening for malnutrition and, if possible, oral nutrition therapy for at least 7–10 days in the case of metabolic risk or manifest deficits [2].

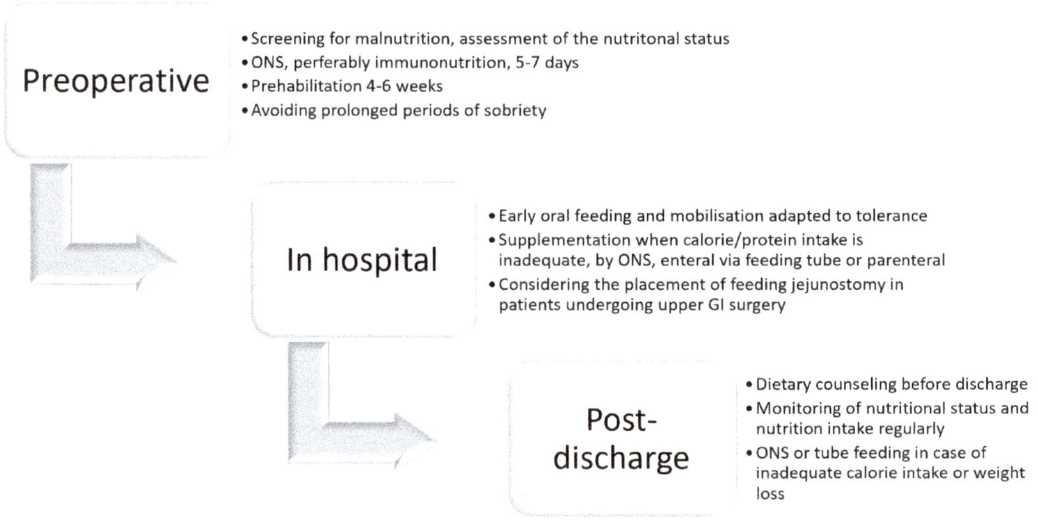

Figure 1. Pathway for nutritional care in patients undergoing gastrointestinal surgery.

4. Prehabilitation

Current "pre-rehabilitation" conditioning concepts include a period of pre-inpatient 4–6 weeks that goes beyond the previously usual inpatient period of 10–14 days. The approach is frequently trimodal: physiotherapy, nutrition therapy, and psychological coaching in order to reduce anxiety and perioperative stress [36–38].

The primary goal is to make the patient "fit for ERAS" and at least to prevent further weight loss. The benefit of perioperative administration of oral nutritional supplements (ONS) has been clearly demonstrated for surgical patients with a reduction in complications and resulting in economic savings [39]. In a systematic review and meta-analysis of nine controlled studies compiled under the aspect of prehabilitation (six exclusively nutritional intervention, three multimodal), a significant reduction in the length of hospital stay by two days was found for a nutritional intervention over at least seven days, with supplementation being continued postoperatively [40].

Currently, however, there are no specific evidence-based recommendations for nutritional therapy in a prehabilitation program [36–38]. Eating a high-protein diet after exercise is recommended regarding the physiology of muscle protein synthesis. Since compliance is often limited for the intake of ONS [41,42], special and repeated motivation is required.

A start of prehabilitation can already be considered with the start of neoadjuvant treatment. It was shown that early nutrition support during neoadjuvant treatment in patients with esophageal cancer leads to less weight loss at 12 months postoperatively [43]. A weight loss of $\geq$10% results in a significantly higher mortality.

5. Immunological Conditioning

The stimulation of the immune system by enriching the diet with suitable immune-enhancing substrates is an interesting and controversially discussed concept as so-called "immunonutrition".

The stimulation of anti-tumor T-cell activity has been shown in vitro for arginine. Anti-inflammatory effects can be expected from the administration of omega-3 fatty acids [44,45]. Numerous randomized clinical studies and their meta-analyses have examined the combination of arginine, omega-3 fatty acids, and ribonucleotides in an enriched oral drink supplement, as well as enterally [45]. Overall, clinical benefits have been shown in reducing the rate of infectious complications and length of hospital stay, as well as cost.

This also applies to an ERAS program [46]. It has been a matter of discussion whether the exclusively preoperative administration offers advantages not only in comparison with a (hospital) diet but also in comparison with standard drinking food [7].

A recent meta-analysis of the available data from 16 randomized studies with 1387 surgical patients with gastrointestinal tumors (immunonutrition $n = 715$, controls $n = 672$) was aimed at this question. Here, the preoperative intake for 5–7 days led to a significant reduction in the incidence of infectious complications in comparison with a normal diet and with an isonitrogenic standard drinking food (OR 0.52; 95% CI 0.38–0.71, $p < 0.0001$). The heterogeneity of the data was low (I = 16%). There was a significant reduction in the length of stay in hospital compared to normal food and a tendency compared to standard drinking food (−1.57 days, 95% CI −2.48–0.66, $p = 0.0007$, I = 34%). Non-infectious complications and mortality were unaffected [47].

The results of this meta-analysis, with a focus on surgical patients with gastrointestinal cancer, show good study quality and acceptable heterogeneity: (1). Oral supplementation carried out exclusively preoperative for 5 days is effective, (2). In this context, immunonutrition is superior, at least in terms of the risk of infectious complications, compared to standard oral nutritional supplements. Focusing on patients undergoing esophagectomy, another recent meta-analysis including six randomized trials of perioperative immunonutrition did not show significant benefits regarding postoperative complication rate [11]. Thus far, the ESPEN guideline recommends ONS before major operations for 5–7 days, with immunomodulating supplements being preferred [7].

6. Postoperative Nutrition

Early Oral Diet

A recent meta-analysis of five randomized controlled studies showed early enteral nutrition after major emergency surgery to be correlated with reduced mortality. Regarding the optimal timing and composition, the authors claimed the need for more high-quality data [48]. In comparison with parenteral nutrition, a recent PRCT showed significant benefits regarding intestinal recovery, hospital length of stay, and immune function for patients with cholangiocarcinoma and obstructive jaundice undergoing surgery [49]. Another randomized controlled trial in patients undergoing pelvic exenteration surgery compared parenteral nutrition and trophic enteral feeding (20 mL/h) via nasogastric tube. No significant difference between the two groups was observed for time to first bowel movement. Postoperative ileus occurred significantly less in enterally fed patients in the per-protocol analysis ($p = 0.036$). In regression analysis, it was shown again that time restriction from an oral diet was significantly associated with the time to first bowel movement and the postoperative complication rate ($p < 0.0005$) [50].

Generally, and even after surgery of the lower gastrointestinal tract, oral food intake can be started within hours (ESPEN guidelines A recommendation) [7]. The diet should be adapted to the individual tolerance and the operation performed; elderly patients require special attention [7].

St. Georg recommendations for postoperative oral diet:

- Very low in fat (max. 30 g fat/day)
- Use of easily digestible, lean protein carriers
- High carbohydrate
- Easily digestible (low fiber content)

7. Early Oral Delivery after Esophagectomy and Gastrectomy

A recent multicenter randomized Dutch study investigated the feasibility and safety of an early oral diet after minimally invasive esophagectomy with intrathoracic anastomosis [51]. In the intervention group ($n = 65$), oral intake took place without delay, while the control group ($n = 67$) was fed only via the enteral tube for 5 days. There was no significant difference in the primary endpoint time of postoperative recovery (7 vs. 8 days) and the secondary parameters of complications with anastomotic leakage (18.5% vs. 16.4%) and pneumonia rate (24.6% vs. 34.3%).

In another retrospective study in patients after gastrectomy, a group with early oral food intake from the first day (EOF $n = 203$) was compared with a historical control group with conventional delayed oral food intake (COF $n = 203$) by means of propensity score matching. The EOF group showed an earlier onset of flatus (2.9 vs. 3.1 days, $p = 0.013$). The length of stay in the hospital was significantly shorter (8.9 ± 5.7 vs. 12.6 ± 10.2 days, $p < 0.01$). No significant differences were observed for morbidity and mortality, with the EOF group showing a lower rate of abdominal infections (3.0% vs. 7.4%, $p = 0.044$) and anastomotic leakages (1.5% vs. 4.9%, $p = 0.048$). A subgroup analysis based on age, gender, surgical method, lymph node dissection, and tumor stage also showed no risk of increased morbidity or anastomotic leakage in the EOF group. The compliance with oral nutrition was the same in both groups [52]. In a prospective study with 50 patients undergoing major abdominal surgery, the protein and calorie intake in the first postoperative week was recorded. In the majority of patients, the energy and protein intake was insufficient (82% and 90%, respectively), leading to more Clavien-Dindo III complications in patients who did not meet their protein targets [53]. In terms of nutritional medicine, it must be emphasized that an early oral diet is feasible but that the oral calorie requirement is not met over a longer period of time comprising the time after discharge, even after post-inpatient, and thus initiates weight loss. This is an argument in favor of oral/enteral supplementation via sip feed/tube feed via feeding jejunostomy (FJT) placed during surgery.

8. Feeding Jejunostomy

Zhuang et al. retrospectively analyzed the outcome of their patients undergoing esophageal resection with and without placing a feeding jejunostomy. The feeding jejunal tube was placed if the patient was considered to be at high risk for developing anastomotic leakage. No significant difference was observed regarding the length of hospital stay, short-term mortality, and overall survival. There was a tendency in FJT patients for the recovery of anastomotic leakage (27.2 vs. 37.4 d, $p = 0.073$). The results suggest FJT being safe and that the placement should be considered in high-risk patients [54]. A meta-analysis for the comparison of jejunostomy versus nasoenteral tube showed benefits for the jejunostomy regarding postoperative pneumonia, length of hospital stay, and dislocation of the tube [55]. Similar results were shown in a Swedish register-based study, which investigated the differences between patients undergoing esophagectomy with and without intraoperative implantation of nutritional jejunostomy. The risk of developing severe complications (≥IIIb Clavien-Dindo) was significantly decreased in jejunostomy patients with an anastomotic leak. There was no increased risk for jejunostomy-related complications [56]. These arguments are for the selective use in high-risk patients undergoing esophagectomy [57], while the preoperative identification may be a matter of debate.

9. Oral Food Intake after Prolonged Intensive Care Treatment

Despite best perioperative management and ERAS protocol as a plan A, serious complications occur, inducing catabolism and subsequently increasing the risk of postoperative deterioration in nutritional status [58]. This needs a plan B, including enteral and even parenteral nutrition. The ESPEN guidelines state: If the energy and nutrient requirements cannot be met by oral and enteral intake alone (<50% of caloric requirement) for more than seven days, a combination of enteral and parenteral nutrition is recommended (GPP).

From a nutritional point of view, the period after transfer from intensive care to a normal ward is very vulnerable. Spontaneous oral food intake is usually inadequate. The main causes may be inappetence and fatigue with limited cooperation. In the anabolic situation, after a longer period of severe catabolism, the undersupply of energy and protein may be a "metabolic catastrophe", increasing the risk for delayed recovery and rehabilitation. In a small cohort study, 32 patients were followed up with their food intake measured by their energy (2000; 1650–2550 kcal/d) and protein (112; 84–129 g/d) requirements after they had been transferred from the intensive care unit. A median of 1238 (869–1813) kcal/d and 60 (35–89) g protein was consumed over 227 days. Most of the patients were fed exclusively orally (55%) or combined enterally (42%). The energy and protein intake were lower than they estimated or even measured by indirect calorimetry. It was shown that energy and protein requirements could only be achieved with a combination of oral and enteral nutrition. The supplementation of an oral diet with ONS only covered about 70% of the requirement [59].

10. Post-Discharge Nutrition

In particular, after resections in the upper gastrointestinal tract, sustained weight loss as a "bariatric effect" has to be expected. Dietary counseling and follow-up monitoring of the nutritional status (minimum: BMI), including documentation of the amount of oral food intake, are essential.

A systematic review of 18 studies revealed a postoperative weight loss of 5% to 12% after 6 months. More than half of the patients lost >10% of their body weight after 12 months [60]. Therefore, these patients have to be considered to be at severe metabolic risk. Koterazawa et al., additionally showed that severe weight loss 3 months after esophagectomy could not be diminished by enteral tube feeding but had a significant impact on the 5-year overall survival rate [61]. The guidelines recommend the implantation of needle-catheter-jejunostomy (FIT) during surgery. Our own results in patients with esophageal and gastric resection, including partial pancreato-duodenectomy, show a weight loss of >10% in 40% of patients after 6 months, even with consistent postoperative continuation of nutritional therapy via feeding jejunostomy [16]. Early postoperative weight decreased up to 3 months, while stabilization occurred between 4–6 months after surgery; a further decline could be prevented by continuing enteral feeding supplementation [16]. In comparison with a control group, Chen et al. demonstrated in elderly patients after esophagectomy significant benefits from home enteral nutrition for at least 8 weeks for the BMI, PG-SGA score, serum albumin, and immune parameters [62].

In a recent meta-analysis, home enteral nutrition (HERN) and oral nutritional supplements were compared in patients with upper gastrointestinal resection for malignancy [63]. A total of 15 RCTs involving 1059 patients were included. Home enteral nutrition seemed to be superior by the significant prevention of weight loss (−3.95 vs. −5.82 kg; SMD: 1.98 kg; 95% CI 1.24–2.73) and reduction of the incidence of malnutrition or latent malnutrition (RR = 0.54; $p < 0.01$). Furthermore, improved levels of albumin, hemoglobin, pre-albumin, and transferrin were observed. Subgroup analysis based on the approach of home nutrition therapy showed that weight loss in the home enteral nutrition subgroup was significantly lower than that of the control group (WMD = 2.69, $p < 0.01$). No significant difference could be observed between the ONS subgroup and the control group. The same results were found in albumin, physical function (WMD: 5.29; 95% CI 1.86–8.73), and fatigue (WMD: −8.59; 95% CI −12.61, −4.58). Furthermore, dimensions in QOL were significantly better in the HERN group.

In a randomized study, significantly less bodyweight loss was observed when ONS was administered until 12 weeks after surgery in patients with total gastrectomy, which was not seen in patients with subtotal resection [64]. In another randomized study in patients after gastrectomy, dietary counseling in combination with ONS [65] showed a significant reduction in weight loss, with a higher BMI and SMI. Fatigue and loss of appetite were also significantly lower than with diet advice alone. Limited compliance in taking sip feeds has to be always considered [17], which may be caused by loss of appetite, taste, bloating, belching, and diarrhea. A recent multicentric randomized trial of 1003 patients after gastrectomy compared the impact of ONS 400 kcal/d on weight loss 1 year after gastrectomy. In the ONS group, only 50.4% of the patients had an intake of more than 200 kcal/day of ONS (average 301 mL) and showed significantly less bodyweight loss (8.2 ± 7.2%) at 1 year compared to the control ($p = 0.0204$) [66].

11. Conclusions

In conclusion, in patients undergoing gastrointestinal surgery with special regard to those of the upper GI tract, nutritional monitoring and therapy may be required even after discharge from the hospital. In line with the guidelines appropriate recommendations are summarized in Table 2. Oral nutritional supplementation and/or enteral nutrition provide clinical benefits for the attenuation of weight loss. From an ethical point of view, randomized studies may be only justified comparing oral versus enteral supplementation. Individualized home nutrition therapy requires a network with the cooperation of the surgeon, general practitioner, dietitian, and home care provider. Regular monitoring of body composition and quantitative registration of oral nutrition intake is mandatory. In the future, patient reporting and dietary counseling may be improved by virtual coaching and assisted digitalized communication by chatbot [67].

Table 2. Perioperative recommendations summary for patients undergoing gastrointestinal surgery.

Recommendations Summary
- Malnutrition and sarcopenia are prognostic factors, a preoperative assessment of the nutritional status is mandatory - If there is a high metabolic risk/severe malnutrition, surgery should be postponed and nutritional therapy should be administered enterally for 10–14 days if possible, and prehabilitation should be considered - Preoperative use of ONS for at least 7 days may diminish the rate of infectious complications and the length of hospital stay - In patients with gastrointestinal cancer, preoperative immunonutrition for 5–7 days may lead to a reduction in infectious complications - Early oral intake is also feasible after esophagectomy and gastrectomy - If the oral calorie intake is insufficient (<50% for more than seven days), ONS and enteral supplementation may be indicated - Metabolic risk patients should receive dietary counseling at the time of discharge and follow-up nutritional care - Follow-up on the nutritional status and nutrition therapy after discharge may be necessary especially after surgery in the upper GI as prolonged weight loss with high metabolic risk is common

As an open question, the impact of perioperative oral and enteral supplementation in the framework of prehabilitation and ERAS on long-term oncological outcomes in patients with gastrointestinal cancer has to be elucidated in the future [68]. Further data from observational and controlled trials, including patient-reported outcomes, are urgently needed.

Author Contributions: Conceptualization, M.W. and A.W.; writing—original draft preparation, A.W. and M.W.; writing—review and editing, A.W. and M.W.; supervision, A.W. All authors have read and agreed to the published version of the manuscript.

Funding: This research received no external funding.

Conflicts of Interest: Arved Weimann received lecture fees from Baxter, B. Braun, Fresenius Kabi, Ethicon, and Falk Foundation, and a research grant from Baxter, Mucos. Maria Wobith declares no conflicts of interest.

References

1. Williams, D.G.A.; Molinger, J.; Wischmeyer, P.E. The malnourished surgery patient: A silent epidemic in perioperative outcome? *Curr. Opin. Anesthesiol.* **2019**, *32*, 405–411. [CrossRef]
2. Gustafsson, U.O.; Scott, M.J.; Hubner, M.; Nygren, J.; Demartines, N.; Francis, N.; Rockall, T.A.; Young-Fadok, T.M.; Hill, A.G.; Soop, M.; et al. Guidelines for Perioperative Care in Elective Colorectal Surgery: Enhanced Recovery After Surgery (ERAS®) Society Recommendations: 2018. *World J. Surg.* **2019**, *43*, 659–695. [CrossRef]

3. Cruz-Jentoft, A.J.; Bahat, G.; Bauer, J.; Boirie, Y.; Bruyère, O.; Cederholm, T.; Cooper, C.; Landi, F.; Rolland, Y.; Sayer, A.A.; et al. Writing Group for the European Working Group on Sarcopenia in Older People 2 (EWGSOP2), and the Extended Group for EWGSOP2. Sarcopenia: Revised European consensus on definition and diagnosis. *Age Ageing* **2019**, *48*, 16–31. [CrossRef] [PubMed]
4. Agarwal, E.; Ferguson, M.; Banks, M.; Vivanti, A.; Batterham, B.; Bauer, J.; Capra, S.; Isenring, E. Malnutrition, poor food intake, and adverse healthcare outcomes in noncritically ill obese acute care hospital patients. *Clin. Nutr.* **2019**, *38*, 759–766. [CrossRef] [PubMed]
5. Jensen, G.L.; Cederholm, T.; Correia, M.T.D.; Gonzalez, M.C.; Fukushima, R.; Higashiguchi, T.; Baptista, G.; Barrazoni, R.; Blaauw, R.; Coats, A.; et al. GLIM criteria for the diagnosis of malnutrition: A consensus report from the global clinical nutrition community. *J. Parenter. Enter. Nutr.* **2019**, *43*, 32–40. [CrossRef]
6. Cao, Y.; Han, D.; Zhou, X.; Han, Y.; Zhang, Y.; Li, H. Effects of preoperative nutrition on postoperative outcomes in esophageal cancer: A systematic review and meta-analysis. *Dis. Esophagus* **2021**. [CrossRef]
7. Weimann, A.; Braga, M.; Carli, F.; Higashiguchi, T.; Hübner, M.; Klek, S.; Laviano, A.; Lobo, D.N.; Ljungqvist, O.; Martindale, R.; et al. ESPEN Guideline Clinical Nutrition in Surgery. *Clin. Nutr.* **2017**, *36*, 623–650. [CrossRef] [PubMed]
8. Kenny, E.; Samavat, H.; Touger-Decker, R.; Parrott, J.S.; Byham-Gray, L.; Agust, D.A. Adverse perioperative outcomes among patients undergoing gastrointestinal cancer surgery: Quantifying attributable risk from malnutrition. *J. Parenter. Enter. Nutr.* **2021**. [CrossRef]
9. Muscaritoli, M.; Arends, J.; Bachmann, P.; Baracos, V.; Barthelemy, N.; Bertz, H.; Bozzetti, F.; Hütterer, E.; Isenring, E.; Kaasa, S.; et al. ESPEN practical guideline: Clinical Nutrition in cancer. *Clin. Nutr.* **2021**, *40*, 2898–2913. [CrossRef]
10. Wischmeyer, P.W.; Carli, F.; Evans, D.C.; Guilbert, S.; Kozar, R.; Pryor, A.; Thiele, R.H.; Everett, S.; Rocott, M.; Gan, T.J.; et al. American Society for Enhanced Recovery and Perioperative Quality Initiative joint consensus statement on nutritional screening and therapy within a surgical enhanced recovery pathway. *Anesth. Analg.* **2018**, *126*, 1883–1895. [CrossRef]
11. Zhang, B.; Najarali, Z.; Ruo, L.; Alhusaini, A.; Solis, N.; Valencia, M.; Sanchez, M.I.P.; Serrano, P.E. Effect of Perioperative Nutritional Supplementation on Postoperative Complications-Systematic Review and Meta-Analysis. *J. Gastrointest. Surg.* **2019**, *23*, 1682–1693. [CrossRef]
12. Simonsen, C.; de Heer, P.; Bjerre, E.D.; Suetta, C.; Hojman, P.; Pedersen, B.K.; Svendsen, L.B.; Christensen, J.F. Sarcopenia and postoperative complication risk in gastrointestinal surgical oncology: A meta-analysis. *Ann. Surg.* **2018**, *268*, 58–69. [CrossRef] [PubMed]
13. Bozzetti, F. Forcing the vicious circle: Sarcopenia increases toxicity, decreases response to chemotherapy and worsens with chemotherapy. *Ann. Oncol.* **2017**, *28*, 2107–2118. [CrossRef] [PubMed]
14. Caillet, P.; Liuu, E.; Raynaud Simon, A.; Bonnefoy, M.; Guerin, O.; Berrut, G.; Lesourd, B.; Jeandel, C.; Ferry, M.; Rolland, Y. Association between cachexia, chemotherapy and outcomes in older cancer patients: A systematic review. *Clin. Nutr.* **2017**, *36*, 1473–1482. [CrossRef]
15. Olotu, C.; Weimann, A.; Bahrs, C.; Schwenk, W.; Scherer, M.; Kiefmann, R. The perioperative care of older patients—Time for a new, interdisciplinary approach. *Dtsch. Ärzteblatt Int.* **2019**, *116*, 63–69.
16. Wobith, M.; Wehle, L.; Haberzettl, D.; Acikgöz, A.; Weimann, A. Needle catheter jejunostomy in patients undergoing surgery for upper gastrointestinal and pancreato-biliary cancer—Impact on nutritional and clinical outcome in the early and late postoperative period. *Nutrients* **2020**, *12*, 2564. [CrossRef]
17. Kondrup, J.; Allison, S.P.; Elia, M.; Vellas, B.; Plauth, M.; Educational and Clinical Practice Committee; European Society of Parenteral and Enteral Nutrition (ESPEN). ESPEN guidelines for nutrition screening 2002. *Clin. Nutr.* **2003**, *22*, 415–421. [CrossRef]
18. Schuetz, P.; Fehr, R.; Baechli, V.; Geiser, M.; Deiss, M.; Gomes, F.; Kutz, A.; Tribolet, F.; Bregenzer, T.; Braun, N.; et al. Individualized nutritional support in medical inpatients at nutritional risk: A randomized clinical trial. *Lancet* **2019**, *393*, 2312–2321. [CrossRef]
19. Martin, L.; Gillis, C.; Atkins, M.; Gillam, M.; Sheppard, C.; Buhler, S.; Hammond, C.B.; Nelson, G.; Gramlich, L. Implementation of an Enhanced Recovery After Surgery Program can change nutrition care practice: A multicenter experience in elective colorectal surgery. *J. Parenter. Enter. Nutr.* **2019**, *43*, 206–219. [CrossRef]
20. Tamandl, D.; Paireder, M.; Asari, R.; Baltzer, P.A.; Schoppmann, S.F.; Ba-Ssalamah, A. Markers of sarcopenia quantified by computed tomography predict adverse long-term outcome in patients with resected oesophageal or gastro-oesophageal junction cancer. *Eur Radiol.* **2016**, *26*, 1359–1367. [CrossRef]
21. Runkel, M.; Diallo, T.D.; Lang, S.A.; Bamberg, F.; Benndorf, M.; Fichtner-Feigl, S. The Role of Visceral Obesity, Sarcopenia and Sarcopenic Obesity on Surgical Outcomes After Liver Resections for Colorectal Metastases. *World J. Surg.* **2021**, *45*, 2218–2226. [CrossRef]
22. Pecorelli, N.; Capretti, G.; Sandini, M.; Damascelli, A.; Cristel, G.; De Cobelli, F.; Gianotti, L.; Zerbi, A.; Braga, M. Impact of Sarcopenic Obesity on Failure to Rescue from Major Complications Following Pancreaticoduodenectomy for Cancer: Results from a Multicenter Study. *Ann. Surg. Oncol.* **2018**, *25*, 308–317. [CrossRef]
23. Feliciano, E.M.C.; Kroenke, C.H.; Meyerhardt, J.A.; Prado, C.M.; Bradshaw, P.T.; Kwan, M.L.; Xiao, J.; Alexeeff, S.; Douglas, D. Association of Systemic Inflammation and Sarcopenia with Survival in Nonmetastatic Colorectal Cancer. *JAMA Oncol.* **2017**, *3*, e172319. [CrossRef]

24. Palmela, C.; Velho, S.; Agostinho, L.; Branco, F.; Santos, M.; Costa Santos, M.P.; Oliveira, M.H.; Strecht, J.; Maio, R.; Cravo, M.; et al. Body composition as a prognostic factor of neoadjuvant chemotherapy toxicity and outcome in patients with locally advanced gastric cancer. *J. Gastric Cancer* **2017**, *17*, 74–87. [CrossRef]
25. Kroenke, C.H.; Prado, C.M.; Meyerhardt, J.A.; Weltzien, E.K.; Xiao, J.; Cespedes Feliciano, E.M.; Caan, B.J. Muscle radiodensity and mortality in patients with colorectal cancer. *Cancer* **2018**, *124*, 3008–3015. [CrossRef] [PubMed]
26. Martin, L.; Hopkins, J.; Malietzis, G.; Jenkins, J.T.; Sawyer, M.B.; Brisebois, R.; MacLean, A.; Nelson, G.; Gramlich, L.; Baracos, V.E. Assessment of Computed Tomography (CT)-Defined Muscle and Adipose Tissue Features in Relation to Short-Term Outcomes After Elective Surgery for Colorectal Cancer: A Multicenter Approach. *Ann. Surg. Oncol.* **2018**, *25*, 2669–2680. [CrossRef] [PubMed]
27. Xiao, J.; Caan, B.J.; Cespedes Feliciano, E.M.; Meyerhardt, J.A.; Kroenke, C.H.; Baracos, V.E.; Weltzien, E.; Kwan, M.L.; Alexeeff, S.E.; Castillo, A.L.; et al. The association of medical and demographic characteristics with sarcopenia and low muscle radiodensity in patients with nonmetastatic colorectal cancer. *Am. J. Clin. Nutr.* **2019**, *109*, 626–634. [CrossRef] [PubMed]
28. Herrod, P.J.J.; Boyd-Carson, H.; Doleman, B.; Trotter, J.; Schlichtemeier, S.; Sathanapally, G.; Somerville, J.; Williams, J.P.; Lund, J.N. Quick and simple; psoas density measurement is an independent predictor of anastomotic leak and other complications after colorectal resection. *Tech. Coloproctol.* **2019**, *23*, 129–134. [CrossRef] [PubMed]
29. Linder, N.; Schaudinn, A.; Langenhan, K.; Krenzien, F.; Hau, H.-M.; Benzing, C.; Atanasov, G.; Schmelzle, M.; Kahn, T.; Busse, H.; et al. Power of computed-tomography-defined sarcopenia for prediction of morbidity after pancreaticoduodenectomy. *BMC Med. Imaging* **2019**, *19*, 32. [CrossRef]
30. Yassaie, S.S.; Keane, C.; French, S.J.H.; Al-Herz, F.A.J.; Young, M.K.; Gordon, A.C. Decreased total psoas muscle area after neoadjuvant therapy is a predictor of increased mortality in patients undergoing oesophageal cancer resection. *ANZ J. Surg.* **2019**, *89*, 515–519. [CrossRef] [PubMed]
31. Martin, L.; Gioulbasanis, I.; Senesse, P.; Baracos, V.E. Cancer-Associated Malnutrition and CT-Defined Sarcopenia and Myosteatosis are Endemic in Overweight and Obese Patients. *J. Parenter. Enter. Nutr.* **2020**, *44*, 227–238. [CrossRef]
32. Xie, H.; Gong, Y.; Kuang, J.; Yan, L.; Ruan, G.; Tang, S.; Gao, F.; Gan, J. Computed Tomography-Determined Sarcopenia is a Useful Imaging Biomarker for Predicting Postoperative Outcomes in Elderly Colorectal Cancer Patients. *Cancer Res. Treat.* **2020**, *52*, 957–972. [CrossRef] [PubMed]
33. Gruber, E.S.; Jomrich, G.; Tamandl, D.; Gnant, M.; Schindl, M.; Sahora, K. Erratum: Sarcopenia and sarcopenic obesity are independent adverse prognostic factors in resectable pancreatic ductal adenocarcinoma. *PLoS ONE* **2020**, *15*, e0244896 1–18. [CrossRef]
34. Ishida, T.; Makino, T.; Yamasaki, M.; Yamashita, K.; Tanaka, K.; Saito, T.; Yamamoto, K.; Takahashi, T.; Kurokawa, Y.; Motoori, M.; et al. Quantity and Quality of Skeletal Muscle as an Important Predictor of Clinical Outcomes in Patients with Esophageal Cancer Undergoing Esophagectomy after Neoadjuvant Chemotherapy. *Ann. Surg. Oncol.* **2021**. [CrossRef]
35. Argillander, T.E.; Spek, D.; van der Zaag-Loonen, H.J.; van Raamt, A.F.; van Duijvendijk, P.; van Munster, B.C. Association between postoperative muscle wasting and survival in older patients undergoing surgery for non-metastatic colorectal cancer. *J. Geriatr. Oncol.* **2021**. [CrossRef]
36. Gillis, C.; Ferton, T.; Sajobi, T.T.; Minnella, E.M.; Awasthi, R.; Loiselle, S.E.; Liberman, A.S.; Stein, B.; Charlebois, P.; Carli, F. Trimodal prehabilitation for colorectal surgery attemuates post-surgical losses in lean body mass: A pooled analysis of randomized controlled trials. *Clin. Nutr.* **2019**, *38*, 1053–1060. [CrossRef] [PubMed]
37. Hughes, M.J.; Hackney, R.J.; Lamb, P.J.; Wigmore, S.J.; Deans, D.C.; Skipworth, R.J.E. Prehabilitation Before Major Abdominal Surgery: A Systematic Review and Meta-analysis. *World J. Surg.* **2019**, *43*, 1661–1668. [CrossRef] [PubMed]
38. Daniels, S.L.; Lee, M.J.; George, J.; Kerr, K.; Moug, S.; Wilson, T.R.; Brown, S.R.; Wyld, L. Prehabilitation in elective abdominal cancer surgery in older patients: Systematic review and meta-analysis. *BJS Open* **2020**, *4*, 1022–1041. [CrossRef]
39. Elia, M.; Normand, C.; Norman, K.; Laviano, A.; Norman, K. A systematic review of the cost and cost effectiveness of using standard oral nutritional supplements in the hospital setting. *Clin. Nutr.* **2016**, *35*, 370–380. [CrossRef]
40. Gillis, C.; Buhler, K.; Bresee, L.; Carli, F.; Gramlich, l.; Culos-Reed, N.; Sajobi, T.T.; Fenton, T.R. Effects of nutritional prehabilitation with and without exercise, on outcomes of patients who undergo colorectal surgery: A systematic review and meta-analysis. *Gastroenterol* **2018**, *155*, 391–410. [CrossRef]
41. Grass, F.; Bertrand, P.C.; Schäfer, M.; Ballabeni, P.; Cerantola, Y.; Demartines, N.; Hübner, M. Compliance with preoperative oral nutritional supplements in patients at nutritional risk-only a question of will? *Eur. J. Clin. Nutr.* **2015**, *69*, 525–529. [CrossRef] [PubMed]
42. Lidoriki, I.; Schizas, D.; Mylonas, K.S.; Frountzas, M.; Mastoraki, A.; Pikoulis, E.; Liakakos, T.; Karavokyros, I. Oral nutritional supplementation following upper gastrointestinal cancer surgery: A prospective analysis exploring potential barriers to compliance. *J. Am. Coll. Nutr.* **2010**, *39*, 650–656. [CrossRef]
43. Davies, S.J.; West, M.A.; Rahman, S.A.; Underwood, T.J.; Marino, L.V. Oesophageal cancer: The effect of early nutrition support on clinical outcomes. *Clin. Nutr. ESPEN* **2021**, *42*, 117–123. [CrossRef]
44. Geiger, R.; Rieckmann, J.C.; Wolf, T.; Basso, C.; Feng, Y.; Fuhrer, T.; Kogadeeva, M.; Picotti, P.; Meissner, F.; Mann, M.; et al. L-Arginine Modulates T Cell Metabolism and Enhances Survival and Anti-tumor Activity. *Cell* **2016**, *167*, 829–842. [CrossRef]

45. Troesch, B.; Eggersdorfer, M.; Laviano, A.; Rolland, Y.; Smith, A.D.; Warnke, I.; Weimann, A.; Calder, P.C. Expert Opinion on Benefits of Long-Chain Omega-3 Fatty Acids (DHA and EPA) in Aging and Clinical Nutrition. *Nutrients* **2020**, *12*, 2555. [CrossRef] [PubMed]
46. Moya, P.; Soriano-Irigiaray, L.; Ramirez, J.M.; Garcea, A.; Blasco, O.; Blanco, F.; Brugiotti, C.; Miranda, E.; Arroyo, A. Perioperative standard oral nutrition supplements versus immunonutrition in patients undergoing colorectal resection in an enhanced recovery(ERAS) protocol: A multicenter randomized Clinical Trial (SONVI Study). *Medicine* **2016**, *95*, e3704. [CrossRef]
47. Adiamah, A.; Skorepa, P.; Weimann, A.; Lobo, D.N. The impact of preoperative immune modulating nutrition on outcomes in patients undergoing durgery for gastrointestinal cancer: A systematic review and metaanalysis. *Ann. Surg.* **2019**, *270*, 247–256. [CrossRef]
48. Burcharth, J.; Falkenberg, A.; Schack, A.; Ekeloef, S.; Gögenur, I. The effects of early enteral nutrition on mortality after major emergency abdominal surgery: A systematic review and meta-analysis with Trial Sequential Analysis. *Clin. Nutr.* **2021**, *40*, 1604–1612. [CrossRef] [PubMed]
49. Ma, B.Q.; Chen, S.Y.; Jiang, Z.B.; Wu, B.; He, Y.; Wang, X.X.; Li, Y.; Gao, P.; Yang, X.J. Effect of postoperative early enteral nutrition on clinical outcomes and immune function of cholangiocarcinoma patients with malignant obstructive jaundice. *World J. Gastroenterol.* **2020**, *26*, 7405–7415. [CrossRef]
50. Hogan, S.; Reece, L.; Solomon, M.; Rangan, A.; Carey, S. Early enteral feeding is beneficial for patients after pelvic exenteration surgery: A randomized controlled trial. *J. Parenter. Enter. Nutr.* **2021**. [CrossRef]
51. Berkelmans, G.H.K.; Fransen, L.F.C.; Dolmans-Zwartjes, A.C.P.; Kouwenhoven, E.A.; van Det, M.J.; Nilsson, M.; Nieuwenhuijzen, G.A.P.; Luyer, M.D.P. Direct oral feeding following minimally invasive esophagectomy (NUTRIENT II trial): An international, multicenter, open-label randomized controlled trial. *Ann. Surg.* **2020**, *271*, 41–47. [CrossRef] [PubMed]
52. Jang, A.; Jeong, O. Early Postoperative Oral Feeding After Total Gastrectomy in Gastric Carcinoma Patients: A Retrospective Before-After Study Using Propensity Score Matching. *J. Parenter. Enter. Nutr.* **2019**, *43*, 649–657. [CrossRef]
53. Constansia, R.D.; Hentzen, J.E.; Hogenbirk, R.N.; van der Plas, W.Y.; Campmans-Kuijpers, M.J.; Buis, C.I.; Kruijff, S.; Klaase, J.M. Actual postoperative protein and calorie intake in patients undergoing major open abdominal cancer surgery: A prospective, observational cohort study. *Nutr. Clin. Pract.* **2021**. [CrossRef]
54. Zhuang, W.; Wu, H.; Liu, H.; Huang, S.; Wu, Y.; Deng, C.; Tian, D.; Zhou, Z.; Shi, R.; Chen, G.; et al. Utility of feeding jejunostomy in patients with esophageal cancer undergoing esophagectomy with a high risk of anastomotic leakage. *J. Gastrointest. Oncol.* **2021**, *12*, 433–445. [CrossRef] [PubMed]
55. Li, H.N.; Chen, Y.; Dai, L.; Wang, Y.-Y.; Chen, M.-W.; Mei, L.-X. A Meta-analysis of Jejunostomy Versus Nasoenteral Tube for Enteral Nutrition Following Esophagectomy. *J. Surg. Res.* **2021**, *264*, 553–561. [CrossRef]
56. Holmén, A.; Hayami, M.; Szabo, E.; Rouvelas, I.; Agustsson, T.; Klevebro, F. Nutritional jejunostomy in esophagectomy for cancer, a national register-based cohort study of associations with postoperative outcomes and survival. *Langenbecks Arch. Surg.* **2020**. [CrossRef]
57. Shen, X.; Zhuo, Z.-G.; Li, G.; Alai, G.-H.; Song, T.-N.; Xu, Z.-J.; Yao, P.; Lin, Y.-D. Is the routine placement of a feeding jejunostomy during esophagectomy worthwhile?—A systematic review and meta-analysis. *Ann. Palliat. Med.* **2021**, *10*, 4232–4241. [CrossRef]
58. Grass, F.; Benoit, M.; Coti Bertrand, P.; Sola, J.; Schäfer, M.; Demartines, N.; Hübner, M. Nutritional status deteriorates postoperatively despite preoperative nutritional support. *Ann. Nutr. Metab.* **2016**, *68*, 291–297. [CrossRef]
59. Ridley, E.J.; Parke, R.L.; Davies, A.R.; Bailey, M.; Hodgson, C.; Deane, A.M.; McGuinness, S.; Cooper, D.J. What happens to nutrition intake in the post-intensive care unit hospitalization period? An observational cohort study in critically ill adults. *J. Parenter. Enter. Nutr.* **2019**, *43*, 88–95. [CrossRef]
60. Baker, M.; Halliday, V.; Williams, R.N.; Bowrey, D.J. A systematic review of the nutritional consequences of esophagectomy. *Clin. Nutr.* **2016**, *35*, 987–994. [CrossRef]
61. Koterazawa, Y.; Oshikiri, T.; Takiguchi, G.; Urakawa, N.; Hasegawa, H.; Yamamoto, M.; Kanaji, S.; Yamashita, K.; Matsuda, T.; Nakamura, T.; et al. Severe weight loss after minimally invasive oesophagectomy is associated with poor survival in patients with oesophageal cancer at 5 years. *BMC Gastroenterol.* **2020**, *20*, 407. [CrossRef] [PubMed]
62. Chen, X.; Zhao, G.; Zhu, L. Home enteral nutrition for postoperative elderly patients with esophageal cancer. *Ann. Palliat. Med.* **2021**, *10*, 278–284. [CrossRef] [PubMed]
63. Xueting, H.; Li, L.; Meng, Y.; Yuqing, C.; Yutong, H.; Lihong, Q.; June, Z. Home enteral nutrition and oral nutritional supplements in postoperative patients with upper gastrointestinal malignancy: A systematic review and meta-analysis. *Clin. Nutr.* **2020**, *40*, 3082–3093. [CrossRef] [PubMed]
64. Hatao, F.; Chen, K.Y.; Wu, J.M.; Wang, M.Y.; Aikou, S.; Onoyama, H.; Shimizu, N.; Fukatsu, K.; Seto, Y.; Lin, M.T. Randomized controlled clinical trial assessing the effects of oral nutritional supplements in postoperative gastric cancer patients. *Langenbecks Arch. Surg.* **2017**, *402*, 203–211. [CrossRef] [PubMed]
65. Meng, Q.; Tan, S.; Jiang, Y.; Han, J.; Xi, Q.; Zhuang, Q.; Wu, G. Post-discharge oral nutritional supplements with dietary advice in patients at nutritional risk after surgery for gastric cancer: A randomized clinical trial. *Clin. Nutr.* **2020**, *40*, 40–46. [CrossRef]
66. Miyazaki, Y.; Omori, T.; Fujitani, K.; Fujita, J.; Kawabata, R.; Imamura, H.; Okada, K.; Moon, J.H.; Hirao, M.; Matsuyama, J.; et al. Oral nutritional supplements versus a regular diet alone for body weight loss after gastrectomy: A phase 3, multicenter, open-label randomized controlled trial. *Gastric Cancer* **2021**. [CrossRef]

67. Tropea, P.; Schlieter, H.; Sterpi, I.; Judica, E.; Gand, K.; Caprino, M.; Gabilondo, I.; Gomez-Esteban, J.C.; Busnatu, S.; Sinescu, C.; et al. Rehabilitation, the great absentee of virtual coaching in medical care: Scoping Review. *J. Med. Internet Res.* **2019**, *21*, e12805. [CrossRef] [PubMed]
68. Aoyama, T.; Nakazono, M.; Nagasawa, S.; Segami, K. Clinical Impact of Perioperative Oral Nutritional Treatment for Body Composition Changes in Gastrointestinal Cancer Treatment. *Anticancer Res.* **2021**, *41*, 1727–1732. [CrossRef]

nutrients

Review

Nutrition in the Intensive Care Unit—A Narrative Review

Aileen Hill [1,*], Gunnar Elke [2] and Arved Weimann [3,*]

1 Department of Intensive Care and Anesthesiology, University Hospital RWTH Aachen University, D-52074 Aachen, Germany
2 Department of Anesthesiology and Intensive Care Medicine, University Medical Center Schleswig-Holstein, Campus Kiel, D-24105 Kiel, Germany; gunnar.elke@uksh.de
3 Department of General, Visceral and Oncological Surgery, Surgical Intensive Care Unit, Klinikum St. Georg, D-04129 Leipzig, Germany
* Correspondence: ahill@ukaachen.de (A.H.); Arved.Weimann@sanktgeorg.de (A.W.); Tel.: +49-(0)241-80-38166 (A.H.); +49-(0)341-909-2200 (A.W.)

Abstract: Background: While consent exists, that nutritional status has prognostic impact in the critically ill, the optimal feeding strategy has been a matter of debate. Methods: Narrative review of the recent evidence and international guideline recommendations focusing on basic principles of nutrition in the ICU and the treatment of specific patient groups. Covered topics are: the importance and diagnosis of malnutrition in the ICU, the optimal timing and route of nutrition, energy and protein requirements, the supplementation of specific nutrients, as well as monitoring and complications of a Medical Nutrition Therapy (MNT). Furthermore, this review summarizes the available evidence to optimize the MNT of patients grouped by primarily affected organ system. Results: Due to the considerable heterogeneity of the critically ill, MNT should be carefully adapted to the individual patient with special focus on phase of critical illness, metabolic tolerance, leading symptoms, and comorbidities. Conclusion: MNT in the ICU is complex and requiring an interdisciplinary approach and frequent reevaluation. The impact of personalized and disease-specific MNT on patient-centered clinical outcomes remains to be elucidated.

Keywords: medical nutrition therapy; critical care; enteral nutrition; parenteral nutrition; micronutrients; energy; protein; review

Citation: Hill, A.; Elke, G.; Weimann, A. Nutrition in the Intensive Care Unit—A Narrative Review. *Nutrients* **2021**, *13*, 2851. https://doi.org/10.3390/nu13082851

Academic Editors: Ina Bergheim, Roland N. Dickerson and Yasuo Tsutsumi

Received: 28 June 2021
Accepted: 17 August 2021
Published: 19 August 2021

1. Introduction

Medical nutrition therapy (MNT) is an essential part of the care for critically ill patients, but the optimal feeding strategy for patients in the intensive care unit (ICU) is still debated and often remains a challenge for the ICU team in clinical practice. Recommendations for MNT in critically ill patients vary between guidelines of the DGEM (German Society for Nutritional Medicine) [1], the ESPEN (European Society of Enteral and Parenteral Nutrition) [2], the A.S.P.E.N (American Society of Enteral and Parenteral Nutrition) [3], and other societies [4,5], and their implementation into clinical practice may be considered a challenge [6]. This article summarizes current recommendations and discusses available evidence. The important questions "why", "who", "when", "how", and "with what" are answered to provide a pragmatic oversight. The optimal MNT for special patient groups in the ICU is presented grouped by organ systems.

This narrative review is a summary of current guideline recommendations and more recent evidence afterwards. For many aspects of the broad topic "nutrition in the ICU", we can only give an overview in this manuscript and refer to other references for greater detail. Pragmatic recommendations are given as a summary at the end of each chapter and represent the author's opinion in the attempt to help the clinician focus on the most important aspects of each section.

2. "Why?"—Nutritional Status as a Prognostic Factor

MNT in the ICU aims at avoiding malnutrition in primarily well-nourished patients and at preventing further deterioration of previously malnourished patients.

Malnutrition is a significant prognostic risk factor for critically ill patients, influencing major outcomes such as mortality, length of stay (LOS), duration of mechanical ventilation, and infection rates [7]. In a current meta-analysis of 20 studies with 1168 patients, the prevalence of malnutrition in ICU patients was 38–78% [7], highlighting the need of optimal and individually tailored MNT in the ICU. In the same vein, the large percentage of malnourished patients makes nutritional risk assessment upon ICU-admission followed by an appropriate MNT mandatory. Special attention and care should be attributed to those patients who will be anticipated to stay for longer than a week in the ICU [8,9].

Feeding protocols should be used and have proven beneficial in the nutrition of ICU patients [10–12]. A clearly defined feeding protocol has shown to decrease the rate of patients who cannot be enterally fed at all and will increase the delivery of calories [13]. Protocols should be locally tailored according to expertise, local barriers, facilities, and patient subpopulation in the ICU [14]

Pragmatic recommendations:

- **Nutrition is a prognostically relevant part of ICU treatment and requires special care**
- **Standard operating protocols improve the local MNT**

3. "Who?"—Assessment of Nutritional Status

Pragmatically, when admitted to the ICU, according to Sundstrom et al., three patient groups may be metabolically differentiated (although the discrimination between them is still debated) [15]:

1. Patients in whom multi-organ insufficiency or -failure cannot be prevented or broken through by MNT
2. Patients who recover quickly and do not require MNT
3. Patients in whom the disease course may be positively influenced by an individualized MNT

The latter may be either patients with pre-existing malnutrition or patients with anticipated complicative courses and prolonged ICU stays. These patients benefit from an individually adjusted MNT, but their distinction from the other patient groups is not trivial. While it may be easy to identify an undernourished patient with low body weight and body mass index (BMI) during clinical routine, in case of normal or even elevated BMI, the detection of malnutrition is less likely [3]. In 2019, the Global Leadership Initiative on Malnutrition (GLIM) developed a new definition of malnutrition based on phenotypic and etiologic criteria [16].

There are no validated and recommended tools to estimate the nutritional status of a critically ill patient. Therefore, in case of (semi-) elective admissions, a screening for malnutrition in standard care before major surgery may be practical and is recommended [12]. Several tools to estimate nutrition risk are the Nutrition Risk Score (NRS 2002), the NUTRIC (Nutrition Risk in the Critically Ill) Score, the Subjective Global Assessment (SGA), or the Malnutrition Universal Screening Tool (MUST). In many of these tools, some of the following factors are included [16,17]:

- Medical history: age, comorbidities, loss of physical function
- Nutrition history: weight loss, reduced food intake, loss of appetite
- Physical examination: BMI, edema, body composition
- Severity of disease: critically ill patients are severely ill by definition

In many critically ill patients, the aforementioned parameters—such as nutrition history and exact BMI—are difficult to obtain. Weighing critically ill patients should be mandatory but feasibility may be limited in the ICU setting. In addition, changes of volume status impede weight measurements and clinical examinations, such as anthropometry [2]. The ESPEN guideline defines every patient who is in the ICU for more than 48 h to be at nutritional risk [2]. The DGEM guideline recommends a combination of low BMI, unintended weight loss and lack of oral food intake, or the SGA for critically ill patients [1].

Regarding technical tools, computed tomography (CT) scan analysis, musculoskeletal ultrasound, and bioelectrical impedance analysis (BIA) may be available for the assessment and monitoring of nutritional status in the ICU, but are not broadly implemented into clinical routine until now [18].

Pragmatic recommendations:

- **There is no single "golden bullet" to diagnose malnutrition, but many helpful tools and criteria**
- **All ICU patients should be regularly screened for risk of malnutrition**

4. "How?"—The Route of Nutrition

4.1. Enteral Nutrition (EN)

For a long time, the issue of "enteral or parenteral" has been controversially and quite emotionally discussed [19,20]. Two large multicenter randomized controlled trials (RCTs) did not show any significant difference in mortality, while EN was associated with a higher risk for gastrointestinal (GI) complications [21,22]. An RCT by Harvey at al. including 2400 patients found no difference in 30-day mortality or infectious complications in ICU patients receiving either EN or PN, while patients receiving PN had significantly less vomiting and hypoglycaemia [21]. A more recent RCT led by Reignier et al. (NUTRIREA-2) recruiting 2410 patients also observed no difference in 28-day mortality and infection rates, but significant more frequent GI complications (vomiting, bowel ischemia, and pseudo-obstruction) [22]

The current international nutrition guidelines uniformly recommend the preferential use of EN in the critically ill patient who is unable to maintain sufficient oral intake [2–4,23]. The physiological advantages, the paradigm of "use the gut or lose it", adverse effects of PN in earlier decades and increased cost effectiveness led to uniform preference of EN. However, EN alone is often insufficient to achieve energy and protein targets particularly in the acute phase of critical illness due to frequent interruptions for procedures and/or GI intolerance [24], which a recent meta-analysis of EN+PN versus EN alone demonstrated [25].

Although clear benefits are lacking regarding the optimal nutritional route, worldwide consensus among experts exists about a cautious individualized approach with 'trophic feeding' in high-risk patients without absolute contraindication followed by a ramp-up strategy until the target is reached [20]. While severe critical illness is frequently associated with considerable GI dysfunction, even severe sepsis or septic shock have not been considered clear contraindications in the guidelines [1–4]. EN can be administered via nasogastric or nasojejunal tube. If the need for EN will potentially exceed four weeks, placement of percutaneus endoscopic gastrostomy/jejunostomy is recommended [23].

Contraindications for EN according to the European Society of Intensive Care Medicine (ESICM) Clinical Practice Guidelines [4] particularly include hemodynamic instability (escalation of or high vasopressor medication and increased lactate) and GI intolerance from minor to major symptoms, e.g., gastric residual volume (GRV) > 500 mL/6 h or acute gastrointestinal injury grade > 2.

4.2. Combination of EN with Parenteral Nutrion (PN)

The progression of EN up to calorie/protein target is often prevented by feeding intolerance or common interruptions of EN [26–28]. Thus, particularly in the patient's first ICU week, EN alone may lead to macronutrient deficiency [29–31]. To avoid large cumulative energy and protein deficits, EN and PN may be combined, either early during the patient's ICU course (combined EN+PN), or after several days once EN is proven to be insufficient or unfeasible (supplementary PN [SPN]) [19].

The debate between early or late PN in addition to EN remains controversial [32–34]. There are strong arguments to start PN latest by day 4 particularly in malnourished patients and those with special risks [34], respecting potential risk of refeeding-syndrome.

A recent meta-analysis by Hill et al. including 12 RCTs with 5543 patients found that treatment with combined EN and PN led to increased delivery of macronutrients in severely ill ICU patients [25]. No statistically significant effect of a combination of EN with PN vs. EN alone on any of the analyzed endpoints were observed: mortality (Risk Ratio [RR] 1.0, 95% Confidence Interval [CI], 0.79 to 1.28 p = 0.99), hospital LOS (Mean Difference [MD]-1.44, CI −5.59 to 2.71, p = 0.50), ICU LOS and ventilation days. Trends toward improved physical outcomes were observed in two of four trials. There was a tendency for reduced mortality in nutritionally at-risk patients in some subgroup analyses, but data were too sparse to draw further conclusions.

The ESPEN guideline recommends as good practice point: "PN should not be started until all strategies to maximize EN tolerance have been attempted" and "In patients who do not tolerate full dose EN during the first week in the ICU, the safety and benefits of initiating should be weighed on a case-by-case basis" [2].

Pragmatic recommendations:

- **EN may be preferred in almost all ICU patients**
- **Macronutrient targets may not be reached with EN alone in the acute phase**
- **The addition of PN or the use of total PN (in the acute phase) needs to be considered on a case-by-case basis**

5. "How Much?"—Energy and Protein Requirements

It remains unclear what the optimal protein energy targets should be and exactly when they should be reached [35–37]. Greater protein and energy intake may be associated with improved mortality in patients at nutritional risk as stated in a recent meta-analysis, but evidence remains controversial [25]. The number of macronutrients for critically ill patients needs to be carefully adapted to the individual patient [1]. Aspects requiring careful consideration and regular re-evaluation are the phase of critical illness (acute vs. post-acute), GI and metabolic tolerance to exogenous substrates, the primary disease, possible comorbidities, macro- and micronutrient deficiencies, and the intraindividual disease trajectory of the patient [1–3]. Tailoring the MNT to the individual patient´s metabolic tolerance is described below.

5.1. Energy

The optimal amount of energy is not yet agreed upon, as the evidence remains conflicting. What the targets should be and when they should be reached is still unclear, especially in the acute phase of critical illness, as targeting only caloric adequacy did not show statistically significant improvements in many studies [19].

The guidelines uniformly recommend using indirect calorimetry to determine energy requirements [1–3]. Recent advances in this technique and development of modern devices may improve feasibility and usefulness in clinical practice [38]. Two meta-analyses from 2021 concluded that patients treated with calorimetry-guided isocaloric nutrition had significantly lower short term mortality rates [39,40]. Aiming at personalized nutrition, these results may further stimulate the use of indirect calorimetry in ICUs worldwide.

Otherwise, weight adapted formulas may be used but may only be considered an alternative. The general recommended calculated energy targets vary between guidelines and are 24–30 kcal/kg/d, see Table A1, Appendix A [1–3].

5.2. Protein

The guidelines currently recommend a protein target of 1.0–2 g/kg/d, see Table A1, Appendix A [1–3], but the influence of protein on the outcome of critically ill patients has also been discussed controversially [41,42]. Increased protein intake, was associated with improved long-term physical recovery and lower mortality in observational trials [43–46] and did not influence duration of renal dysfunction [47]. However, a systematic review and meta-analysis of 14 RCTs did not show any impact of different amounts of protein delivery on outcomes mortality, mechanical ventilation, infections, and length of stay [48]. It must be noted that the protein delivery of the included trials was below the guideline recommendations (mean: 67.15 g vs. 42.95 g/d). Suppressed autophagy has been discussed to explain the discrepancy between observational studies and the meta-analysis [49].

The most recent evidence from RCTs regarding protein showed no significant differences in clinical outcomes. Targeting full energy and protein delivery a RCT (FEED) including 60 patients had shown significantly attenuated muscle loss and a lower number of malnourished patients [50]. In an RCT from 2021, patients receiving higher protein delivery (1.5 ± 0.5 g vs. 1.0 ± 0.5 g/kg_{BW}/d) did not show different clinical outcomes or changes in quadriceps muscle layer thickness [51]. For comparison of protein delivery according to the international guidelines vs. usual care, a recent multicenter RCT including 120 patients investigated the feasibility of a high-protein enteral nutrition formula (100 g/L vs. 63 g/L). The two groups were comparable in energy delivery as well. Protein delivery in the intervention group was higher (1.52 vs. 0.99 g/kg ideal body weight per day). No differences in the clinical outcomes including 90-day mortality were observed [52].

Even in nutrition trials targeting the adequate provision of protein, EN failed to provide more than 1.5 g/kg/d in all the mentioned trials [53]. A very recent meta-analysis concluded that a higher versus a lower protein delivery (1.31 ± 0.48 vs. 0.90 ± 0.30 g/kg) did not significantly influence overall mortality (RR 0.91, 95% CI -0.75 to 1.10, $p = 0.34$) or other outcomes [54]. Large trials, such as the EFFORT Trial (NCT03160547) are currently ongoing to evaluate the influence of high and low protein dosages in critically ill [41].

5.3. Non-Nutritional Calories

Calculation of total calorie intake should include intake of non-nutritional calories. Depending on the dose, propofol used for sedation patients may represent a significant portion of the total calorie intake. 1% and 2% propofol contains 0.1 g fat/mL; at a propofol infusion rate of 20 mL/h, fat intake would be 48 g fat/d; thus, about 450 kcal/d would be additionally provided. In a retrospective analysis of 687 critically ill patients, sedation with propofol resulted in an additional calorie intake of 146 ± 117 kcal/d, corresponding to 17% of total calorie intake [55].

Trisodium citrate is commonly used for regional anti-coagulation during renal replacement therapy (RRT). The number of effective calories provided by citrate depends on the citrate concentration /infusion rate, the blood flow rate, the filtration fraction of the ultrafiltrate per unit time and type of filter. For example, a trisodium citrate solution may contain 0.59 kcal/mmol. An infusion rate of 11–20 mmol/h according to 3 kcal/g would result in a caloric intake of 150–280 kcal/d.

A retrospective study of 146 critically ill patients showed that the median propofol and citrate contribution to total calorie intake was 6–18% during the first seven days after ICU admission. In individual cases, however, this portion may increase up to one-third of total calorie intake [56].

Pragmatic recommendations:

- **Optimal macronutrient targets are controversially debated**
- **Prefer indirect calorimetry to measure energy expenditure**
- **Weight-adapted formulas may be used alternatively:**
    - **Energy: 24–30 kcal/kg/d**
    - **Protein: 1.0–2 g/kg/d**
- **Do not neglect non-nutritional calories**

6. "When?"—The Timing of Nutrition

6.1. Early EN (EEN)

The early initiation of EEN within 24–48 h is uniformly recommended by the guidelines in the critically ill patient who is unable to maintain sufficient oral intake [2–4,23]. Bearing in mind the traditional concept that the gut may be the "motor" for multi-organ dysfunction, EN should be started at a low feeding rate (e.g. 5–10 mL) and increased carefully and individually adapted to hemodynamic stability and tolerance [12].

Regarding the effect of EEN, a meta-analysis by Tian et al. [57] (8 RCTs, 1895 patients) calculated a significantly decreased lower rate of mortality (RR 0.68; 95% CI, 0.51 to 0.92; $p = 0.01$) and GI intolerance (RR 0.65; 95% CI 0.43 to 0.99; $p = 0.05$) in a subgroup of patients with low energy intake (33.3–66.6% of energy target). However, GI intolerance was only reported in three studies, including 452 patients.

Comparing EEN and PN four recent meta-analyses [35,57–59] including up to 25 studies with 3816 patients provided these results:

- In general, EEN has no significant impact on mortality, but this may be reconsidered in high-risk subgroups.
- EEN significantly decreases the risk for infectious complications.

6.2. How Early Is Too Early to Add PN?

Regarding the exact timing of initiating PN, the recommendations are contradictory. The A.S.P.E.N guideline recommends that PN should be withheld in patients at low nutrition risk during the first 7 days following ICU admission [3]. The ESPEN guideline advises implementing PN within 3–7 days in these patients if EN is contraindicated [2]. The DGEM does not address a particular recommendation regarding the timing of starting PN but recommends to already use PN in the acute phase if calorie-protein targets cannot be reached by EN alone [1].

However, in severely malnourished patients or patients at high nutrition risk, ESPEN and A.S.P.E.N guidelines state that early and progressive PN should be provided to patients with contraindications for EN [2,3]. The DGEM guideline states that PN may be the better route for malnourished patients, because they frequently experience GI dysfunctions [1].

The above-mentioned meta-analysis from Hill et al. did not detect any clinical differences between an early combined EN+PN or the addition of SPN several days after ICU admission regarding mortality, ICU and hospital LOS or mechanical ventilation [25].

Therefore, it is obvious that the indication for PN in the critically ill has become more critical and individualized. The ESPEN guideline states "In patients who do not tolerate full dose EN during the first week in the ICU, the safety and benefits of initiating should be weighed on a case-by-case basis" and adds a practical point: "PN should not be started until all strategies to maximize EN tolerance have been attempted" [2].

Pragmatic recommendations:

- **Initiate EN within 24–48 h in patients without contraindications**
- **Optimize EN delivery**
- **The addition of PN needs individual adaptation**

7. "What?"—Formula Considerations

In most critically ill patients, standard polymeric formulas for EN should be used according to the guidelines [3,12]. For PN, all-in-one bags should be preferred [12].

7.1. Energy-Dense versus Standard Formula

Due to impaired GI tolerance in the critically ill, energy-dense EN may theoretically improve nutrient delivery. Two multicenter RCTs concluded that energy dense EN (1.5 kcal/mL) increased macronutrient delivery, while adverse effects were not increased [60,61]. However, in the latter trial [61], the need for insulin was higher in the 1.5 kcal group and the 90-day survival was not significantly different. In a recent RCT, scintigraphic measurement of gastric retention at 120 min was greater in the group with the energy-dense formula, intestinal energy delivery and glucose absorption were not improved [52].

7.2. Special Enteral Diets: Synbiotics

Synbiotics refer to the combination of both probiotics and prebiotics, containing Lactobacillus organisms alongside fiber. Synbiotics reveal trophic effects in the colon, focusing on the preservation of the microbiome, promoting mucosal regeneration in balance with the microenvironment. Their administration is a challenging concept with a potential impact on diarrhea and infectious complications [62,63].

Results from current evidence are equivocal. Including nine studies, a meta-analysis from Batra et al., showed a reduction of the incidence of pneumonia (RR: 0.70, CI 0.56 to 0.88; $p = 0.002$), the duration of mechanical ventilation (MD −3.75, CI −6.93 to −0.58; $p = 0.02$), ICU LOS (MD −4.20, CI −6.73 to −1.66; $p = 0.001$), and in-hospital mortality (OR 0.73, CI 0.54 to 0.98; $p = 0.04$) [64]. A more recent meta-analysis emphasized the considerable heterogeneity of the studies. While no benefits for clinical outcome parameters were found, reduced prevalence or duration of diarrhea episodes were observed [65]. A very recent RCT including 218 patients did not confirm a significant difference in infectious complications [66].

The A.S.P.E.N. guidelines suggest that a commercial mixed fiber formula should not be used routinely in the adult critically ill patient prophylactically to promote bowel regularity or prevent diarrhea [3]. The German DGEM guidelines recommend probiotics for patients with severe trauma and those undergoing liver transplantation [1].

Pragmatic recommendations:

- **Standard formulas should be chosen for most ICU patients**
- **Consider energy dense formulas in patients with fluid restrictions and in malnourished patients**
- **Intensified blood sugar monitoring may be beneficial in patients with energy dense formulas**

8. Supplementation with Specific Nutrients

The supplementation of EN or PN with immune-enhancing and anti-inflammatory nutrients is a challenging and controversial issue.

8.1. Arginine, Glutamine, and Omega-3 Fatty Acids

The guidelines currently do not recommend the use of arginine, glutamine, and omega-3 fatty acids in the general critically ill patient population.

Regarding arginine, in theory, the availability is reduced in sepsis, but as arginine supplementation could induce the formation of nitric oxide and increase hypotension in patients with septic shock, there is limited data [67,68].

In critical illness, the glutamine concentration in plasma may be reduced and considered an expression of the severity of the disease and infection [69]. Supplementation of glutamine may possibly lead to a reduction in bacterial translocation from the gut, improved immune cell function, decreased proinflammatory cytokine production, and increased antioxidant capacity. The clinical impact of these findings, however, has not yet been clearly established and in most meta-analyses, no statistical significance was reached. Because a high dosage in patients with (multi-)organ dysfunction may be detrimental [70,71], in unstable and complex ICU patients, particularly in those suffering from liver and renal failure, parenteral glutamine shall not be administered [2]. An "umbrella" overview of 22 meta-analyses and a current meta-analysis of 15 randomized studies have shown again benefits for glutamine supplementation in ICU patients with regard to the rate of infectious complications and the hospital LOS [72–74]. Reference was also made to the considerable and statistically significant heterogeneity of the studies and meta-analyses [72]. The new ESPEN practical guideline "clinical nutrition in surgery" states parenteral glutamine may be considered in patients requiring exclusive PN [12].

Fish oil or ω-3 fatty acids may have a high anti-inflammatory potential due to the shift in inflammatory mediator synthesis but may intensify an already existing immunosuppression. A recent meta-analysis of 49 RCTs, risk of infection was 40% lower with ω-3 fatty-acid enriched PN than standard PN (RR 0.60, 95% CI 0.49 to 0.72; $p < 0.00001$). Patients given ω-3 fatty-acid enriched PN had reduced ICU LOS ($p = 0.01$) and reduced length of hospital LOS ($p < 0.00001$). Risk of sepsis (9 RCTs) was reduced by 56% in those given ω-3 fatty-acid enriched PN (RR 0.44, 95% CI 0.28–0.70; $p = 0.0004$) [75]. Focusing on enrichment with omega-3 fatty acids in EN and supplemental PN in a recent RCT with 100 mechanically ventilated ICU patients, neither improvement in lung function, nor decrease of complications were observed. However, the supplemented group could be weaned earlier from catecholamine treatment and PN [76]. In patients requiring PN, the new ESPEN practical guideline "clinical nutrition in surgery" states postoperative PN including ω-3 fatty acids should be considered [12].

8.2. Selenium

Selenium is a known antioxidant and decreases in septic patients and after major surgery. Current evidence for the use of intravenous selenium has been inconclusive. An RCT (SISPCT) administering high-dose selenium to 1089 patients with severe sepsis or septic shock, the 28-day mortality rate in the selenium group was not significantly affected ($p = 0.30$) [77]. A recent meta-analysis focusing on eight RCTs with low risk of bias did not find effects of antioxidant micronutrient supplementation on the reduction of mortality [78]. Currently, an international multicenter RCT of selenium in cardiac surgery (SUSTAIN-CSX, NCT02002247) is undergoing statistical evaluation [79].

Currently the guidelines do not recommend pharmacological supplementation with selenium [1,2].

8.3. Vitamin D

Vitamin D exerts pleiotropic effects with actions far beyond its classic role in mineral homeostasis. Tissue actions require two enzymatic conversions to 25-hydroxyvitamin D and 1,25-dihydroxy (OH) vitamin D after which vitamin D has been shown to modulate the immune response amongst others. Vitamin D deficiency is highly prevalent across all age groups and countries [80]. Although vitamin D deficiency was further determined to be associated with greater illness severity (also in patients with COVID 19), a causal relationship between vitamin D deficiency and multiple organ dysfunction has not been established [81–84]. Thus, the efficacy of vitamin D as a therapeutic in critically ill patients remains controversial [85].

While in a previous phase 2 trial (underpowered for mortality as endpoint [VITdAL-ICU], involving 475 patients), vitamin D supplementation administered to vitamin D–deficient, critically ill patients was associated with lower observed mortality than placebo at 28 days (21.9% vs. 28.6%, $p = 0.14$) and at 6 months (35.0% vs. 42.9%, $p = 0.09$) [86]. The subsequent randomized, double-blind, placebo-controlled, phase 3 (VIOLET) trial including 1360 patients (1078 vitamin D deficient at baseline defined as 25-hydroxyvitamin D level, <20 ng/mL) receiving a single enteral dose of 540,000 IU of vitamin D or matched placebo did not reveal clinically important differences between the groups with respect to secondary clinical, physiological, or safety end points. The severity of vitamin D deficiency at baseline did not affect the association between the treatment assignment and mortality.

In a recent sub-study of the VIOLET trial, long-term cognitive outcomes were measured. The adjusted median score at follow-up (median 443 days) was not significantly different (adjusted OR, 0.83; 95% CI, 0.50 to 1.38) [87].

Another randomized, placebo-controlled, double-blind, multicenter, international trial, is ongoing with planned recruitment of 2400 adult patients with severe vitamin D deficiency (25-OH Vit D $\leq$ 12 ng/mL) receiving a enteral loading dose of 540,000 IU cholecalciferol within 72 h after ICU admission, followed by 4000 IU daily for 90 days or placebo [88].

So far, there is no clear evidence for pharmacological vitamin D supplementation in patients with established deficiency.

Given the controversial evidence, different recommendations can be found among the guidelines (1, 2) (see Table 1).

Table 1. Current guideline recommendation regarding micronutrients, vitamins, and antioxidants.

Substrate	DGEM	ESPEN	A.S.P.E.N.
Micro-nutrients	A patient should receive vitamins and trace elements, if EN cannot meet daily needs, and if supplemental PN is required to ensure the desired calorie and protein intake according to the disease phase and individual metabolic tolerance.	To enable substrate metabolism, micronutrients (i.e., trace elements and vitamins) should be provided daily with PN.	N/A
Vitamin D	Patients may receive a pharmacotherapy with vitamin D, when they have a severe vitamin D deficiency (25 [OH] D $\leq$ 30 nmol/L corresponding to 12 ng/mL).	In critically ill patients with measured low plasma levels (25-hydroxy-vitamin D < 12.5 ng/mL, or 50 nmol/L) a high dose of vitamin D3 (500,000 UI) as a single dose can be administered within a week after admission.	N/A
Antioxidants/ Selenium	• Patients shall not receive a pharmacotherapy with selenium. • Patients should not receive routinely a pharmacotherapy with zinc, alpha-tocopherol, vitamins A and C, or with a combination of those.	Antioxidants as high dose monotherapy should not be administered without proven deficiency.	• We suggest that a combination of antioxidant vitamins and trace minerals in doses reported to be safe in critically ill patients be provided to those patients who require specialized nutrition therapy. • We cannot make a recommendation regarding selenium, zinc, and antioxidant supplementation in sepsis at this time due to conflicting studies.

DGEM, German Society for Nutritional Medicine; ESPEN, European Society of Enteral and Parenteral Nutrition; A.S.P.E.N, American Society of Enteral and Parenteral Nutrition.

8.4. Vitamin C

Vitamin C is a pleiotropic nutrient and powerful antioxidant. While suboptimal vitamin C status is common among critically ill patients, vitamin C is currently not recommended as pharmacotherapy for these patients. Nevertheless, a very recent meta-analysis from Patel et al. (acceped for publication, registered at PROSPERO: CRD42021244074) could demonstrate with a trend towards reduction in overall mortality (RR 0.88, 95% CI 0.75 to 1.02, $p = 0.09$), which became significant when comparing high dose vitamin C supplementation ($\geq$10,000 mg/d) with placebo (RR = 0.70, 95%, CI 0.52 to 0.96, $p = 0.03$). In the CITRIS-ALI Trial from Fowler et al., in 2019, vitamin C–infused patients exhibited a significant reduction in 28-day all-cause mortality ($\chi^2 = 4.84$; $p = 0.03$) [89].

Regarding the "Marik cocktail", a treatment with intravenous hydrocortisone, ascorbic acid (vitamin C) and thiamine (HAT) [90], a current meta-analysis from 2021 in septic patients found no significant difference between both groups in long term mortality, ICU mortality, or incidence of acute kidney injury, hospital and ICU LOS, and ICU free days on day 28 between the intervention and control groups [91]. There was, however, a significant difference in the reduction of SOFA score on day 3 from baseline (MD −0.92; 95% CI −1.43 to −0.41; $p < 0.05$).

In the before-mentioned meta-analysis (PROSPERO: CRD42021244074), vitamin C monotherapy was associated with a significant reduction in overall mortality (RR 0.66,95% CI 0.49 to 0.89, p = 0.006), while there was no effect on overall mortality in the trials administering vitamin C in combination with thiamine and hydrocortisone (RR 0.99, 95% CI 0.82 to 1.19, p = 0.91, test for subgroup differences was significant, p = 0.02).

Pragmatic recommendations:

- **The guidelines currently do not recommend pharmacotherapy with micronutrients in the general critically ill patient population.**
- **Proven micronutrient deficits need to be treated.**
- **Micronutrients must be supplemented whenever PN is administered to a patient.**
- **While arginine and glutamine have no indication in the critically ill with special regard to those with organ failure, the enteral and parenteral supplementation of fish oil remains a matter of debate.**

9. Monitoring and Complications

In addition to the clinical examination of the patient's abdomen, blood chemistry includes serum glucose, triglycerides, lactate, and procalcitonin should be measured. Phosphate should be controlled to detect and treat a potential refeeding syndrome. To avoid measurement of nitrogen balance, urea excretion rate per 24 h will help to estimate the extent of catabolism. For the estimation of intestinal function, the potential impact of biomarkers as citrulline and fatty acid binding protein plasma level have to be validated in clinical studies.

9.1. Gastrointestinal Intolerance

There is a variety of definitions for feeding intolerance including a lot of uncertainty for the clinical management [92]. Many critically ill patients experience feeding intolerance, motility disorders, which include delayed passage with slow gastric emptying and constipation, and accelerated passage with impaired small intestinal nutrient absorption or nutrition-related diarrhea [92,93]. According to the ERAS Society's (Enhanced Recovery After Surgery) various guidelines [94–96], causes of GI intolerance include opioid analgesia, sedation, edema, and insufficient stimulation. Treatment options are oral/EN, prokinetics, mobilization and physical therapy, prokinetics and optimization of sedation and volume status. Clinical observation of the patient's abdomen, bowel motility, and gastric reflux is mandatory.

While measurement of GRV is not standardized the optimal threshold is uncertain. Therefore, the impact of the measurement of GRV has been controversial. The monitoring of GRV is considered still relevant in surgical ICU patients and severely critically ill patients with a high risk for GI dysfunction. A GRV of more than 500 mL/6 h may be considered critical [93]. In two controlled studies non-monitoring of GRV was without significant effect on the risk of ventilator-associated pneumonia in adults receiving mechanical ventilation and EEN [97,98].

Additional possibilities to monitor GI tolerance include sonography and CT scans. If the latter shows bowel loops with intramural air accumulation, bowel ischemia must be considered. Intramucosal phi tonometry as a tool for the measurement of splanchnic perfusion has not been used frequently in clinical practice.

A research agenda has been proposed by the working group GI failure of the European Society of Intensive Care Medicine (ESCIM) including the core set of monitoring and outcomes [99].

9.2. Metabolic Intolerance

Hyperglycemia may be due either to hyperalimentation in the (acute) phase of insulin resistance or underlying subclinical diabetes. Hyperglycemia is associated with increased mortality, which has been the rationale for intensified insulin therapy [100]. To avoid any life-threatening hypoglycemia, nowadays, glucose levels up to 180 mg/dL (10 mmol/L) are accepted [5,101]. Reduction of glucose calories should be considered before insulin is administered in nondiabetic patients with moderate dosages of 0–4 IU/h. An intensive insulin therapy is not recommended to avoid the potential consequences of hypoglycemia [5].

To circumvent hypertriglyceridemia, bolus application of lipids should be avoided. During continuous lipid infusion serum triglyceride levels should not exceed 400 mg/dL [101]. Reversible liver steatosis or hypertriglyceridemia-induced acute pancreatitis may develop if metabolic control cannot be achieved within several days.

In severely malnourished patients and patients after long starving periods, a refeeding syndrome may occur within the first few days after starting MNT. Supply of carbohydrates and fluid will stimulate insulin secretion and intracellular shift of glucose and electrolytes. The decrease of serum potassium, magnesium, and phosphate levels will lead to impaired neuromuscular transmission causing life-threatening arrhythmias, and convulsions. Calories and fluid administration should be increased slowly under electrolyte- and preferably ECG monitoring [102]. In high-risk patients, a preventive thiamine administration 200 mg once daily for 2 days is necessary.

Pragmatic recommendations:

- **Clinical examination, bowel motility, and gastric reflux and GRV are helpful to assess gastrointestinal tolerance of EN.**
- **Optimization of analgosedation, fluid status, and GI-stimulation can improve the tolerance of EN**
- **Markers of metabolic tolerance include blood sugar, lipids, and electrolytes with special focus on potassium, magnesium, and phosphate**

10. Tailoring MNT to the Individual Patient's Needs

The heterogeneity of the patients in the ICU complicates MNT and many different factors such as primary disease, comorbidities, phase of critical illness influence the patient's individual requirements. The most important aspects of the different patient groups are summarized in Figure 1 to give a pragmatic overview.

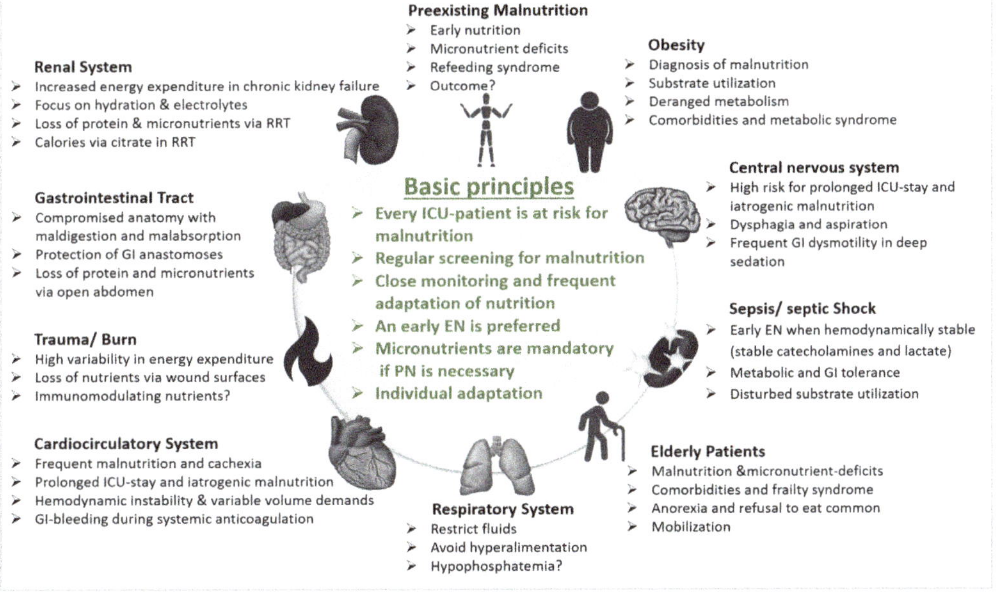

Figure 1. Overview important aspects of medical nutrition therapy in different patient groups. Abbreviations: EN: enteral nutrition, GI: gastrointestinal, PN: parenteral nutrition, RRT: renal replacement therapy.

10.1. Adapting MNT to the Phase of Critical Illness

MNT must be adjusted to the metabolic tolerance which may considerably alter in the different phases of critical illness and resulting catabolic response [1,2]. While metabolic tolerance may be extremely limited by severe inflammation in the early acute phase bearing the high risk of overfeeding, it is different in the post-acute phase according to chronical inflammation or beginning resolution and recovery—the shift to anabolism includes the risk of underfeeding. Therefore, the guidelines recommend an individualized approach due to specific pathophysiologic and resulting metabolic changes, the guidelines recommend an individual adaptation of MNT to the different phases of critical illness [1,2].

Immediately after onset of the critical illness, the 'acute' phase begins that can be divided into an 'early acute' phase (about 1–3 days post-onset with the possibility of fatality due to the most severe illness entity) and a 'late acute' phase (approximately lasting for 2–4 days if the patient survives the early acute phase). The post-acute phase can be described as a 'recovery' phase (duration >7 days), which is usually spent in the primary care hospital. During this phase of recovery from catabolic critical illness, substrate tolerance has been normalized with a metabolic shift to anabolism. From a nutritional point of view, sufficient macronutrient supply in this period may be considered pivotal to the patient's recovery and long-term outcome. The amount of administered calories are considered to be 1.2–1.5 fold the calculated energy requirement. Because these patients are weaned from ventilation and able to eat, it will be frequently—and falsely—assumed that they will manage to cover these requirements by the oral route. It has been recently shown that this may be achieved only by a combination of oral and enteral feeding [103]. It has been shown that after extubating, many patients will receive no more than 700 kcal/d [104]. Reasons may be early discontinuation of MNT in favor of oral nutrition (especially in case of discharge to standard care), limited oral intake due to post-critical weakness, fatigue, anorexia, and isolation. Documentation of the amount of oral intake is mandatory, supplementation with oral nutritional supplements (ONS), in some cases enteral or even SPN should be considered.

After the recovery phase, the 'rehabilitation' phase (lasting several months) follows, in which, among others, the metabolic damage suffered initially is repaired slowly. Usually, patients will not go through this phase in the primary care hospital. Alternatively, the 'post-acute' phase may merge into a 'chronic' phase (of uncertain duration) characterized by persistent organ dysfunction and an uncertain prognosis. This particular course may be described as a "persistent inflammatory immunosuppressed catabolism syndrome" [105,106]. If there is a new disturbance of homeostasis, the process will start again with the acute phase.

Pragmatic recommendations:

- **Adapt MNT to the phase of illness**
- **In the early acute phase, a lower supply of macronutrients may be beneficial**
- **In the postacute phase, reaching nutritional targets may be of great importance for the patient's long-term outcome**

10.2. Critically Ill Patients with Altered Nutritional Status

Malnutrition is a common phenomenon and includes under- as well as overnutrition. While undernourished patients with low weight and BMI can be easily recognized during clinical routine, malnutrition may be underestimated in patients with normal or elevated BMI [3]. An obese patient with weigh loss in the past months or with reduced intake of nutrition is at risk for malnutrition, which may not be diagnosed at first sight. Therefore, a combination of clinical evaluation and the use of a validated screening tool may be recommended [1–3,107]. If the patient is at nutritional risk, a detailed nutritional assessment should follow.

10.2.1. Undernourished Patients

In patients with preexisting malnutrition, according to the DGEM guideline, the same energy and protein targets may be used, as in other patients, as there is still a lack of data concerning the therapeutic relevance of MNT in this patient cohort although the prognostic relevance of malnutrition is clear [1]. In these patients, an MNT should be commenced early to avoid large cumulative macronutrient debts. This may be achieved through ONS or though SPN. Should an EN not be feasible, an early hypocaloric PN should be administered (75% of caloric target, protein $\geq$ 1 g/kg/d) [1].

In severely malnourished patients, according to the DGEM guideline, a more aggressive MNT is not recommended to avoid GI and metabolic intolerance and potential complications such as acute hyperalimentation, refeeding syndrome, or increased rates of infection [1].

Contrasting this recommendation, the A.S.P.E.N. suggests a rapid progression of preferably EN with the aim to reach the target within 24–48 h under careful monitoring. Within 48–72 h, >80% of energy and protein goals should be achieved [3]. However, these recommendations of the A.S.P.E.N. rely on debatable evidence.

While the essential role of micronutrients for many biological processes in the human body is accentuated in current literature, reliable evidence remains sparse. Therefore, micronutrients in pharmacological dosages should only be administered after clinically raised assumptions or measured micronutrient deficiencies as stated by the DGEM [1].

Special attention should be paid to the occurrence of a refeeding syndrome in severely malnourished patients.

10.2.2. Obese Patients

Obese patients do not have 'reserves', as often assumed, but in contrast they frequently suffer from disturbances in substrate utilization, leading this patient group to be predisposed to loss of muscle mass during an ICU stay. Increased attention should be paid to the monitoring of metabolism, markers of metabolic syndrome and possible comorbidities [3]. While there is only weak evidence, the goal of an MNT is avoidance of muscle catabolism, improvement in body composition, reduction of insulin resistance and hyperglycemia and reduction of infection rates.

For obese patients with BMI $\geq$ 30 kg/m^2, similar guideline recommendations were issued as for other ICU patients [1–3]. The best MNT for obese patients is still debated particularly because evidence remains weak from smaller and quite older observational studies, as are formulas to calculate energy demand in these patients, but indirect calorimetry should be used to measure EE [1,2]. Otherwise, obese patients should be nourished weight-adapted hypocalorically and high in protein (1.5 g protein (1.8 g amino acids)/$kg_{ideal\ BW}$/d) [1–3,108]. For calculation of ideal BW, the so-called Peterson formula is recommended in a slightly different modification by the ESPEN and DGEM guideline (DGEM: ideal BW = 48.4 + 77.0 $\times$ (height-1.50 m) related to a BMI of 22 kg/m^2; ESPEN: adjusted BW = (current BW-ideal BW) $\times$ 0.33 + ideal BW, ideal BW = 2.2 $\times$ BMI + 3.5 $\times$ BMI $\times$ (height-1.50 m) related to an "overweight" BMI of 25 kg/m^2) [109]. In clinical routine, a hypocaloric high-protein nutrition may be achieved via enteral or parenteral protein supplements. Additionally, patients who experienced weight loss in the past months or who underwent bariatric surgery, should receive vitamin and trace element supplements with special focus on thiamine [1,3].

Pragmatic recommendations:

- **Diagnosis of malnutrition is not trivial and needs care**
- **In undernourished patients, an early MNT including ONS, EN, and PN may be considered with special attention to metabolic and GI feeding tolerance**
- **In overweight patients, an iso-proteinic/lower calorie approach might avoid hyperalimentation or muscle-catabolism**

10.3. Elderly Critically Ill Patients

In elderly patients, the most important goal is to optimize the functional capabilities to achieve the best possible quality of life and autonomy [110]. Elderly patients have higher incidences of comorbidities, malnutrition, sarcopenia, and cachexia, which may be regarded as the frailty syndrome [110]. Because the multidimensional syndrome "frailty" and other reasons for malnutrition, such as depression, anorexia, polypharmacy, low activity and catabolism of chronic diseases, and inflammation are so common in elderly patients and affects nutritional status, all elderly patients should be screened for malnutrition and be treated accordingly [111,112]. If the elderly critically ill patient is able to feed orally, sedative measures and dietary restrictions should be minimized to avoid a reduced food intake, which may negatively affect patient outcome [110]. In obese elderly patients, a weight-reducing diet should be circumvented, to avoid muscle catabolism [110]. Possibilities to optimize an oral nutrition include enriched nutrition, several small meals, ONS, monitoring of food intake and company during meals. If oral nutrition is insufficient, EN and or PN should be administered with the same indications as in other critically ill patients, while minding an early start of a MNT if indicated [110].

In the absence of methodologically sound clinical evidence, the DGEM chose not to give special recommendations for this patient group [1]. The ESPEN assumes that an adequate macronutrient supply may lower the incidence of frailty in elderly patients and recommends higher protein target (1.2–1.5 g/kg/d) for elderly malnourished patients [2]. In elderly critically ill patients a hypercaloric and high-protein supplementation (30 kcal/kg/d energy and $\geq$1 g/kg/d protein) is recommended by the ESPEN as well [110]. Micronutrients should be supplied as for healthy elderly patients, special micronutrient deficits shall be counterbalanced [110]. The A.S.P.E.N. does not provide special recommendations for elderly ICU patients.

In addition to an effective MNT, the focus should be on adequate hydration, potential refeeding syndrome [113], as well as on preservation of muscle mass physical activity to allow for an independent living after hospital discharge [110].

Pragmatic recommendations:

- **Malnutrition and dehydration are common among this group**
- **A critical (re-)evaluation of nutritional status is necessary**
- **Focus on the most physiologic route of nutrition, patient's preferences, and function**
- **Consider a higher protein target to counteract anabolic resistance**

10.4. Critically Ill Patients in Shock

Physiologic advantages or an EN are the maintenance of the gut mucosa and the GI barrier, the modulation of an inflammatory reaction and the reduction of insulin resistance. In case of compromised hemodynamics, the concern for mesenterial ischemia due to the increased demands for the gut system is voiced in the guidelines [2,3]. Cautious EN in patients with the need for catecholamines or vasopressors may be considered [114], but there is still a risk for nonocclusive bowel disease and a lack of data from controlled randomized trials [115–117]. In the NUTRIREA-2 trial including 2410 patients, early EN did not reduce mortality or the risk of secondary infections but was associated with a greater risk of digestive complications compared with early isocaloric PN [22].

In the concept of "minimal trophic nutrition" it is recommended to start EN with a low flow rate (10–20 mL/h) [118]. In case of hemodynamic instability and the administration of catecholamines, limited enteral tolerance should be anticipated. Careful clinical examination of the abdomen must be performed and keeping in mind that the energy target cannot be achieved in such patients via the EN in most cases within the first ICU week. Complete stop of enteral supply should be avoided whenever possible [23].

The guidelines recommend EEN in septic patients after achieving hemodynamic stability [1–5]. The MNT should be progressed slowly to achieve more than 80% of the nutritional target within the first week [3]. If EN is contraindicated, PN should be used. An SPN shall be administered according to the ESPEN after day 3 [2]. Contrasting the ESPEN guideline, the A.S.P.E.N. and Surviving Sepsis Campaign (SSC) do not recommend SPN or PN in the acute phase of a severe sepsis due to the limited substrate utilization [3,5]. Because septic patients in the hypermetabolic phase may have an increased need for vitamins and trace elements, these should be supplemented in physiological dosages after a proven deficit or of PN is needed [1]. Immonutrition, as well as micronutrients—such as selenium, glutamine, arginine, or carnitine—should not be administered [3,5].

Attention should be paid to the macronutrient balance in the first week of sepsis, because EN alone is often hypocaloric due to limited GI and metabolic tolerance. Whether this is advantageous or not remains controversial. The GI tolerance may be improved through prokinetics and postpyloric nutrition tubes, according to the SSC guideline [5].

The guidelines furthermore state, that patients with septic shock can receive EEN after stable hemodynamics are ensured (mean arterial pressure ≥60 mmHg and stable/falling lactate/need for vasopressors) but should progress slowly and be adapted to the patient's tolerance [2,3]. In patients with uncontrolled shock, no EN should be given [2,3].

Pragmatic recommendations:

- **In hemodynamically stable patients, EN is preferable**
- **Careful monitoring of clinical and laboratory parameters is necessary**

10.5. Critically Ill Patients after Trauma and Burn Injury

Due to insufficient data in these patient groups, they should be treated according to the general recommendations for ICU patients [1]. The A.S.P.E.N. recommends EN should be preferred to PN [3].

Patients with burn injury and exudative losses through wound surfaces have an increased need for vitamins and trace elements. A highly variable EE leads to inaccuracy of estimation formulas regarding macronutrient needs. Therefore, indirect calorimetry shall be preferred [1,3]. The ESPEN recommends—contrasting the DGEM—enteral glutamine to be supplemented (0.3–0.5 g/kg/d) for 10–15 days [2] and the A.S.P.E.N. a higher dosage of protein (1.5–2 g/kg/d) [3]. According to DGEM, protein losses via drains and dressings should be compensated [1].

For trauma patients, the ESPEN recommends glutamine (0.2–0.3 g/kg/d) for the first 5 days (10–15 days if wound healing is complicated) [2]. Immunomodulating solutions with fish oil and arginine may be considered for patients after severe trauma according to the A.S.P.E.N. [3]. The DGEM advises against the use of these substrates in the sense of pharmacotherapy [1].

Pragmatic recommendations:

- **EN should be preferred**
- **Use indirect calorimetry as energy needs may be highly variable**
- **Consider supplementing more protein and micronutrients**

10.6. Critically Ill Patients with Central Nervous Diseases

The guidelines state that patients with head trauma, ischemic or hemorrhagic stroke and spinal trauma should receive EEN [2–4].

Patients with **stroke** are especially vulnerable for malnutrition, dehydration, and aspiration pneumonia due to impaired consciousness, swallowing problems as well as cognitive and perceptive deficits [119]. Therefore, all stroke-patients should be screened for swallowing problems and malnutrition and treated accordingly [120]. ONS may support the individualized therapy [119,120]. In patients, where impaired consciousness or dysphagia prohibit oral nutrition, an EN should be started within 72 h via nasogastric tube [119].

Young patients with **head trauma** are usually not malnourished at ICU admission but have an increased nutritional risk due to long ICU stays, variable EE (up to 200%) and frequently experienced a profound muscle catabolism. A higher protein supplementation 1.5–2.5 g/kg/d may therefore be considered [2,3]. The A.S.P.E.N. suggests the administration of formulas including arginine and omega-3-fatty-acids [3].

Patients with a diagnosed **swallowing disorder**, which make up to 60% of the general ICU population [121], should receive a logopedic therapy and an optimization of food texture to optimize oral food intake as stated in the ESPEN guideline "clinical nutrition in neurology" [119]. Special attention should be paid to oral and pharyngeal residue of food, sufficient food and fluid intake and aspiration [119,120]. If no safe and sufficient nutrition can be ensured, EN via nasogastric route shall be added to the oral nutrition [2, 119,120]. If there is additional risk of aspiration, EN boles should be avoided, a postpyloric tube should be placed, the head should be elevated by 30°, and prokinetics should be considered [2,3,120]. PN is another option for these patients, especially in cases with preexisting malnutrition or if an EN is not sufficient to ensure adequate nutrition and hydration [2,120].

Pragmatic recommendations:

- **Increased risk of malnutrition during prolonged ICU stay, frequent re-assessment necessary**
- **EN is preferred, but often difficult due to paralytic ileus and increased risk of aspiration**

10.7. Critically Ill Patients with Cardiac Diseases

Malnutrition, cachexia, and sarcopenia are common comorbidities in patients with cardiac failure [17,122,123]. In contrast to often quickly recovering patients after elective "simple" cardiac surgery, patients with preexisting malnutrition, complex heart surgery and elevated risk for prolonged ICU stays should receive regular screenings for nutritional risk [123,124]. At the same time, these patients are vulnerable for fluctuation of the volume status, which is why the use energy-dense formulae seems reasonable.

For patients with mechanical assist devices (extracorporeal membrane oxygenation (ECMO)/ extracorporeal life support (ECLS), or ventricular assist devices (VAD)), the same recommendations apply as for other critically ill patients. According to the guidelines, in patients with stable hemodynamics and intact GI system, EN should be preferred [1,2,4]. Regarding monitoring, special attention should be paid to GI bleeding during therapeutic anticoagulation. In patients on ECMO/ECLS, diagnostic measures and therefore interruptions of EN are frequent. Although evidence remains sparse, SPN seems safe and should be infused via central venous catheter (as opposed to directly into the ECMO circuit [1]. Indirect calorimetry is not reliable in this patient group due to the extracorporeal CO_2-elimination, therefore weight-based formulas should be used to estimate energy targets [1].

Pragmatic recommendations:

- **Heterogeneous patient group, focus MNT on patients with increased risk of malnutrition**
- **EN is preferred**
- **Special attention should be paid to hemodynamic stability and fluid status**

10.8. Critically Ill Patients with Respiratory Diseases

In patients with respiratory failure, in theory, formulae with increased fat content and reduced carbohydrates may be useful to influence the respiratory quotient and decrease CO_2-production. The A.S.P.E.N. advises against these formulae in patients with acute respiratory failure due to the lack of evidence, but recommends avoiding hyperalimentation, because lipogenesis increases CO_2-production [3].

According to the guidelines, EN should not be administered to patients with life-threatening hypoxia, hypercapnia, and acidosis as sign of respiratory decompensation. In patients with stable and compensated respiratory failure, EN can be commenced [2,4]. A restriction of fluids seems reasonable to avoid aggravating overhydration and edema. Therefore, energy-dense formulae (1.5–2 kcal/mL) are recommended [3]. Hypophosphatemia is a common (and commonly unrecognized) problem and may lead to weakness of respiratory muscles and weaning-failure. Therefore, phosphate should be monitored closely and replaced as necessary [3]. A functional GI-System should be used, so EEN should be performed in patients managed in prone position [2]. The hypothesis that the abdominal compression leads to problems with transport and resorption has not been proven [1,2,4]. One possibility to increase tolerance and avoid aspiration of EN is to bring the entire bed in a 30° head elevation. Measurements of intraabdominal pressure may aid in the early detection of an abdominal/GI problem (not only) in patients in prone position.

Critically Ill COVID-19 Patient

Most COVID-19 patients admitted to the ICU are at high risk of or have preexisting malnutrition [125]. According to Ochoa et al., COVID-19 patients present with three different phenotypes of nutrition risk: (1) the frail older patient, (2) the patient with severe ongoing chronic illness, and (3) the patient with severe and morbid obesity [126]. The measurement of the upper waist circumference of COVID-19 patients has recently shown that with every centimeter increase there is a 1.13-fold higher probability of intensive treatment and a 1.25-fold higher probability of mechanical ventilation [127].

Regarding MNT in this patient cohort, no RCTs exist so far. Instead, the ESPEN and A.S.P.E.N. have published expert statements as an adaptation of their existing guidelines [128,129].

Before admission to the ICU, anorexia secondary to infection, dyspnea, dysosmia, dysgeusia, and impaired meal preparation during quarantine may have reduced food intake. Therefore, upon ICU admission, nutritional assessment is mandatory. An individualized approach including indirect calorimetry is recommended, because persistent hypermetabolism was observed in these patients [130,131].

While EN may be performed even in prone position, Martindale et al. recommend a lower threshold for switching to PN in cases of intolerance, high risk of aspiration, or escalating vasopressor support [129]. In case of ARDS, Thibault et al. have recommended the use of EN enriched with omega-3 fatty acids, and for PN fish oil-enriched intravenous fat emulsions [125].

In a multicenter, double-blind, RCT conducted at two sites in Sao Paulo, Brazil, 240 hospitalized patients with COVID-19 being moderately to severely ill were randomized to receive a single oral dose of 200,000 IU of vitamin D3 or placebo [132]. LOS defined as the primary endpoint was not significantly different between the vitamin D3 and placebo groups (log-rank $p = 0.59$; unadjusted hazard ratio for hospital discharge, 1.07 [95% CI, 0.82 to 1.39]; $p = 0.62$). Furthermore, there were no significant differences for in-hospital mortality, admission to the ICU ($p = 0.30$), or need for mechanical ventilation ($p = 0.09$). Mean serum levels of 25-hydroxyvitamin D significantly increased after a single dose of vitamin D3 vs. placebo (44.4 ng/mL vs. 19.8 ng/mL; difference, 24.1 ng/mL [95% CI, 19.5 to 28.7]; $p < 0.001$).

Pragmatic recommendations:

- **EN is preferred in patients with respiratory problems including COVID-19 and ARDS patients, as well as patients in prone position**
- **Do not use EN in patients with respiratory decompensation**

10.9. Critically Ill Patients with Abdominal Diseases

Critically ill patients requiring abdominal surgery often present anatomical and functional characteristics that require a critical evaluation and adaptation of MNT. Compromised functions are GI-motility, digestion, and absorption of nutrients, which often leads to a reduced tolerance of EN. On the other hand, EN also nurtures the gut mucosa, increases intestinal perfusion and peristalsis [3,17].

10.9.1. Patients after GI-Surgery

If the GI system is functional, it shall be used in patients after **GI surgery** as recommended uniformly by all guidelines [1–4]. Therefore, in most abdominal-surgery patients an EEN within 24–48 h is recommended [1,12]. Even if the nutrition of these patients remains a controversially debated topic, it is not necessary to withhold EN as per standard or administer clear liquids only [3]. The MNT should be discussed in the interdisciplinary teams to optimize the nutrition for each individual patient [12].

In patients after **upper GI-surgery,** an intraoperatively placed nasojejunal or percutaneous postpyloric tube can allow for EN of the distal GI parts without risking injury of a fresh anastomosis during feeding tube placement and without risk of regurgitation and aspiration. In general, fresh anastomoses *per se* are not a contraindication for EN, whereas the individual therapy needs to be decided in the interdisciplinary team [2].

Abdominal surgery patients with **complicative courses** often accumulate great energy debts. Therefore, the ESPEN recommends to consider an early SPN [2]. In patients with leaky anastomoses, with internal or external fistulas, an access to the distal part of the gut shall be used for EN. Enteroclysis shall be considered and re-evaluated on a regular basis, to increase resorption of nutrients and prevent mucosal atrophy as well as bacterial overgrowth with the risk of bacterial translocation and bacteriemia [2]. If EN is insufficient, the patient needs PN [2].

Patients with **open abdomen** should receive an EEN (24–48 h postinjury) in the absence of bowel injury [3,4]. Protein losses via drains and dressings should be compensated in the form of enteral protein supplements or parenteral albumin (15–30 g protein/ liter exudate) [1,3]. In addition to the above-mentioned guideline recommendations, an algorithm was proposed by Friese et al. in 2012 [133] and Moore and Burlew in 2016 [134].

10.9.2. Patients with Liver Failure

According to the ESPEN guideline on clinical nutrition in liver disease [135] and the A.S.P.E.N. [3], patients with liver failure have an increased risk for malnutrition and may develop severe disturbances regarding carbohydrate, protein and fat metabolism with impeded hepatic gluconeogenesis and lactate clearance, as well as catabolism [3,135]. Therefore, all patients with liver disease should be screened for malnutrition and treated accordingly [135]. EE may be highly variable in these patients; therefore, indirect calorimetry should be used [3,135]. If the latter is not possible, the patient's dry weight shall be used to estimate energy and protein targets and 1.3× resting EE should be supplied [135]. Patients with acute liver failure should receive oral nutrition as long as possible, or in case of hepatic encephalopathy, EEN [135]. A generalized protein restriction is not recommended prevent muscle degradation, which contributes to the development of a hepatic encephalopathy [3]. In patients with encephalopathy and high ammonia, the protein supplementation can be delayed for 24–48 h [135]. A normal EN formula shall be used, because of insufficient evidence for the use of branched chained amino acids [3,135].

10.9.3. Patients with Acute Pancreatitis

In patients with acute pancreatitis, there is a great range of severity. Therefore, the MNT should be re-evaluated frequently. In mild cases, patients can receive oral nutrition ad libitum. In medium or severe cases, EEN should be commenced with a low infusion rate via gastral or jejunal path. If the patient undergoes surgery for necrosectomy, placement of needle catheter jejunostomy should be considered [136]. Before PN is considered one week after onset of pancreatitis, measures to increase GI-tolerance can be performed, according to A.S.P.E.N. [3].

Pragmatic recommendations:

- **Patients with abdominal diseases have a high risk for malnutrition**
- **MNT in patients with abdominal diseases often requires an interdisciplinary team**
- **EN is the preferred route in patients with liver failure, acute pancreatitis, after GI surgery and may be used even in patients with enteric fistulas and/or open abdomen.**

10.10. Critically Ill Patients with Renal Diseases

Patients with renal failure represent a heterogeneous group with different needs regarding macro- and micronutrients. EN should be preferred in this patient group. If an

EEN is not possible, SPN or TPN should be started early according to the German guideline for Patients with Kidney Disease [137].

A patient with acute kidney failure has the same needs for energy and protein and should be treated according to the standard of other critically ill patients—compared to patients with chronic kidney failure, who have increased energy demands as stated in the guidelines [1,3,137]. Even when deranged values for potassium and phosphate are rare in patients with acute-on-chronic kidney failure, a careful electrolyte monitoring is important. In case of electrolyte derangements and no indication for RRT, special renal formulae can be used [3]. These formulae contain less fluid and protein, are high in calories and have lower potassium and phosphate content and can contain additional substances such as carnitine [137].

A RRT increases losses of energy, water soluble molecules—such as amino acids, electrolytes, trace elements, and vitamins—and induces systemic inflammation and protein catabolism. On the other hand, substances sch as lactate and citrate are added excessively with the dialysis or hemofiltration solution. This needs to be minded for calorie calculations. The guidelines vary in their recommendations regarding energy supplementation for ICU patients with kidney failure during RRT. The DGEM guideline recommends macronutrient calculation as per the standard in other ICU patients, which should be ramped up slowly while considering macronutrient losses. The A.S.P.E.N guideline suggests administration of high protein dosages of 2.5 g/kg/d, to achieve nitrogen balance [1,3,137]. To cope with the increased need of vitamins and trace elements, an increased supply is recommended, with special regard to vitamin C, folate and thiamine [1,137].

Pragmatic recommendations:

- **Use the same principles as in other ICU patients regarding macronutrient requirements, timing, and route**
- **In patients on RRT, an increased supply of amino acids, electrolytes vitamin, and trace elements should be considered**
- **Electrolytes need to be monitored closely**

11. Summary and Conclusions

Despite ongoing research activities, the level of evidence remains often low due to a lack of data from large RCTs.

Evidence based international guidelines are available. Nevertheless, the implementation of the different recommendations into clinical routine remains often insufficient.

It should be kept in mind that every ICU-patient is at risk for malnutrition, regular screenings for malnutrition should be performed, close monitoring and frequent adaptation of nutrition is necessary. An early enteral nutrition with a standard formula is preferred in almost all critically ill patients. The addition of parenteral nutrition and micronutrients should be considered individually. Feeding protocols which are tailored to the treating units will improve the nutritional performance.

Due to the heterogeneity of the patients, MNT should be carefully adapted to the individual patient with special focus on phase of critical illness, metabolic tolerance, leading symptoms, and comorbidities. Nutrition in the ICU is a complex therapy requiring an interdisciplinary approach and frequent reevaluation. This article can only be regarded as an introduction and summary for the complex topic of nutrition in the critically ill patient and may be of help for the clinical routine.

Appreciating recent advances, the long-lasting open questions will remain:

- The optimal nutritional assessment and monitoring
- The impact of personalized and disease-specific MNT on patient-centered clinical outcomes
- The impact of high vs. low protein and energy intake according to the phase of illness

Author Contributions: Conceptualization, A.H. and A.W. together with G.E.; Writing—original draft preparation, A.H. and A.W. together with G.E.; Writing—review and editing, A.H. and A.W. together with G.E.; Visualization, A.H.; Supervision, A.W. All authors have read and agreed to the published version of the manuscript.

Funding: This research received no external funding.

Institutional Review Board Statement: Not applicable.

Informed Consent Statement: Not applicable.

Data Availability Statement: Not applicable.

Acknowledgments: A.H. is currently supported by a stipend from the Ministry of Culture and Science of the State of North Rhine-Westphalia "Landesprogramm für chancengerechte Hochschulmedizin".

Conflicts of Interest: G.E. received lecture fees, travel support and advisor honoraria from Fresenius Kabi within the last three years, A.W. received lecture fees from Baxter, B. Braun, Fresenius Kabi, Ethicon, and Falk Foundation, Research grant from Baxter, Mucos. A.H. received a research grant from Fresenius Kabi.

Abbreviations

A.S.P.E.N	American Society of Enteral and Parenteral Nutrition
BIA	Bioelectrical Impedance Analysis
BMI	Body Mass Index
CT	Computed Tomography
DGEM	Deutsche Gesellschaft für Ernährungsmedizin (German Society for Nutritional Medicine)
EN	Enteral Nutrition
EN+PN	Combined Enteral and Parenteral Nutrition
ESICM	European Society of Intensive Care Medicine
ESPEN	European Society of Enteral and Parenteral Nutrition
GI	Gastrointestinal
GLIM	Global Leadership Initiative on Malnutrition
ICU	Intensive Care Unit
LOS	Length of Stay
MNT	Medical Nutrition Therapy
MUST	Malnutrition Universal Screening Tool
NRS 2002	Nutrition Risk Score 2002
NUTRIC	Nutrition Risk in the Critically Ill Score
ONS	Oral Nutritional Supplements
PN	Parenteral Nutrition
RCT	Randomized Controlled Trial
RRT	Renal Replacement Therapy
SPN	Supplementary Parenteral Nutrition
SGA	Subjective Global Assessment

Appendix A

Table A1. Weight based calculation of macronutrient requirements, *, recommendation for acute phase, use actual body weight for calculation.

Guideline	Energy in kcal/kg/d	Protein in g/kg/d
DGEM [1]	24 *	1.0–1.2 *
ESPEN [2]	20–25	1.3
ASPEN [3]	25–30	1.2–2

References

1. Elke, G.; Hartl, W.H.; Kreymann, K.G.; Adolph, M.; Felbinger, T.W.; Graf, T.; de Heer, G.; Heller, A.R.; Kampa, U.; Mayer, K.; et al. Clinical Nutrition in Critical Care Medicine—Guideline of the German Society for Nutritional Medicine (DGEM). *Clin. Nutr. ESPEN* **2019**, *33*, 220–275. [CrossRef]
2. Singer, P.; Blaser, A.R.; Berger, M.M.; Alhazzani, W.; Calder, P.C.; Casaer, M.; Hiesmayr, M.; Mayer, K.; Montejo, J.C.; Pichard, C.; et al. ESPEN guideline on clinical nutrition in the intensive care unit. *Clin. Nutr.* **2018**, *38*, 48–79. [CrossRef] [PubMed]
3. McClave, S.A.; Taylor, B.E.; Martindale, R.G.; Warren, M.M.; Johnson, D.R.; Braunschweig, C.; McCarthy, M.S.; Davanos, E.; Rice, T.W.; Cresci, G.A.; et al. Guidelines for the Provision and Assessment of Nutrition Support Therapy in the Adult Critically Ill Patient: Society of Critical Care Medicine (SCCM) and American Society for Parenteral and Enteral Nutrition (A.S.P.E.N.). *J. Parenter. Enter. Nutr.* **2016**, *40*, 159–211. [CrossRef] [PubMed]
4. Reintam Blaser, A.; Starkopf, J.; Alhazzani, W.; Berger, M.M.; Casaer, M.P.; Deane, A.M.; Fruhwald, S.; Hiesmayr, M.; Ichai, C.; Jakob, S.M.; et al. Early enteral nutrition in critically ill patients: ESICM clinical practice guidelines. *Intensive Care Med.* **2017**, *43*, 380–398. [CrossRef]
5. Rhodes, A.; Evans, L.E.; Alhazzani, W.; Levy, M.M.; Antonelli, M.; Ferrer, R.; Kumar, A.; Sevransky, J.E.; Sprung, C.L.; Nunnally, M.E.; et al. Surviving Sepsis Campaign: International Guidelines for Management of Sepsis and Septic Shock: 2016. *Crit. Care Med.* **2017**, *45*, 486–552. [CrossRef] [PubMed]
6. Mooi, N.M.; Ncama, B.P. Evidence on nutritional therapy practice guidelines and implementation in adult critically ill patients: A systematic scoping review. *Curationis* **2019**, *42*, e1–e13. [CrossRef]
7. Lew, C.C.H.; Yandell, R.; Fraser, R.J.L.; Chua, A.P.; Chong, M.F.F.; Miller, M. Association Between Malnutrition and Clinical Outcomes in the Intensive Care Unit: A Systematic Review. *J. Parenter. Enter. Nutr.* **2017**, *41*, 744–758. [CrossRef] [PubMed]
8. Lambell, K.J.; Tatucu-Babet, O.A.; Chapple, L.A.; Gantner, D.; Ridley, E.J. Nutrition therapy in critical illness: A review of the literature for clinicians. *Crit. Care* **2020**, *24*, 1–11. [CrossRef]
9. Kondrup, J. Nutritional-risk scoring systems in the intensive care unit. *Curr. Opin. Clin. Nutr. Metab. Care* **2014**, *17*, 177–182. [CrossRef]
10. Barr, J.; Hecht, M.; Flavin, K.E.; Khorana, A.; Gould, M.K. Outcomes in critically ill patients before and after the implementation of an evidence-based nutritional management protocol. *Chest* **2004**, *125*, 1446–1457. [CrossRef]
11. Heyland, D.K.; Cahill, N.E.; Dhaliwal, R.; Sun, X.; Day, A.G.; McClave, S.A. Impact of enteral feeding protocols on enteral nutrition delivery: Results of a multicenter observational study. *J. Parenter. Enter. Nutr.* **2010**, *34*, 675–684. [CrossRef]
12. Weimann, A.; Braga, M.; Carli, F.; Higashiguchi, T.; Hubner, M.; Klek, S.; Laviano, A.; Ljungqvist, O.; Lobo, D.N.; Martindale, R.G.; et al. ESPEN practical guideline: Clinical nutrition in surgery. *Clin. Nutr.* **2021**, *40*, 4745–4761. [CrossRef] [PubMed]
13. Heyland, D.K.; Dhaliwal, R.; Lemieux, M.; Wang, M.; Day, A.G. Implementing the PEP uP Protocol in Critical Care Units in Canada: Results of a Multicenter, Quality Improvement Study. *J. Parenter. Enter. Nutr.* **2015**, *39*, 698–706. [CrossRef]
14. Kozeniecki, M.; McAndrew, N.; Patel, J.J. Process-Related Barriers to Optimizing Enteral Nutrition in a Tertiary Medical Intensive Care Unit. *Nutr. Clin. Pract.* **2016**, *31*, 80–85. [CrossRef]
15. Sundstrom Rehal, M.; Tjader, I.; Wernerman, J. Nutritional needs for the critically ill in relation to inflammation. *Curr. Opin. Clin. Nutr. Metab. Care* **2016**, *19*, 138–143. [CrossRef]
16. Cederholm, T.; Jensen, G.L.; Correia, M.I.T.D.; Gonzalez, M.C.; Fukushima, R.; Higashiguchi, T.; Baptista, G.; Barazzoni, R.; Blaauw, R.; Coats, A.; et al. GLIM criteria for the diagnosis of malnutrition—A consensus report from the global clinical nutrition community. *J. Cachexia Sarcopenia Muscle* **2019**, *10*, 207–217. [CrossRef] [PubMed]
17. Hill, A.; Goetzenich, A.; Marx, G.; Stoppe, C. Role of Nutrition Support in Cardiac Surgery Patients—An Overview. *Anasthesiol. Intensivmed. Notf. Schmerzther. AINS* **2018**, *53*, 466–479.
18. Looijaard, W.; Molinger, J.; Weijs, P.J.M. Measuring and monitoring lean body mass in critical illness. *Curr. Opin. Crit. Care* **2018**, *24*, 241–247. [CrossRef] [PubMed]
19. Hill, A.; Heyland, D.K.; Elke, G.; Schaller, S.J.; Stocker, R.; Haberthur, C.; von Loeffelholz, C.; Suchner, U.; Puthucheary, Z.A.; Bear, D.E.; et al. Meeting nutritional targets of critically ill patients by combined enteral and parenteral nutrition: Review and rationale for the EFFORTcombo trial. *Nutr. Res. Rev.* **2020**, *33*, 312–320. [CrossRef]
20. Preiser, J.-C.; van Zanten, A.R.H.; Berger, M.M.; Biolo, G.; Casaer, M.P.; Doig, G.S.; Griffiths, R.D.; Heyland, D.K.; Hiesmayr, M.; Iapichino, G.; et al. Metabolic and nutritional support of critically ill patients: Consensus and controversies. *Crit. Care* **2015**, *19*, 1–11. [CrossRef]
21. Harvey, S.E.; Parrott, F.; Harrison, D.A.; Bear, D.E.; Segaran, E.; Beale, R.; Bellingan, G.; Leonard, R.; Mythen, M.G.; Rowan, K.M.; et al. Trial of the route of early nutritional support in critically ill adults. *N. Engl. J. Med.* **2014**, *371*, 1673–1684. [CrossRef] [PubMed]
22. Reignier, J.; Boisrame-Helms, J.; Brisard, L.; Lascarrou, J.-B.; Ait Hssain, A.; Anguel, N.; Argaud, L.; Asehnoune, K.; Asfar, P.; Bellec, F.; et al. Enteral versus parenteral early nutrition in ventilated adults with shock: A randomised, controlled, multicentre, open-label, parallel-group study (NUTRIREA-2). *Lancet* **2018**, *391*, 133–143. [CrossRef]
23. Weimann, A.; Braga, M.; Carli, F.; Higashiguchi, T.; Hübner, M.; Klek, S.; Laviano, A.; Ljungqvist, O.; Lobo, D.N.; Martindale, R.; et al. ESPEN guideline: Clinical nutrition in surgery. *Clin. Nutr.* **2017**, *36*, 623–650. [CrossRef]
24. Heyland, D.K.; Schroter-Noppe, D.; Drover, J.W.; Jain, M.; Keefe, L.; Dhaliwal, R.; Day, A. Nutrition support in the critical care setting: Current practice in canadian ICUs–opportunities for improvement? *J. Parenter. Enter. Nutr.* **2003**, *27*, 74–83. [CrossRef]

25. Hill, A.; Heyland, D.K.; Ortiz Reyes, L.A.; Laaf, E.; Wendt, S.; Elke, G.; Stoppe, C. A combination of enteral and parenteral nutrition in the acute phase of critical illness: An updated systematic review and meta-analysis. *J. Parenter. Enter. Nutr.* **2021**, 1–16. [CrossRef]
26. Heyland, D.; Cook, D.J.; Winder, B.; Brylowski, L.; Van deMark, H.; Guyatt, G. Enteral nutrition in the critically ill patient: A prospective survey. *Crit. Care Med.* **1995**, *23*, 1055–1060. [CrossRef] [PubMed]
27. Adam, S.; Batson, S. A study of problems associated with the delivery of enteral feed in critically ill patients in five ICUs in the UK. *Intensive Care Med.* **1997**, *23*, 261–266. [CrossRef]
28. McClave, S.A.; Sexton, L.K.; Spain, D.A.; Adams, J.L.; Owens, N.A.; Sullins, M.B.; Blandford, B.S.; Snider, H.L. Enteral tube feeding in the intensive care unit: Factors impeding adequate delivery. *Crit. Care Med.* **1999**, *27*, 1252–1256. [CrossRef]
29. Berger, M.M.; Chiolero, R.L. Enteral nutrition and cardiovascular failure: From myths to clinical practice. *J. Parenter. Enter. Nutr.* **2009**, *33*, 702–709. [CrossRef] [PubMed]
30. Berger, M.M.; Revelly, J.-P.; Cayeux, M.-C.; Chiolero, R.L. Enteral nutrition in critically ill patients with severe hemodynamic failure after cardiopulmonary bypass. *Clin. Nutr.* **2005**, *24*, 124–132. [CrossRef]
31. Villet, S.; Chiolero, R.L.; Bollmann, M.D.; Revelly, J.P.; Cayeux, M.C.; Delarue, J.; Berger, M.M. Negative impact of hypocaloric feeding and energy balance on clinical outcome in ICU patients. *Clin. Nutr.* **2005**, *24*, 502–509. [CrossRef]
32. Casaer, M.P.; Hermans, G.; Wilmer, A.; Van den Berghe, G. Impact of early parenteral nutrition completing enteral nutrition in adult critically ill patients (EPaNIC trial): A study protocol and statistical analysis plan for a randomized controlled trial. *Trials* **2011**, *12*, 1–11. [CrossRef]
33. Wischmeyer, P.E.; Hasselmann, M.; Kummerlen, C.; Kozar, R.; Kutsogiannis, D.J.; Karvellas, C.J.; Besecker, B.; Evans, D.K.; Preiser, J.-C.; Gramlich, L.; et al. A randomized trial of supplemental parenteral nutrition in underweight and overweight critically ill patients: The TOP-UP pilot trial. *Crit. Care* **2017**, *21*, 1–14. [CrossRef] [PubMed]
34. Heidegger, C.P.; Berger, M.M.; Graf, S.; Zingg, W.; Darmon, P.; Costanza, M.C.; Thibault, R.; Pichard, C. Optimisation of energy provision with supplemental parenteral nutrition in critically ill patients: A randomised controlled clinical trial. *Lancet* **2013**, *381*, 385–393. [CrossRef]
35. Elke, G.; van Zanten, A.R.H.; Lemieux, M.; McCall, M.; Jeejeebhoy, K.N.; Kott, M.; Jiang, X.; Day, A.G.; Heyland, D.K. Enteral versus parenteral nutrition in critically ill patients: An updated systematic review and meta-analysis of randomized controlled trials. *Crit. Care* **2016**, *20*, 1–14. [CrossRef] [PubMed]
36. Arabi, Y.M.; Aldawood, A.S.; Al-Dorzi, H.M.; Tamim, H.M.; Haddad, S.H.; Jones, G.; McIntyre, L.; Solaiman, O.; Sakkijha, M.H.; Sadat, M.; et al. Permissive Underfeeding or Standard Enteral Feeding in High- and Low-Nutritional-Risk Critically Ill Adults. Post Hoc Analysis of the PermiT Trial. *Am. J. Respir. Crit. Care Med.* **2017**, *195*, 652–662. [CrossRef] [PubMed]
37. Petros, S.; Horbach, M.; Seidel, F.; Weidhase, L. Hypocaloric vs Normocaloric Nutrition in Critically Ill Patients: A Prospective Randomized Pilot Trial. *J. Parenter. Enter. Nutr.* **2016**, *40*, 242–249. [CrossRef]
38. De Waele, E.; Jonckheer, J.; Wischmeyer, P. Indirect calorimetry in critical illness: A new standard of care? *Curr. Opin. Crit. Care* **2021**, *27*, 334–343. [CrossRef]
39. Duan, J.Y.; Zheng, W.H.; Zhou, H.; Xu, Y.; Huang, H.B. Energy delivery guided by indirect calorimetry in critically ill patients: A systematic review and meta-analysis. *Crit. Care* **2021**, *25*, 1–10. [CrossRef]
40. Pertzov, B.; Bar-Yoseph, H.; Menndel, Y.; Bendavid, I.; Kagan, I.; Glass, Y.D.; Singer, P. The effect of indirect calorimetry guided isocaloric nutrition on mortality in critically ill patients-a systematic review and meta-analysis. *Eur. J. Clin. Nutr.* **2021**, 1–11. [CrossRef]
41. Heyland, D.K.; Patel, J.; Bear, D.; Sacks, G.; Nixdorf, H.; Dolan, J.; Aloupis, M.; Licastro, K.; Jovanovic, V.; Rice, T.W.; et al. The Effect of Higher Protein Dosing in Critically Ill Patients: A Multicenter Registry-Based Randomized Trial: The EFFORT Trial. *J. Parenter. Enter. Nutr.* **2018**, *43*, 326–334. [CrossRef]
42. Heyland, D.K.; Stapleton, R.; Compher, C. Should We Prescribe More Protein to Critically Ill Patients? *Nutrients* **2018**, *10*, 462. [CrossRef] [PubMed]
43. Alberda, C.; Gramlich, L.; Jones, N.; Jeejeebhoy, K.; Day, A.G.; Dhaliwal, R.; Heyland, D.K. The relationship between nutritional intake and clinical outcomes in critically ill patients: Results of an international multicenter observational study. *Intensive Care Med.* **2009**, *35*, 1728–1737. [CrossRef]
44. Heyland, D.K.; Stephens, K.E.; Day, A.G.; McClave, S.A. The success of enteral nutrition and ICU-acquired infections: A multicenter observational study. *Clin. Nutr.* **2011**, *30*, 148–155. [CrossRef] [PubMed]
45. Heyland, D.K.; Cahill, N.; Day, A.G. Optimal amount of calories for critically ill patients: Depends on how you slice the cake! *Crit. Care Med.* **2011**, *39*, 2619–2626. [CrossRef]
46. Ferrie, S.; Allman-Farinelli, M.; Daley, M.; Smith, K. Protein Requirements in the Critically Ill: A Randomized Controlled Trial Using Parenteral Nutrition. *J. Parenter. Enter. Nutr.* **2016**, *40*, 795–805. [CrossRef] [PubMed]
47. Doig, G.S.; Simpson, F.; Bellomo, R.; Heighes, P.T.; Sweetman, E.A.; Chesher, D.; Pollock, C.; Davies, A.; Botha, J.; Harrigan, P.; et al. Intravenous amino acid therapy for kidney function in critically ill patients: A randomized controlled trial. *Intensive Care Med.* **2015**, *41*, 1197–1208. [CrossRef]
48. Davies, M.L.; Chapple, L.-A.S.; Chapman, M.J.; Moran, J.L.; Peake, S.L. Protein delivery and clinical outcomes in the critically ill: A systematic review and meta-analysis. *Crit. Care Resusc.* **2017**, *19*, 117–127.

49. Gunst, J.; Vanhorebeek, I.; Thiessen, S.E.; Van den Berghe, G. Amino acid supplements in critically ill patients. *Pharmacol. Res.* **2018**, *130*, 127–131. [CrossRef] [PubMed]
50. Fetterplace, K.; Deane, A.M.; Tierney, A.; Beach, L.J.; Knight, L.D.; Presneill, J.; Rechnitzer, T.; Forsyth, A.; Gill, B.M.T.; Mourtzakis, M.; et al. Targeted Full Energy and Protein Delivery in Critically Ill Patients: A Pilot Randomized Controlled Trial (FEED Trial). *J. Parenter. Enter. Nutr.* **2018**, *42*, 1252–1262. [CrossRef]
51. Dresen, E.; Weissbrich, C.; Fimmers, R.; Putensen, C.; Stehle, P. Medical high-protein nutrition therapy and loss of muscle mass in adult ICU patients: A randomized controlled trial. *Clin. Nutr.* **2021**, *40*, 1562–1570. [CrossRef]
52. Chapple, L.S.; Summers, M.J.; Bellomo, R.; Chapman, M.J.; Davies, A.R.; Ferrie, S.; Finnis, M.E.; Hurford, S.; Lange, K.; Little, L.; et al. Use of a High-Protein Enteral Nutrition Formula to Increase Protein Delivery to Critically Ill Patients: A Randomized, Blinded, Parallel-Group, Feasibility Trial. *J. Parenter. Enter. Nutr.* **2021**, *45*, 699–709. [CrossRef] [PubMed]
53. van Zanten, A.R.H.; Petit, L.; De Waele, J.; Kieft, H.; de Wilde, J.; van Horssen, P.; Klebach, M.; Hofman, Z. Very high intact-protein formula successfully provides protein intake according to nutritional recommendations in overweight critically ill patients: A double-blind randomized trial. *Crit. Care* **2018**, *22*, 1–12. [CrossRef] [PubMed]
54. Lee, Z.Y.; Yap, C.S.L.; Hasan, M.S.; Engkasan, J.P.; Barakatun-Nisak, M.Y.; Day, A.G.; Patel, J.J.; Heyland, D.K. The effect of higher versus lower protein delivery in critically ill patients: A systematic review and meta-analysis of randomized controlled trials. *Crit. Care* **2021**, *25*, 1–15. [CrossRef]
55. Charriere, M.; Ridley, E.; Hastings, J.; Bianchet, O.; Scheinkestel, C.; Berger, M.M. Propofol sedation substantially increases the caloric and lipid intake in critically ill patients. *Nutrition* **2017**, *42*, 64–68. [CrossRef] [PubMed]
56. Bousie, E.; van Blokland, D.; Lammers, H.J.; van Zanten, A.R. Relevance of non-nutritional calories in mechanically ventilated critically ill patients. *Eur. J. Clin. Nutr.* **2016**, *70*, 1443–1450. [CrossRef]
57. Tian, F.; Heighes, P.T.; Allingstrup, M.J.; Doig, G.S. Early Enteral Nutrition Provided Within 24 Hours of ICU Admission: A Meta-Analysis of Randomized Controlled Trials. *Crit. Care Med.* **2018**, *46*, 1049–1056. [CrossRef]
58. Lewis, S.R.; Schofield-Robinson, O.J.; Alderson, P.; Smith, A.F. Enteral versus parenteral nutrition and enteral versus a combination of enteral and parenteral nutrition for adults in the intensive care unit. *Cochrane Database Syst. Rev.* **2018**, *6*, CD012276. [CrossRef] [PubMed]
59. Zhang, G.; Zhang, K.; Cui, W.; Hong, Y.; Zhang, Z. The effect of enteral versus parenteral nutrition for critically ill patients: A systematic review and meta-analysis. *J. Clin. Anesth.* **2018**, *51*, 62–92. [CrossRef]
60. Peake, S.L.; Davies, A.R.; Deane, A.M.; Lange, K.; Moran, J.L.; O'Connor, S.N.; Ridley, E.J.; Williams, P.J.; Chapman, M.J.; TARGET investigators and the Australian and New Zealand Intensive Care Society Clinical Trials Group. Use of a concentrated enteral nutrition solution to increase calorie delivery to critically ill patients: A randomized, double-blind, clinical trial. *Am. J. Clin. Nutr.* **2014**, *100*, 616–625. [CrossRef]
61. Chapman, M.; Peake, S.L.; Bellomo, R.; Davies, A.; Deane, A.; Horowitz, M.; Hurford, S.; Lange, K.; Little, L.; Mackle, D.; et al. Energy-Dense versus Routine Enteral Nutrition in the Critically Ill. *N. Engl. J. Med.* **2018**, *379*, 1823–1834. [CrossRef]
62. Manzanares, W.; Langlois, P.L.; Wischmeyer, P.E. Restoring the Microbiome in Critically Ill Patients: Are Probiotics Our True Friends When We Are Seriously Ill? *J. Parenter. Enter. Nutr.* **2017**, *41*, 530–533. [CrossRef]
63. Didari, T.; Solki, S.; Mozaffari, S.; Nikfar, S.; Abdollahi, M. A systematic review of the safety of probiotics. *Expert Opin. Drug Saf.* **2014**, *13*, 227–239. [CrossRef] [PubMed]
64. Batra, P.; Soni, K.D.; Mathur, P. Efficacy of probiotics in the prevention of VAP in critically ill ICU patients: An updated systematic review and meta-analysis of randomized control trials. *J. Intensive Care* **2020**, *8*, 1–14. [CrossRef]
65. Seifi, N.; Jafarzadeh Esfahani, A.; Sedaghat, A.; Rezvani, R.; Khadem-Rezaiyan, M.; Nematy, M.; Safarian, M. Effect of gut microbiota modulation on feeding tolerance of enterally fed critically ill adult patients: A systematic review. *Syst. Rev.* **2021**, *10*, 1–11. [CrossRef] [PubMed]
66. Litton, E.; Anstey, M.; Broadhurst, D.; Chapman, A.; Currie, A.; Ferrier, J.; Gummer, J.; Higgins, A.; Lim, J.; Manning, L.; et al. Early and sustained Lactobacillus plantarum probiotic therapy in critical illness: The randomised, placebo-controlled, restoration of gut microflora in critical illness trial (ROCIT). *Intensive Care Med.* **2021**, *47*, 307–315. [CrossRef]
67. Bertolini, G.; Iapichino, G.; Radrizzani, D.; Facchini, R.; Simini, B.; Bruzzone, P.; Zanforlin, G.; Tognoni, G. Early enteral immunonutrition in patients with severe sepsis: Results of an interim analysis of a randomized multicentre clinical trial. *Intensive Care Med.* **2003**, *29*, 834–840. [CrossRef] [PubMed]
68. Luiking, Y.C.; Poeze, M.; Deutz, N.E. Arginine infusion in patients with septic shock increases nitric oxide production without haemodynamic instability. *Clin. Sci.* **2015**, *128*, 57–67. [CrossRef] [PubMed]
69. Buter, H.; Bakker, A.J.; Kingma, W.P.; Koopmans, M.; Boerma, E.C. Plasma glutamine levels in patients after non-elective or elective ICU admission: An observational study. *BMC Anesthesiol.* **2016**, *16*, 1–5. [CrossRef]
70. Heyland, D.; Muscedere, J.; Wischmeyer, P.E.; Cook, D.; Jones, G.; Albert, M.; Elke, G.; Berger, M.M.; Day, A.G.; Canadian Critical Care Trials, G. A randomized trial of glutamine and antioxidants in critically ill patients. *N. Engl. J. Med.* **2013**, *368*, 1489–1497. [CrossRef] [PubMed]
71. Heyland, D.; Wischmeyer, P.E.; Day, A.G.; Canadian Clinical Care Trials, G. Glutamine and antioxidants in critically ill patients. *N. Engl. J. Med.* **2013**, *369*, 484–485. [CrossRef] [PubMed]
72. McRae, M.P. Therapeutic benefits of glutamine: An umbrella review of meta-analyses. *Biomed. Rep.* **2017**, *6*, 576–584. [CrossRef] [PubMed]

73. Stehle, P.; Ellger, B.; Kojic, D.; Feuersenger, A.; Schneid, C.; Stover, J.; Scheiner, D.; Westphal, M. Glutamine dipeptide-supplemented parenteral nutrition improves the clinical outcomes of critically ill patients: A systematic evaluation of randomised controlled trials. *Clin. Nutr. ESPEN* **2017**, *17*, 75–85. [CrossRef]
74. Sun, Y.; Zhu, S.; Li, S.; Liu, H. Glutamine on critical-ill patients: A systematic review and meta-analysis. *Ann. Palliat. Med.* **2021**, *10*, 1503–1520. [CrossRef]
75. Pradelli, L.; Klek, S.; Mayer, K.; Omar Alsaleh, A.J.; Rosenthal, M.D.; Heller, A.R.; Muscaritoli, M. Omega-3 fatty acid-containing parenteral nutrition in ICU patients: Systematic review with meta-analysis and cost-effectiveness analysis. *Crit. Care* **2020**, *24*, 634. [CrossRef]
76. Singer, P.; Bendavid, I.; Mesilati-Stahy, R.; Green, P.; Rigler, M.; Lev, S.; Schif-Zuck, S.; Amiram, A.; Theilla, M.; Kagan, I. Enteral and supplemental parenteral nutrition enriched with omega-3 polyunsaturated fatty acids in intensive care patients—A randomized, controlled, double-blind clinical trial. *Clin. Nutr.* **2021**, *40*, 2544–2554. [CrossRef]
77. Bloos, F.; Trips, E.; Nierhaus, A.; Briegel, J.; Heyland, D.K.; Jaschinski, U.; Moerer, O.; Weyland, A.; Marx, G.; Grundling, M.; et al. Effect of Sodium Selenite Administration and Procalcitonin-Guided Therapy on Mortality in Patients With Severe Sepsis or Septic Shock: A Randomized Clinical Trial. *JAMA Intern. Med.* **2016**, *176*, 1266–1276. [CrossRef]
78. Gudivada, K.K.; Kumar, A.; Shariff, M.; Sampath, S.; Varma, M.M.; Sivakoti, S.; Krishna, B. Antioxidant micronutrient supplementation in critically ill adults: A systematic review with meta-analysis and trial sequential analysis. *Clin. Nutr.* **2021**, *40*, 740–750. [CrossRef] [PubMed]
79. Stoppe, C.; McDonald, B.; Rex, S.; Manzanares, W.; Whitlock, R.; Fremes, S.; Fowler, R.; Lamarche, Y.; Meybohm, P.; Haberthür, C.; et al. SodiUm SeleniTe Adminstration IN Cardiac Surgery (SUSTAIN CSX-trial): Study design of an international multicenter randomized double-blinded controlled trial of high dose sodium-selenite administration in high-risk cardiac surgical patients. *Trials* **2014**, *15*, 1–10. [CrossRef]
80. Palacios, C.; Gonzalez, L. Is vitamin D deficiency a major global public health problem? *J. Steroid Biochem. Mol. Biol.* **2014**, *144*, 138–145. [CrossRef]
81. McNally, J.D.; Nama, N.; O'Hearn, K.; Sampson, M.; Amrein, K.; Iliriani, K.; McIntyre, L.; Fergusson, D.; Menon, K. Vitamin D deficiency in critically ill children: A systematic review and meta-analysis. *Crit. Care* **2017**, *21*, 1–13. [CrossRef]
82. Munshi, R.; Hussein, M.H.; Toraih, E.A.; Elshazli, R.M.; Jardak, C.; Sultana, N.; Youssef, M.R.; Omar, M.; Attia, A.S.; Fawzy, M.S.; et al. Vitamin D insufficiency as a potential culprit in critical COVID-19 patients. *J. Med. Virol.* **2021**, *93*, 733–740. [CrossRef] [PubMed]
83. Herrera-Quintana, L.; Gamarra-Morales, Y.; Vazquez-Lorente, H.; Molina-Lopez, J.; Castano-Perez, J.; Machado-Casas, J.F.; Coca-Zuniga, R.; Perez-Villares, J.M.; Planells, E. Bad Prognosis in Critical Ill Patients with COVID-19 during Short-Term ICU Stay regarding Vitamin D Levels. *Nutrients* **2021**, *13*, 1988. [CrossRef]
84. Vassiliou, A.G.; Jahaj, E.; Orfanos, S.E.; Dimopoulou, I.; Kotanidou, A. Vitamin D in infectious complications in critically ill patients with or without COVID-19. *Metabol. Open* **2021**, *11*, 100106. [CrossRef] [PubMed]
85. Langlois, P.L.; Szwec, C.; D'Aragon, F.; Heyland, D.K.; Manzanares, W. Vitamin D supplementation in the critically ill: A systematic review and meta-analysis. *Clin. Nutr.* **2018**, *37*, 1238–1246. [CrossRef]
86. Amrein, K.; Schnedl, C.; Holl, A.; Riedl, R.; Christopher, K.B.; Pachler, C.; Urbanic Purkart, T.; Waltensdorfer, A.; Munch, A.; Warnkross, H.; et al. Effect of high-dose vitamin D3 on hospital length of stay in critically ill patients with vitamin D deficiency: The VITdAL-ICU randomized clinical trial. *JAMA* **2014**, *312*, 1520–1530. [CrossRef]
87. Han, J.H.; Ginde, A.A.; Brown, S.M.; Baughman, A.; Collar, E.M.; Ely, E.W.; Gong, M.N.; Hope, A.A.; Hou, P.C.; Hough, C.L.; et al. Effect of Early High-Dose Vitamin D3 Repletion on Cognitive Outcomes in Critically Ill Adults. *Chest* **2021**, *152*, 1070–1086. [CrossRef]
88. Amrein, K.; Parekh, D.; Westphal, S.; Preiser, J.C.; Berghold, A.; Riedl, R.; Eller, P.; Schellongowski, P.; Thickett, D.; Meybohm, P.; et al. Effect of high-dose vitamin D3 on 28-day mortality in adult critically ill patients with severe vitamin D deficiency: A study protocol of a multicentre, placebo-controlled double-blind phase III RCT (the VITDALIZE study). *BMJ Open* **2019**, *9*, e031083. [CrossRef]
89. Truwit, J.D.; Hite, R.D.; Morris, P.E.; DeWilde, C.; Priday, A.; Fisher, B.; Thacker, L.R.; Natarajan, R.; Brophy, D.F.; Sculthorpe, R.; et al. Effect of Vitamin C Infusion on Organ Failure and Biomarkers of Inflammation and Vascular Injury in Patients With Sepsis and Severe Acute Respiratory Failure: The CITRIS-ALI Randomized Clinical Trial. *JAMA* **2019**, *322*, 1261–1270. [CrossRef]
90. Marik, P.E.; Khangoora, V.; Rivera, R.; Hooper, M.H.; Catravas, J. Hydrocortisone, Vitamin C, and Thiamine for the Treatment of Severe Sepsis and Septic Shock: A Retrospective Before-After Study. *Chest* **2017**, *151*, 1229–1238. [CrossRef]
91. Zayed, Y.; Alzghoul, B.N.; Banifadel, M.; Venigandla, H.; Hyde, R.; Sutchu, S.; Khasawneh, M.; Borok, Z.; Urbine, D.; Jantz, M.; et al. Vitamin C, Thiamine, and Hydrocortisone in the Treatment of Sepsis: A Meta-Analysis and Trial Sequential Analysis of Randomized Controlled Trials. *J. Intensive Care Med.* **2021**. [CrossRef]
92. Reintam Blaser, A.; Starkopf, J.; Malbrain, M.L. Abdominal signs and symptoms in intensive care patients. *Anaesthesiol. Intensive Ther.* **2015**, *47*, 379–387. [CrossRef]
93. Elke, G.; Felbinger, T.W.; Heyland, D.K. Gastric residual volume in critically ill patients: A dead marker or still alive? *Nutr. Clin. Pract.* **2015**, *30*, 59–71. [CrossRef] [PubMed]
94. ERAS Society. Enhanced Recovery after Surgery Society (ERAS). Available online: https://erassociety.org/guidelines/list-of-guidelines/ (accessed on 18 August 2021).

95. Ljungqvist, O. ERAS—Enhanced recovery after surgery: Moving evidence-based perioperative care to practice. *J. Parenter. Enter. Nutr.* **2014**, *38*, 559–566. [CrossRef] [PubMed]
96. Engelman, D.T.; Ben Ali, W.; Williams, J.B.; Perrault, L.P.; Reddy, V.S.; Arora, R.C.; Roselli, E.E.; Khoynezhad, A.; Gerdisch, M.; Levy, J.H.; et al. Guidelines for Perioperative Care in Cardiac Surgery: Enhanced Recovery After Surgery Society Recommendations. *JAMA Surg.* **2019**, *154*, 755–766. [CrossRef]
97. Reignier, J.; Mercier, E.; Le Gouge, A.; Boulain, T.; Desachy, A.; Bellec, F.; Clavel, M.; Frat, J.P.; Plantefeve, G.; Quenot, J.P.; et al. Effect of not monitoring residual gastric volume on risk of ventilator-associated pneumonia in adults receiving mechanical ventilation and early enteral feeding: A randomized controlled trial. *JAMA* **2013**, *309*, 249–256. [CrossRef]
98. Poulard, F.; Dimet, J.; Martin-Lefevre, L.; Bontemps, F.; Fiancette, M.; Clementi, E.; Lebert, C.; Renard, B.; Reignier, J. Impact of not measuring residual gastric volume in mechanically ventilated patients receiving early enteral feeding: A prospective before-after study. *J. Parenter. Enter. Nutr.* **2010**, *34*, 125–130. [CrossRef]
99. Reintam Blaser, A.; Preiser, J.C.; Fruhwald, S.; Wilmer, A.; Wernerman, J.; Benstoem, C.; Casaer, M.P.; Starkopf, J.; van Zanten, A.; Rooyackers, O.; et al. Gastrointestinal dysfunction in the critically ill: A systematic scoping review and research agenda proposed by the Section of Metabolism, Endocrinology and Nutrition of the European Society of Intensive Care Medicine. *Crit. Care* **2020**, *24*, 1–17. [CrossRef]
100. van den Berghe, G.; Wouters, P.; Weekers, F.; Verwaest, C.; Bruyninckx, F.; Schetz, M.; Vlasselaers, D.; Ferdinande, P.; Lauwers, P.; Bouillon, R. Intensive insulin therapy in critically ill patients. *N. Engl. J. Med.* **2001**, *345*, 1359–1367. [CrossRef] [PubMed]
101. Hartl, W.H.; Jauch, K.W. Metabolic self-destruction in critically ill patients: Origins, mechanisms and therapeutic principles. *Nutrition* **2014**, *30*, 261–267. [CrossRef] [PubMed]
102. CNSG East Cheshire NHS Trust: Guidelines for Prevention and Management of Refeeding Syndrome in Adults. 2015. Available online: https://docplayer.net/45404167-Guidelines-for-prevention-and-management-of-refeeding-syndrome-in-adults.html (accessed on 18 August 2021).
103. Ridley, E.J.; Davies, A.R.; Parke, R.; Bailey, M.; McArthur, C.; Gillanders, L.; Cooper, D.J.; McGuinness, S.; Investigators, S.P.N.C. Supplemental parenteral nutrition versus usual care in critically ill adults: A pilot randomized controlled study. *Crit. Care* **2018**, *22*, 1–11. [CrossRef] [PubMed]
104. Peterson, S.J.; Tsai, A.A.; Scala, C.M.; Sowa, D.C.; Sheean, P.M.; Braunschweig, C.L. Adequacy of oral intake in critically ill patients 1 week after extubation. *J. Am. Diet. Assoc.* **2010**, *110*, 427–433. [CrossRef]
105. Gentile, L.F.; Cuenca, A.G.; Efron, P.A.; Ang, D.; Bihorac, A.; McKinley, B.A.; Moldawer, L.L.; Moore, F.A. Persistent inflammation and immunosuppression: A common syndrome and new horizon for surgical intensive care. *J. Trauma Acute Care Surg.* **2012**, *72*, 1491–1501. [CrossRef] [PubMed]
106. Rosenthal, M.D.; Moore, F.A. Persistent Inflammation, Immunosuppression, and Catabolism: Evolution of Multiple Organ Dysfunction. *Surg. Infect.* **2016**, *17*, 167–172. [CrossRef] [PubMed]
107. Valentini, L.; Volkert, D.; Schütz, T.; Ockenga, J.; Pirlich, M.; Druml, W.; Schindler, K.; Ballmer, P.E.; Bischoff, S.C.; Weimann, A.; et al. Leitlinie der Deutschen Gesellschaft fur Ernahrungsmedizin (DGEM). *Aktuelle Ernährungsmedizin* **2013**, *38*, 97–111. [CrossRef]
108. Hill, A.; Stoppe, C.; Elke, G.; Marx, G. Nutrition in Critically Ill Patients—Challenges in Special Patient Groups. *Anästhesiol Intensivmed. Notf. Schmerzther.* **2020**, *55*, 91–105. [CrossRef]
109. Peterson, C.M.; Thomas, D.M.; Blackburn, G.L.; Heymsfield, S.B. Universal equation for estimating ideal body weight and body weight at any BMI. *Am. J. Clin. Nutr.* **2016**, *103*, 1197–1203. [CrossRef] [PubMed]
110. Volkert, D.; Beck, A.M.; Cederholm, T.; Cruz-Jentoft, A.; Goisser, S.; Hooper, L.; Kiesswetter, E.; Maggio, M.; Raynaud-Simon, A.; Sieber, C.C.; et al. ESPEN guideline on clinical nutrition and hydration in geriatrics. *Clin. Nutr.* **2018**, *38*, 10–47. [CrossRef]
111. Fried, L.P.; Tangen, C.M.; Walston, J.; Newman, A.B.; Hirsch, C.; Gottdiener, J.; Seeman, T.; Tracy, R.; Kop, W.J.; Burke, G.; et al. Frailty in older adults: Evidence for a phenotype. *J. Gerontology. Ser. A Biol. Sci. Med Sci.* **2001**, *56*, M146–M156. [CrossRef]
112. Buttery, A.K.; Busch, M.A.; Gaertner, B.; Scheidt-Nave, C.; Fuchs, J. Prevalence and correlates of frailty among older adults: Findings from the German health interview and examination survey. *BMC Geriatr.* **2015**, *15*, 1–9. [CrossRef]
113. Aubry, E.; Friedli, N.; Schuetz, P.; Stanga, Z. Refeeding syndrome in the frail elderly population: Prevention, diagnosis and management. *Clin. Exp. Gastroenterol.* **2018**, *11*, 255–264. [CrossRef]
114. Zaloga, G.P.; Roberts, P.R.; Marik, P. Feeding the hemodynamically unstable patient: A critical evaluation of the evidence. *Nutr. Clin. Pract.* **2003**, *18*, 285–293. [CrossRef]
115. Bruns, B.R.; Kozar, R.A. Feeding the Postoperative Patient on Vasopressor Support: Feeding and Pressor Support. *Nutr. Clin. Pract.* **2016**, *31*, 14–17. [CrossRef]
116. Khalid, I.; Doshi, P.; DiGiovine, B. Early enteral nutrition and outcomes of critically ill patients treated with vasopressors and mechanical ventilation. *Am. J. Crit. Care* **2010**, *19*, 261–268. [CrossRef]
117. Piton, G.; Cypriani, B.; Regnard, J.; Patry, C.; Puyraveau, M.; Capellier, G. Catecholamine use is associated with enterocyte damage in critically ill patients. *Shock* **2015**, *43*, 437–442. [CrossRef]
118. Weimann, A.; Felbinger, T.W. Gastrointestinal dysmotility in the critically ill: A role for nutrition. *Curr. Opin. Clin. Nutr. Metab. Care* **2016**, *19*, 353–359. [CrossRef] [PubMed]
119. Burgos, R.; Bretón, I.; Cereda, E.; Desport, J.C.; Dziewas, R.; Genton, L.; Gomes, F.; Jésus, P.; Leischker, A.; Muscaritoli, M.; et al. ESPEN guideline clinical nutrition in neurology. *Clin. Nutr.* **2018**, *37*, 354–396. [CrossRef] [PubMed]

120. Volkert, D.; Bauer, J.M.; Frühwald, T.; Gehrke, I.; Lechleitner, M.; Lenzen-Großimlinghaus, R.; Wirth, R.; Sieber, C.; das DGEM Steering Committee. Leitlinie der Deutschen Gesellschaft für Ernährungsmedizin (DGEM) in Zusammenarbeit mit der GESKES, der AKE, der DGN und der DGG. *Aktuelle Ernährungsmedizin* **2013**, *38*, e1–e48.
121. Zuercher, P.; Moret, C.S.; Dziewas, R.; Schefold, J.C. Dysphagia in the intensive care unit: Epidemiology, mechanisms, and clinical management. *Crit. Care* **2019**, *23*, 1–11. [CrossRef]
122. Hill, A.; Arora, R.C.; Engelman, D.T.; Stoppe, C. Preoperative Treatment of Malnutrition and Sarcopenia in Cardiac Surgery: New Frontiers. *Crit. Care Clin.* **2020**, *36*, 593–616. [CrossRef] [PubMed]
123. Hill, A.; Nesterova, E.; Lomivorotov, V.; Efremov, S.; Goetzenich, A.; Benstoem, C.; Zamyatin, M.; Chourdakis, M.; Heyland, D.; Stoppe, C. Current Evidence about Nutrition Support in Cardiac Surgery Patients-What Do We Know? *Nutrients* **2018**, *10*, 597. [CrossRef] [PubMed]
124. Stoppe, C.; Goetzenich, A.; Whitman, G.; Ohkuma, R.; Brown, T.; Hatzakorzian, R.; Kristof, A.; Meybohm, P.; Mechanick, J.; Evans, A.; et al. Role of nutrition support in adult cardiac surgery: A consensus statement from an International Multidisciplinary Expert Group on Nutrition in Cardiac Surgery. *Crit. Care* **2017**, *21*, 131. [CrossRef]
125. Thibault, R.; Seguin, P.; Tamion, F.; Pichard, C.; Singer, P. Nutrition of the COVID-19 patient in the intensive care unit (ICU): A practical guidance. *Crit. Care* **2020**, *24*, 1–8. [CrossRef]
126. Ochoa, J.B.; Cardenas, D.; Goiburu, M.E.; Bermudez, C.; Carrasco, F.; Correia, M. Lessons Learned in Nutrition Therapy in Patients With Severe COVID-19. *J. Parenter. Enter. Nutr.* **2020**, *44*, 1369–1375. [CrossRef]
127. Petersen, A.; Bressem, K.; Albrecht, J.; Thiess, H.M.; Vahldiek, J.; Hamm, B.; Makowski, M.R.; Niehues, A.; Niehues, S.M.; Adams, L.C. The role of visceral adiposity in the severity of COVID-19: Highlights from a unicenter cross-sectional pilot study in Germany. *Metabolism* **2020**, *110*, 154317. [CrossRef] [PubMed]
128. Barazzoni, R.; Bischoff, S.C.; Breda, J.; Wickramasinghe, K.; Krznarić, Ž; Nitzan, D.; Pirlich, M.; Singer, P. ESPEN expert statements and practical guidance for nutritional management of individuals with SARS-CoV-2 infection. *Clin. Nutr.* **2020**, *39*, 1631–1638.
129. Martindale, R.; Patel, J.J.; Taylor, B.; Arabi, Y.M.; Warren, M.; McClave, S.A. Nutrition Therapy in Critically Ill Patients With Coronavirus Disease 2019. *J. Parenter. Enter. Nutr.* **2020**, *44*, 1174–1184. [CrossRef]
130. Whittle, J.; Molinger, J.; MacLeod, D.; Haines, K.; Wischmeyer, P.E.; Group, L.-C.S. Persistent hypermetabolism and longitudinal energy expenditure in critically ill patients with COVID-19. *Crit. Care* **2020**, *24*, 581. [CrossRef] [PubMed]
131. Lakenman, P.L.M.; van der Hoven, B.; Schuijs, J.M.; Eveleens, R.D.; van Bommel, J.; Olieman, J.F.; Joosten, K.F.M. Energy expenditure and feeding practices and tolerance during the acute and late phase of critically ill COVID-19 patients. *Clin. Nutr. ESPEN* **2021**, *43*, 383–389. [CrossRef] [PubMed]
132. Murai, I.H.; Fernandes, A.L.; Sales, L.P.; Pinto, A.J.; Goessler, K.F.; Duran, C.S.C.; Silva, C.B.R.; Franco, A.S.; Macedo, M.B.; Dalmolin, H.H.H.; et al. Effect of a Single High Dose of Vitamin D3 on Hospital Length of Stay in Patients With Moderate to Severe COVID-19: A Randomized Clinical Trial. *JAMA* **2021**, *325*, 1053–1060. [CrossRef]
133. Friese, R.S. The open abdomen: Definitions, management principles, and nutrition support considerations. *Nutr. Clin. Pract.* **2012**, *27*, 492–498. [CrossRef] [PubMed]
134. Moore, S.M.; Burlew, C.C. Nutrition Support in the Open Abdomen. *Nutr. Clin. Pract.* **2016**, *31*, 9–13. [CrossRef]
135. Plauth, M.; Bernal, W.; Dasarathy, S.; Merli, M.; Plank, L.D.; Schütz, T.; Bischoff, S.C. ESPEN guideline on clinical nutrition in liver disease. *Clin. Nutr.* **2019**, *38*, 485–521. [CrossRef] [PubMed]
136. Weimann, A.; Braunert, M.; Muller, T.; Bley, T.; Wiedemann, B. Feasibility and safety of needle catheter jejunostomy for enteral nutrition in surgically treated severe acute pancreatitis. *J. Parenter. Enter. Nutr.* **2004**, *28*, 324–327. [CrossRef] [PubMed]
137. Druml, W.; Contzen, B.; Joannidis, M.; Kierdorf, H.; Das, M.K.; Und, K. S1-Guideline of the German Society for Nutritional Medicine (DGEM) in Cooperation with the GESKES, the AKE and the DGfN Enteral and Parenteral Nutrition in Patients with Kidney Disease. *Aktuelle Ernährungsmedizin* **2015**, *40*, 21–37.

nutrients

Review

Impact of Dietary Factors on Brugada Syndrome and Long QT Syndrome

Sara D'Imperio [1], Michelle M. Monasky [1], Emanuele Micaglio [1], Gabriele Negro [1] and Carlo Pappone [1,2,*]

1 Department of Arrhythmology, IRCCS Policlinico San Donato, Piazza E. Malan 1, San Donato Milanese, 20097 Milan, Italy; Sara.DImperio@grupposandonato.it (S.D.); michelle.monasky@grupposandonato.it (M.M.M.); Emanuele.micaglio@grupposandonato.it (E.M.); gabriele.negro87@gmail.com (G.N.)
2 Vita-Salute San Raffaele University, 20132 Milan, Italy
* Correspondence: carlo.pappone@af-ablation.org; Tel.: +39-0252774260; Fax: +39-0252774306

Abstract: A healthy regime is fundamental for the prevention of cardiovascular diseases (CVD). In inherited channelopathies, such as Brugada syndrome (BrS) and Long QT syndrome (LQTS), unfortunately, sudden cardiac death could be the first sign for patients affected by these syndromes. Several known factors are used to stratify the risk of developing cardiac arrhythmias, although none are determinative. The risk factors can be affected by adjusting lifestyle habits, such as a particular diet, impacting the risk of arrhythmogenic events and mortality. To date, the importance of understanding the relationship between diet and inherited channelopathies has been underrated. Therefore, we describe herein the effects of dietary factors on the development of arrhythmia in patients affected by BrS and LQTS. Modifying the diet might not be enough to fully prevent arrhythmias, but it can help lower the risk.

Keywords: Brugada syndrome; long QT syndrome; diet; ingredients; glucose; ketone bodies; ROS; sudden cardiac death

Citation: D'Imperio, S.; Monasky, M.M.; Micaglio, E.; Negro, G.; Pappone, C. Impact of Dietary Factors on Brugada Syndrome and Long QT Syndrome. *Nutrients* **2021**, *13*, 2482. https://doi.org/10.3390/nu13082482

Academic Editor: Ina Bergheim

Received: 24 June 2021
Accepted: 19 July 2021
Published: 21 July 2021

1. Introduction

A healthy regime is fundamental for the prevention of cardiovascular diseases (CVD). CVD are widely known to be responsible for 33% of all deaths worldwide [1]. One of the multiple types of cardiac disorders is arrhythmia, which refers to a group of several conditions that interfere with heart rhythm. Among different types of cardiac arrhythmias, there are inherited channelopathies (IC), including long QT syndrome (LQTS), short QT syndrome (SQTS), Asian sudden unexplained nocturnal death syndrome (SUNDS), catecholaminergic polymorphic ventricular tachycardia (CPVT), and Brugada syndrome (BrS). Unfortunately, sudden cardiac death (SCD) could be the first symptom of patients affected by IC. Risk stratification and current diagnosis have been primarily focused on clinically detectable changes and abnormalities in the heart structure and function [2,3]. Several known markers are used to predict cardiac arrhythmias, although none are specific for patients affected by inherited channelopathies.

Several risks factors, such as obesity, diabetes, sleep apnea, anorexia nervosa, electrolyte imbalances, and unhealthy food consumption, are associated with these inherited channelopathies (Figure 1). Therefore, it is very important to address these risk factors in order to manage and prevent adverse outcomes. Most of the correlations are well established, and, indeed, these risks factors can be affected by adjusting lifestyle habits, such as a particular diet, impacting the risk of arrhythmogenic events and mortality. Thus, the aim of this review is to present objective insights into different daily diets. Specifically, here, we highlight the effects of various dietary factors and their suspected roles in the development of arrhythmias in BrS and LQTS, providing a summary of current literature and presenting questions that need to be the subject of future studies. This information is desperately needed to advise the patients diagnosed with these syndromes in the clinic.

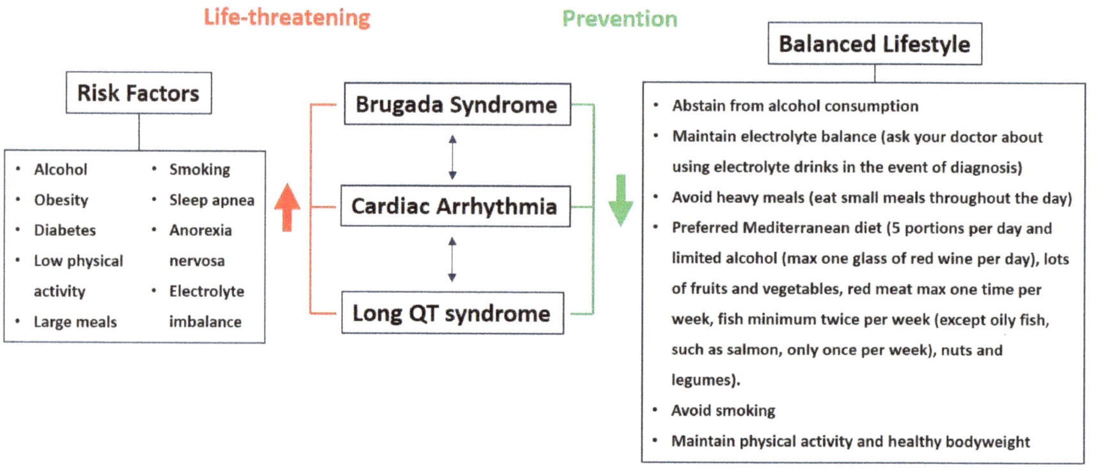

Figure 1. Cardiac arrhythmias, such as Brugada syndrome and long QT syndrome, can be triggered by a variety of factors, but maintaining a certain lifestyle may reduce these risks.

2. Brugada Syndrome and Long QT Syndrome

Brugada syndrome and long QT syndrome are among the most common inherited cardiac arrhythmias, with high risks of malignant arrhythmias, and they present with abnormalities in the 12-lead electrocardiogram (ECG): ST-segment elevation and prolonged QT interval, respectively.

BrS is associated with right ventricular conduction abnormalities and coved-type ST-segment elevation in the right precordial leads of the ECG [4]. The syndrome is clinically characterized by syncope episodes and SCD due to ventricular fibrillation [5]. The majority of the patients are asymptomatic, and they are usually diagnosed by chance due to the dynamicity of the BrS ECG pattern, since it fluctuates throughout the day [6]. Moreover, these fluctuations can be induced by several factors, such as fever, hypercalcemia, diabetes, excessive consumption of food or alcohol, hyperkalemia, and certain drugs [7]. The coved-type ST-segment elevation is diagnostic when the "Type 1" Brugada pattern is seen, which can be found either on a spontaneous ECG or after a drug challenge with a class Ia anti-arrhythmic drug, such as ajmaline, which can provoke the Type 1 pattern so that the syndrome can be discovered [8]. The true incidence of BrS is currently unknown, due to the lack of availability of centers that can perform the drug challenge to unmask the syndrome. This is because these tests must be performed in specialized centers capable of specialized resuscitation due to the risks associated with the drug ajmaline [9], which can potentially induce life-threatening ventricular arrhythmias (VAs). For this reason, although used widely in Europe, the drug ajmaline is not used in the United States of America, due to risks associated with this drug. The BrS is believed to be genetic in nature, described as being oligogenic in nature, although it may occur in a minority of families as a Mendelian condition [10,11].

LQTS is defined by a variable degree of QT prolongation, absence of structural heart diseases, and it can manifest in three different subtypes (LQTS1, LQTS2, and LQTS3). The syndrome is clinically characterized by syncope episodes and SCD due to ventricular tachyarrhythmias [12]. LQTS can be congenital or acquired, therefore, a prolonged QT interval may result from genetic abnormalities, mineral imbalances, or certain medications [13]. It manifests as ECG abnormalities, including prolonged QT-interval, Torsade de Pointes (TdP), and ventricular fibrillation (VF) [14]. LQTS is usually diagnosed by measuring the QT interval on the ECG and it can include adrenalin and isoproterenol test. Moreover, epinephrine test is able to unmask the prolongation of the QT intervals in patients with 'concealed LQTS' [15]. The ECG may result normal at rest but based on the subtype, ECG abnormalities can be triggered by physical stress, emotional stress, dietary changes, certain drugs, or during sleep or at rest [16,17], and it has also been associated with metabolic syndromes and eating disorders [13]. LQTS primarily affects young people and is one of the main causes of SCD in this population [18].

3. BrS ECG Pattern Triggered by Food or Alcohol Intake

Several studies have repeatedly stressed that the BrS ECG pattern can appear after the consumption of alcohol [19,20] or caused by glucose-induced insulin secretion after the ingestion of a meal [21]. A 29-year-old BrS patient experienced several episodes of palpitation and syncope after alcohol consumption [22]. Another case report described a 25-year-old man who experienced three syncope events while consuming alcohol [23]. In yet another study, a 21-year-old man was diagnosed with BrS after he developed two episodes of syncope after the consumption of a high quantity of alcohol; specifically, his ECG, during the syncope, revealed ventricular fibrillation and the elevation of the ST-segment at high intercostal space [24]. In addition, a 56-year-old male experienced cardiac arrest while asleep after the consumption of one bottle of wine. The patient, five years previously, already had an abnormal ECG showing an elevation of the ST-segment, however, unfortunately, this abnormality was not considered important as a diagnostic criteria for BrS [25]. A nine-year-old boy, who was diagnosed with BrS by an ajmaline test, experienced syncope and cardiopulmonary arrest after the ingestion of a large hot dog [26]. Another case report of a 53-year-old man, who had been diagnosed with BrS after four weeks of Ramadan-fasting, had syncope events and several sudden cardiac arrests after the ingestion of a large meal [27].

Glucose and insulin intravenous infusion in patients affected by BrS results in a significant accentuation of the abnormal J-ST configuration [26]. In one study, 75% of BrS patients, compared to the controls, had a higher incidence of ECG fluctuation after an oral glucose tolerance test (OGTT) [28]. Although it is widely understood that large meals can trigger the BrS ECG pattern, the cellular mechanism responsible for the appearance of this ECG pattern by high levels of glucose and/or insulin concentration in BrS is still unclear.

In conclusion, the relationship between large meal intake, alcohol, and the development of the BrS ECG pattern is still unclear. However, it seems that the interplay between them is very relevant and should be investigated further.

4. Cortisol and Sudden Death

The majority of sudden cardiac deaths seem to occur from 4 am to 6 am, so in a period of nocturnal sleep in which cortisol concentration tends to be lower than in other periods of the day [29]. This observation suggests a role for abnormal sympathetic activity, which has been observed in a SUNDS cohort (sudden unexplained death during nocturnal sleep), and SUNDS is still considered to share common genetic causes with BrS [30,31].

Moreover, it is very well known that cortisol can affect the incidence and the clinical manifestations of sudden cardiac death. More recently, the role of the enzyme 11β-hydroxysteroid dehydrogenase (11β-HSD1) has been considered [32]. This enzyme is one of the most promising molecular targets to treat Type 2 diabetes mellitus and its complications [33]. Basically, 11β-HSD1 is able to catalyze the production of cortisol, and the levels of 11β-HSD1 correspond to the levels of cortisol concentration [29]. High concentrations of cortisol have already been demonstrated to be associated with both cardiac arrhythmias and diabetes mellitus [34]. The drug-mediated 11β-HSD1 inhibition alleviates most metabolic abnormalities associated with both diabetes and a cortisol concentration beyond normal levels [32]. The expression of 11β-HSD1 can be induced by high-fat diet in mouse models [35], suggesting the usefulness to assess the activity of 11β-HSD1 among human patients with BrS. Therefore, these mechanisms should be further explored to better understand why the majority of sudden deaths occur during this time period.

5. The Mechanisms behind Food and Alcohol Intake as a Trigger for BrS ECG Pattern Manifestation

The mechanism behind the manifestation of ventricular arrhythmias after ingestion of alcohol or large amounts of food is still uncertain. The main product produced by the metabolism of fatty acids (FA), carbohydrates, ketones, and amino acids [36] is adenosine triphosphate (ATP), which is essential as energy source for cardiac work. The energy metabolic pathways include (1) glycolysis, where ATP is produced by glucose oxidation, (2) the citric acid cycle (Krebs cycle or TCA cycle), where guanosine-triphosphate (GTP), nicotinamide adenine dinucleotide (NADH) and flavin adenine dinucleotide ($FADH_2$) are produced by acetyl-CoA oxidation, (3) the electron-transport chain (ETC), where most of the ATP is produced, and (4) fatty acid beta-oxidation, where FAs breakdown into acetyl-CoA and they are used by the TCA cycle. Acetyl-coenzyme A (CoA) is produced by either oxidative decarboxylation of pyruvate from glycolysis, beta-oxidation of long chain fatty acids, or oxidative degradation of some amino acids [37]. CoA is synthesized within the mitochondria, and it is the main substrate for the TCA cycle, which is a series of enzyme catalyzed reactions that generate energy sources via oxidative reactions. NADH and $FADH_2$ are generated, then they enter into the ETC and, together with oxygen, they generate ATP through a redox reaction. The ATP level in the cells is maintained constant via two mechanisms: the production of ATP by oxidative phosphorylation and the hydrolysis of ATP [38]. In order to sustain normal cardiac activity in a healthy human heart, if there is an increase of fatty acid oxidation, there is also a decrease of pyruvate oxidation, and thus a glucose oxidation, and vice versa [39].

However, dysfunction of glucose metabolism, whether hypoglycemia or hyperglycemia, could have detrimental effects on cardiomyocytes. Several studies describe the deleterious effect of high glucose and alcohol on cardiomyocytes. In addition, the deprivation of glucose results in cardiac myocyte apoptosis [40–42]. A study conducted on rat cardiomyocytes demonstrated that high glucose-induced mitochondrial hyperpolarization increases cell injury [43]. Moreover, an in vitro induction of high glucose resulted in cardiomyocyte apoptosis [44]. Furthermore, several studies have demonstrated that ethanol depresses myocardial contractility both in humans and animals [45,46]. The first enzyme involved in alcohol metabolism is alcohol dehydrogenase (ADH), which catalyzes the conversion of ethanol into acetaldehyde, which is a reactive and toxic product that contributes to the formation of reactive oxygen species (ROS) and reduces the oxidation process in liver cells; the second enzyme involved is aldehyde dehydrogenase 2 (ALDH2), which converts acetaldehyde into acetate, which is then ready to be incorporated into Acetyl CoA and then entered into the TCA cycle. Indeed, a study conducted on 198 Japanese patients affected by BrS demonstrated that arrhythmic events caused by the consumption of alcohol were associated with the increased activity of the alcohol-metabolizing enzyme ADH1B in BrS patients [47].

BrS is a complex disease that has been described to have an oligogenic model of inheritance [10,48], and several possible genetic causes have been proposed, reviewed elsewhere [10,49]. It has also been hypothesized that, when people are overstressed by large meals and alcohol, especially during festivities, possible mutations in genes encoding for SULT1A enzymes result in the inability of those enzymes to deactivate catecholamines in the intestine, possibly inducing cardiac arrhythmia [50]. Finally, another hypothesis could be that glucose metabolism dysfunction can interfere with the homeostasis of ATP and ROS within the cardiomyocytes [51], possibly leading to arrhythmic events via mitochondria defects and impaired intracellular cation homeostasis.

Adhering to the "Mediterranean diet" may reduce the risk of cardiovascular disease. However, unfortunately, this diet is not clearly defined. However, generally, it consists of eating smaller portions throughout the day, such as eating five smaller meals instead of three larger ones. The Mediterranean diet could allow for max one glass of red wine per day, but patients with cardiac arrhythmias are probably best to abstain completely from alcohol, especially if even low amounts of alcohol make them feel unwell. In our experience, even one glass of alcohol can make some patients with BrS fell unwell. In these cases, it is best to avoid alcohol completely. The Mediterranean diet includes a lot of plant-based foods and olive oil, but tends to be low in saturated fat, meat, and dairy products. Red meat could be eaten once per week. Fish are generally included minimum twice per week, but oily fish, such as salmon, which are larger fish and could contain a higher amount of mercury, should be limited to only once per week, although it is still good because it contains a high amount of omega-3 fatty acids. Nuts are also good for omega fatty acids and legumes are good for proteins.

6. Sudden Cardiac Death and QT Prolongation Triggered by Ketogenic Diet

While carbohydrates provide a readily available fuel for our body, fats and oils (lipids) are considered our primarily source of stored energy. Fat enters into the body through food and breaks down into triglycerides, and then into fatty acids and glycerol. Mitochondria provide the main source of energy, and, therefore, the dysfunction of the two metabolic pathways of β-oxidation of fatty acids (FAO) and oxidative phosphorylation (OXPHOS) can lead to mitochondrial remodeling and the manifestation of heart failure, arrhythmias, and ventricular hypertrophy [52,53].

Acute arrhythmic events associated with a ketogenic diet and calorie-restricted diets can include QT prolongation, leading to sudden cardiac death, and these phenomena have been described in children, adolescents, and adults [54–56]. A ketogenic diet is characterized by very low carbohydrate intake, with 75% of calories derived from fat. This carbohydrate (CHO) restriction helps to reduce blood glucose and insulin [57]. However, it is known that, among several complications related to a ketogenic diet, there is selenium deficiency, QT prolongation, and SCD [58–60]. A case report described a Torsades de Point (TdP) event in a patient with a dual-chamber implantable cardioverter-defibrillator (ICD) affected by LQT2 triggered by an uncommon factor: the ketogenic diet. Specifically, while the patient was following a ketogenic diet, she experienced four episodes of ventricular fibrillation, due to TdP, over the course of only three weeks. Selenium, ketone bodies, and alcohol levels were all within normal limits. One month later, she stopped the ketogenic diet, and the ICD showed no further arrhythmic episodes [17].

The majority of SCD cases of pediatrics patients are associated with QT interval prolongation and ketosis. In a report about two cases of death in two children on a ketogenic diet for seizure control, both patients experienced QT prolongation and suffered from selenium deficiency [58]. However, another case described a correlation between a QT interval prolongation and ketogenic diets in the absence of electrolyte imbalance in children [60]. Specifically, a direct correlation was observed between QT interval prolongation and β-hydroxybutyrate concentrations, and between QT interval prolongation and systemic acidosis [60]. Interestingly, in a study of 70 children with drug-resistance epilepsy, receiving a ketogenic diet for a 12-month period, no deleterious effects on corrected QT interval, QT dispersion, and Tp-e interval were reported [61]. Finally, another case described a five-year-old boy who developed selenium deficiency, acute reversible cardiomyopathy and ventricular tachycardia with prolonged QT interval after following a ketogenic diet to treat refractory epilepsy. Then, his clinical status improved and got back to normal after selenium supplementation [62].

In conclusion, the extended association between QT interval prolongation and/or SCD and ketosis conditions/ketogenic diet has been described, and it raises the question of whether ketosis may directly affect cardiac repolarization.

7. The Mechanisms behind Ketosis as a Trigger for the Manifestation of QT Prolongation

The relationship behind the manifestation of ventricular arrhythmias and high-fat and low-carbohydrate diet is still unclear. However, it is known that the heart uses fatty acids as a main substrate for source energy, but it can also use ketone bodies. Ketone bodies are produced by breaking down fatty acids and ketogenic amino acids in a process called ketogenesis, previously reviewed elsewhere [63]. Briefly, ketogenesis involves the anabolic hormone insulin, and the catabolic hormones glucagon, cortisol, catecholamines, and growth hormone, of which insulin and glucagon are considered the most crucial for this pathway [64,65] (Figure 2). Ketone body concentration is lowered by insulin, which promotes glucose uptake and oxidation. The reduction of circulating insulin levels is the principal triggering event for accelerating ketogenesis [63]. Insulin acts on the adipose tissue, liver, and the periphery; specifically, the low amount of insulin and a high amount of glucagon in our blood stream trigger the augmentation of free fatty acids (FFAs), increase uptake of FFAs into the mitochondria, and increase production of ketones in the liver, by activating the acyltransferase system through the inhibition on malonyl-CoA synthesis. Indeed, the ketogenesis pathway occurs primarily in the mitochondria of hepatocytes. FFAs are converted into fatty acyl CoA (Acyl CoA), which enters into the hepatic mitochondria through CPT1-mediated transport. Then Acyl CoA undergoes β-oxidation to produce Acetyl CoA, which is only employed to generate ATP if there is enough oxaloacetate. When carbohydrate intake is limited, such as in the ketogenic diet, the liver uses the majority of oxaloacetate to produce glucose through gluconeogenesis; therefore, the liver diverts the acetyl CoA to form ketone bodies. Thiolase enzyme (acetyl coenzyme A acetyltransferase (ACAT)) catalyzes the reversible reaction where two molecules of acetyl CoA are combined to generate acetoacetyl CoA. At this point, mitochondrial β-Hydroxy β-methylglutaryl-CoA (*HMG*-CoA) synthase catalyzes a condensation reaction by adding an extra acetyl CoA molecule onto the acetoacetyl CoA. Then, the enzyme *HMG*-CoA lyase cleaves the *HMG*-CoA, which releases CoA and forms acetoacetate, the metabolized ketone body. Within the mitochondrial matrix, 3-hydroxybutyrate dehydrogenase can reduce the acetoacetate into two other ketone bodies: acetone and β-hydroxybutyrate (β-HB), through non-enzymatic decarboxylation or by beta-hydroxybutyrate dehydrogenase, respectively. Moreover, the ratio of $NADH/NAD^+$ helps to maintain an equilibrium between acetoacetate and β-HB within the mitochondria matrix. Both acetoacetate and β-HB are considered fuel molecules normally found in the heart and renal cortex. Acetone cannot be metabolized. At this point, due to the fact that the liver does not have enzyme beta ketoacyl-CoA transferase, and it cannot utilize ketone bodies [66], acetoacetate and β-HB reach the extrahepatic tissues. β-HB is converted into acetoacetate, which is then converted back to acetyl-CoA. Acetyl-CoA goes through the TCA and produces 22 ATP by oxidative phosphorylation. Therefore, due to the acid nature of ketone bodies, this causes an anion gap metabolic acidosis. This condition usually results in electrolyte imbalances, especially a reduction in K^+, Mg^{2+}, and P.

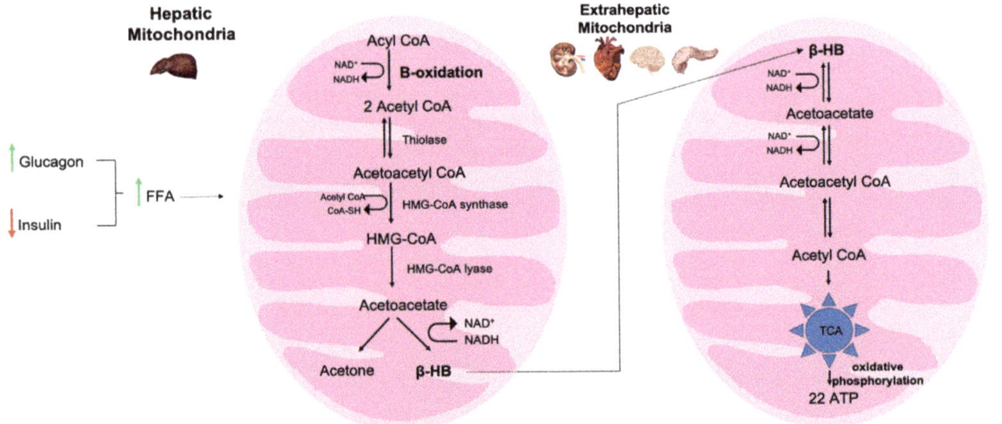

Figure 2. Ketogenesis pathway in hepatic mitochondria. Low amounts of insulin and high amounts of glucagon in our blood stream trigger the augmentation of free fatty acids (FFAs), and increase uptake of FFAs into the mitochondria. Acetyl CoA is generated from beta-oxidation of Acyl CoA, which derives from FFAs. The production of ketones (acetoacetate, acetone, β-HB) in the liver is then increased.

A metabolomic study on individuals with arrhythmogenic cardiomyopathy (AC) identified as a possible biomarker β-HB, due to its elevated amount in the plasma and hearts [67]. Specifically, the β-HB produced by cardiomyocytes of AC patients is released into the blood, and its levels are significantly higher compared to controls [67]. Therefore, it was demonstrated that cardiac ketogenesis occurs in CA, and β-HB may be used as a potential metabolic marker to predict CA.

8. Oxidative Stress

Oxidative stress is usually considered a state in which the production of ROS and antioxidant defenses are not balanced [68]. ROS are derivatives of molecular oxygen, such as superoxide (O_2^-), hydrogen peroxide (H_2O_2), peroxynitrite ($ONOO^-$), and hydroxyl radicals (OH) [69]. ROS are mainly produced by mitochondria, and their homeostasis is maintained by the enzyme glutathione peroxidase (GSH-Px). In damaged mitochondria, the calcium ions are overloaded and drive the augmentation of ROS concentration, which leads to excitotoxicity damage [70]. Indeed, excessive generation of ROS, impaired calcium homeostasis, and diminished ATP production directly impact mitochondrial function [71].

Repetitive or prolonged oxidative stress can damage proteins and lipids within the cell, and it might result in a contractile dysfunction, a downregulation of gene expression, and also a disrupted energy transfer, which could induce cardiomyocyte apoptosis, followed by heart failure [72]. There have been limited studies into the mechanisms linking oxidative stress and arrhythmias, but it has been shown that cardiac conditions with increased arrhythmic risks are associated with an unbalanced production of ROS [73]. ROS, in addition to their role as a messenger in cell signal transduction and the cell cycle, regulate both cellular metabolism and ion homeostasis in excitable cells. An elevated presence of ROS within the cells can be highly toxic, and they can lead to arrhythmogenic triggers, such as alterations of ion channels (Na, Ca^{2+}, and K^+), dysfunction of the mitochondria, and gap junction remodeling [74]. Therefore, an excessive production of ROS and ineffective ROS scavenging can culminate in cell death [52].

It is well known that ketone bodies are always present in the blood, and their concentration increases during prolonged exercise and fasting. In vitro studies demonstrated that KBs stimulate insulin release [75,76], cause lipid peroxidation, and generate oxygen radicals [77]. It is also known that KBs are able to reduce oxidative stress, due to the activation of multiple protective antioxidant pathways. However, we hypothesize that patients affected by inherited channelopathies could have a potential dysfunction of these pathways, and they are not able to reduce the levels of ROS. To sustain our hypothesis, there are some studies suggesting that exposure to high concentrations of KBs could provoke oxidative stress. A study conducted on calf hepatocytes suggested that both β-HB and acetoacetic acid decrease the activity of antioxidant enzymes superoxide dismutase (SOD), catalase (CAT), and glutathione peroxidase (GSH-Px), and increase malondialdehyde (MDA) and nitric oxide (NO), which are markers of oxidative stress [78,79]. Another study attributed acetoacetate treatment to the activation of mitogen-activated protein kinase (MAPK) pathway, which is known to be activated by oxidative stress in rat hepatocyte cells [80]. Therefore, even if there are several studies associating KBs with the inhibition of oxidative stress and ROS production, there are still some studies showing a correlation between KBs and the induction of oxidative stress.

It is known that a nutrient overload helps to release free fatty acids and can also induce damages to the mitochondria [81]. Therefore, free fatty acids might be related to the excess production of oxidative stress. Indeed, there is a lot of evidence suggesting that conditions of high levels of glucose, lipids, or their combination can interfere with mitochondria metabolism, and they may modulate mitochondrial ATP synthesis capacity and increase ROS production [71]. Moreover, in patients with inherited channelopathies, malignant ventricular arrhythmias and SCD occur before overt structural changes of the heart. To prevent the arrhythmogenic substrate progression, more studies to understand the electrical instability of cardiomyocytes are needed. At present, a lot of studies are focused on the connection between metabolic disease and the manifestation of arrhythmic events. Specifically, the attention is focused on the mitochondrial dysfunction, which can drive the manifestation of arrhythmic events by interfering with the electrical activity of the cardiomyocytes. The role of mitochondrial dysfunction in inherited channelopathies is still unclear and will be the subject of future studies.

9. Vitamin D

Vitamin D, also known as calciferol or hydroxyvitamin D (25-OHD), is a fat-soluble vitamin that is naturally found in a few foods and is also produced endogenously by the exposure of the skin to sunlight ultraviolet rays [82]. Vitamin D is responsible to increase the intestinal absorption of calcium and to maintain the serum calcium and phosphate concentration [82]. Vitamin D is also involved in the reduction of inflammation, cell growth, neuromuscular, immune function, and glucose metabolism [83–85].

Vitamin D deficiency has been related to different cardiovascular disorders, including SCD [86]. Moreover, a decreased level of 25-OHD has been linked to structural and ionic channel remodeling, which may increase arrhythmic events [87]. Indeed, prolonged QTc is commonly induced by hypocalcemia, which can be caused by vitamin D inadequacy or resistance.

In a case of a patient with severe vitamin D deficiency, hypocalcemia and prolonged QTc resulted in TdP and cardiac arrest. After the administration of vitamin D and calcium supplements, the QTc interval became normal, and the patient did not experience additional arrhythmic events [88]. A hypocalcemic teenage girl affected by hypoparathyroidism experienced a few episodes of syncope during exercise, and the ECG on admission showed prolonged QTc. After the treatment with alphacalcidol, which is an analog of vitamin D, and calcium supplements, the QTs of the patient became normal [89]. Moreover, in a case of a 40-year-old woman, who followed a vegan diet, she was affected by hypocalcemia, due to severe deficiency of vitamin D. The patient manifested symptoms of palpitations, presyncope, and a long QT (556 ms). Therefore, she was first treated with calcium gluconate, then by vitamin D and calcium oral supplementation. After the treatment, the QTc normalized, and the symptoms disappeared [90].

Vitamin D deficiency can not only induce the prolongation of the QTc, but it can also promote inflammatory reactions. It is known that cardiac contraction is affected by an overload of Ca^{2+} ions in myocardial cells. Studies have hypothesized that lack of vitamin D could interfere with the function of Ca^{2+} in myocardial cells; specifically, it can induce hypertrophy [91], increase anti-inflammatory cytokines [92], increase fibrosis [93], and impact the production of ROS in the atria [94].

The correlation between atrial fibrillation and vitamin D is poorly understood. Some studies suggest a direct correlation between atrial fibrillation and vitamin D deficiency [95–97]. However, other studies did not find a connection between vitamin D levels and the incidence of atrial fibrillation [98–100]. It would be important to better understand the link between vitamin D and calcium, because it is known that calcium overload causes myocytes apoptosis and cardiac failure [101,102]. Oxidative stress associated with cardiovascular risks factors is real, and, therefore, further studies are needed to investigate this aspect.

In conclusion, vitamin D deficiency appears to play an important role in CVD, and it is important to investigate the possible role of vitamin D deficiency in the development of arrhythmic events.

10. Omega-3 Fatty Acids

Omega-3 fatty acids, also called ω-3 (n-3) fatty acids, are polyunsaturated fatty acids (PUFAs), and they are essential nutrients involved in lipid metabolism. It has been demonstrated that n-3 PUFAs, including eicosapentaenoic acid (EPA), docosahexaenoic acid (DHA), and α-linolenic acid (ALA), play an important role in the human diet and cellular physiology, and they may have beneficial effects against CVD and risks factors, including arrhythmias, probably by the modulation of cardiac ion channels [103]. DHA and EPA fatty acids are mostly found in seafood, seaweed, and algae, while ALA is in nuts and seeds. ALA might improve cardiac function by inhibiting apoptosis through anti-inflammatory and anti-oxidative stress effects in diabetic, but not normal, rats [104].

An investigation into whether a diet enriched with fish and PUFAs could be associated with changes in QT duration on a resting ECG showed that long-term consumption of fish could positively influence the duration of QTc by lowering it; therefore, fish intake may be considered an antiarrhythmic protection [105]. A retrospective study of men affected by BrS evaluated the correlation between the serum levels of EPA and DHA and the risks factors for SCD. A multivariate logistic regression analysis showed that low levels of EPA and DHA were linked to the incidence of syncope in patients affected by BrS. This same study suggested that all levels of omega-3 PUFAs may play an important role in preventing ventricular fibrillation in BrS [106]. Furthermore, a study on 123 Langendorff-perfused rabbit hearts, used to mimic LQT2 and LQT3 syndrome, showed that PUFAs were able to prevent TdP by reverting the AP prolongation. This effect was stronger in LQT2 than in LQT3 syndrome, and the antitorsadogenic effect was more distinctive with DHA and EPA compared with ALA [107].

A recent review demonstrated that n-3 PUFAs have no significant effect on mortality or cardiovascular health [108]. For example, in a study investigating the correlation between n-3 PUFAs from fish and risks of CVD, including SCD, PUFAs were inversely related to QTc and JTc intervals. However, QTc and JTc did not reduced the inverse relationship between n-3 PUFAs and SCD risks, suggesting that this association cannot explain the prevention of prolonged ventricular repolarization [109]. Moreover, in a study investigating the association between mercury (Hg), EPA, and DHA, and large seafood consumption and heart rate variability (HRV) and QT interval duration, the authors found a possible association between specific seafood types and arrhythmias, such as tuna steak with QTc and anchovies with HRV [110]. Finally, data from four trials suggested that a high dose (4.0 g/d) of n-3 PUFAs could increase the risk of atrial fibrillation development [111].

In conclusion, data on the effects of omega-3 are still inconsistent, requiring further studies to assess their beneficial effects on preventing arrhythmic events.

11. Arrhythmogenic Ingredients

Several factors can increase the risk of the development of arrhythmias in BrS and LQTS. Although not well studied in these two specific syndromes, one of the factors that could result in arrhythmias is the consumption of specific foods, which are considered to be arrhythmogenic [112,113], as described in Table 1, which lists both arrhythmogenic and anti-arrhythmogenic dietary factors.

Several case reports have described patients experiencing arrhythmic events and cardiac arrests after the consumption of energy drinks [114–117]. Taurine has been shown to modulate ion channel activity by suppressing the activity of sodium, calcium, and potassium channels [118]. Moreover, it shortens the action potential duration and decelerates the rate of terminal repolarization of the cardiac action potential, inducing atrial and ventricular arrhythmias or cardiac arrest [116]. Furthermore, atrial and ventricular arrhythmias can also be induced by caffeine, which is able to interfere with calcium homeostasis [115,119] by increasing intracellular calcium concentration [115]. Indeed, it was demonstrated that caffeine is able to stimulate calcium release from the sarcoplasmic reticulum [120], and calcium imbalances, particularly sarcoplasmic reticulum calcium stores, may be altered in BrS [121–123]. A 24-year-old male developed arrhythmias and collapsed after the ingestion, for the first time, of a small quantity of an energy drink combined with alcohol; specifically, the drink contained 80 mg of caffeine and 1000 mg of taurine, plus Vodka. He was then diagnosed with BrS [116]. Another case report showed that energy drinks could have triggered an abnormal QT response in a 13-year-old female affected by LQTS1. Moreover, a 22-year-old female affected by LQTS1 experienced cardiac arrest after the consumption of an elevated quantity of energy drink [124].

Grapefruit has been identified as an inhibitor of drug-induced TdP and QTc prolongation, thus enhancing their pharmacokinetics [113]. Specifically, naringenin, which is a flavonoid found in grapefruit, is able to block the hERG channel and induce TdP and/or QTc prolongation [113,125]. These hERG channels are known receptors of the drug ajmaline, which is used to provoke the diagnostic type-1 BrS ECG pattern [126]. Therefore, other ingredients could be suspected to be arrhythmogenic, because they contain flavonoids: citrus fruit, parsley, onions, berries, bananas, red wine, chocolate, grains, nuts, tea, coffee, and various other fruits and vegetables [127–129], including spinach, cauliflower, broccoli, black beans, and chickpeas [130–134]. Moreover, additional ingredients like lemon, lime, clementine, oranges, and bergamot oil could be considered a risk, due to the same organic compounds of grapefruit: furanocoumarins [129]. However, other studies have suggested that flavonoids could actually be beneficial for cardiovascular health by reducing inflammation. Flavonoids, polyphenolic compound derivatives, reduce inflammation and risk of cardiovascular disease by reducing NFκB and its resulting transcription factors involved in the inflammatory pathway [128]. Many other natural products, which have polyphenols as their major compound, have been shown to have anti-inflammatory effects, such as mushrooms, honey, plant extracts, plant juices, plant powders, and essential oils [128,135,136]. Thus, further studies are needed to understand these foods and the resulting effects, in order to better advise the patients about dietary supplements or restraints.

Table 1. Arrhythmogenic and anti-arrhythmogenic dietary factors.

Possible Dietary Triggers Involved in Cardiac Arrhythmia	
Pro-Arrhythmic	**Anti-Arrhythmic**
Alcohol [22–24,137]	Vitamin D, E, C [138]
Heavy Meals [21,27,139]	Carotenoids [138]
Energy Drinks [114–117]	Resveratrol [140]
Flavonoids (found in grapefruit [125], citrus fruit [113,125,129], parsley [129], onions, berries [129], bananas, red wine [129], chocolate [129], grains, nuts, tea, coffee, spinach, cauliflower, and broccoli)	Electrolytes [141]
Licorice [142]	Omega–3 [103,105,106]
Honey [113,135] §	Honey [127,128,135,136] §
Mushrooms [113] §	Mushrooms [136] §

§ Honey and mushrooms have been described as both arrhythmogenic and anti-arrhythmogenic, depending on the study.

12. Ingredients That Suppress Cardiac Arrhythmogenesis

An interesting review showed that 18 active ingredients such as alkaloids, flavonoids, saponins, quinones, and terpenes, Wenxin-Keli, and Shensongyangxin could have antiarrhythmic effects [143]. In particular, the Chinese herb extract, Wenxin-Keli (WK), has been reported as an effective treatment of atrial and ventricular fibrillation by inhibiting the transient K^+ outward current (I_{to}) [144]. The experiments were conducted in a canine experimental model of BrS, and the authors observed an inhibition of I_{to} and indirect adrenergic sympathomimetic effects using WK in combination with quinidine [144]. However, the 18 ingredients described in the review, with the exception of omega 3, have been tested in vitro or in animal models as natural drugs or in combination with other antiarrhythmic treatments, and not as ingredients for regular meals. Therefore, further studies are needed.

Another ingredient with the potential to prevent antiarrhythmic events, in moderate doses, is resveratrol, a stilbenoid polyphenol found in grapes. It has been described as a potential inhibitor of intracellular calcium release able to eliminate calcium overload in AF, and, therefore, able to preserve the cardiomyocyte contractile function [140]. Finally, based on an interesting review on antioxidant therapies for atrial fibrillation [138], it would be interesting to further study the role of vitamin C, vitamin E, and carotenoids, found in fruits and vegetables, and their role as suppressors of ROS, and, therefore, their role as anti-arrhythmic nutrients.

13. Electrolytes

Food is the main source of electrolytes, essential minerals for life. Electrolytes are required for maintaining osmotic pressure in cells and generating action potentials in nerves and muscles. In particular, sodium, calcium, potassium, and magnesium play an important role in the heart. Indeed, it has been demonstrated that electrolyte imbalance can have detrimental effects on the heart, such as triggering cardiac arrhythmias or cardiac arrest [141], including what has been coined "BrS phenocopies" [145], which may provide important insights into the mechanisms involved in BrS [11]. It is important to understand better the connection between food intake, electrolyte imbalance, and BrS or LQTS.

14. Vagal Tone Activity and Arrhythmic Events

The heart rhythm is regulated by cardiac parasympathetic (vagal) nerves, the sympathetic nerves, and the pacemaker cells [121]. Autonomic activity can influence the elevation of the ST-segment [146–148]. Indeed, late in the 1990s, it was shown that the nocturnal vagal activity may be involved in the cardiac arrhythmic events of BrS [149]. Moreover, the high vagal activity could lead to the manifestation of ventricular tachyarrhythmias in patients with BrS or LQTS [150,151].

The relationship between the autonomic modulation and arrhythmic events is very complex and still unclear. However, based on the studies presented in this review, the association between the vagal activity and the consumption of large meals may contribute to the manifestation of arrhythmic events.

15. Conclusions

We describe herein the effects of dietary factors in patients affected by cardiac arrhythmias, specifically BrS and LQTS. To date, the importance of understanding the relationship between diet and inherited channelopathies has been underrated. It is evident that dietary factors can influence the risk of the development of arrhythmic events. Therefore, we recommend eating and drinking small portions throughout the day and to try to limit certain types of ingredients in order to prevent arrhythmic events. Modifying the diet might not be enough to fully prevent arrhythmias, but it can help lower the risk.

Author Contributions: S.D. conceived and drafted the paper. M.M.M. revised the paper. M.M.M., E.M. and G.N. provided useful feedback. C.P. secured funding for the project. All authors have read and agreed to the published version of the manuscript.

Funding: This study was partially supported by Ricerca Corrente funding from Italian Ministry of Health to IRCCS Policlinico San Donato.

Conflicts of Interest: The authors declare no conflict of interest. The funders had no role in the design of the study; in the collection, analyses, or interpretation of data; in the writing of the manuscript, or in the decision to publish the results.

References

1. Joseph, P.; Leong, D.; McKee, M.; Anand, S.S.; Schwalm, J.-D.; Teo, K.; Mente, A.; Yusuf, S. Reducing the Global Burden of Cardiovascular Disease, Part 1: The Epidemiology and Risk Factors. *Circ. Res.* **2017**, *121*, 677–694. [CrossRef]
2. Calkins, H.; Corrado, D.; Marcus, F. Risk Stratification in Arrhythmogenic Right Ventricular Cardiomyopathy. *Circulation* **2017**, *136*, 2068–2082. [CrossRef] [PubMed]
3. Calkins, H. Arrhythmogenic right ventricular cardiomyopathy: Evaluation of the current diagnostic criteria and differential diagnosis. *Eur. Heart J.* **2019**. [CrossRef]
4. Brugada, J.; Campuzano, O.; Arbelo, E.; Sarquella-Brugada, G.; Brugada, R. Present Status of Brugada Syndrome: JACC State-of-the-Art Review. *J. Am. Coll. Cardiol.* **2018**, *72*, 1046–1059. [CrossRef] [PubMed]

5. Wilde, A.A.M.; Antzelevitch, C.; Borggrefe, M.; Brugada, J.; Brugada, R.; Brugada, P.; Corrado, D.; Hauer, R.N.W.; Kass, R.S.; Nademanee, K.; et al. Proposed Diagnostic Criteria for the Brugada Syndrome: Consensus Report. *Circulation* **2002**, *106*, 2514–2519. [CrossRef] [PubMed]
6. Antzelevitch, C.; Fish, J.M. Therapy for the Brugada syndrome. *Handb. Exp. Pharmacol.* **2006**, *171*, 305–330. [CrossRef]
7. Hedley, P.; Jãrgensen, P.; Schlamowitz, S.; Moolman-Smook, J.; Kanters, J.; Corfield, V.A.; Christiansen, M. The genetic basis of Brugada syndrome: A mutation update. *Hum. Mutat.* **2009**, *30*, 1256–1266. [CrossRef]
8. Pappone, C.; Brugada, J.; Vicedomini, G.; Ciconte, G.; Manguso, F.; Saviano, M.; Vitale, R.; Cuko, A.; Giannelli, L.; Calovic, Z.; et al. Electrical Substrate Elimination in 135 Consecutive Patients with Brugada Syndrome. *Circ. Arrhythm. Electrophysiol.* **2017**, *10*, e005053. [CrossRef]
9. Ciconte, G.; Monasky, M.M.; Vicedomini, G.; Borrelli, V.; Giannelli, L.; Pappone, C. Unusual response to ajmaline test in Brugada syndrome patient leads to extracorporeal membrane oxygenator support. *Eur. Eur. Pacing Arrhythm. Card. Electrophysiol. J. Work. Groups Card. Pacing Arrhythm. Card. Cell. Electrophysiol. Eur. Soc. Cardiol.* **2019**, *21*, 1574. [CrossRef]
10. Monasky, M.M.; Micaglio, E.; Ciconte, G.; Pappone, C. Brugada Syndrome: Oligogenic or Mendelian Disease? *Int. J. Mol. Sci.* **2020**, *21*, 1687. [CrossRef]
11. Monasky, M.M.; Micaglio, E.; Locati, E.T.; Pappone, C. Evaluating the Use of Genetics in Brugada Syndrome Risk Stratification. *Front. Cardiovasc. Med.* **2021**, *8*, 652027. [CrossRef]
12. Moss, A.J.; Robinson, J. Clinical features of the idiopathic long QT syndrome. *Circulation* **1992**, *85*, I140–I144. [PubMed]
13. D'Imperio, S.; Monasky, M.M.; Micaglio, E.; Negro, G.; Pappone, C. Early Morning QT Prolongation During Hypoglycemia: Only a Matter of Glucose? *Front. Cardiovasc. Med.* **2021**, *8*, 688875. [CrossRef] [PubMed]
14. Crotti, L.; Celano, G.; Dagradi, F.; Schwartz, P.J. Congenital long QT syndrome. *Orphanet J. Rare Dis.* **2008**, *3*, 18. [CrossRef]
15. Kim, Y.G.; Oh, S.-K.; Choi, H.Y.; Choi, J.-I. Inherited arrhythmia syndrome predisposing to sudden cardiac death. *Korean J. Intern. Med.* **2021**, *36*, 527–538. [CrossRef] [PubMed]
16. Marzuillo, P.; Benettoni, A.; Germani, C.; Ferrara, G.; D'Agata, B.; Barbi, E. Acquired Long QT Syndrome. *Pediatr. Emerg. Care* **2014**, *30*, 257–261. [CrossRef]
17. Sudhakaran, S.; Yazdani, L.; Wheelan, K.R.; Rao, P.K. The ketogenic diet and the QT interval. *Bayl. Univ. Med Cent. Proc.* **2019**, *33*, 77–79. [CrossRef]
18. Saul, P.J.P.; Gajewski, K.K. Sudden cardiac death in children and adolescents (excluding Sudden Infant Death Syndrome). *Ann. Pediatr. Cardiol.* **2010**, *3*, 107–112. [CrossRef]
19. Antzelevitch, C.; Brugada, P.; Borggrefe, M.; Brugada, J.; Brugada, R.; Corrado, D.; Gussak, I.; Lemarec, H.; Nademanee, K.; Pérez-Riera, A.R.; et al. Brugada Syndrome: Report of the Second Consensus Conference. *Heart Rhythm.* **2005**, *2*, 429–440. [CrossRef]
20. Antzelevitch, C.; Brugada, P.; Borggrefe, M.; Brugada, J.; Brugada, R.; Corrado, D.; Gussak, I.; Lemarec, H.; Nademanee, K.; Pérez-Riera, A.R.; et al. Brugada syndrome: Report of the second consensus conference: Endorsed by the Heart Rhythm Society and the European Heart Rhythm Association. *Circulation* **2005**, *111*, 659–670. [CrossRef]
21. Nishizaki, M.; Sakurada, H.; Mizusawa, Y.; Niki, S.; Hayashi, T.; Tanaka, Y.; Maeda, S.; Fujii, H.; Ashikaga, T.; Yamawake, N.; et al. Influence of Meals on Variations of ST Segment Elevation in Patients with Brugada Syndrome. *J. Cardiovasc. Electrophysiol.* **2007**, *19*, 62–68. [CrossRef] [PubMed]
22. Shimada, M.; Miyazaki, T.; Miyoshi, S.; Soejima, K.; Hori, S.; Mitamura, H.; Ogawa, S. Sustained Monomorphic Ventricular Tachycardia in a Patient with Brugada Syndrome. *Jpn. Circ. J.* **1996**, *60*, 364–370. [CrossRef] [PubMed]
23. Ohkubo, K.; Nakai, T.; Watanabe, I. Alcohol-induced ventricular fibrillation in a case of Brugada syndrome. *EP Eur.* **2013**, *15*, 1058. [CrossRef] [PubMed]
24. Iizuka, C.; Sato, M.; Kitazawa, H.; Ikeda, Y.; Okabe, M.; Kugiyama, K.; Aizawa, Y. A case of brugada syndrome presenting with ventricular fibrillation storm and prominent early repolarization. *J. Electrocardiol.* **2016**, *49*, 564–568. [CrossRef]
25. Achaiah, A.; Andrews, N. Intoxication with alcohol: An underestimated trigger of Brugada syndrome? *JRSM Open* **2016**, *7*. [CrossRef]
26. Ozyilmaz, I.; Akyol, B.; Ergul, Y. Sudden Cardiac Arrest While Eating a Hot Dog: A Rare Presentation of Brugada Syndrome in a Child. *Pediatrics* **2017**, *140*, e20162485. [CrossRef]
27. Talib, S.; Van De Poll, S.W. Brugada syndrome diagnosed after Ramadan. *Lancet* **2013**, *382*, 100. [CrossRef]
28. Nishizaki, M.; Sakurada, H.; Ashikaga, T.; Yamawake, N.; Fujii, H.; Arita, M.; Isobe, M.; Hiraoka, M. Effects of glucose-induced insulin secretion on ST segment elevation in the Brugada syndrome. *J. Cardiovasc. Electrophysiol.* **2003**, *14*, 243–249. [CrossRef]
29. Kanki, M.; Young, M.J. Corticosteroids and circadian rhythms in the cardiovascular system. *Curr. Opin. Pharmacol.* **2021**, *57*, 21–27. [CrossRef]
30. Mongkhonsiri, P.; Tong-Un, T.; Wyss, J.M.; Roysommuti, S. Blunted Nighttime Sympathetic Nervous System Response to Stress among Thai Men with Positive Family History of Sudden Unexplained Nocturnal Death Syndrome. *Int. Hear. J.* **2019**, *60*, 55–62. [CrossRef]
31. Vatta, M.; Dumaine, R.; Varghese, G.; Richard, T.A.; Shimizu, W.; Aihara, N.; Nademanee, K.; Brugada, R.; Brugada, J.; Veerakul, G.; et al. Genetic and biophysical basis of sudden unexplained nocturnal death syndrome (SUNDS), a disease allelic to Brugada syndrome. *Hum. Mol. Genet.* **2002**, *11*, 337–345. [CrossRef]
32. Azmi, N.M.; Juliana, N.; Azmani, S.; Effendy, N.M.; Abu, I.; Teng, N.M.F.; Das, S. Cortisol on Circadian Rhythm and Its Effect on Cardiovascular System. *Int. J. Environ. Res. Public Health* **2021**, *18*, 676. [CrossRef] [PubMed]
33. Gregory, S.; Hill, D.; Grey, B.; Ketelbey, W.; Miller, T.; Muniz-Terrera, G.; Ritchie, C.W. 11β-hydroxysteroid dehydrogenase type 1 inhibitor use in human disease-a systematic review and narrative synthesis. *Metabolism* **2020**, *108*, 154246. [CrossRef] [PubMed]

34. Koracevic, G.; Mićić, S.; Stojanović, M.; Zdravković, M.; Kostić, T.; Koraćević, M.; Kutlešić, M.; Pavlović, M.; Dakić, S. High likelihood for atrial fibrillation in Cushing's syndrome. *Eur. Rev. Med. Pharmacol. Sci* **2020**, *24*, 1391–1397. [CrossRef] [PubMed]
35. Densmore, V.S.; Morton, N.M.; Mullins, J.J.; Seckl, J.R. 11 beta-hydroxysteroid dehydrogenase type 1 induction in the arcuate nucleus by high-fat feeding: A novel constraint to hyperphagia? *Endocrinology* **2006**, *147*, 4486–4495. [CrossRef] [PubMed]
36. Kolwicz, S.C.; Purohit, S.; Tian, R. Cardiac Metabolism and its Interactions with Contraction, Growth, and Survival of Cardiomyocytes. *Circ. Res.* **2013**, *113*, 603–616. [CrossRef] [PubMed]
37. Bhagavan, N.V.; Ha, C.-E. *Essentials of Medical Biochemistry: With Clinical Cases*, 2nd ed.; Elsevier: Boston, MA, USA, 2015.
38. Stanley, W.C.; Recchia, F.A.; Lopaschuk, G.D. Myocardial Substrate Metabolism in the Normal and Failing Heart. *Physiol. Rev.* **2005**, *85*, 1093–1129. [CrossRef]
39. Lopaschuk, G.D.; Ussher, J.R.; Folmes, C.; Jaswal, J.S.; Stanley, W.C. Myocardial Fatty Acid Metabolism in Health and Disease. *Physiol. Rev.* **2010**, *90*, 207–258. [CrossRef]
40. Bialik, S.; Cryns, V.L.; Drincic, A.; Miyata, S.; Wollowick, A.L.; Srinivasan, A.; Kitsis, R.N. The Mitochondrial Apoptotic Pathway Is Activated by Serum and Glucose Deprivation in Cardiac Myocytes. *Circ. Res.* **1999**, *85*, 403–414. [CrossRef]
41. Zhang, Z.; Yu, B.; Tao, G.-Z. Apelin protects against cardiomyocyte apoptosis induced by glucose deprivation. *Chin. Med. J.* **2009**, *122*, 2360–2365.
42. Sun, X.; Momen, A.; Wu, J.; Noyan, H.; Li, R.-K.; von Harsdorf, R.; Husain, M. p27 Protein Protects Metabolically Stressed Cardiomyocytes from Apoptosis by Promoting Autophagy. *J. Biol. Chem.* **2014**, *289*, 16924–16935. [CrossRef]
43. Sedlic, F.; Muravyeva, M.Y.; Sepac, A.; Sedlic, M.; Williams, A.M.; Yang, M.; Bai, X.; Bosnjak, Z.J. Targeted Modification of Mitochondrial ROS Production Converts High Glucose-Induced Cytotoxicity to Cytoprotection: Effects on Anesthetic Preconditioning. *J. Cell. Physiol.* **2017**, *232*, 216–224. [CrossRef]
44. Cheng, G.; Li, L. High-glucose-induced apoptosis, ROS production and pro-inflammatory response in cardiomyocytes is attenuated by metformin treatment via PP2A activation. *J. Biosci.* **2020**, *45*, 126. [CrossRef]
45. Danziger, R.S.; Sakai, M.; Capogrossi, M.C.; Spurgeon, H.A.; Hansford, R.G.; Lakatta, E. Ethanol acutely and reversibly suppresses excitation-contraction coupling in cardiac myocytes. *Circ. Res.* **1991**, *68*, 1660–1668. [CrossRef] [PubMed]
46. Fernández-Solà, J. The Effects of Ethanol on the Heart: Alcoholic Cardiomyopathy. *Nutrients* **2020**, *12*, 572. [CrossRef] [PubMed]
47. Wu, Q.; Hayashi, H.; Hira, D.; Sonoda, K.; Ueshima, S.; Ohno, S.; Makiyama, T.; Terada, T.; Katsura, T.; Miura, K.; et al. Genetic variants of alcohol-metabolizing enzymes in Brugada syndrome: Insights into syncope after drinking alcohol. *J. Arrhythmia* **2019**, *35*, 752–759. [CrossRef] [PubMed]
48. Aiba, T. Recent understanding of clinical sequencing and gene-based risk stratification in inherited primary arrhythmia syndrome. *J. Cardiol.* **2019**, *73*, 335–342. [CrossRef]
49. Campuzano, O.; Sarquella-Brugada, G.; Cesar, S.; Arbelo, E.; Brugada, J.; Brugada, R. Update on Genetic Basis of Brugada Syndrome: Monogenic, Polygenic or Oligogenic? *Int. J. Mol. Sci.* **2020**, *21*, 7155. [CrossRef]
50. Eagle, K. Hypothesis Holiday sudden cardiac death: Food and alcohol inhibition of SULT1A enzymes as a precipitant. *J. Appl. Toxicol.* **2012**, *32*, 751–755. [CrossRef]
51. Song, S.; Hwang, E. A Rise in ATP, ROS, and Mitochondrial Content upon Glucose Withdrawal Correlates with a Dysregulated Mitochondria Turnover Mediated by the Activation of the Protein Deacetylase SIRT1. *Cells* **2018**, *8*, 11. [CrossRef]
52. Ii, G.W.D.; Vega, R.B.; Kelly, D.P. Mitochondrial biogenesis and dynamics in the developing and diseased heart. *Genes Dev.* **2015**, *29*, 1981–1991. [CrossRef]
53. Nsiah-Sefaa, A.; McKenzie, M. Combined defects in oxidative phosphorylation and fatty acid β-oxidation in mitochondrial disease. *Biosci. Rep.* **2016**, *36*. [CrossRef]
54. Fisler, J.S. Cardiac effects of starvation and semistarvation diets: Safety and mechanisms of action. *Am. J. Clin. Nutr.* **1992**, *56*, 230S–234S. [CrossRef]
55. Sours, H.E. Sudden death associated with very low calorie weight reduction regimens. *Am. J. Clin. Nutr.* **1981**, *34*, 453–461. [CrossRef]
56. Stevens, A.; Robinson, D.P.; Turpin, J.; Groshong, T.; Tobias, J.D. Sudden cardiac death of an adolescent during dieting. *South. Med. J.* **2002**, *95*, 1047–1049. [CrossRef]
57. Volek, J.S.; Feinman, R.D. Carbohydrate restriction improves the features of Metabolic Syndrome. Metabolic Syndrome may be defined by the response to carbohydrate restriction. *Nutr. Metab.* **2005**, *2*, 31. [CrossRef] [PubMed]
58. Bank, I.M.; Shemie, S.D.; Rosenblatt, B.; Bernard, C.; Mackie, A.S. Sudden Cardiac Death in Association with the Ketogenic Diet. *Pediatr. Neurol.* **2008**, *39*, 429–431. [CrossRef]
59. Bergqvist, A.G.C.; Chee, C.M.; Lutchka, L.; Rychik, J.; Stallings, V.A. Selenium Deficiency Associated with Cardiomyopathy: A Complication of the Ketogenic Diet. *Epilepsia* **2003**, *44*, 618–620. [CrossRef] [PubMed]
60. Best, T.H.; Franz, D.N.; Gilbert, D.L.; Nelson, D.P.; Epstein, M.R. Cardiac complications in pediatric patients on the ketogenic diet. *Neurology* **2000**, *54*, 2328–2330. [CrossRef] [PubMed]
61. Güzel, O.; Küçük, M.; Karadeniz, C.; Yılmaz, Ü.; Calik, T.; Meşe, T.; Özdemir, R. The Impact of 3:1 Ketogenic Diet on Cardiac Repolarization Changes in Children with Refractory Seizures: A Prospective Follow-Up Study. *Neuropediatrics* **2016**, *47*, 157–161. [CrossRef] [PubMed]
62. Sirikonda, N.S.; Patten, W.D.; Phillips, J.R.; Mullett, C.J. Ketogenic Diet: Rapid Onset of Selenium Deficiency-Induced Cardiac Decompensation. *Pediatr. Cardiol.* **2012**, *33*, 834–838. [CrossRef]

63. Longo, R.; Peri, C.; Cricrì, D.; Coppi, L.; Caruso, D.; Mitro, N.; De Fabiani, E.; Crestani, M. Ketogenic Diet: A New Light Shining on Old but Gold Biochemistry. *Nutrients* **2019**, *11*, 2497. [CrossRef]
64. McGarry, J.D.; Foster, D.W. Ketogenesis and its regulation. *Am. J. Med.* **1976**, *61*, 9–13. [CrossRef]
65. Alberti, K.G.; Johnston, D.G.; Gill, A.; Barnes, A.J.; Orskov, H. Hormonal regulation of ketone-body metabolism in man. *Biochem. Soc. Symp.* **1978**, *43*, 163–182.
66. Grabacka, M.; Pierzchalska, M.; Dean, M.; Reiss, K. Regulation of Ketone Body Metabolism and the Role of PPARα. *Int. J. Mol. Sci.* **2016**, *17*, 2093. [CrossRef]
67. Song, J.-P.; Chen, L.; Chen, X.; Ren, J.; Zhang, N.-N.; Tirasawasdichai, T.; Hu, Z.-L.; Hua, W.; Hu, Y.-R.; Tang, H.-R.; et al. Elevated plasma β-hydroxybutyrate predicts adverse outcomes and disease progression in patients with arrhythmogenic cardiomyopathy. *Sci. Transl. Med.* **2020**, *12*, eaay8329. [CrossRef] [PubMed]
68. Schieber, M.; Chandel, N.S. ROS Function in Redox Signaling and Oxidative Stress. *Curr. Biol.* **2014**, *24*, R453–R462. [CrossRef] [PubMed]
69. Van Opbergen, C.J.M.; Braven, L.D.; Delmar, M.; Van Veen, T.A.B. Mitochondrial Dysfunction as Substrate for Arrhythmogenic Cardiomyopathy: A Search for New Disease Mechanisms. *Front. Physiol.* **2019**, *10*, 1496. [CrossRef] [PubMed]
70. Balaban, R.S.; Nemoto, S.; Finkel, T. Mitochondria, Oxidants, and Aging. *Cell* **2005**, *120*, 483–495. [CrossRef] [PubMed]
71. Bhatti, J.S.; Bhatti, G.K.; Reddy, P.H. Mitochondrial dysfunction and oxidative stress in metabolic disorders—A step towards mitochondria based therapeutic strategies. *Biochim. Biophys. Acta (BBA) Mol. Basis Dis.* **2017**, *1863*, 1066–1077. [CrossRef] [PubMed]
72. Depre, C.; Taegtmeyer, H. Metabolic aspects of programmed cell survival and cell death in the heart. *Cardiovasc. Res.* **2000**, *45*, 538–548. [CrossRef]
73. Sovari, A.A. Cellular and Molecular Mechanisms of Arrhythmia by Oxidative Stress. *Cardiol. Res. Pr.* **2016**, *2016*, 1–7. [CrossRef]
74. Saadeh, K.; Fazmin, I.T. Mitochondrial Dysfunction Increases Arrhythmic Triggers and Substrates; Potential Anti-arrhythmic Pharmacological Targets. *Front. Cardiovasc. Med.* **2021**, *8*, 646932. [CrossRef] [PubMed]
75. Ikeda, T.; Yoshida, T.; Ito, Y.; Murakami, I.; Mokuda, O.; Tominaga, M.; Mashiba, H. Effect of β-hydroxybutyrate and acetoacetate on insulin and glucagon secretion from perfused rat pancreas. *Arch. Biochem. Biophys.* **1987**, *257*, 140–143. [CrossRef]
76. Malaisse, W.J.; Lebrun, P.; Yaylali, B.; Càmara, J.; Valverde, I.; Sener, A. Ketone bodies and islet function: 45Ca handling, insulin synthesis, and release. *Am. J. Physiol. Metab.* **1990**, *259*, E117–E122. [CrossRef] [PubMed]
77. Jain, S.; Kannan, K.; Lim, G. Ketosis (acetoacetate) can generate oxygen radicals and cause increased lipid peroxidation and growth inhibition in human endothelial cells. *Free Radic. Biol. Med.* **1998**, *25*, 1083–1088. [CrossRef]
78. Shi, X.; Li, D.; Deng, Q.; Peng, Z.; Zhao, C.; Li, X.; Wang, Z.; Li, X.; Liu, G. Acetoacetic acid induces oxidative stress to inhibit the assembly of very low density lipoprotein in bovine hepatocytes. *J. Dairy Res.* **2016**, *83*, 442–446. [CrossRef] [PubMed]
79. Shi, X.; Li, X.; Li, D.; Li, Y.; Song, Y.; Deng, Q.; Wang, J.; Zhang, Y.; Ding, H.; Yin, L.; et al. Hydroxybutyrate Activates the NF-κB Signaling Pathway to Promote the Expression of Pro-Inflammatory Factors in Calf Hepatocytes. *Cell. Physiol. Biochem.* **2014**, *33*, 920–932. [CrossRef]
80. Abdelmegeed, M.A.; Kim, S.K.; Woodcroft, K.J.; Novak, R.F. Acetoacetate Activation of Extracellular Signal-Regulated Kinase 1/2 and p38 Mitogen-Activated Protein Kinase in Primary Cultured Rat Hepatocytes: Role of Oxidative Stress. *J. Pharmacol. Exp. Ther.* **2004**, *310*, 728–736. [CrossRef]
81. Liesa, M.; Shirihai, O.S. Mitochondrial Dynamics in the Regulation of Nutrient Utilization and Energy Expenditure. *Cell Metab.* **2013**, *17*, 491–506. [CrossRef] [PubMed]
82. Faghih, S.; Abdolahzadeh, M.; Mohammadi, M.; Hassanzadeh, J. Prevalence of Vitamin D Deficiency and Its Related Factors Among University Students in Shiraz, Iran. *Int. J. Prev. Med.* **2014**, *5*, 796–799.
83. Scott, J.F.; Das, L.M.; Ahsanuddin, S.; Qiu, Y.; Binko, A.M.; Traylor, Z.P.; Debanne, S.M.; Cooper, K.D.; Boxer, R.; Lu, K.Q. Oral Vitamin D Rapidly Attenuates Inflammation from Sunburn: An Interventional Study. *J. Investig. Dermatol.* **2017**, *137*, 2078–2086. [CrossRef]
84. Sassi, F.; Tamone, C.; D'Amelio, P. Vitamin D: Nutrient, Hormone, and Immunomodulator. *Nutrients* **2018**, *10*, 1656. [CrossRef]
85. Matsui, M.S. Vitamin D Update. *Curr. Dermatol. Rep.* **2020**, *9*, 323–330. [CrossRef]
86. Bagrul, D.; Atik, F. Association of vitamin D deficiency with ventricular repolarization abnormalities. *Kardiol. Pol.* **2019**, *77*, 853–858. [CrossRef]
87. Yetkin, D.O.; Kucukkaya, B.; Turhan, M.; Ören, M. The effect of 25-hydroxyvitamin D levels on QT interval duration and dispersion in type 2 diabetic patients. *Croat. Med. J.* **2015**, *56*, 525–530. [CrossRef] [PubMed]
88. Halawa, A.; Dave, I.; Gautam, S. Torsade de Pointes with severe vitamin D deficiency, an unusual presentation of a common problem. *J. Cardiol. Cases* **2019**, *20*, 132–134. [CrossRef] [PubMed]
89. Mangat, J.S.; Till, J.; Bridges, N. Hypocalcaemia mimicking long QT syndrome: Case report. *Eur. J. Nucl. Med. Mol. Imaging* **2007**, *167*, 233–235. [CrossRef] [PubMed]
90. Gupta, P.; Agstam, S.; Yadav, A.; Ghosh, S. Malignant prolongation of the QTc interval due to severe vitamin D deficiency: An unusual presentation. *BMJ Case Rep.* **2020**, *13*, e237157. [CrossRef] [PubMed]
91. Chen, S.; Law, C.S.; Grigsby, C.; Olsen, K.; Hong, T.-T.; Zhang, Y.; Yeghiazarians, Y.; Gardner, D.G. Cardiomyocyte-Specific Deletion of the Vitamin D Receptor Gene Results in Cardiac Hypertrophy. *Circulation* **2011**, *124*, 1838–1847. [CrossRef] [PubMed]
92. Fanari, Z.; Hammami, S.; Hammami, M.B.; Hammami, S.; Abdellatif, A. Vitamin D deficiency plays an important role in cardiac disease and affects patient outcome: Still a myth or a fact that needs exploration? *J. Saudi Hear. Assoc.* **2015**, *27*, 264–271. [CrossRef]

93. Latic, N.; Erben, R.G. Vitamin D and Cardiovascular Disease, with Emphasis on Hypertension, Atherosclerosis, and Heart Failure. *Int. J. Mol. Sci.* **2020**, *21*, 6483. [CrossRef] [PubMed]
94. Thompson, J.; Nitiahpapand, R.; Bhatti, P.; Kourliouros, A. Vitamin D deficiency and atrial fibrillation. *Int. J. Cardiol.* **2015**, *184*, 159–162. [CrossRef] [PubMed]
95. Demir, M.; Uyan, U.; Melek, M. The Effects of Vitamin D Deficiency on Atrial Fibrillation. *Clin. Appl. Thromb.* **2012**, *20*, 98–103. [CrossRef] [PubMed]
96. Liu, X.; Wang, W.; Tan, Z.; Zhu, X.; Liu, M.; Wan, R.; Hong, K. The relationship between vitamin D and risk of atrial fibrillation: A dose-response analysis of observational studies. *Nutr. J.* **2019**, *18*, 73. [CrossRef]
97. Trevisan, C.; Piovesan, F.; Lucato, P.; Zanforlini, B.M.; De Rui, M.; Maggi, S.; Noale, M.; Corti, M.C.; Perissinotto, E.; Manzato, E.; et al. Parathormone, vitamin D and the risk of atrial fibrillation in older adults: A prospective study. *Nutr. Metab. Cardiovasc. Dis.* **2019**, *29*, 939–945. [CrossRef] [PubMed]
98. Rienstra, M.; Cheng, S.; Larson, M.G.; McCabe, E.L.; Booth, S.L.; Jacques, P.F.; Lubitz, S.A.; Yin, X.; Levy, D.; Magnani, J.W.; et al. Vitamin D status is not related to development of atrial fibrillation in the community. *Am. Heart J.* **2011**, *162*, 538–541. [CrossRef]
99. Bie, L. The Status and Research Progress on Vitamin D Deficiency and Atrial Fibrillation. *Braz. J. Cardiovasc. Surg.* **2019**, *34*, 605–609. [CrossRef]
100. Zhang, Z.; Yang, Y.; Ng, C.Y.; Wang, D.; Wang, J.; Li, G.; Liu, T. Meta-analysis of Vitamin D Deficiency and Risk of Atrial Fibrillation. *Clin. Cardiol.* **2016**, *39*, 537–543. [CrossRef]
101. Vassalle, M.; Lin, C.-I. Calcium overload and cardiac function. *J. Biomed. Sci.* **2004**, *11*, 542–565. [CrossRef]
102. Xia, P.; Liu, Y.; Cheng, Z. Signaling Pathways in Cardiac Myocyte Apoptosis. *BioMed Res. Int.* **2016**, *2016*, 1–22. [CrossRef] [PubMed]
103. Parikh, M.; Netticadan, T.; Pierce, G.N. Flaxseed: Its bioactive components and their cardiovascular benefits. *Am. J. Physiol. Circ. Physiol.* **2018**, *314*, H146–H159. [CrossRef] [PubMed]
104. Xie, N.; Zhang, W.; Li, J.; Liang, H.; Zhou, H.; Duan, W.; Xu, X.; Yu, S.; Zhang, H.; Yi, D. α-Linolenic Acid Intake Attenuates Myocardial Ischemia/Reperfusion Injury through Anti-inflammatory and Anti-oxidative Stress Effects in Diabetic But Not Normal Rats. *Arch. Med. Res.* **2011**, *42*, 171–181. [CrossRef]
105. Chrysohoou, C.; Panagiotakos, D.B.; Pitsavos, C.; Skoumas, J.; Krinos, X.; Chloptsios, Y.; Nikolaou, V.; Stefanadis, C. Long-term fish consumption is associated with protection against arrhythmia in healthy persons in a Mediterranean region—The ATTICA study. *Am. J. Clin. Nutr.* **2007**, *85*, 1385–1391. [CrossRef] [PubMed]
106. Yagi, S.; Soeki, T.; Aihara, K.-I.; Fukuda, D.; Ise, T.; Kadota, M.; Bando, S.; Matsuura, T.; Tobiume, T.; Yamaguchi, K.; et al. Low Serum Levels of Eicosapentaenoic Acid and Docosahexaenoic Acid are Risk Factors for Cardiogenic Syncope in Patients with Brugada Syndrome. *Int. Heart J.* **2017**, *58*, 720–723. [CrossRef]
107. Milberg, P.; Frommeyer, G.; Kleideiter, A.; Fischer, A.; Osada, N.; Breithardt, G.; Fehr, M.; Eckardt, L. Antiarrhythmic effects of free polyunsaturated fatty acids in an experimental model of LQT2 and LQT3 due to suppression of early afterdepolarizations and reduction of spatial and temporal dispersion of repolarization. *Heart Rhythm.* **2011**, *8*, 1492–1500. [CrossRef]
108. Jaca, A.; Durão, S.; Harbron, J. Omega-3 fatty acids for the primary and secondary prevention of cardiovascular disease. *S. Afr. Med. J.* **2020**, *110*, 1158–1159. [CrossRef]
109. Tajik, B.; Kurl, S.; Tuomainen, T.-P.; Virtanen, J.K. Associations of the serum long-chain omega-3 polyunsaturated fatty acids and hair mercury with heart rate-corrected QT and JT intervals in men: The Kuopio Ischaemic Heart Disease Risk Factor Study. *Eur. J. Nutr.* **2016**, *56*, 2319–2327. [CrossRef]
110. Miller, C.; Karimi, R.; Silbernagel, S.; Kostrubiak, D.; Schiavone, F.; Zhang, Q.; Yang, J.; Rashba, E.; Meliker, J.R. Mercury, omega-3 fatty acids, and seafood intake are not associated with heart rate variability or QT interval. *Arch. Environ. Occup. Health* **2017**, *73*, 251–257. [CrossRef]
111. Curfman, G. Omega-3 Fatty Acids and Atrial Fibrillation. *JAMA* **2021**, *325*, 1063. [CrossRef]
112. Enriquez, A.; Frankel, D.S. Arrhythmogenic effects of energy drinks. *J. Cardiovasc. Electrophysiol.* **2017**, *28*, 711–717. [CrossRef]
113. Raymond, L.W. Arrhythmogenic foods—A growing medical problem. *Trends Cardiovasc. Med.* **2020**, *30*, 310–312.
114. Berger, A.J.; Alford, K. Cardiac arrest in a young man following excess consumption of caffeinated "energy drinks. " *Med. J. Aust.* **2009**, *190*, 41–43. [CrossRef]
115. Cannon, E.M.; Cooke, C.T.; McCarthy, J. Caffeine-induced cardiac arrhythmia: An unrecognised danger of healthfood products. *Med. J. Aust.* **2001**, *174*, 520–521. [CrossRef] [PubMed]
116. Rutledge, M.; Witthed, A.; Khouzam, R.N. It took a Redbull to unmask Brugada syndrome. *Int. J. Cardiol.* **2012**, *161*, e14–e15. [CrossRef] [PubMed]
117. Gray, B.; Ingles, J.; Medi, C.; Driscoll, T.; Semsarian, C. Cardiovascular Effects of Energy Drinks in Familial Long QT Syndrome: A Randomized Cross-Over Study. *Int. J. Cardiol.* **2017**, *231*, 150–154. [CrossRef] [PubMed]
118. Satoh, H. Cardiac Actions of Taurine as a Modulator of the Ion Channels. In *Taurine 3*; Schaffer, S., Lombardini, J.B., Huxtable, R.J., Eds.; Springer: Boston, MA, USA, 1998; Volume 442, pp. 121–128. [CrossRef]
119. Myers, M.G. Caffeine and Cardiac Arrhythmias. *Chest* **1988**, *94*, 4–5. [CrossRef]
120. Bogdanovic, E.; Potet, F.; Marszalec, W.; Iyer, H.; Galiano, R.; Hong, S.J.; Leung, K.P.; Wasserstrom, J.A.; Mustoe, T.A. The sodium channel Na X: Possible player in excitation–contraction coupling. *IUBMB Life* **2020**, *72*, 601–606. [CrossRef]
121. Monasky, M.M.; Pappone, C.; Piccoli, M.; Ghiroldi, A.; Micaglio, E.; Anastasia, L. Calcium in Brugada Syndrome: Questions for Future Research. *Front. Physiol.* **2018**, *9*, 1088. [CrossRef]

122. Antzelevitch, C. Brugada syndrome: Historical perspectives and observations. *Eur. Heart J.* **2002**, *23*, 676–678. [CrossRef]
123. Antzelevitch, C.; Yan, G.-X.; Ackerman, M.J.; Borggrefe, M.; Corrado, D.; Guo, J.; Gussak, I.; Hasdemir, C.; Horie, M.; Huikuri, H.; et al. J-Wave syndromes expert consensus conference report: Emerging concepts and gaps in knowledge. *J. Arrhythmia* **2016**, *32*, 315–339. [CrossRef] [PubMed]
124. Rottlaender, D.; Motloch, L.; Reda, S.; Larbig, R.; Hoppe, U. Cardiac arrest due to long QT syndrome associated with excessive consumption of energy drinks. *Int. J. Cardiol.* **2012**, *158*, e51–e52. [CrossRef] [PubMed]
125. Scholz, E.P.; Zitron, E.; Kiesecker, C.; Lück, S.; Thomas, D.; Kathöfer, S.; Kreye, V.A.W.; Katus, H.A.; Kiehn, J.; Schoels, W.; et al. Inhibition of cardiac HERG channels by grapefruit flavonoid naringenin: Implications for the influence of dietary compounds on cardiac repolarisation. *Naunyn-Schmiedeberg's Arch. Pharmacol.* **2005**, *371*, 516–525. [CrossRef]
126. Kiesecker, C.; Zitron, E.; Bloehs, R.; Scholz, E.P.; Thomas, D.; Kreye, V.A.W.; Katus, H.A.; Schoels, W.; Karle, C.A.; Kiehn, J. Class Ia anti-arrhythmic drug ajmaline blocks HERG potassium channels: Mode of action. *Naunyn-Schmiedeberg's Arch. Pharmacol.* **2004**, *370*, 423–435. [CrossRef]
127. Zeinali, M.; Rezaee, S.A.; Hosseinzadeh, H. An overview on immunoregulatory and anti-inflammatory properties of chrysin and flavonoids substances. *Biomed. Pharmacother.* **2017**, *92*, 998–1009. [CrossRef]
128. Choy, K.W.; Murugan, D.; Leong, X.-F.; Abas, R.; Alias, A.; Mustafa, M.R. Flavonoids as Natural Anti-Inflammatory Agents Targeting Nuclear Factor-Kappa B (NFκB) Signaling in Cardiovascular Diseases: A Mini Review. *Front. Pharmacol.* **2019**, *10*, 1295. [CrossRef]
129. Havakuk, O.; Schwartz, A.L.; Rosso, R.; Viskin, S. Editorial commentary: A question on proarrhythmic food: Is grapefruit "the forbidden fruit" for patients with long QT syndrome? *Trends Cardiovasc. Med.* **2020**, *30*, 313–314. [CrossRef] [PubMed]
130. Ahmed, F.A.; Ali, R. Bioactive Compounds and Antioxidant Activity of Fresh and Processed White Cauliflower. *BioMed Res. Int.* **2013**, *2013*, 1–9. [CrossRef] [PubMed]
131. Bergquist, S.Å.M.; Gertsson, U.E.; Knuthsen, P.; Olsson, M.E. Flavonoids in Baby Spinach (Spinacia oleraceaL.): Changes during Plant Growth and Storage. *J. Agric. Food Chem.* **2005**, *53*, 9459–9464. [CrossRef] [PubMed]
132. Rao, S.; Gou, Y.; Yu, T.; Cong, X.; Gui, J.; Zhu, Z.; Zhang, W.; Liao, Y.; Ye, J.; Cheng, S.; et al. Effects of selenate on Se, flavonoid, and glucosinolate in broccoli florets by combined transcriptome and metabolome analyses. *Food Res. Int.* **2021**, *146*, 110463. [CrossRef] [PubMed]
133. Chavez-Santoscoy, R.A.; Gutiérrez-Uribe, J.A.; Granados, O.; Torre-Villalvazo, I.; Serna-Saldívar, S.O.; Torres, N.; Palacios-González, B.; Tovar, A.R. Flavonoids and saponins extracted from black bean (Phaseolus vulgaris L.) seed coats modulate lipid metabolism and biliary cholesterol secretion in C57BL/6 mice. *Br. J. Nutr.* **2014**, *112*, 886–899. [CrossRef] [PubMed]
134. Bhagyawant, S.S.; Gautam, A.K.; Narvekar, D.T.; Gupta, N.; Bhadkaria, A.; Srivastava, N.; Upadhyaya, H.D. Biochemical diversity evaluation in chickpea accessions employing mini-core collection. *Physiol. Mol. Biol. Plants* **2018**, *24*, 1165–1183. [CrossRef] [PubMed]
135. Khalil, M.I.; Sulaiman, S.A. The potential role of honey and its polyphenols in preventing heart diseases: A review. *Afr. J. Tradit. Complement. Altern. Med. AJTCAM* **2010**, *7*, 315–321. [CrossRef]
136. Azab, A.; Nassar, A.; Azab, A.N. Anti-Inflammatory Activity of Natural Products. *Molecules* **2016**, *21*, 1321. [CrossRef]
137. Karetnikova, V.; Gruzdeva, O.; Uchasova, E.; Osokina, A.; Barbarash, O. Glucose levels as a prognostic marker in patients with ST-segment elevation myocardial infarction: A case–control study. *BMC Endocr. Disord.* **2016**, *16*, 31. [CrossRef]
138. Orenes-Piñero, E.; Valdés, M.; Lip, G.Y.H.; Marín, F. A comprehensive insight of novel antioxidant therapies for atrial fibrillation management. *Drug Metab. Rev.* **2015**, *47*, 388–400. [CrossRef]
139. Nogami, A.; Nakao, M.; Kubota, S.; Sugiyasu, A.; Doi, H.; Yokoyama, K.; Yumoto, K.; Tamaki, T.; Kato, K.-I.; Hosokawa, N.; et al. Enhancement of J-ST-segment elevation by the glucose and insulin test in Brugada syndrome. *Pacing Clin. Electrophysiol.* **2003**, *26*, 332–337. [CrossRef]
140. Stephan, L.S.; Almeida, E.D.; Markoski, M.M.; Garavaglia, J.; Marcadenti, A. Red Wine, Resveratrol and Atrial Fibrillation. *Nutrients* **2017**, *9*, 1190. [CrossRef]
141. Fisch, C. Relation of Electrolyte Disturbances to Cardiac Arrhythmias. *Circulation* **1973**, *47*, 408–419. [CrossRef] [PubMed]
142. Omar, H.R.; Komarova, I.; El-Ghonemi, M.; Fathy, A.; Rashad, R.; Abdelmalak, H.D.; Yerramadha, M.R.; Ali, Y.; Helal, E.; Camporesi, E.M. Licorice abuse: Time to send a warning message. *Ther. Adv. Endocrinol. Metab.* **2012**, *3*, 125–138. [CrossRef] [PubMed]
143. Li, J.; Hu, D.; Song, X.; Han, T.; Gao, Y.; Xing, Y. The Role of Biologically Active Ingredients from Natural Drug Treatments for Arrhythmias in Different Mechanisms. *BioMed Res. Int.* **2017**, *2017*, 1–10. [CrossRef]
144. Minoura, Y.; Panama, B.K.; Nesterenko, V.V.; Betzenhauser, M.; Barajas-Martínez, H.; Hu, D.; Di Diego, J.M.; Antzelevitch, C. Effect of Wenxin Keli and quinidine to suppress arrhythmogenesis in an experimental model of Brugada syndrome. *Heart Rhythm.* **2013**, *10*, 1054–1062. [CrossRef] [PubMed]
145. Manne, J.R.R.; Garg, J. Hyperkalemia induced Brugada phenocopy. *J. Arrhythmia* **2021**, *37*, 249–250. [CrossRef]
146. Kawaguchi, T.; Nomura, M.; Tujikawa, T.; Nakaya, Y.; Ito, S. 123I-metaiodo-benzylguanidine myocardial scintigraphy in the Brugada-type ECG. *J. Med. Investig.* **2006**, *53*, 95–102. [CrossRef]
147. Miyazaki, T.; Mitamura, H.; Miyoshi, S.; Soejima, K.; Aizawa, Y.; Ogawa, S. Autonomic and antiarrhythmic drug modulation of ST segment elevation in patients with Brugada syndrome. *J. Am. Coll. Cardiol.* **1996**, *27*, 1061–1070. [CrossRef]
148. Kasanuki, H.; Ohnishi, S.; Ohtuka, M.; Matsuda, N.; Nirei, T.; Isogai, R.; Shoda, M.; Toyoshima, Y.; Hosoda, S. Idiopathic ventricular fibrillation induced with vagal activity in patients without obvious heart disease. *Circulation* **1997**, *95*, 2277–2285. [CrossRef]

149. Matsuo, K.; Kurita, T.; Inagaki, M.; Kakishita, M.; Aihara, N.; Shimizu, W.; Taguchi, A.; Suyama, K.; Kamakura, S.; Shimomura, K. The circadian pattern of the development of ventricular fibrillation in patients with Brugada syndrome. *Eur. Heart J.* **1999**, *20*, 465–470. [CrossRef]
150. Lane, R.D.; Zareba, W.; Reis, H.T.; Peterson, D.R.; Moss, A.J. Changes in Ventricular Repolarization Duration During Typical Daily Emotion in Patients with Long QT Syndrome. *Psychosom. Med.* **2011**, *73*, 98–105. [CrossRef] [PubMed]
151. Ikeda, T.; Abe, A.; Yusu, S.; Nakamura, K.; Ishiguro, H.; Mera, H.; Yoshino, H. The Full Stomach Test as a Novel Diagnostic Technique for Identifying Patients at Risk of Brugada Syndrome. *J. Cardiovasc. Electrophysiol.* **2006**, *17*, 602–607. [CrossRef] [PubMed]

nutrients

MDPI

Review

Relationship between Dietary Patterns with Benign Prostatic Hyperplasia and Erectile Dysfunction: A Collaborative Review

Giorgio Ivan Russo [1,*], Giuseppe Broggi [2], Andrea Cocci [3], Paolo Capogrosso [4], Marco Falcone [5], Ioannis Sokolakis [6], Murat Gül [7], Rosario Caltabiano [2] and Marina Di Mauro [8,†]

1 Urology Section, University of Catania, 95123 Catania, Italy
2 Department of Medical, Surgical Sciences and Advanced Technologies "G.F. Ingrassia", Anatomic Pathology, University of Catania, 95123 Catania, Italy; giuseppe.broggi@gmail.com (G.B.); rosario.caltabiano@unict.it (R.C.)
3 Department of Urology, University of Florence, 50100 Florence, Italy; cocci.andrea@gmail.com
4 Department of Urology and Andrology, Ospedale di Circolo and Macchi Foundation, 21100 Varese, Italy; paolo.capogrosso@gmail.com
5 Department of Urology, Molinette Hospital, A.O.U. Città della Salute e della Scienza di Torino, 10100 Turin, Italy; marco.falcone@unito.it
6 Department of Urology, Martha-Maria Hospital Nuremberg, 90491 Nuremberg, Germany; sokolakisi@gmail.com
7 School of Medicine, Department of Urology, Selcuk University, 42005 Konya, Turkey; drmuratgul@hotmail.com
8 Urology Section, San Marco Hospital, 95100 Catania, Italy; marinadimauro@live.it
* Correspondence: giorgioivan.russo@unict.it
† Membership of the EAU-YAU Sexual and Reproductive Health Group is provided in the Acknowledgments.

Citation: Russo, G.I.; Broggi, G.; Cocci, A.; Capogrosso, P.; Falcone, M.; Sokolakis, I.; Gül, M.; Caltabiano, R.; Di Mauro, M. Relationship between Dietary Patterns with Benign Prostatic Hyperplasia and Erectile Dysfunction: A Collaborative Review. *Nutrients* **2021**, *13*, 4148. https://doi.org/10.3390/nu13114148

Academic Editor: Lynnette Ferguson

Received: 7 November 2021
Accepted: 17 November 2021
Published: 19 November 2021

Abstract: Interest in the role of dietary patterns has been consistently emerging in recent years due to much research that has documented the impact of metabolism on erectile dysfunction (ED) and/or benign prostatic hyperplasia (BPH). We conducted a non-systematic review of English articles published from 1964 to September 2021. The search terms were: ("dietary patterns" OR "diet") AND/OR ("erectile dysfunction") AND/OR ("benign prostatic hyperplasia"). In the present review, we have highlighted how the association between dietary patterns and two of the most frequent pathologies in urology, namely erectile dysfunction and benign prostatic hyperplasia, is present in the literature. The data suggested that a diet that is more adherent to the Mediterranean diet or that emphasizes the presence of vegetables, fruits, nuts, legumes, and fish or other sources of long-chain (n-3) fats, in addition to reduced content of red meat, may have a beneficial role on erectile function. At the same time, the same beneficial effects can be transferred to BPH as a result of the indirect regulatory effects on prostatic growth and smooth muscle tone, thus determining an improvement in symptoms. Certainly, in-depth studies and translational medicine are needed to confirm these encouraging data.

Keywords: prostate; diet; metabolism; benign prostatic hyperplasia; erectile dysfunction

1. Introduction

The interest in the role of dietary patterns has been consistently emerging in recent years due to much research that has documented the impact of metabolism on erectile dysfunction (ED) [1] and benign prostatic hyperplasia (BPH) [2,3].

Although much research has already demonstrated a significant association between dietary patterns and prostate cancer, including polyphenols, phytoestrogens or phenolic acids [4–6], the evidence regarding ED and BPH have not been fully investigated.

It is estimated that the worldwide prevalence of erectile dysfunction will likely increase to 322 million men by 2025 [7]. These results also explain the increase in interest over the years in Google trend searches for many terms such as prosthetic surgical treatment (+1.7%), for prostaglandins (+0.7%), for traction (+0.6%), and for shock wave therapy (+1.8%) [8].

Although many modifiable risk factors for ED have been investigated, such as cardiovascular disease (CVD), smoking, obesity, sedentary behavior, diabetes, hypertension, hyperlipidemia, and metabolic syndrome [9], it is still unknown whether healthy dietary patterns are associated with a reduced risk (Figure 1).

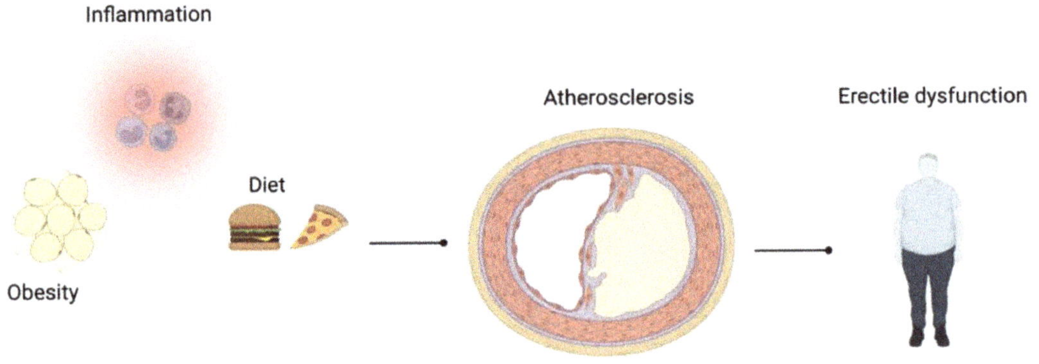

Figure 1. Overview of mechanisms associated with erectile dysfunction.

In addition to these concepts, it is interesting to state that men with ED are 1.33–6.24 times more likely to have BPH than men without ED [10].

In fact, data from the literature have shown a close relationship between BPH and ED, suggesting that pathophysiological mechanisms involved in the metabolic syndrome are key factors in both disorders [11].

In particular, several possible pathophysiological mechanisms have been discussed, including the NOS/NO (the nitric oxide synthase) and the Rho-kinase activation pathways, autonomic hyperactivity, pelvic ischemia and microvascular dysfunction, inflammatory pathways, sex hormones, and psychological factors [11].

All these premises lay the groundwork for the hypothesis that dietary patterns may have a role in exacerbating or even preventing ED and BPH [12].

For these reasons, in this non-systematic review, we aimed to address current evidence on the role of dietary patterns on ED and BPH. We also addressed the impact of diabetes, atherosclerosis, and metabolic syndrome on ED and BPH.

2. Materials and Methods

We conducted a non-systematic review of English articles published from 1964 to September 2021. The search terms were: ("dietary patterns" OR "diet") AND/OR ("erectile dysfunction") AND/OR ("benign prostatic hyperplasia").

3. Erectile Dysfunction

In recent years, interest in the quality of food has grown, paying attention not only to the individual food, but also to the micronutrients contained in the food [13,14].

3.1. Studies in Animal Models

In recent years, there have been important advances in the study of the molecular pathways that regulate the association between diet, exercise, and endothelial function in the penis. Authors have demonstrated that pigs fed with a high-fat diet had significantly reduced cGMP levels, an increase in eNOS uncoupling, and eNOS binding to caveolin-1 (indicating reduced NO availability) in the penis of sedentary pigs. They also reported that exercise of pigs counteracted these abnormalities [15].

These data demonstrated that unhealthy diets determine a detrimental impact in endothelial cells by the increase in oxidative stress and significant reduction in nitric oxide, an essential molecule that maintains erectile function [16].

Akomolafe et al. evaluated the effect of a diet supplemented with raw and toasted pumpkin seeds on some key biochemical parameters relevant to erectile function in the body cavernous tissues of male rats. Animals were divided into six groups (10 animals per group) for the evaluation of adenosine deaminase (ADA), phosphodiesterase-5 (PDE-5), arginase, and acetylcholinesterase (AChE) activity, including nitric oxide level and malondialdehyde (MDA). Group I: normal control rats fed with a basal diet; Groups II and III: (5% and 10% pumpkin seeds [PS]) rats fed with diet supplemented with 5% and 10% raw pumpkin seeds, respectively; Groups IV and V: (5% and 10% roasted pumpkin seeds [PS]) rats fed with diet supplemented with 5% and 10% roasted pumpkin seeds, respectively; Group VI: rats treated with Sildenafil citrate (5 mg/kg). The diet supplemented with roasted pumpkin seeds showed better PDE-5, ADA, and arginase activities, as well as NO and MDA levels. No significant differences were observed in the AChE activities of rats treated with raw and roasted pumpkin seeds. The authors concluded that the modulatory effects of raw and roasted pumpkin seeds on enzymes associated with erectile dysfunction suggest the biochemical rationale for their therapeutic role in improving erectile function. However, it appears that roasted pumpkin seeds (10%, w/w of the diet) have more beneficial effects than raw seeds [17]. In addition, a 2016 study by Akomolafe revealed that pumpkin seeds contain phenolic acids and flavonoids and prevent oxidative damage to testicular tissues [18].

Recent studies have suggested that the treatment of ED may benefit from the modulation of other enzymes such as ectonucleotidases (E-NTPDase) and ADA [19–21], which are involved in the regulation of biomolecules such as cGMP (cyclic guanosine monophosphate), NO (nitrogen oxide), ATP, and adenosine that are involved in penile erection [22,23].

In a study conducted in rats with ED, the authors investigated the effects of quercetin as a promising source of dietary phytochemicals for ED management.

The authors divided the rats into different groups by the administration of normal saline, cyclosporine-induced hypertension, rats administered with sildenafil (5 mg kg^{-1} day^{-1}), and rats administered quercetin 25 mg kg^{-1} day^{-1} or Q 50 mg kg^{-1} day^{-1} (50 Q). They demonstrated that quercetin improved the activities of enzymes associated with better ATP bioavailability (E-NTPDase and 5′-nucleotidase). Its effects were able to decrease ADA activity and increase NO levels [24].

These findings highlighted the concept that polyphenols are phytochemicals and can promote good health and improved erectile dysfunction [25].

3.2. Studies in Humans

The importance of diet on ED arises from the assumption that this pathology is often the first symptom of coronary heart disease (CHD) [9,26,27]. In fact, the pathophysiology of ED is similar to that of atherosclerosis [28,29]. The correlation between diet and diseases of the cardiovascular system has been known for years [1]. Therefore, a possible correlation between erectile dysfunction and dietary factors is conceivable.

In the study by Nicolosi et al., it is shown that 31.8% of men with a below-average level of physical activity demonstrate ED, compared to 17.5% of men with an average activity level and only 13.9% of men who have an above-average level of physical activity, thus demonstrating a linear association between the level of physical activity and ED [30].

Lu et al. recently studied the relationship between a plant-based diet and erectile dysfunction in 184 Chinese patients. The ED group (92 subjects) and the free ED group (92 subjects) were similar in terms of baseline characteristics ($p > 0.05$), with the exception of lifestyle ($p < 0.05$). The plant diet index (PDI) and the healthy plant diet index (hPDI) in the ED group were significantly lower than those of the control group ($p < 0.001$). Adjusted multivariate analysis indicated that the presence of ED was negatively associated with nitric oxide, PDI, and hPDI levels (all $p < 0.05$) and was positively related to body mass index, metabolic syndrome and E-selectin levels. Additionally, both PDI and hPDI significantly increased with increasing International Erectile Function Index (IIEF-5) scores within the ED group ($p < 0.05$). Finally, multimodal multivariate analysis was performed, which indicated the robustness of the results [31].

From previous studies, we know that ED is less frequent in patients who adhere to the Mediterranean diet model, characterized by the prevalent presence of fish, vegetables, fruit, whole grains, and nuts, compared to those who mainly consumed a diet containing red and processed meats and refined cereals [16,32].

Giugliano et al. assessed the relationship between adherence to the Mediterranean diet and sexual function among 555 men with type 2 diabetes, showing that men with the highest scores of adherences to the Mediterranean diet had lower overall prevalence of erectile dysfunction. Additionally, men in the middle and highest tertiles had a lower prevalence of severe ED compared to men in the lowest adhesion tertile [33].

Sticking to an unhealthy diet pattern can cause early endothelial damage through oxidative stress, which results in a reduction in the availability of nitric oxide, indispensable in the physiology of erection [34].

Food antioxidants have been shown to improve erectile dysfunction in men. A rich source of polyphenols is represented by the standardized French maritime pine bark (Pinus pinaster) Pycnogenol (PYC) extract. The main constituent of the polyphenols in PYC is made up of 70% procyanidins. PYC has significant antioxidant and multiple power biomodulatory effects, such as transcription factor inhibition NF-κB, cyclooxygenase, NO stimulation of the hypertensive effect by inhibition of the angiotensin converting enzyme, antimutagen effects, and alleviation of allergic and anti-glycemic asthma symptoms. Trebaticky enrolled 53 patients with ED who were divided into two groups (32 with diabetes mellitus, 21 non-diabetes mellitus) in a randomized, blinded, placebo-controlled study. During the 3-month intervention with Pycnogenol or placebo and one month after the end of the intervention, the patients were investigated for ED with a validated International Index of Erectile Function-5 (IIEF-5) questionnaire. Lipid profile and blood glucose were analyzed in each group. The results showed that of the natural polyphenols extracted, Pycnogenol improved erectile function in the DM group by 45% compared to the NDM group, where the improvement was also significant, but only 22%. The levels of total cholesterol, LDL cholesterol and glucose were lowered by Pycnogenol in patients with DM. Glucose level was not affected by Pycnogenol in non-DMs. The placebo showed no effect on the monitored parameters in both groups [35].

In a recent study, Salas-Huetos et al. (2019) reported that the consumption of walnuts increased sexual function. Indeed, a secondary outcome analysis of the FERTINUTS study, a 14-week randomized controlled trial with 83 subjects consuming a Western-style diet, reported that consuming 60 g/day of nuts was associated with increased orgasmic function and sexual desire compared with the control group (avoiding nuts), but no significant differences in erectile function were observed [36].

Previously, Mykoniatis et al. enrolled 350 adult men who were asked to complete an anonymous web-based questionnaire. Erectile dysfunction was diagnosed with the International Erectile Function Index (IIEF) and flavonoid intake was recorded using food-frequency questionnaires, with a focus on flavonoid-rich foods such as coffee, fruit, etc. Participants were divided into two groups based on the IIEF scores: control group without ED (IIEF score 26; n = 264) and the case group with ED (IIEF score < 26; n = 86). Men with erectile dysfunction reported a lower median monthly intake of total flavonoids ($p < 0.001$) and all flavonoid subclasses ($p < 0.001$) than the controls. Adjusting the intake for age and body mass index, it was found that the consumption of 50 mg/day of flavonoids reduced the risk of erectile dysfunction by 32% (odds ratio = 0.68, $p < 0.001$). Of all the flavonoids recorded, flavones appeared to contribute the most to healthy erectile function. Controls reported higher consumption of fruit and vegetables, lower consumption of dairy products and alcoholic beverages, and less intense smoking than the cases ($p < 0.001$) [37].

Cassidy et al. were among the first authors to evaluate the intake of flavonoids in erectile dysfunction. The results of the Health Professions Follow-up study showed that a reduced incidence of ED was associated with increased habitual intake of specific foods rich in flavonoids. The greatest benefits from the increase in the consumption of flavonones, flavones, and anthocyanins were observed in overweight or obese young men. Of all the flavonoids recorded, flavones appeared to contribute the most to healthy erectile function [38], according to Mykoniatis's study [37].

Bauer also conducted a population-based prospective cohort study which included men from the Health Professionals Follow-up Study with follow-up from January 1998 through January 2014, demonstrating that a higher-quality diet based on adherence to a Mediterranean diet or alternative to the Healthy Eating Index diet, which emphasize eating vegetables, fruits, nuts, legumes, and fish or other sources of long-chain (n-3) fats, as well as avoiding red and processed meats, was found to be associated with a lower risk of developing erectile dysfunction [39].

Taking together all these considerations, we postulate that dietary patterns have a significant role in ED severity (Figure 2) and should be investigated further in future studies.

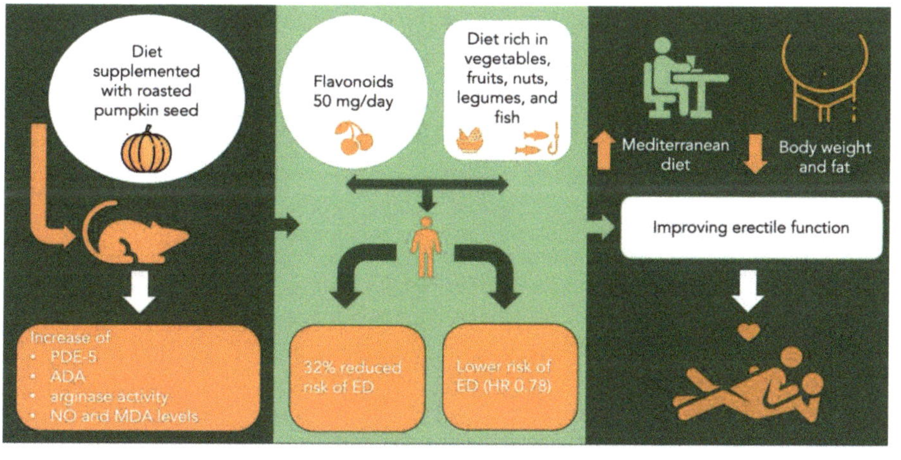

Figure 2. Relationship between dietary patterns and erectile dysfunction. Relationship between dietary patterns and erectile dysfunction. PDE5 = phosphodiesterase-5 inhibitors; ADA = adenosine deaminase; NO = nitrogen oxide; MDA = Malondialdehyde; ED = erectile dysfunction.

4. Benign Prostatic Hyperplasia

The histopathology of BPH characteristically consists of a dual hyperplasia of the epithelial and stromal compartment of the transitional zone of the prostate. Epithelial hyperplastic features include nodules composed of variably sized and sometimes cystically dilated prostatic glands with a retained basal cell layer, often exhibiting corpora amylacea and/or calcifications; stromal hyperplasia consists of nodular proliferation of bland-looking spindle cells with rounded to ovoid nuclei, frequently resembling smooth muscle cells [40].

Several biological factors, including oxidative stress, inflammation, androgens, and enhanced expression of multiple growth factors, have been associated with benign and malignant prostatic disorders [41–45] (Figure 3).

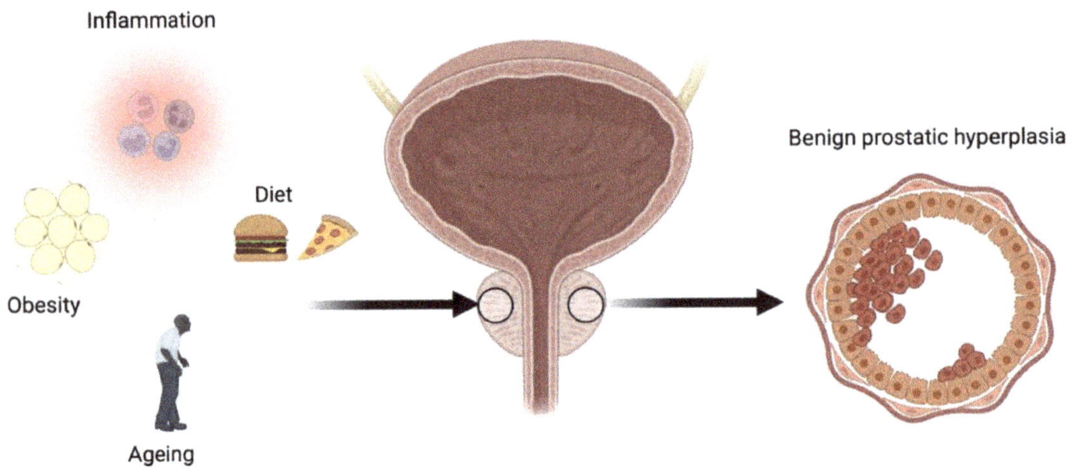

Figure 3. Overview of mechanisms associated with benign prostatic hyperplasia.

In this regard, as some evidence has suggested that a high-fat diet is intrinsically correlated with BPH by stimulating inflammation and oxidative stress [46], in recent decades, the potential association between different dietary patterns, including both macro- and micronutrients, and incidence of BPH has become one of the most debated topics in scientific literature [47–49].

4.1. Studies in Animal Models

Numerous advances have been made in the comprehension of the molecular basis of the relationship between diet and BPH in animals.

Zhang et al. found that Vitamin D (VD) deficiency in early life promoted BPH in middle-aged mice [50]; male pups, whose dams were fed with VD-deficient diets during pregnancy and lactation and continued to receive VD-deficient diets after weaning. Higher incidences of BPH were observed in this group of mice, compared to a control group of dams and male pups that received a standard diet. In addition, the authors [50] found that VD-deficient diets induced prostatic inflammation and fibrosis through the activation of the NF-κB-mediated pathway and the production of IL-6, as well as upregulation of the STAT3-mediated pathway that stimulates cell proliferation and growth. Interestingly, these prostatic effects were partially reversible if the standard diet was restored.

Li et al. showed that the combination of androgens and a high-fat diet-induced hyperinsulinemia promoted BPH in rats and the activation of p-ERK1/2 could be implicated in this process [51]. In particular, higher immunohistochemical and Western blot levels of p-ERK1/2 were observed both in rats with BPH plus a high-fat diet and rats with BPH compared to those found in rats with a high-fat diet and in a control group. However, it has also been reported that the administration of flaxseed reduced epithelial cell proliferative activity in rats with BPH [52,53]; this finding led to the supposition that different fat types and contents are involved in BPH onset and maintenance. As further evidence for this hypothesis, Kayode et al. emphasized that a ketogenic diet, consisting of high fat, moderate protein and low carbohydrate consumption, ameliorated testosterone propionate-induced BPH in male Wistar rats [54]. Similarly, epigallocatechin-3-gallate (EGCG), a green tea component, has been found to play an antioxidative and anti-BPH role in a metabolic syndrome rat model [55].

Recently, Aljehani et al. investigated the role of icariin (ICA), a flavonol glycoside with marked phytoestrogenic activity in rats with metabolic syndrome (MS)-induced

BPH [56]. Animals were divided into five groups (two out of five groups were fed a standard diet and MS was induced in the remaining three groups). MS rat groups were given vehicle, 25 mg/kg of ICA and 50 mg/kg of ICA, respectively. The authors found that the administration of both ICA doses had positive effects on prostate weight, prostate index, and histopathologic features of BPH. Furthermore, ICA seemed to play antiproliferative, proapoptotic, antioxidant, and anti-inflammatory functions by regulating cyclin D1, Bax, Bcl2, and tumor necrosis factor-α expression.

Mangosteen pericarp powder (MPP), which originates from Mangosteen, a tropical fruit from the Malay islands and the Indonesian Moluccas, has been traditionally used for wounds and cutaneous infections. Recently, the consumption of MMP, whose xanthones are the main polyphenol compounds, has been found implicated in the decrease in prostate weight, serum testosterone, and attenuation of BPH in F344 male rats [57].

4.2. Studies in Human

BPH is a very frequent and age-related disease as it is estimated that about 50% of men over the age of 50, and 80% of those older than 70 suffer from it [58,59]. Patients with BPH often exhibit acute urinary symptoms deriving from urethra compression and/or lower urinary tract symptoms (LUTSs) [59]. Due to the huge impact of BPH on populations, multiple studies have investigated its relationship with environmental factors, including the effects of several macro- and micronutrients [60,61].

A statistically significant correlation between higher risk of BPH and high consumption of fats and red meat or low consumption of proteins and vegetables has been found by the Prostate Cancer Prevention Trial (PCPT) within a cohort of 18,800 patients aged more than 50 years. A slight relationship between lower risk of BPH and multiple nutrients, such as lycopene, VD and zinc was also established for this group; conversely, no association between this disease and antioxidant consumption was identified. Total but not dietary vitamin D was associated with reduced risk. Compared with men in the lowest quintile of total vitamin D intake, those in the highest quintile had an 18% reduced BPH risk (p-trend = 0.032). Compared with men eating red meat less than once per week, men eating red meat at least daily had a 38% increased BPH risk ($p = 0.044$) and, compared with men eating fewer than one serving of vegetables per day, men eating four or more servings had a 32% decreased BPH risk ($p = 0.011$) [62]. The protective role of lycopene supplementation on BPH has been also shown by Schwarz et al. who enrolled 40 patients with histologically proven BPH. The authors divided their patients into two groups: (i) lycopene at a dose of 15 mg/d for 6 months, (ii) placebo for 6 months, and found that patients who received lycopene had decreased PSA levels. No evidence of further prostate enlargement nor higher amelioration in symptoms of the disease were found, as assessed by the International Prostate Symptom Score (IPSS) questionnaire, compared to those who were given the placebo [63]. Further evidence of the anti-BPH effect of lycopene has been previously provided by Kim et al. from a clinical study on prostate cancer patients in which lycopene was found capable of inducing apoptosis in tumor-free prostatic tissue exhibiting the histologic features of BPH [64].

Rohrmann et al. prospectively investigated the effects of fruits, vegetables and micronutrients on BPH and found an inverse correlation between intake of vegetables, especially those rich in beta-carotene, lutein, vitamin C (VC), and BPH. Interestingly, fruit intake resulted to be unrelated to the onset of the disease [65]. Conversely, the study by Lagiou et al., through the administration of a food frequency questionnaire, reported for a cohort of 420 patients, all permanent residents in Athens area, that fruit consumption with high levels of beta-carotene, lutein and VC were inversely correlated with BPH risk, while a high-fat diet, especially with increased intake of butter and margarine, had a positive correlation with the disease [66]. Based on these findings, the exact effects of a fruit-rich diet on BPH is still to be elucidated. It may be hypothesized that fruit consumption, according to the type and quantity, influences BPH onset and progression in a diversified manner. Similarly, little is known about the usefulness of polyphenols contained in green tea; in this regard, it has been suggested that they could be used for the treatment of BPH-related symptoms, due to their positive effects on LUTS [67].

Even with regard to the role of a high-fat diet, the evidence from the literature is not unique; although some of the above-mentioned studies stated an increased risk of BPH in patients with high fat consumption [62,66], Suzuki et al. found that BPH risk was modestly associated with intake of eicosapentaenoic, docosahexaenoic, and arachidonic acids, but not with energy-adjusted total fat intake [68].

Similar to what has been reported on animal models [52,53], a strong utility of dietary flaxseed lignan extract in improving BPH-related LUTSs, compared to that of alpha1A-adrenoceptor blockers and 5alpha-reductase inhibitors, has been reported by Zhang et al. [69]. These authors conducted a randomized clinical trial in which a placebo, 300, or 600 mg/day of secoisolariciresinol diglucoside, a flaxseed extract, were administered to 87 patients affected by BPH and found a decrease in IPSS and improvement of quality of life score and LUTSs, respectively [69]. Conversely, it has been shown that pumpkin seed extract did not have any benefits on BPH compared to the placebo over a 1-year period [70].

The above-mentioned findings led us to emphasize that the exact relationship between different dietary patterns and BPH has not yet been fully elucidated; this is probably due to the fact that BPH is a multifactorial disease, whose pathogenesis seems to be correlated with different biological factors, including oxidative stress, androgenic stimulation and inflammatory and growth factors. Finally, recent advances have also highlighted the potential role of statins in reverting BPH symptoms by the improvement of hypercholesterolemia and metabolic syndrome [71]. As far as we are aware, as most macro- and micronutrients that have been associated with BPH risk also influence steroid concentrations, oxidative stress level and inflammation, it is reliable to suppose that they also have positive effects on BPH, regulating prostatic growth and smooth muscle tone (Figure 4).

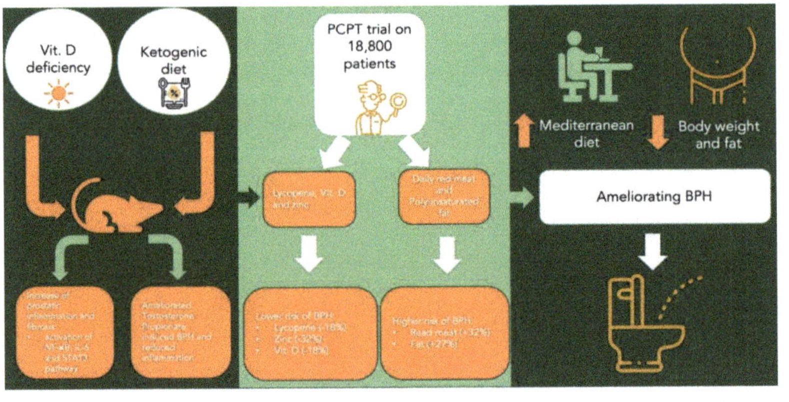

Figure 4. Relationship between dietary patterns and BPH. Relationship between dietary patterns and BPH. STAT3 = Signal Transducer And Activator Of Transcription 3; BPH = benign prostatic hyperplasia.

5. Conclusions

In the present review, we have highlighted how the association between dietary patterns and two of the most frequent pathologies in urology, namely, erectile dysfunction and benign prostatic hyperplasia, is present in the literature. Evidence comes from both animal studies and in part from human studies. The data suggested that a diet that is more adherent to the Mediterranean diet or that emphasizes the presence of vegetables, fruits, nuts, legumes, and fish or other sources of long-chain (n-3) fats, in addition to reduced content of red meat, may have a beneficial role on erectile function. At the same time, the same beneficial effects can be transferred to the BPH side due to the indirect regulatory effects on prostatic growth and smooth muscle tone, thus determining an improvement in symptoms. Certainly, in-depth studies and translational medicine are needed to confirm these encouraging data. Studies could address the relationship of dietary patterns and tissue expression as a marker of disease severity, such as NO and cAMP for ED and markers of inflammation for BPH.

Finally, clinical studies investigating the role of specific drugs for metabolic syndromes, such as statins or hypoglycemic drugs, together with investigations of dietary patterns could be beneficial in better understanding how to counteract ED and BPH.

Author Contributions: Conceptualization, G.I.R.; methodology, G.I.R.; writing—original draft preparation, G.I.R., G.B., M.D.M.; writing—review and editing, A.C., M.F., P.C., I.S., M.G., R.C., EAU-YAU Sexual, Reproductive Health Group. All authors have read and agreed to the published version of the manuscript.

Funding: This research received no external funding.

Institutional Review Board Statement: Not applicable.

Informed Consent Statement: Not applicable.

Acknowledgments: We would like to thank Giorgio Ivan Russo, Paolo Capogrosso, Marco Falcone, Ioannis Sokolakis, Murat Gül as member of the EAU-YAU Sexual, Reproductive Health Group.

Conflicts of Interest: The authors declare no conflict of interest.

References

1. Fung, T.T.; Rexrode, K.M.; Mantzoros, C.S.; Manson, J.E.; Willett, W.C.; Hu, F.B. Mediterranean diet and incidence of and mortality from coronary heart disease and stroke in women. *Circulation* **2009**, *119*, 1093–1100. [CrossRef]
2. Russo, G.I.; Cimino, S.; Fragalà, E.; Privitera, S.; La Vignera, S.; Condorelli, R.; Calogero, A.E.; Castelli, T.; Favilla, V.; Morgia, G. Insulin resistance is an independent predictor of severe lower urinary tract symptoms and of erectile dysfunction: Results from a cross-sectional study. *J. Sex. Med.* **2014**, *11*, 2074–2082. [CrossRef]
3. Russo, G.I.; Castelli, T.; Urzì, D.; Privitera, S.; La Vignera, S.; Condorelli, R.A.; Calogero, A.E.; Favilla, V.; Cimino, S.; Morgia, G. Emerging links between non-neurogenic lower urinary tract symptoms secondary to benign prostatic obstruction, metabolic syndrome and its components: A systematic review. *Int. J. Urol.* **2015**, *22*, 982–990. [CrossRef]
4. Russo, G.I.; Campisi, D.; Mauro, M.D.; Regis, F.; Reale, G.; Marranzano, M.; Ragusa, R.; Solinas, T.; Madonia, M.; Cimino, S.; et al. Dietary consumption of phenolic acids and prostate cancer: A case-control study in sicily, Southern Italy. *Molecules* **2017**, *22*, 2159. [CrossRef] [PubMed]
5. Russo, G.I.; Di Mauro, M.; Regis, F.; Reale, G.; Campisi, D.; Marranzano, M.; Giudice, A.L.; Solinas, T.; Madonia, M.; Cimino, S.; et al. Association between dietary phytoestrogens intakes and prostate cancer risk in Sicily. *Aging Male* **2018**, *21*, 48–54. [CrossRef]
6. Reale, G.; Russo, G.I.; Di Mauro, M.; Regis, F.; Campisi, D.; Giudice, A.L.; Marranzano, M.; Ragusa, R.; Castelli, T.; Cimino, S.; et al. Association between dietary flavonoids intake and prostate cancer risk: A case-control study in Sicily. *Complement Ther. Med.* **2018**, *39*, 14–18. [CrossRef]
7. McKinlay, J.B. The worldwide prevalence and epidemiology of erectile dysfunction. *Int. J. Impot. Res.* **2000**, *12* (Suppl. S4), S6–S11. [CrossRef] [PubMed]
8. Russo, G.I.; di Mauro, M.; Cocci, A.; Cacciamani, G.; Cimino, S.; Serefoglu, E.C.; Albersen, M.; Capogrosso, P.; Fode, M.; Verze, P. Consulting "Dr Google" for sexual dysfunction: A contemporary worldwide trend analysis. *Int. J. Impot. Res.* **2019**, *32*, 455–461. [CrossRef]
9. Salonia, A.; Bettocchi, C.; Boeri, L.; Capogrosso, P.; Carvalho, J.; Cilesiz, N.C.; Cocci, A.; Corona, G.; Dimitropoulos, K.; Gül, M.; et al. European Association of Urology Guidelines on Sexual and Reproductive Health—2021 Update: Male Sexual Dysfunction. *Eur. Urol.* **2021**, *80*, 333–357. [CrossRef]
10. Kessler, A.; Sollie, S.; Challacombe, B.; Briggs, K.; Van Hemelrijck, M. The global prevalence of erectile dysfunction: A review. *BJU Int.* **2019**, *124*, 587–599. [CrossRef]
11. De Nunzio, C.; Roehrborn, C.G.; Andersson, K.-E.; McVary, K.T. Erectile Dysfunction and Lower Urinary Tract Symptoms. *Eur. Urol. Focus* **2017**, *3*, 352–363. [CrossRef] [PubMed]
12. ElJalby, M.; Thomas, D.; Elterman, D.; Chughtai, B. The effect of diet on BPH.; LUTS and ED. *World J. Urol.* **2019**, *37*, 1001–1005. [CrossRef]
13. Kalter-Leibovici, O.; Wainstein, J.; Ziv, A.; Harman-Bohem, I.; Murad, H.; Raz, I. Clinical, socioeconomic, and lifestyle parameters associated with erectile dysfunction among diabetic men. *Diabetes Care* **2005**, *28*, 1739–1744. [CrossRef] [PubMed]
14. Bacon, C.G.; Mittleman, M.A.; Kawachi, I.; Giovannucci, E.; Glasser, D.B.; Rimm, E.B. Sexual Function in Men Older Than 50 Years of Age: Results from the Health Professionals Follow-up Study. *Ann. Intern. Med.* **2003**, *139*, 161–168. [CrossRef]

15. Musicki, B.; Liu, T.; Strong, T.; Jin, L.; Laughlin, M.H.; Turk, J.R.; Burnett, A.L. Low-fat diet and exercise preserve eNOS regulation and endothelial function in the penis of early atherosclerotic pigs: A molecular analysis. *J. Sex. Med.* **2008**, *5*, 552–561. [CrossRef]
16. Esposito, K.; Giugliano, F.; Maiorino, M.I.; Giugliano, D. Dietary Factors, Mediterranean Diet and Erectile Dysfunction. *J. Sex. Med.* **2010**, *7*, 2338–2345. [CrossRef]
17. Akomolafe, S.F.; Olasehinde, T.A.; Aluko, B.T. Diets supplemented with raw and roasted pumpkin (Cucurbita pepo L) seeds improved some biochemical parameters associated with erectile function in rats. *J. Food Biochem.* **2021**, *45*, e13629. [CrossRef]
18. Akomolafe, S.F.; Oboh, G.; Oyeleye, S.I.; Molehin, O.R.; Ogunsuyi, O.B. Phenolic composition and inhibitory ability of methanolic extract from pumpkin (Cucurbita pepo L) seeds on Fe-induced thiobarbituric acid reactive species in albino rat's testicular tissue in-vitro. *J. Appl. Pharm. Sci.* **2016**, *6*, 115–120. [CrossRef]
19. Robson, S.C.; Sévigny, J.; Zimmermann, H. The E-NTPDase family of ectonucleotidases: Structure function relationships and pathophysiological significance. *Purinergic Signal* **2006**, *2*, 409–430. [CrossRef]
20. Sträter, N. Ecto-5′-nucleotidase: Structure function relationships. *Purinergic Signal* **2006**, *2*, 343–350. [CrossRef]
21. Blackburn, M.R.; Kellems, R.E. Adenosine deaminase deficiency: Metabolic basis of immune deficiency and pulmonary inflammation. *Adv. Immunol.* **2005**, *86*, 1–41. [CrossRef]
22. Olabiyi, A.A.; Carvalho, F.B.; Bottari, N.B.; Lopes, T.F.; da Costa, P.; Stefanelo, N.; Morsch, V.M.; Akindahunsi, A.A.; Oboh, G.; Schetinger, M.R. Dietary supplementation of tiger nut alters biochemical parameters relevant to erectile function in L-NAME treated rats. *Food Res. Int.* **2018**, *109*, 358–367. [CrossRef]
23. Wen, J.; Jiang, X.; Dai, Y.; Zhang, Y.; Tang, Y.; Sun, H.; Mi, T.; Kellems, R.E.; Blackburn, M.R.; Xia, Y. Adenosine Deaminase Enzyme Therapy Prevents and Reverses the Heightened Cavernosal Relaxation in Priapism. *J. Sex. Med.* **2010**, *7*, 3011–3022. [CrossRef] [PubMed]
24. Olabiyi, A.A.; AlliSmith, Y.R.; Ukwenya, V.O. Quercetin enhances sexual behavior and improves ectonucleotidases activity in the hypothalamus of rats treated with cyclosporine. *J. Food Biochem.* **2021**, *45*, e13864. [CrossRef]
25. Eleazu, C.; Obianuju, N.; Eleazu, K.; Kalu, W. The role of dietary polyphenols in the management of erectile dysfunction–Mechanisms of action. *Biomed. Pharmacother.* **2017**, *88*, 644–652. [CrossRef] [PubMed]
26. Cheitlin, M.D. Erectile dysfunction: The earliest sign of generalized vascular disease? *J. Am. Coll. Cardiol.* **2004**, *43*, 185–186. [CrossRef] [PubMed]
27. Minhas, S.; Bettocchi, C.; Boeri, L.; Capogrosso, P.; Carvalho, J.; Cilesiz, N.C.; Cocci, A.; Corona, G.; Dimitropoulos, K.; Gül, M.; et al. European Association of Urology Guidelines on Male Sexual and Reproductive Health: 2021 Update on Male Infertility. *Eur. Urol.* **2021**, *80*, 603–620. [CrossRef]
28. Djordjevic, D.; Vukovic, I.; Milenkovic Petronic, D.; Radovanovic, G.; Seferovic, J.; Micic, S.; Tepavcevic, D.K. Erectile dysfunction as a predictor of advanced vascular age. *Andrology* **2015**, *3*, 1125–1131. [CrossRef]
29. Blumentals, W.A.; Gomez-Caminero, A.; Joo, S.; Vannappagari, V. Should erectile dysfunction be considered as a marker for acute myocardial infarction? Results from a retrospective cohort study. *Int. J. Impot. Res.* **2004**, *16*, 350–353. [CrossRef] [PubMed]
30. Nicolosi, A.; Glasser, D.B.; Moreira, E.D.; Villa, M. Prevalence of erectile dysfunction and associated factors among men without concomitant diseases: A population study. *Int. J. Impot. Res.* **2003**, *15*, 253–257. [CrossRef] [PubMed]
31. Lu, Y.; Kang, J.; Li, Z.; Wang, X.; Liu, K.; Zhou, K.; Wang, W.; Shen, C. The association between plant-based diet and erectile dysfunction in Chinese men. *Basic Clin. Androl.* **2021**, *31*, 1–10. [CrossRef]
32. Esposito, K.; Giugliano, F.; De Sio, M.; Carleo, D.; Di Palo, C.; D'Armiento, M. Dietary factors in erectile dysfunction. *Int. J. Impot. Res.* **2006**, *18*, 370–374. [CrossRef] [PubMed]
33. Giugliano, F.; Maiorino, M.I.; Bellastella, G.; Autorino, R.; De Sio, M.; Giugliano, D.; Esposito, K. Adherence to mediterranean diet and erectile dysfunction in men with type 2 diabetes. *J. Sex. Med.* **2010**, *7*, 1911–1917. [CrossRef]
34. Lopez-Garcia, E.; Hu, F.B. Nutrition and the endothelium. *Curr. Diab. Rep.* **2004**, *4*, 253–259. [CrossRef]
35. Trebaticky, B.; Muchova, J.; Ziaran, S.; Bujdak, P.; Breza, J.; Durackova, Z. Natural polyphenols improve erectile function and lipid profile in patients suffering from erectile dysfunction. *Bratisl. Med. J.* **2019**, *120*, 941–944. [CrossRef] [PubMed]
36. Huetos, A.S.; Muralidharan, J.; Galiè, S.; Salas-Salvadó, J.; Bulló, M. Effect of nut consumption on erectile and sexual function in healthy males: A Secondary outcome analysis of the fertinuts randomized controlled trial. *Nutrients* **2019**, *11*, 1372. [CrossRef]
37. Mykoniatis, I.; Grammatikopoulou, M.G.; Bouras, E.; Karampasi, E.; Tsionga, A.; Kogias, A.; Vakalopoulos, I.; Haidich, A.-B.; Chourdakis, M. Sexual Dysfunction Among Young Men: Overview of Dietary Components Associated With Erectile Dysfunction. *J. Sex. Med.* **2018**, *15*, 176–182. [CrossRef]
38. Cassidy, A.; Franz, M.; Rimm, E.B. Dietary flavonoid intake and incidence of erectile dysfunction. *Am. J. Clin. Nutr.* **2016**, *103*, 534–541. [CrossRef]
39. Bauer, S.R.; Breyer, B.N.; Stampfer, M.J.; Rimm, E.B.; Giovannucci, E.L.; Kenfield, S.A. Association of Diet with Erectile Dysfunction among Men in the Health Professionals Follow-up Study. *JAMA Netw. Open* **2020**, *3*, e2021701. [CrossRef]
40. McNeal, J.E. Normal Histology of the Prostate. *Am. J. Surg. Pathol.* **1988**, *12*, 619–633. [CrossRef]
41. Broggi, G.; Lo Giudice, A.; Di Mauro, M.; Asmundo, M.G.; Pricoco, E.; Piombino, E.; Caltabiano, R.; Morgia, G.; Russo, G.I. SRSF-1 and microvessel density immunohistochemical analysis by semi-automated tissue microarray in prostate cancer patients with diabetes (DIAMOND study). *Prostate* **2021**, *81*, 882–892. [CrossRef]

42. Broggi, G.; Lo Giudice, A.; Di Mauro, M.; Pricoco, E.; Piombino, E.; Ferro, M.; Caltabiano, R.; Morgia, G.; Russo, G.I. Insulin signaling, androgen receptor and PSMA immunohistochemical analysis by semi-automated tissue microarray in prostate cancer with diabetes (DIAMOND study). *Transl. Res.* **2021**, *238*, 25–35. [CrossRef] [PubMed]
43. Adaramoye, O.A.; Akanni, O.O.; Abiola, O.J.; Owumi, S.E.; Akinloye, O.; Olapade-olaopa, E.O. Methyl jasmonate reduces testosterone-induced benign prostatic hyperplasia through regulation of inflammatory and apoptotic processes in rats. *Biomed. Pharmacother.* **2017**, *95*, 1493–1503. [CrossRef]
44. Quintero-García, M.; Delgado-González, E.; Sánchez-Tusie, A.; Vázquez, M.; Aceves, C.; Anguiano, B. Iodine prevents the increase of testosterone-induced oxidative stress in a model of rat prostatic hyperplasia. *Free Radic. Biol. Med.* **2018**, *115*, 298–308. [CrossRef]
45. Scott Lucia, M.; Lambert, J.R. Growth factors in benign prostatic hyperplasia: Basic science implications. *Curr. Urol. Rep.* **2008**, *9*, 272–278. [CrossRef]
46. Shankar, E.; Bhaskaran, N.; MacLennan, G.T.; Liu, G.; Daneshgari, F.; Gupta, S. Inflammatory Signaling Involved in High-Fat Diet Induced Prostate Diseases. *J. Urol. Res.* **2015**, *2*, 1018.
47. Denis, L.; Morton, M.S.; Griffiths, K. Diet and Its Preventive Role in Prostatic Disease. *Eur. Urol.* **1999**, *35*, 377–387. [CrossRef] [PubMed]
48. Penson, D.F.; Albertsen, P.C. Lessons learnt about early prostate cancer from large scale databases: Population-based pearls of wisdom. *Surg. Oncol.* **2002**, *11*, 3–11. [CrossRef]
49. Sebastiano, C. Dietary patterns and prostatic diseases. *Front. Biosci.* **2012**, *E4*, 195. [CrossRef]
50. Zhang, Z.-H.; Luo, B.; Xu, S.; Fu, L.; Chen, Y.-H.; Zhang, C.; Wang, H.; Xie, D.-D.; Xu, D.-X. Vitamin D deficiency promotes prostatic hyperplasia in middle-age mice through exacerbating local inflammation. *J. Steroid Biochem. Mol. Biol.* **2018**, *182*, 14–20. [CrossRef]
51. Li, Y.-Z.; Shi, B.-K.; Li, J.-Y.; Zhu, X.-W.; Liu, J.; Liu, Y.-L. Role of p-ERK1/2 in Benign Prostatic Hyperplasia during Hyperinsulinemia. *Urol. J.* **2020**, *18*, 225–229. [CrossRef]
52. de Amorim Ribeiro, I.C.; da Costa, C.A.S.; da Silva, V.A.P.; Côrrea, L.B.N.S.; Boaventura, G.T.; Chagas, M.A. Flaxseed reduces epithelial proliferation but does not affect basal cells in induced benign prostatic hyperplasia in rats. *Eur. J. Nutr.* **2017**, *56*, 1201–1210. [CrossRef]
53. Said, M.M.; Hassan, N.S.; Schlicht, M.J.; Bosland, M.C. Flaxseed Suppressed Prostatic Epithelial Proliferation in a Rat Model of Benign Prostatic Hyperplasia. *J. Toxicol. Environ. Health Part A* **2015**, *78*, 453–465. [CrossRef] [PubMed]
54. Kayode, O.T.; Owolabi, A.V.; Kayode, A.A.A. Biochemical and histomorphological changes in testosterone propionate-induced benign prostatic hyperplasia in male Wistar rats treated with ketogenic diet. *Biomed. Pharmacother.* **2020**, *132*, 110863. [CrossRef]
55. Chen, J.; Song, H. Protective potential of epigallocatechin-3-gallate against benign prostatic hyperplasia in metabolic syndrome rats. *Environ. Toxicol. Pharmacol.* **2016**, *45*, 315–320. [CrossRef]
56. Aljehani, A.A.; Albadr, N.A.; Nasrullah, M.Z.; Neamatallah, T.; Eid, B.G.; Abdel-Naim, A.B. Icariin ameliorates metabolic syndrome-induced benign prostatic hyperplasia in rats. *Environ. Sci. Pollut. Res.* **2021**, 1–9. [CrossRef] [PubMed]
57. Tsai, H.-H.; Chen, C.-W.; Yu, P.-L.; Lin, Y.-L.; Hsieh, R-H. Mangosteen pericarp components alleviate progression of prostatic hyperplasia and mitochondrial dysfunction in rats. *Sci. Rep.* **2020**, *10*, 322. [CrossRef] [PubMed]
58. Berry, S.J.; Coffey, D.S.; Walsh, P.C.; Ewing, L.L. The Development of Human Benign Prostatic Hyperplasia with Age. *J. Urol.* **1984**, *132*, 474–479. [CrossRef]
59. Chughtai, B.; Forde, J.C.; Thomas, D.D.M.; Laor, L.; Hossack, T.; Woo, H.H.; Te, A.E.; Kaplan, S.A. Benign prostatic hyperplasia. *Nat. Rev. Dis. Prim.* **2016**, *2*, 16031. [CrossRef] [PubMed]
60. Parsons, J.K. Modifiable Risk Factors for Benign Prostatic Hyperplasia and Lower Urinary Tract Symptoms: New Approaches to Old Problems. *J. Urol.* **2007**, *178*, 395–401. [CrossRef] [PubMed]
61. Pashootan, P.; Ploussard, G.; Cocaul, A.; De Gouvello, A.; Desgrandchamps, F. Association between metabolic syndrome and severity of lower urinary tract symptoms (LUTS): An observational study in a 4666 European men cohort. *BJU Int.* **2015**, *116*, 124–130. [CrossRef] [PubMed]
62. Kristal, A.R.; Arnold, K.B.; Schenk, J.M.; Neuhouser, M.L.; Goodman, P.; Penson, D.F.; Thompson, I.M. Dietary Patterns, Supplement Use, and the Risk of Symptomatic Benign Prostatic Hyperplasia: Results from the Prostate Cancer Prevention Trial. *Am. J. Epidemiol.* **2008**, *167*, 925–934. [CrossRef] [PubMed]
63. Schwarz, S.; Obermuller-Jevic, U.C.; Hellmis, E.; Koch, W.; Jacobi, G.; Biesalski, H.-K. Lycopene Inhibits Disease Progression in Patients with Benign Prostate Hyperplasia. *J. Nutr.* **2008**, *138*, 49–53. [CrossRef] [PubMed]
64. Kim, H.-S.; Bowen, P.; Chen, L.; Duncan, C.; Ghosh, L.; Sharifi, R.; Christov, K. Effects of Tomato Sauce Consumption on Apoptotic Cell Death in Prostate Benign Hyperplasia and Carcinoma. *Nutr. Cancer* **2003**, *47*, 40–47. [CrossRef] [PubMed]
65. Rohrmann, S.; Giovannucci, E.; Willett, W.C.; Platz, E.A. Fruit and vegetable consumption, intake of micronutrients, and benign prostatic hyperplasia in US men. *Am. J. Clin. Nutr.* **2007**, *85*, 523–529. [CrossRef]
66. Lagiou, P.; Wuu, J.; Trichopoulou, A.; Hsieh, C.-C.; Adami, H.-O.; Trichopoulos, D. Diet and benign prostatic hyperplasia: A study in Greece. *Urology* **1999**, *54*, 284–290. [CrossRef]
67. Bettuzzi, S.; Brausi, M.; Rizzi, F.; Castagnetti, G.; Peracchia, G.; Corti, A. Chemoprevention of Human Prostate Cancer by Oral Administration of Green Tea Catechins in Volunteers with High-Grade Prostate Intraepithelial Neoplasia: A Preliminary Report from a One-Year Proof-of-Principle Study. *Cancer Res.* **2006**, *66*, 1234–1240. [CrossRef]

68. Suzuki, S.; Platz, E.A.; Kawachi, I.; Willett, W.C.; Giovannucci, E. Intakes of energy and macronutrients and the risk of benign prostatic hyperplasia. *Am. J. Clin. Nutr.* **2002**, *75*, 689–697. [CrossRef]
69. Zhang, W.; Wang, X.; Liu, Y.; Tian, H.; Flickinger, B.; Empie, M.W.; Sun, S.Z. Effects of Dietary Flaxseed Lignan Extract on Symptoms of Benign Prostatic Hyperplasia. *J. Med. Food* **2008**, *11*, 207–214. [CrossRef]
70. Vahlensieck, W.; Theurer, C.; Pfitzer, E.; Patz, B.; Banik, N.; Engelmann, U. Effects of Pumpkin Seed in Men with Lower Urinary Tract Symptoms due to Benign Prostatic Hyperplasia in the One-Year, Randomized, Placebo-Controlled GRANU Study. *Urol. Int.* **2015**, *94*, 286–295. [CrossRef] [PubMed]
71. Russo, G.I.; Larganà, G.; Sebastianelli, A.; Cocci, A.; Mauro, M.; Di Rapallo, I.; Morgia, G.; Morgia, M.; La Vignera, S.; Condorelli, R.; et al. The Investigative Role of Statins in Ameliorating Lower Urinary Tract Symptoms (LUTS): A Systematic Review. *J. Clin. Med.* **2021**, *10*, 416. [CrossRef] [PubMed]

nutrients

Review

Nutrition and Kidney Stone Disease

Roswitha Siener

University Stone Center, Department of Urology, University Hospital Bonn, Venusberg-Campus 1, 53127 Bonn, Germany; Roswitha.Siener@ukbonn.de; Tel.: +49-228-287-19034

Abstract: The prevalence of kidney stone disease is increasing worldwide. The recurrence rate of urinary stones is estimated to be up to 50%. Nephrolithiasis is associated with increased risk of chronic and end stage kidney disease. Diet composition is considered to play a crucial role in urinary stone formation. There is strong evidence that an inadequate fluid intake is the major dietary risk factor for urolithiasis. While the benefit of high fluid intake has been confirmed, the effect of different beverages, such as tap water, mineral water, fruit juices, soft drinks, tea and coffee, are debated. Other nutritional factors, including dietary protein, carbohydrates, oxalate, calcium and sodium chloride can also modulate the urinary risk profile and contribute to the risk of kidney stone formation. The assessment of nutritional risk factors is an essential component in the specific dietary therapy of kidney stone patients. An appropriate dietary intervention can contribute to the effective prevention of recurrent stones and reduce the burden of invasive surgical procedures for the treatment of urinary stone disease. This narrative review has intended to provide a comprehensive and updated overview on the role of nutrition and diet in kidney stone disease.

Keywords: calcium oxalate stone formation; diet; dietary assessment; fatty acids; fluid; oxalate; protein; sodium; uric acid; water

Citation: Siener, R. Nutrition and Kidney Stone Disease. *Nutrients* **2021**, *13*, 1917. https://doi.org/10.3390/nu13061917

Academic Editor: Ina Bergheim

Received: 2 May 2021
Accepted: 31 May 2021
Published: 3 June 2021

1. Introduction

The prevalence of urolithiasis of the general population has increased worldwide in the past decades and was reported to be 4.7% in Germany and up to 10.1% in the United States [1–3]. The recurrence rate of urinary stones is high and is estimated to be approximately 50% at 10 years [4]. Nephrolithiasis is associated with an increased risk of chronic and end-stage kidney disease, probably due to kidney injury from obstructive nephropathy [5,6]. The most common stone type is calcium oxalate (67%) followed by calcium phosphate (17%), uric acid (8%), struvite (3%), and cystine (0.4%) [7]. Urinary stone formation is a multifactorial process to which metabolic derangements, genetic factors, anatomical and functional abnormalities may contribute, whereby nutrition plays a crucial role. Diet composition can affect urine risk profile and the supersaturation with the stone-forming salt, which can modify the risk of urinary stone formation [8,9].

A reliable stone analysis is an essential precondition for specific treatment regimens, because urinary risk factors for stone formation vary with the type of stone [10,11]. Moreover, a thorough metabolic evaluation of the stone patient is required, including detailed medical history, dietary assessment as well as blood and urine analysis [11–17]. To reduce the risk of recurrent stone formation, targeted dietary treatment should be individually adjusted to the metabolic risk profile of each patient. The collection of two consecutive 24 h urines is recommended to detect frequent metabolic disturbances such as hypercalciuria, hypocitraturia, hyperoxaluria, and hyperuricosuria and to identify dietary risk factors for kidney stone formation. Specific nutritional therapy, based on dietary assessment and metabolic evaluation, has been demonstrated to be more effective than general dietary measures in preventing recurrent stone formation [18]. The aim of this narrative review is to summarize the current knowledge of the role of nutrition in kidney stone formation.

2. Fluid Intake

A low urine volume caused by insufficient fluid consumption or excessive fluid loss is one of the most crucial risk factors for kidney stone formation [19]. Numerous physiological and pathological conditions may lead to dehydration that modifies water needs. These factors include excessive sweating due to heat exposure, mental stress, high physical activity level, and occupation as well as chronic diarrhea in the setting of fat malabsorption due to different gastrointestinal disorders [19–21]. A study of 100 steel plant workers reported a history of stone disease in 16%, with a urine osmolality of more than 700 mOsm, defined as dehydration, in more than 50% of workers [22]. Occupations at elevated risk for stone formation also include health care professionals with limited access to water. A survey of employees found that physicians working in the operating room had the highest prevalence of nephrolithiasis (17.4% vs. 9.7%) and reported higher stress levels and lower fluid intake compared to those who work at other locations [23]. Poor access to fluids or to bathroom facilities are other circumstances, in which occupation can affect urolithiasis, such as, for example, in professional drivers, airplane pilots or schoolteachers [24].

A randomized controlled study of 199 first-time idiopathic calcium oxalate stone patients evaluated the impact of a high fluid intake in preventing stone recurrence [25]. Patients were randomly assigned to a high water intake to achieve a urine volume of at least 2 L/day or to a control group without specific instructions. After the five-year follow-up period, the intervention group was found to have a 2.5-fold higher urine volume, a lower recurrence rate (12% vs. 27%; $p = 0.008$) and a longer time to recurrence (38.7 months vs. 25.1 months; $p = 0.016$) than the control group. Several systematic reviews and meta-analyses on the role of fluid intake in the secondary prevention of urolithiasis concluded that high total fluid intake to achieve a urine volume of greater than 2.0 to 2.5 L/day decreases the risk of stone recurrence [26–31]. While fluid intake was reported to be a protective factor in the primary prevention of kidney stone formation in cohort studies [30], no evidence was found from randomized controlled trials [31].

An adequate fluid intake is the most important nutritional measure to prevent kidney stone recurrence, regardless of urine stone composition and individual risk factors for stone formation [19]. The intake of an adequate amount of fluid increases urine dilution, thereby reducing the concentration of lithogenic constituents and encouraging the expulsion of crystals by decreasing the renal intratubular transit time [32]. According to the guidelines on urolithiasis, a copious fluid intake to maintain a urine volume of at least 2.0 to 2.5 L/24 h is recommended for most types of stones [15,16]. For the recurrence prevention of patients with cystinuria, excessive urine dilution is of the utmost importance. In adult cystine stone patients, urine volume of at least 3.0 L/24 h should be achieved to decrease urinary cystine concentration to below the solubility limit of 1.3 mmol/L at pH 6.0 [14,15,33]. A study of 27 adult patients with cystinuria demonstrated that maintaining a urine volume of more than 3.0 L/day significantly reduced recurrent stone formation [34]. The specific recommendations regarding fluid intake are summarized in Table 1.

Table 1. Recommendations for fluid intake—adapted from [14–16].

General Measures
Urine volume: at least 2.0 to 2.5 L/24 h Urine density: <1.010 g/cm^3 Fluid intake evenly distributed throughout the day Fluid intake before going to bed Replacement of extrarenal fluid losses caused by extensive physical activity, hot and/or dry environments, occupation, mental stress, and diarrhea
Neutral Beverages
Fruit tea, herbal tea, kidney tea, bladder tea Tap water (attention must be paid to the sterility of water) Mineral water with a low content of calcium, bicarbonate, and sulfate

Table 1. *Cont.*

Alkalizing Beverages
Mineral water with high bicarbonate (≥1500 mg/L) and low calcium content (<150 mg/L)
Orange juice
Unsuitable Beverages
Green tea, black tea, caffeinated coffee (maximum 0.5 L/day)
Sugar-sweetened soft drinks, including cola
Alcoholic beverages, including wine and beer

While the benefit of high fluid intake has been confirmed, the effect of different beverages is still debated. The constituents of different beverages can affect urine composition and consequently the risk of stone formation.

2.1. Tap Water and Mineral Water

The impact of the composition of tap water and mineral water on kidney stone formation is still a matter of debate. The composition of drinking water, predominantly the content of the divalent cations calcium and magnesium, varies widely between geographic regions within the same country [35]. Hard tap water, defined as a calcium carbonate concentration above 120 mg/L [36], can contribute to the daily dietary calcium intake. A study on 2295 patients from two regions in the United States found an increased risk of urinary stone formation for individuals consuming tap water from a private well [37]. However, the cause of elevated risk by private well users is unknown. Other cohort studies have not identified an association between the hardness of water from public water suppliers and urinary stone disease [37–39].

In contrast to tap water, bicarbonate is a natural constituent of mineral water besides calcium, magnesium and other ions. The intake of bicarbonate increases the buffering capacity of the body and is a strong alkalizing agent. Mineral water rich in bicarbonate can support alkalinization therapy and contribute to inhibitory potential of urine by increasing urine pH and citrate excretion [40]. A randomized crossover trial of healthy subjects compared the effect of an equimolar alkali load in form of bicarbonate-rich mineral water or potassium citrate on the urinary risk profile for calcium oxalate and uric acid stone formation [41]. The intake of 2 L/day of mineral water containing 1715 mg/L bicarbonate or 2.55 g/day potassium citrate significantly increased urine pH and citrate excretion and decreased oxalate excretion. The relative supersaturation of calcium oxalate and uric acid declined significantly in both groups. A study of healthy individuals conducted under controlled dietary conditions demonstrated that a mineral water rich in bicarbonate, calcium and magnesium increased urine pH, as well as the excretion of the urinary inhibitors of calcium oxalate stone formation citrate and magnesium [42]. Despite the significant increase in calcium excretion, no change in the relative supersaturation of calcium oxalate occurred.

A study of 22 idiopathic calcium oxalate stone patients evaluated the impact of three mineral waters with different contents of bicarbonate and calcium. The intake of the mineral water rich in bicarbonate and calcium resulted in a significant increase in urinary citrate and decrease in urinary oxalate excretion, while urinary calcium excretion and the supersaturation of calcium oxalate did not change [43]. A study by Rodgers [44] found that the intake of 1.5 L/day of a mineral water with high calcium content (202 mg/L) compared to 1.5 L/day of a tap water with low calcium content (13 mg/L) significantly increased calcium excretion in 20 healthy individuals of each sex as well as 20 female calcium oxalate stone formers but not in male patients. Urinary oxalate excretion and the relative supersaturation of calcium oxalate did not change in any of the four groups. A double-blind crossover study of 34 recurrent calcium oxalate stone patients under a regular diet examined the effect of 1.5 L/day of a bicarbonate-rich mineral water (2673 mg/L) vs. 1.5 L/day of a mineral water with a low bicarbonate content (98 mg/L) on the risk of urinary stone formation [45]. The intake of the bicarbonate-rich mineral water resulted in a significant rise in urine pH, citrate and magnesium excretion compared to the control. The relative supersaturation of calcium oxalate decreased to a similar extent in both groups.

The impact of mineral water on the risk of calcium oxalate and uric acid stone formation is mainly determined by the presence of bicarbonate [40,41,45]. The effect of bicarbonate-rich mineral water corresponds to that of sodium bicarbonate in galenic form [46]. If calcium in high concentration is present in the water, the positive effect of increased urine pH and citrate excretion is neutralized by an increased urinary calcium excretion [42,44]. A randomized trial of 129 healthy women and men found that the consumption of at least 1.5 to 2.0 L/d of mineral water rich in bicarbonate (>1800 mg/L) can decrease the net acid excretion by reducing the dietary acid load [47]. Urine alkalinization is an important nutritional measure in the treatment of patients with calcium oxalate, uric acid and cystine stones, but it is not indicated in calcium phosphate and struvite stone disease [14]. Comparisons of the composition of commercial bottled 'still', 'carbonated' and 'sparkling' water from 10 European countries found a wide variation in the calcium content, with concentrations reaching up to 581.6 mg/L [48,49]. As the dietary reference intake of calcium of a total of 1000 to 1200 mg per day are already achieved by consuming 2 L of calcium-rich water, patients should be aware of the calcium content of the water.

2.2. Fruit Juices and Fruit Juice Beverages

The metabolic effect of fruit juices is primarily determined by their alkali citrate content. Dietary citrate is absorbed in the gastrointestinal tract and metabolized to bicarbonate, which may then increase urine pH and citrate excretion [50]. Citrus juices, such as lemon, orange and grapefruit juice, supply large quantities of citric acid and could be a dietary alternative to the pharmacotherapy with alkalizing agents. Orange juice is among the most popular fruit juices consumed worldwide. Findings from studies on the impact of orange juice intake on urinary risk factors for stone formation are mixed.

Three cohort studies reported that the intake of orange juice is associated with a reduced risk of kidney stone formation [51]. Interventional studies under a controlled dietary regimen showed that orange juice provided an alkali load that increased urine pH and citrate excretion [52–54]. Despite the alkalizing effect, orange juice did not change the calculated risk for calcium oxalate in the majority of studies. Although the oxalate concentration of orange juice is very low (Table 2) [55,56], two of the three studies reported a significant rise in urinary oxalate excretion, which could be due to in vivo conversion of ascorbate to oxalate [52,54]. Due to concerns over the high sugar and energy content and the lack of dietary fiber of orange juice, the nutritional advice is to opt for the whole fruit over the juice, to limit daily consumption of fruit juice to one serving and to dilute the juice with water [14]. A randomized crossover study of 10 healthy subjects under regular diet comparing a crystal light lemonade beverage and two low-calorie orange juice beverages noted a higher urine pH as the only significant change between the groups when participants consumed Kroger low-calorie orange juice beverage [57]. However, citrate excretion, a major urinary inhibitor of calcium oxalate stone formation, did not significantly differ between the groups. Moreover, concerns over health risks associated especially with overconsumption and with certain artificial colorings, preservatives, sweeteners and additives, such as ascorbic acid and calcium, severely limit the health benefit of these beverages for kidney stone formers [57,58].

Table 2. Oxalate content of beverages.

Beverage	Description	Oxalate Content (mg/100 mL)	References
Vegetable Juices			
Rhubarb nectar	60% juice	198	[56]
Beetroot juice	100% juice	60–70	[56]
Tomato juice	100% juice	4.1–8.1	[55,56]
Multi-vegetable juice	100% juice	3.6–8.5	[56]
Carrot juice	100% juice	4.6–5.8	[55,56]
Soybean drink	62% soymilk	4.4	[56]

Table 2. *Cont.*

Beverage	Description	Oxalate Content (mg/100 mL)	References
Fruit Juices			
Grape juice, red	100% juice	2.1–3.9	[55,56]
Grape juice, white	100% juice	1.5	[55,56]
Apple juice	100% juice	0.9	[55,56]
Grapefruit juice	100% juice	0.1–0.3	[55,56]
Orange juice	100% juice	<d.l.–0.2	[55,56]
Tea and Coffee			
Green tea	Brewed	0.8–14.0	[59,60]
Black tea	Brewed	3.9–6.3	[60]
Iced tea	Ready-to-drink	0.3–2.0	[60]
Coffee	Filtered	0.6	[55]
Beer and Wine			
Malt beer		1.8	[60]
Wheat beer		1.3–1.8	[55,60]
Pils		1.3	[60]
Red wine		0.7–1.3	[60]
White wine		0.3	[60]

d.l., detection limit.

Compared to orange juice, lemon and lime juices were found to have higher citrate concentrations [61]. However, findings from studies on the impact of lemon juice and lemonade on urinary risk factors for stone formation are likewise inconclusive. Although several studies in stone patients with or without hypocitraturia found a significant rise in urinary citrate excretion after the administration of lemonade [62–65], others failed to confirm the citraturic effects of regular lemonade [54,66,67]. Reasons for the variability in urinary citrate excretion with lemonade could be different degrees of dilution of pure lemon juice and the higher proportion of non-alkaline citric acid, which could have neutralized the alkali potential of citrate [57,68].

Studies of other fruit juices, such as grapefruit, apple, cranberry, and blackcurrant juice, have also provided inconsistent results. The findings of two studies of healthy subjects on the impact of grapefruit juice yielded no or only a partial effect on the risk of calcium oxalate stone formation (500 mL/day: $p < 0.05$; 720 mL/day: n.s.; 1000 mL/day: n.s.), although urinary citrate excretion increased significantly [53,69]. The administration of 0.5 or 1.0 L/day of apple juice likewise increased urinary citrate excretion but did not change the relative supersaturation of calcium oxalate [53]. In healthy volunteers, the consumption of 330 mL/day of blackcurrant juice significantly increased urine pH and citrate excretion [70]. However, a simultaneous rise of urinary oxalate excretion was observed, probably due to in vivo conversion of ascorbate to oxalate. Interventional studies on the effect of cranberry juice reported that oxalate excretion decreased [71], remained unchanged [72] or increased [70] in healthy subjects and increased in calcium oxalate stone patients [72]. Overall, the relative risk of calcium oxalate stone formation was unaffected [70,72] or reduced [71] in healthy volunteers but increased in stone patients after the intake of cranberry juice [72]. Finally, a study of healthy subjects showed that the consumption of 1.9 L/day of coconut water compared to tap water significantly raised urinary citrate, potassium, and chloride excretion, without affecting urine pH [73]. Although fruit and vegetable juices could be useful in the dietary therapy of kidney stone disease, the oxalate concentration has to be taken into account.

2.3. *Soft Drinks*

One large randomized controlled trial in stone patients with a soft drink intake of at least 160 mL/day randomly assigned men to refrain from soft drink consumption or to the control group [74]. The study showed that the consumption of soft drinks, especially those acidified by phosphoric acid, significantly increased the risk for recurrent stone formation. The analysis of data from 194,095 participants in the Health Professionals Follow-Up Study (HPFS), and Nurses' Health Study (NHS) I and II with a median follow-up of more than eight years indicated significant positive associations with the risk of stone formation for sugar-sweetened cola and sugar-sweetened non-cola [51]. According to findings from a cross-sectional trial based on the Third National Health and Nutrition Examination Survey (NHANES-III), sugar-sweetened soft drink consumption was positively associated with serum uric acid concentration and frequency of hyperuricemia [75]. Accordingly, cohort studies also indicated a strong positive association between sugar-sweetened soft drinks and the risk of gout in men [76]. These results could at least partly be explained by the fructose content of sugar-sweetened soft drinks, which has been associated with an increased risk of incident kidney stone formation [30,77].

2.4. Tea and Coffee

Tea and coffee are among the most commonly consumed beverages. A systematic review and large cohort studies supported in general a potentially preventive role for both coffee and tea consumption against stone formation [30,51,78–80]. It is assumed that the beneficial effect of tea and coffee could be attributed primarily to the diuretic action of the consumption of large amounts of caffeine, which could offset, at least partially, the hypercalciuric effect [78,81,82]. The European Food Safety Authority considers the habitual caffeine consumption of up to 400 mg/day, corresponding to about four cups of brewed coffee, a safe amount for healthy adults, except pregnant women [83]. An increased total fluid intake and the antioxidative effect of phytochemicals, such as polyphenols, could be other explanations for the preventive effect of tea consumption [78,84].

A limitation of the cohort studies is that no distinction was made between different coffee and tea types, such as black, green and herbal tea. Moreover, there has been concern regarding the oxalate content of coffee and tea. While the oxalate content of coffee is low [55], black tea and green tea contain varying amounts of oxalate depending on the origin, quality, time of harvest and preparation [59,60,85] (Table 2). The highest oxalate concentration was detected in black and green tea [59,60,86], whereas other types of tea such as herbal and fruit tea were found to be low in oxalate [60,86]. Therefore, the exact mechanism for the protective effect of black and green tea against stone formation remains to be elucidated.

3. Protein

The recommended intake of protein for adults is between 0.8 and 1.0 g per kg normal body weight per day [87]. High dietary intake of protein was reported to exert potential detrimental effects on urinary risk factors of stone formation. The acid load provided by a high protein intake may increase urinary calcium and reduce urine pH and citrate excretion [88,89]. A study of 18 hypercalciuric stone patients showed that a protein restriction to 0.8 g/kg body weight/day decreased urinary calcium and uric acid and increased urinary citrate excretion [90].

However, the evidence from systematic reviews regarding the relationship between protein intake and the risk of kidney stone formation is inconsistent [91,92]. The two cohort studies [93,94] included in the systematic review by Pedersen [91] found no association between dietary protein intake and stone formation. However, one systematic review confirmed that high-protein diets were associated with increased urinary calcium excretion, which is a risk factor for calcium stone formation [95]. In healthy subjects, the administration of 1.5 g per day L-methionine did not raise urinary calcium excretion [96], whereas the intake of 3 g per day L-methionine significantly increased calcium excretion by about 1 mmol per day [97]. To date, there is no randomized controlled trial comparing the isolated effect of a high versus low protein intake on the risk of urinary stone formation.

While findings on the association between dietary protein consumption and the risk of stone formation are inconclusive, large observational studies found that a higher dietary net acid load was associated with higher risk of stone formation [98]. These data suggested that the proportion of consumed vegetables and fruits compared to ingested protein intake, rather than total protein per se, could be a more reliable indicator for the risk of urinary stone formation. Vegetables and fruits have a distinct alkalizing potential and can neutralize the proton load, metabolically generated from ingested protein [99,100]. In hypocitraturic stone patients, the introduction of vegetables and fruits increased urine pH and citrate excretion and reduced the relative supersaturation of calcium oxalate and uric acid [101]. Reduced urine pH and citrate excretion, resulting from high nutritional proton load or high dietary acidity, are risk factors for several types of urinary stones, particularly for the most common, i.e., calcium oxalate and uric acid. The higher the urine pH, the higher the stone-inhibiting citrate excretion and calcium-binding capacity and the lower the urinary calcium excretion [102].

4. Carbohydrates

Findings from studies on the impact of carbohydrates on the risk of kidney stone formation are inconclusive. While some studies reported a similar intake of carbohydrates in stone patients and controls [103,104], others noted a higher intake of carbohydrates in stone patients than in controls [105,106]. A major limitation of these trials is that they did not distinguish between different types of carbohydrates, especially the most common disaccharide sucrose and its monomers glucose and fructose.

Prospective cohort studies found a positive relationship between sucrose consumption and the risk for stone formation in women but not in men [93,94,107,108]. A previous study reported a higher rate of urinary calcium excretion after oral ingestion of 100 g glucose or sucrose in normal subjects and calcium oxalate stone formers, where the response was even more pronounced in the latter group [109]. The rise in calcium excretion after an oral glucose load has been attributed to an increased intestinal absorption and reduced renal tubular reabsorption of calcium [110–112]. It was suggested that this effect could be, at least partially, mediated by an increase in serum insulin. A study of calcium stone patients with idiopathic hypercalciuria and healthy controls conducted using a fixed metabolic diet concluded that hyperinsulinemia is unlikely to play a significant role in the pathogenesis of calcium stone formation among patients with idiopathic hypercalciuria [113].

Consumption of fructose has clearly increased over the last decades as fructose is used as sweetener in beverages or food as a replacement for sucrose or glucose. A systematic review and meta-analysis noted a positive relationship between fructose intake and the risk of incident stone formation [30], but the underlying mechanisms are not well-understood [114]. Fructose consumption was assumed to increase the risk of stone formation in part via effects on the urinary excretion of calcium [115], oxalate [115,116], urine pH [116], and by effects on uric acid metabolism [115–118]. A cohort study of men showed a positive association between fructose intake and the risk of incident gout [76]. Studies on fixed metabolic diets are required to evaluate the effect of sucrose, glucose and fructose on metabolism and urinary risk factors for uric acid and calcium oxalate stone formation.

5. Fat

Data regarding the association between dietary fat consumption and the risk of urinary stone formation are scarce and inconsistent. While several studies reported a similar fat intake in stone patients and controls [103,104], others found higher dietary fat intake in stone formers [105,106].

Studies have suggested that the dietary fatty acid pattern, especially the ratio of n-6 to n-3 polyunsaturated fatty acids, may affect the risk of calcium oxalate stone formation through various complex mechanisms [119]. An abnormal concentration of arachidonic acid (C20:4n-6) in plasma and erythrocyte membrane phospholipids was found in idiopathic calcium oxalate stone patients compared to healthy controls [120]. Arachidonic acid in cell membrane phospholipids can be released by phospholipase enzymes and subsequently serve as a precursor of PGE_2 [121,122]. An elevated PGE_2 production is assumed to induce hypercalciuria by increasing intestinal calcium absorption and bone resorption [123,124], and by decreasing renal tubular calcium reabsorption [125,126]. Increased phospholipid arachidonic acid levels may also induce hyperoxaluria by activating anion carriers and consequently the intestinal and renal transport activity of oxalate [120,127]. A study of 20 healthy volunteers reported that the supplementation of n-3 polyunsaturated fatty acids DHA (22:6n-3) and EPA (20:5n-3) led to their incorporation into cell membrane phospholipids, which was partly at the expense of arachidonic acid [128]. A change in the membrane fatty acid pattern due to increasing dietary intake of n-3 polyunsaturated fatty acids, therefore, was assumed to decrease the urinary excretion of calcium and oxalate.

Analyses of 24 h urine samples and dietary records of 58 idiopathic calcium oxalate stone patients showed a positive association between the dietary arachidonic acid content and urinary oxalate excretion [129]. Several studies investigated the role of fish oil administration in the dietary management of stone formation. Fish oil supplementation was found to reduce oxalate excretion in healthy subjects [130], and to decrease urinary excretion of calcium and/or oxalate in calcium stone patients in most trials [119]. Further studies should identify those patients who benefit most from n-3 polyunsaturated fatty acid supplementation.

6. Oxalate

Urinary oxalate is regarded as an essential risk factor for calcium oxalate stone formation. Changes in urinary oxalate concentration can significantly increase the urinary supersaturation of calcium oxalate [131,132]. A prospective study of 134 recurrent calcium oxalate stone patients identified the rise in oxalate excretion as the major urinary determinant for relapse after a two-year follow-up [133]. Oxalate is an end product of metabolism. Urinary oxalate is derived from endogenous oxalate synthesis and dietary oxalate intake [134]. Endogenous oxalate metabolism occurs predominantly in the liver and is affected by dietary intake of precursors, such as ascorbic acid and hydroxyproline [135,136].

The effect of dietary oxalate intake on urinary oxalate excretion and the risk of stone formation has been examined in several interventional trials. In a study of healthy subjects, the mean contribution of dietary oxalate to urinary oxalate excretion ranged from 24% (10 mg/day dietary oxalate) to 42% (250 mg/day dietary oxalate) [137]. A study of 20 healthy women and men showed that a controlled oxalate-rich diet (600 mg/day dietary oxalate) compared to a diet normal in oxalate (100 mg/day dietary oxalate) significantly increased oxalate excretion from 0.354 to 0.542 mmol/24 h by 0.188 mmol/24 h, i.e., >50%, corresponding to 35% of total urinary oxalate excretion [138]. This study also showed that the supersaturation of calcium oxalate increases significantly with a high dietary oxalate intake. However, a prospective cohort study reported only a modest positive association between dietary oxalate intake and the risk for incident stone formation [139]. Several reasons could be responsible for these inconsistencies, including the use of food frequency questionnaires, which are prone to errors, to evaluate dietary oxalate intake in large cohort studies, the daily variation in the oxalate ingestion and the variability of the oxalate content due to growth conditions, preparation and processing of food [59,140–142]. Therefore, studies using diets strictly controlled in their oxalate and nutrient content and the use of comprehensive data on the oxalate content of raw and processed foods are required.

Dietary oxalate is mainly derived from plant foods. Estimates of dietary oxalate intake are in a wide range, depending on the ingestion of oxalate-rich foods. Therefore, it is essential to consider sources of excess dietary oxalate. Detailed knowledge of the oxalate content of foods is important in the dietary therapy of calcium oxalate stone patients. However, the previous lack of accurate and complete data on the oxalate concentration of foods hindered the elucidation of the role of dietary oxalate in urinary oxalate excretion and the risk of stone formation. Analysis of the oxalate content of a wide variety of foods by a HPLC enzyme-reactor method provided a comprehensive database of the oxalate content of foods and beverages and detected a considerable number of foods with high or extremely high oxalate concentrations [55,56,59,60,143–145]. An overview of foods rich in oxalate, including vegetables, legumes, fruits, cereals and pseudocereals, nuts, herbs and spices, is presented in Table 3.

Table 3. Oxalate-rich foods.

Food	Description	Oxalate Content (mg/100 g)	References
Vegetables			
Spinach	Raw	1959	[143]
Sorrel	Raw	1391	[143]
Rhubarb	Raw	1235	[143]
Mangold	Raw	874	[143]
Sweet potato	Raw	496	[145]
Okra	Raw	317	[145]
Beetroot	Raw	160	[143]
Olive, green	Canned	45.7	[55]
Legumes			
Beans, white	Seeds, dry	548	[145]
Soybeans	Seeds, dry	277	[145]
Quail beans	Seeds, dry	177	[145]
Kidney beans	Seeds, dry	74.6	[145]
Green beans	Raw	65.2	[145]
Fruits			
Star fruit	Raw	295	[55]
Elderberry, black	Raw	72.1	[55]
Blackberry	Raw	29.2	[55]
Gooseberry, green	Raw	27.0	[55]
Kiwi fruit	Raw	23.0	[55]
Fig	Raw	20.5	[55]
Pseudocereals			
Amaranth	Nuts	232	[143]
Quinoa	Nuts	184	[143]
Buckwheat	Nuts	143	[143]
Cereals			
Wheat	Bran	457	[144]
Wheat	Wholegrain flour	70.0	[144]
Bulgur		59.4	[144]
Couscous		65.2	[144]
Nuts and Cocoa			
Sesame		3800	[55]
Almond		383	[55]
Hazel nut		167	[55]
Pistachio		56.5	[55]
Cocoa powder		567–619	[55,145]
Herbs and Spices			
Licorice	Root	3569	[145]
Blue fenugreek	Powder	1246	[145]
Pepper, black	Grinded	623	[55]
Parsley	Raw	136	[55]

As boiling may lead to considerable losses of oxalate into the cooking water, food processing and preparation methods are important determinants for the oxalate content [141]. For example, the oxalate concentration of raw spinach was found to be more than five times higher in raw (1959 mg/100 g) compared to cooked spinach (364 mg/100 g) [55,143]. To avoid losses of water-soluble vitamins and minerals, it is recommended to shortly steam spinach, which, however, also preserves most of the oxalate content of raw spinach. Moreover, beverages such as vegetable and fruit juices as well as black, green and iced teas were found to contain considerable amounts of highly bioavailable oxalate (Table 2) [56,59,60]. Knowledge of the oxalate concentration of beverages is particularly crucial as a high fluid intake is the most important nutritional measure for recurrence prevention of kidney stone disease. Kidney stone formers should be advised to avoid oxalate-rich foods and beverages.

A case-control study of 186 calcium oxalate stone patients found a significant positive association between dietary ascorbic acid intake and urinary oxalate excretion [134]. The association between ascorbic acid intake and the risk of urinary stone formation has been noted in several large cohort studies [146,147]. A study under controlled dietary conditions noted a significant increase in urinary oxalate in both calcium oxalate stone patients and healthy controls after oral supplementation of 2 g ascorbic acid daily [148]. Dietary hydroxyproline, mainly present in collagen/gelatin, may also contribute to endogenous oxalate synthesis and urinary oxalate excretion in healthy subjects and in primary hyperoxaluria [135,149,150]. A study using infusions of [^{15}N,^{13}C$_5$]-hydroxyproline found that hydroxyproline contributed at least 15% to urinary oxalate excretion in healthy volunteers and could be a major source of the oxalate produced in patients with primary hyperoxaluria [149,150].

Intestinal hyperabsorption of oxalate may additionally contribute to urinary oxalate, even in the absence of intestinal diseases associated with fat malabsorption. A study under standardized conditions using [^{13}C$_2$]oxalate demonstrated that the mean intestinal oxalate absorption was significantly higher in 120 idiopathic calcium oxalate stone patients compared to 120 healthy subjects (10.2% vs. 8.0%; $p < 0.001$) [151]. Intestinal hyperabsorption, defined as oxalate absorption exceeding 10%, was noted in 28% of healthy volunteers and in 46% of patients. The amount of dietary oxalate that is intestinally absorbed is affected by dietary constituents, especially the calcium intake [152]. Moreover, enteric hyperoxaluria in the setting of fat malabsorption due to different gastrointestinal disorders is a crucial risk factor contributing to kidney stone formation [153–155]. Unabsorbed fatty acids bind to calcium, decreasing the intraluminal calcium concentration for complexation with oxalate [156]. With the depletion of free calcium, a larger percentage of unbound oxalate is available for absorption in the gut. A study of 51 patients with intestinal fat malabsorption found that the resection status is a major risk factor for hyperoxaluria and kidney stone formation [157].

Finally, the intestinal colonization with the Gram-negative anaerobic oxalate-degrading bacterium *Oxalobacter formigenes* can be inversely associated with calcium oxalate stone formation [158]. Although the administration of *Oxalobacter formigenes* or other probiotic preparations was suggested to reduce urinary oxalate excretion and the lithogenic risk, findings from several interventional studies are conflicting [138,159]. Future studies should include functional and nutritional aspects of the interaction between nutrients and the gut microbiota composition.

7. Calcium

Hypercalciuria is a crucial risk factor for calcium stone formation. Balanced dietary calcium intake from both dairy and non-dairy sources has been demonstrated to exert a preventive effect against urinary stone formation [160]. Dietary calcium restriction should be avoided as it may induce bone loss and result in the hyperabsorption and hyperexcretion of oxalate. Dietary calcium restriction reduces intestinal calcium concentration, which enhances the absorption of uncomplexed oxalate and subsequently urinary oxalate excretion [152]. A study of healthy volunteers with a standardized [^{13}C$_2$]oxalate absorption test showed that intestinal oxalate absorption was inversely associated with dietary calcium intake within the range of 200 (17% oxalate absorption) to 1200 mg/day calcium (2.6% oxalate absorption) [152]. Dietary calcium intake above 1200 mg per day had only a minor impact on oxalate absorption. Epidemiologic studies found an inverse relationship between dietary calcium intake and the risk of stone formation in women and men [93,107,108]. A five-year prospective randomized study of 120 men with calcium oxalate stone formation and hypercalciuria showed that recurrences were less frequent on a normal calcium (1200 mg/day), normal-protein, and low-salt diet compared to a low-calcium diet (400 mg/day) [161]. For idiopathic calcium stone formers, a total dietary calcium intake of 1000 to 1200 mg/day is recommended [14–16].

8. Sodium Chloride

Dietary sodium chloride consumption increases the risk of stone formation due to its propensity to enhance urinary calcium excretion [162]. High sodium chloride intake may promote calcium excretion by inhibiting renal tubular calcium reabsorption from the sodium-induced expansion of extracellular fluid volume [163]. Interventional studies in normal adults showed that every 100 mmol (2300 mg) increase in sodium intake per day enhances daily calcium excretion by approximately 1 mmol [164]. A correlation between dietary sodium intake and kidney stone formation was observed in one cohort study [108], but was not confirmed by others [93,94,107]. These discrepancies may reflect the difficulties in obtaining reliable estimates of dietary salt intake based on food frequency questionnaires. A randomized controlled trial of 210 idiopathic calcium oxalate stone patients demonstrated that a low-salt diet could effectively decrease urinary calcium excretion compared to a control diet [165]. The recommended dietary sodium intake is <100 mmol (2300 mg) or 6 g of salt (sodium chloride) per day [14,16,17].

9. Dietary Management

Diet modification is an effective method to correct urinary risk factors for kidney stone formation, particularly of the most common stone type calcium oxalate. Dietary therapy should be tailored to every single patient according to the patient-specific biochemical and dietary risk profile. A detailed nutritional assessment is an essential component of the evaluation and the main prerequisite for a successful dietary therapy of the stone-forming patient. Seven-day dietary records are regarded as the most accurate technique for assessing habitual dietary intake. A variety of dietary factors, including fluid intake, dietary protein, carbohydrates, oxalate, calcium, and sodium chloride can modulate urinary risk profile and contribute to the risk of kidney stone formation. The specific dietary recommendations for calcium oxalate stone formers are included in Table 4.

Table 4. Dietary recommendations for calcium oxalate stone patients.

Urinary Risk Factor	Limit	Recommendation
Urine volume	Urine volume < 2.0 L/24 h	Fluid intake that achieves urine volume ≥ 2.0 to 2.5 L/24 h Neutral and alkalizing beverages
Hypercalciuria	Calcium > 5 mmol/24 h	Calcium intake: 1000 to 1200 mg/day Protein intake: 0.8 to 1.0 g/kg normal body weight/day Sodium chloride intake: <6 g/day Increased intake of vegetables and fruits
Hyperoxaluria	Oxalate > 0.5 mmol/24 h	Low dietary oxalate intake Calcium intake: 1000 to 1200 mg/day (IH) Calcium supplementation (EH)
Hyperuricosuria	Uric acid > 4 mmol/24 h	Protein intake: 0.8 to 1.0 g/kg normal body weight/day Reduced dietary purine intake Increased intake of vegetables and fruits
Hypocitraturia	Citrate < 1.7 mmol/24 h	Protein intake: 0.8 to 1.0 g/kg normal body weight/day Increased intake of vegetables and fruits

IH, idiopathic hyperoxaluria; EH, enteric hyperoxaluria.

10. Conclusions

Nutritional factors play an important role in kidney stone formation. A careful dietary assessment should obtain from the patient any dietary habits that predispose them to urinary stone formation. An appropriate diet can modulate the urinary risk profile and contribute to the reduction of the risk of urinary stone formation. Specific dietary therapy, based on nutritional assessment and metabolic evaluation, has been demonstrated to be more effective than general dietary measures in preventing recurrent stone formation.

Funding: This research was supported in part by a grant from the Association of German Mineral Water Bottlers (VDM).

Institutional Review Board Statement: Not applicable.

Informed Consent Statement: Not applicable.

Data Availability Statement: Not applicable.

Conflicts of Interest: The author declares no conflict of interest.

References

1. Hesse, A.; Brändle, E.; Wilbert, D.; Köhrmann, K.U.; Alken, P. Study on the prevalence and incidence of urolithiasis in Germany comparing the years 1979 vs. 2000. *Eur. Urol.* **2003**, *44*, 709–713. [CrossRef]
2. Romero, V.; Akpinar, H.; Assimos, D.G. Kidney stones: A global picture of prevalence, incidence, and associated risk factors. *Rev. Urol.* **2010**, *12*, e86–e96. [PubMed]
3. Chewcharat, A.; Curhan, G. Trends in the prevalence of kidney stones in the United States from 2007 to 2016. *Urolithiasis* **2021**, *49*, 27–39. [CrossRef] [PubMed]
4. Uribarri, J.; Oh, M.S.; Carroll, H.J. The first kidney stone. *Ann. Intern. Med.* **1989**, *111*, 1006–1009. [CrossRef] [PubMed]
5. Shoag, J.; Halpern, J.; Goldfarb, D.S.; Eisner, B.H. Risk of chronic and end stage kidney disease in patients with nephrolithiasis. *J. Urol.* **2014**, *192*, 1440–1445. [CrossRef] [PubMed]
6. Dhondup, T.; Kittanamongkolchai, W.; Vaughan, L.E.; Mehta, R.A.; Chhina, J.K.; Enders, F.T.; Hickson, L.J.; Lieske, J.C.; Rule, A.D. Risk of ESRD and mortality in kidney and bladder stone formers. *Am. J. Kidney Dis.* **2018**, *72*, 790–797. [CrossRef]
7. Lieske, J.C.; Rule, A.D.; Krambeck, A.E.; Williams, J.C.; Bergstralh, E.J.; Mehta, R.A.; Moyer, T.P. Stone composition as a function of age and sex. *Clin. J. Am. Soc. Nephrol.* **2014**, *9*, 2141–2146. [CrossRef]
8. Siener, R.; Hesse, A. The effect of different diets on urine composition and the risk of calcium oxalate crystallization in healthy subjects. *Eur. Urol.* **2002**, *42*, 289–296. [CrossRef]
9. Siener, R.; Schade, N.; Nicolay, C.; von Unruh, G.E.; Hesse, A. The efficacy of dietary intervention on urinary risk factors for stone formation in recurrent calcium oxalate stone patients. *J. Urol.* **2005**, *173*, 1601–1605. [CrossRef]
10. Siener, R.; Buchholz, N.; Daudon, M.; Hess, B.; Knoll, T.; Osther, P.J.; Reis-Santos, J.; Sarica, K.; Traxer, O.; Trinchieri, A. Quality assessment of urinary stone analysis: Results of a multicenter study of laboratories in Europe. *PLoS ONE* **2016**, *11*, e0156606. [CrossRef]
11. Williams, J.C.; Gambaro, G.; Rodgers, A.; Asplin, J.; Bonny, O.; Costa-Bauzá, A.; Ferraro, P.M.; Fogazzi, G.; Fuster, D.G.; Goldfarb, D.S.; et al. Urine and stone analysis for the investigation of the renal stone former: A consensus conference. *Urolithiasis* **2021**, *49*, 1–16. [CrossRef]
12. Yagisawa, T.; Chandhoke, P.S.; Fan, J. Comparison of comprehensive and limited metabolic evaluations in the treatment of patients with recurrent calcium urolithiasis. *J. Urol.* **1999**, *161*, 1449–1452. [CrossRef]
13. Parks, J.H.; Goldfisher, E.; Asplin, J.R.; Coe, F.L. A single 24-hour urine collection is inadequate for the medical evaluation of nephrolithiasis. *J. Urol.* **2002**, *167*, 1607–1612. [CrossRef]
14. Hesse, A.; Tiselius, H.G.; Siener, R.; Hoppe, B. *Urinary Stones: Diagnosis, Treatment and Prevention of Recurrence*, 3rd ed.; Karger: Basel, Switzerland, 2009.
15. Türk, C.; Neisius, A.; Petrik, A.; Seitz, C.; Skolarikos, A.; Somani, B.; Thomas, K.; Gambaro, G.; Davis, N.F.; Donaldson, J.F.; et al. *EAU Guidelines on Urolithiasis*; European Association of Urology, EAU Guidelines Office: Arnhem, The Netherlands, 2021.
16. Pearle, M.S.; Goldfarb, D.S.; Assimos, D.G.; Curhan, G.; Denu-Ciocca, C.J.; Matlaga, B.R.; Monga, M.; Penniston, K.L.; Preminger, G.M.; Turk, T.M.T.; et al. Medical management of kidney stones: AUA guideline. *J. Urol.* **2014**, *192*, 316–324. [CrossRef]
17. Gambaro, G.; Croppi, E.; Coe, F.; Lingeman, J.; Moe, O.; Worcester, E.; Buchholz, N.; Bushinsky, D.; Curhan, G.C.; Ferraro, P.M.; et al. Metabolic diagnosis and medical prevention of calcium nephrolithiasis and its systemic manifestations: A consensus statement. *J. Nephrol.* **2016**, *29*, 715–734. [CrossRef] [PubMed]
18. Kocvara, A.; Plasgura, P.; Petrik, A.; Louzensky, G.; Bartonickova, K.; Dvoracek, J. A prospective study of nonmedical prophylaxis after a first kidney stone. *BJU Int.* **1999**, *84*, 393–398. [CrossRef] [PubMed]
19. Siener, R.; Hesse, A. Fluid intake and epidemiology of urolithiasis. *Eur. J. Clin. Nutr.* **2003**, *57*, S47–S51. [CrossRef]
20. Caudarella, R.; Rizzoli, E.; Pironi, L.; Malavolta, N.; Martelli, G.; Poggioli, G.; Gozzetti, G.; Miglioli, M. Renal stone formation in patients with inflammatory bowel disease. *Scanning Microsc.* **1993**, *7*, 371–380.
21. Parks, J.H.; Worcester, E.M.; O´Connor, R.C.; Coe, F.L. Urine stone risk factors in nephrolithiasis patients with and without bowel disease. *Kidney Int.* **2003**, *63*, 255–265. [CrossRef]
22. Lotan, Y.; Antonelli, J.; Jiménez, I.B.; Gharbi, H.; Herring, R.; Beaver, A.; Dennis, A.; Merveldt, D.V.; Carter, S.; Cohen, A.; et al. The kidney stone and increased water intake trial in steel workers: Results from a pilot study. *Urolithiasis* **2017**, *45*, 177–183. [CrossRef]

23. Linder, B.J.; Rangel, L.J.; Krambeck, A.E. The effect of work location on urolithiasis in health care professionals. *Urolithiasis* **2013**, *41*, 327–331. [CrossRef]
24. Malieckal, D.A.; Goldfarb, D.S. Occupational kidney stones. *Curr. Opin. Nephrol. Hypertens.* **2020**, *29*, 232–236. [CrossRef]
25. Borghi, L.; Meschi, T.; Amato, F.; Briganti, A.; Novarini, A.; Giannini, A. Urinary volume, water and recurrences in idiopathic calcium nephrolithiasis: A 5-year randomized prospective study. *J. Urol.* **1996**, *155*, 839–843. [CrossRef]
26. Fink, H.A.; Akornor, J.W.; Garimella, P.S.; MacDonald, R.; Cutting, A.; Rutks, I.R.; Monga, M.; Wilt, T.J. Diet, fluid, or supplements for secondary prevention of nephrolithiasis: A systematic review and meta-analysis of randomized trials. *Eur. Urol.* **2009**, *56*, 72–80. [CrossRef]
27. Fink, H.A.; Wilt, T.J.; Eidman, K.E.; Garimella, P.S.; MacDonald, R.; Rutks, I.R.; Brasure, M.; Kane, R.L.; Ouellette, J.; Monga, M. Medical management to prevent nephrolithiasis in adults: A systematic review for an American College of Physicians clinical guideline. *Ann. Intern. Med.* **2013**, *158*, 535–543. [CrossRef] [PubMed]
28. Qaseem, A.; Dallas, P.; Forciea, M.A.; Starkey, M.; Denberg, T.D. Dietary and pharmacologic management to prevent recurrent nephrolithiasis in adults: A clinical practice guideline from the American College of Physicians. *Ann. Intern. Med.* **2014**, *161*, 659–667. [CrossRef] [PubMed]
29. Cheungpasitporn, W.; Rossetti, S.; Friend, K.; Erickson, S.B.; Lieske, J.C. Treatment effect, adherence, and safety of high fluid intake for the prevention of incident and recurrent kidney stones: A systematic review and meta-analysis. *J. Nephrol.* **2016**, *29*, 211–219. [CrossRef] [PubMed]
30. Lin, B.B.; Lin, M.E.; Huang, R.H.; Hong, Y.K.; Lin, B.L.; He, X.J. Dietary and lifestyle factors for primary prevention of nephrolithiasis: A systematic review and meta-analysis. *BMC Nephrol.* **2020**, *21*, 267. [CrossRef]
31. Bao, Y.; Tu, X.; Wei, Q. Water for preventing urinary stones. *Cochrane Database Syst. Rev.* **2020**, 1–26. [CrossRef]
32. Borghi, L.; Meschi, T.; Maggiore, U.; Prati, B. Dietary therapy in idiopathic nephrolithiasis. *Nutr. Rev.* **2006**, *64*, 301–312. [CrossRef]
33. Servais, A.; Thomas, K.; Strologo, L.D.; Sayer, J.A.; Bekri, S.; Bertholet-Thomas, A.; Bultitude, M.; Capolongo, G.; Cerkauskiene, R.; Daudon, M.; et al. Cystinuria: Clinical practice recommendation. *Kidney Int.* **2021**, *99*, 48–58. [CrossRef]
34. Barbey, F.; Joly, D.; Rieu, P.; Méjean, A.; Daudon, M.; Jungers, P. Medical treatment of cystinuria: Critical reappraisal of long-term results. *J. Urol.* **2000**, *163*, 1419–1423. [CrossRef]
35. Willis, S.; Goldfarb, D.S.; Thomas, K.; Bultitude, M. Comment—Water to prevent kidney stones: Tap vs. bottled; soft vs. hard—Does it matter? *BJU Int.* **2019**, *124*, 905–906. [CrossRef]
36. World Health Organization (WHO). Hardness in Drinking-Water. In *Background Document for Development of WHO Guidelines for Drinking-Water Quality*; WHO Press: Geneva, Switzerland, 2011.
37. Shuster, J.; Finlayson, B.; Scheaffer, R.; Sierakowski, R.; Zoltek, J.; Dzegede, S. Water hardness and urinary stone disease. *J. Urol.* **1982**, *128*, 422–425. [CrossRef]
38. Schwartz, B.F.; Schenkman, N.S.; Bruce, J.E.; Leslie, S.W.; Stoller, M.L. Calcium nephrolithiasis: Effect of water hardness on urinary electrolytes. *Urology* **2002**, *60*, 23–27. [CrossRef]
39. Basiri, A.; Shakhssalim, N.; Koshdel, A.R.; Pakmanesh, H.; Radfar, M.H. Drinking water composition and incidence of urinary calculus: Introducing an new index. *Iran. J. Kidney Dis.* **2011**, *5*, 15–20. [PubMed]
40. Siener, R. Can the manipulation of urinary pH by beverages assist with the prevention of stone recurrence? *Urolithiasis* **2016**, *44*, 51–56. [CrossRef]
41. Keßler, T.; Hesse, A. Cross-over study of the influence of bicarbonate-rich mineral water on urinary composition in comparison with sodium potassium citrate in healthy male subjects. *Br. J. Nutr.* **2000**, *84*, 865–871. [CrossRef] [PubMed]
42. Siener, R.; Jahnen, A.; Hesse, A. Influence of a mineral water rich in calcium, magnesium and bicarbonate on urine composition and the risk of calcium oxalate crystallization. *Eur. J. Clin. Nutr.* **2004**, *58*, 270–276. [CrossRef] [PubMed]
43. Caudarella, R.; Rizzoli, E.; Buffa, A.; Bottura, A.; Stefoni, S. Comparative study of the influence of 3 types of mineral water in patients with idiopathic calcium lithiasis. *J. Urol.* **1998**, *159*, 658–663. [CrossRef]
44. Rodgers, A.L. Effect of mineral water containing calcium and magnesium on calcium oxalate urolithiasis risk factors. *Urol. Int.* **1997**, *58*, 93–99. [CrossRef]
45. Karagülle, O.; Smorag, U.; Candir, F.; Gundermann, G.; Jonas, U.; Becker, A.J.; Gehrke, A.; Gutenbrunner, C. Clinical study on the effect of mineral waters containing bicarbonate on the risk of urinary stone formation in patients with multiple episodes of CaOx-urolithiasis. *World J. Urol.* **2007**, *25*, 315–323. [CrossRef] [PubMed]
46. Pinheiro, V.B.; Baxmann, A.C.; Tiselius, H.G.; Heilberg, I.P. The effect of sodium bicarbonate upon urinary citrate excretion in calcium stone formers. *Urology* **2013**, *82*, 33–37. [CrossRef] [PubMed]
47. Wasserfurth, P.; Schneider, I.; Ströhle, A.; Nebl, J.; Bitterlich, N.; Hahn, A. Effects of mineral waters on acid-base status in healthy adults: Results of a randomized trial. *Food Nutr. Res.* **2019**, *63*, 3515. [CrossRef] [PubMed]
48. Stoots, S.J.M.; Geraghty, R.; Kamphuis, G.M.; Jamnadass, E.; Henderickx, M.M.E.L.; Ventimiglia, E.; Traxer, O.; Keller, E.X.; DeConinck, V.; Talso, M.; et al. Variations in the mineral content of bottled 'still' water across Europe: Comparison of 182 brands across 10 countries. *J. Endourol.* **2021**, *35*, 206–214. [CrossRef] [PubMed]
49. Stoots, S.J.M.; Geraghty, R.; Kamphuis, G.M.; Jamnadass, E.; Henderickx, M.M.E.L.; Ventimiglia, E.; Traxer, O.; Keller, E.X.; DeConinck, V.; Talso, M.; et al. Variations in the mineral content of bottled 'carbonated or sparkling' water across Europe: Comparison of 126 brands across 10 countries. *Cent. Eur. J. Urol.* **2021**, *74*. [CrossRef]

50. Hamm, L.L.; Hering-Smith, K.S. Pathophysiology of hypocitraturic nephrolithiasis. *Endocrinol. Metab. Clin. N. Am.* **2002**, *31*, 885–893. [CrossRef]
51. Ferraro, P.M.; Taylor, E.N.; Gambaro, G.; Curhan, G.C. Soda and other beverages and the risk of kidney stones. *Clin. J. Am. Soc. Nephrol.* **2013**, *8*, 1389–1395. [CrossRef] [PubMed]
52. Wabner, C.L.; Pak, C.Y.C. Effect of orange juice consumption on urinary stone risk factors. *J. Urol.* **1993**, *149*, 1405–1408. [CrossRef]
53. Hönow, R.; Laube, N.; Schneider, A.; Keßler, T.; Hesse, A. Influence of grapefruit-, orange- and apple-juice consumption on urinary variables and risk of crystallization. *Br. J. Nutr.* **2003**, *90*, 295–300. [CrossRef]
54. Odvina, C.V. Comparative value of orange juice versus lemonade in reducing stone-forming risk. *Clin. J. Am. Soc. Nephrol.* **2006**, *1*, 1269–1274. [CrossRef]
55. Hönow, R.; Hesse, A. Comparison of extraction methods for the determination of soluble and total oxalate in foods by HPLC-enzyme-reactor. *Food Chem.* **2002**, *78*, 511–521. [CrossRef]
56. Siener, R.; Seidler, A.; Voss, S.; Hesse, A. The oxalate content of fruit and vegetable juices, nectars and drinks. *J. Food Compos. Anal.* **2016**, *45*, 108–112. [CrossRef]
57. Large, T.; Williams, J.; Asplin, J.R.; Krambeck, A. Using low-calorie orange juice as a dietary alternative to alkali therapy. *J. Endourol.* **2020**, *34*, 1082–1087. [CrossRef] [PubMed]
58. Kregiel, D. Health safety of soft drinks: Contents, containers, and microorganisms. *Biomed Res. Int.* **2015**, *2015*, 1–15. [CrossRef] [PubMed]
59. Hönow, R.; Gu, K.L.R.; Hesse, A.; Siener, R. Oxalate content of green tea of different origin, quality, preparation and time of harvest. *Urol. Res.* **2010**, *38*, 377–381. [CrossRef]
60. Siener, R.; Seidler, A.; Voss, S.; Hesse, A. Oxalate content of beverages. *J. Food Compos. Anal.* **2017**, *63*, 184–188. [CrossRef]
61. Penniston, K.L.; Nakada, S.Y.; Holmes, R.P.; Assimos, D.G. Quantitative assessment of citric acid in lemon juice, lime juice, and commercially-available fruit juice products. *J. Endourol.* **2008**, *22*, 567–570. [CrossRef] [PubMed]
62. Seltzer, M.A.; Low, R.K.; McDonald, M.; Shami, G.S.; Stoller, M.L. Dietary manipulation with lemonade to treat hypocitraturic calcium nephrolithiasis. *J. Urol.* **1996**, *156*, 907–909. [CrossRef]
63. Penniston, K.L.; Steele, T.H.; Nakada, S.Y. Lemonade therapy increases urinary citrate and urine volumes in patients with recurrent calcium oxalate stone formation. *Urology* **2007**, *70*, 856–860. [CrossRef]
64. Kang, D.E.; Sur, R.L.; Haleblian, G.E.; Fitzsimons, N.J.; Borawski, K.M.; Preminger, G.M. Long-term lemonade based dietary manipulation in patients with hypocitraturic nephrolithiasis. *J. Urol.* **2007**, *177*, 1358–1362. [CrossRef]
65. Aras, B.; Kalfazade, N.; Tugcu, V.; Kemahli, E.; Özbay, B.; Polat, H.; Tasci, A.I. Can lemon juice be an alternative to potassium citrate in the treatment of urinary calcium stones in patients with hypocitraturia? A prospective randomized study. *Urol. Res.* **2008**, *36*, 313–317. [CrossRef]
66. Koff, S.G.; Paquette, E.L.; Cullen, J.; Gancarczyk, K.K.; Tucciarone, P.R.; Schenkman, N.S. Comparison between lemonade and potassium citrate and impact on urine pH and 24-hour urine parameters in patients with kidney stone formation. *Urology* **2007**, *69*, 1013–1016. [CrossRef]
67. Cheng, J.W.; Wagner, H.; Asplin, J.R.; Hodgkin, G.; Schlaifer, A.; Fargusson, M.; Fargusson, J.; Baldwin, D.D. The effect of lemonade and diet lemonade upon urinary parameters affecting calcium urinary stone formation. *J. Endourol.* **2019**, *33*, 160–166. [CrossRef]
68. Eisner, B.H.; Asplin, J.R.; Goldfarb, D.S.; Ahmad, A.; Stoller, M.L. Citrate, malate and alkali content in commonly consumed diet sodas: Implications for nephrolithiasis treatment. *J. Urol.* **2010**, *183*, 2419–2423. [CrossRef] [PubMed]
69. Goldfarb, D.S.; Asplin, J.R. Effect of grapefruit juice on urinary lithogenicity. *J. Urol.* **2001**, *166*, 263–267. [CrossRef]
70. Keßler, T.; Jansen, B.; Hesse, A. Effect of blackcurrant-, cranberry- and plum juice consumption on risk factors associated with kidney stone formation. *Eur. J. Clin. Nutr.* **2002**, *56*, 1020–1023. [CrossRef] [PubMed]
71. McHarg, T.; Rodgers, A.; Charlton, K. Influence of cranberry juice on the urinary risk factors for calcium oxalate kidney stone formation. *BJU Int.* **2003**, *92*, 765–768. [CrossRef]
72. Gettman, M.T.; Ogan, K.; Brinkley, L.J.; Adams-Huet, B.; Pak, C.Y.C.; Pearle, M.S. Effect of cranberry juice consumption on urinary stone risk factors. *J. Urol.* **2005**, *174*, 590–594. [CrossRef]
73. Patel, R.M.; Jiang, P.; Asplin, J.; Granja, I.; Capretz, T.; Osann, K.; Okhunov, Z.; Landman, J.; Clayman, R.V. Coconut water: An unexpected source of urinary citrate. *Biomed. Res. Int.* **2018**, *2018*, 3061742. [CrossRef]
74. Shuster, J.; Jenkins, A.; Logan, C.; Barnett, T.; Riehle, R.; Zackson, D.; Wolfe, H.; Dale, R.; Daley, M.; Malic, I.; et al. Soft drink consumption and urinary stone recurrence: A randomized prevention trial. *J. Clin. Epidemiol.* **1992**, *45*, 911–916. [CrossRef]
75. Choi, J.W.J.; Ford, E.S.; Gao, X.; Choi, H.K. Sugar-sweetened soft drinks, diet soft drinks, and serum uric acid level: The Third National Health and Nutrition Examination Survey. *Arthritis Rheum.* **2008**, *59*, 109–116. [CrossRef] [PubMed]
76. Choi, H.K.; Curhan, G. Soft drinks, fructose consumption, and the risk of gout in men: Prospective cohort study. *Br. Med. J.* **2008**, *336*, 309–312. [CrossRef] [PubMed]
77. Taylor, E.N.; Curhan, G.C. Fructose consumption and the risk of kidney stones. *Kidney Int.* **2008**, *73*, 207–212. [CrossRef] [PubMed]
78. Barghouthy, Y.; Corrales, M.; Doizi, S.; Somani, B.K.; Traxer, O. Tea and coffee consumption and pathophysiology related to kidney stone formation: A systematic review. *World J. Urol.* **2020**, 1–10. [CrossRef] [PubMed]
79. Chen, H.Y.; Wu, J.S.; Chang, Y.F.; Sun, Z.J.; Chang, C.J.; Lu, F.H.; Yang, Y.C. Increased amount and duration of tea consumption may be associated with decreased risk of renal stone disease. *World J. Urol.* **2019**, *37*, 379–384. [CrossRef] [PubMed]

80. Littlejohns, T.J.; Neal, N.L.; Bradbury, K.E.; Heers, H.; Allen, N.E.; Turney, B.W. Fluid intake and dietary factors and the risk of incident kidney stones in UK biobank: A population-based prospective cohort study. *Eur. Urol. Focus* **2020**, *6*, 752–761. [CrossRef]
81. Ferraro, P.M.; Taylor, E.N.; Gambaro, G.; Curhan, G.C. Caffeine intake and the risk of kidney stones. *Am. J. Clin. Nutr.* **2014**, *100*, 1596–1603. [CrossRef]
82. Massey, L.K.; Wise, K.J. Impact of gender and age on urinary water and mineral excretion responses to acute caffeine doses. *Nutr. Res.* **1992**, *12*, 605–612. [CrossRef]
83. European Food Safety Authority (EFSA) Panel on Dietetic Products, Nutrition and Allergies. Scientific Opinion on the Safety of Caffeine. *EFSA J.* **2015**, *13*, 4102. [CrossRef]
84. Shu, X.; Cai, H.; Xiang, Y.B.; Li, H.; Lipworth, L.; Miller, N.L.; Zheng, W.; Shu, X.O.; His, R.S. Green tea intake and risk of incident kidney stones: Prospective cohort studies in middle-aged and elderly Chinese individuals. *Int. J. Urol.* **2019**, *26*, 241–246. [CrossRef]
85. Mahdavi, R.; Yagin, N.L.; Liebman, M.; Nikniaz, Z. Effect of different brewing times on soluble oxalate content of loose-packed black teas and tea bags. *Urolithiasis* **2013**, *41*, 15–19. [CrossRef]
86. McKay, D.W.; Seviour, J.P.; Comerford, A.; Vasdev, S.; Massey, L.K. Herbal tea: An alternative to regular tea for those who form calcium oxalate stones. *J. Am. Diet. Assoc.* **1995**, *95*, 360–361. [CrossRef]
87. Richter, M.; Baerlocher, K.; Bauer, J.M.; Elmadfa, I.; Heseker, H.; Leschik-Bonnet, E.; Stangl, G.; Volkert, D.; Stehle, P. Revised reference values for the intake of protein. *Ann. Nutr. Metab.* **2019**, *74*, 242–250. [CrossRef] [PubMed]
88. Kok, D.J.; Iestra, J.A.; Doorenbos, C.J.; Papapoulos, S.E. The effects of dietary excesses in animal protein and in sodium on the composition and the crystallization kinetics of calcium oxalate monohydrate in urines of healthy men. *J. Clin. Endocrinol. Metab.* **1990**, *71*, 861–867. [CrossRef] [PubMed]
89. Reddy, S.T.; Wang, C.Y.; Sakhaee, K.; Brinkley, L.; Pak, C.Y.C. Effect of low-carbohydrate high-protein diets on acid-base balance, stone-forming propensity, and calcium metabolism. *Am. J. Kidney Dis.* **2002**, *40*, 265–274. [CrossRef]
90. Giannini, S.; Nobile, M.; Sartori, L.; Carbonare, L.D.; Ciuffreda, M.; Corrò, P.; D'Angelo, A.; Calò, L.; Crepaldi, G. Acute effects of moderate dietary protein restriction in patients with idiopathic hypercalciuria and calcium nephrolithiasis. *Am. J. Clin. Nutr.* **1999**, *69*, 267–271. [CrossRef]
91. Pedersen, A.N.; Kondrup, J.; Børsheim, E. Health effects of protein intake in healthy adults: A systematic literature review. *Food Nutr. Res.* **2013**, *57*, 21245. [CrossRef] [PubMed]
92. Van Elswyk, M.E.; Weatherford, C.A.; McNeill, S.H. A systematic review of renal health in healthy individuals associated with protein intake above the US Recommended Daily Allowance in randomized controlled trials and observational studies. *Adv. Nutr.* **2018**, *9*, 404–418. [CrossRef]
93. Curhan, G.C.; Willett, W.C.; Knight, E.L.; Stampfer, M.J. Dietary factors and the risk of incident kidney stones in younger women. *Arch. Intern. Med.* **2004**, *164*, 885–891. [CrossRef]
94. Taylor, E.N.; Stampfer, M.J.; Curhan, G.C. Dietary factors and the risk of incident kidney stones in men: New insights after 14 years of follow-up. *J. Am. Soc. Nephrol.* **2004**, *15*, 3225–3232. [CrossRef]
95. Schwingshackl, L.; Hoffmann, G. Comparison of high vs. normal/low protein diets on renal function in subjects without chronic kidney disease: A systematic review and meta-analysis. *PLoS ONE* **2014**, *9*, e97656. [CrossRef] [PubMed]
96. Siener, R.; Struwe, F.; Hesse, A. Effect of L-methionine on the risk of phosphate stone formation. *Urology* **2016**, *98*, 39–43. [CrossRef] [PubMed]
97. Remer, T. Influence of diet on acid-base balance. *Semin. Dial.* **2000**, *13*, 221–226. [CrossRef] [PubMed]
98. Ferraro, P.M.; Mandel, E.I.; Curhan, G.C.; Gambaro, G.; Taylor, E.N. Dietary protein and potassium, diet–dependent net acid load, and risk of incident kidney stones. *Clin. J. Am. Soc. Nephrol.* **2016**, *11*, 1834–1844. [CrossRef]
99. Remer, T.; Manz, F. Estimation of the renal net acid excretion by adults consuming diets containing variable amounts of protein. *Am. J. Clin. Nutr.* **1994**, *59*, 1356–1361. [CrossRef]
100. Remer, T.; Manz, F. Potential renal acid load of foods and its influence on urine pH. *J. Am. Diet. Assoc.* **1995**, *95*, 791–797. [CrossRef]
101. Meschi, T.; Maggiore, U.; Fiaccadori, E.; Schianchi, T.; Bosi, S.; Adorni, G.; Ridolo, E.; Guerra, A.; Allegri, F.; Novarini, A.; et al. The effect of fruits and vegetables on urinary stone risk factors. *Kidney Int.* **2004**, *66*, 2402–2410. [CrossRef]
102. Moe, O.W.; Huang, C.L. Hypercalciuria from acid load: Renal mechanisms. *J. Nephrol.* **2006**, *19*, S53–S61.
103. Power, C.; Barker, D.J.P.; Nelson, M.; Winter, P.D. Diet and renal stones: A case-control study. *Br. J. Urol.* **1984**, *56*, 456–459. [PubMed]
104. Fellström, B.; Danielson, B.G.; Karlström, B.; Lithell, H.; Ljunghall, S.; Vessby, B. Dietary habits in renal stone patients compared with healthy subjects. *Br. J. Urol.* **1989**, *63*, 575–580. [CrossRef]
105. Al Zahrani, H.; Norman, R.W.; Thompson, C.; Weerasinghe, S. The dietary habits of idiopathic calcium stone-formers and normal control subjects. *BJU Int.* **2000**, *85*, 616–620. [CrossRef] [PubMed]
106. Trinchieri, A.; Mandressi, A.; Luongo, P.; Longo, G.; Pisani, E. The influence of diet on urinary risk factors for stones in healthy subjects and idiopathic renal calcium stone formers. *Br. J. Urol.* **1991**, *67*, 230–236. [CrossRef]
107. Curhan, G.C.; Willett, W.C.; Rimm, E.B.; Stampfer, M.J. A prospective study of dietary calcium and other nutrients and the risk of symptomatic kidney stones. *N. Engl. J. Med.* **1993**, *328*, 833–838. [CrossRef] [PubMed]
108. Curhan, G.C.; Willett, W.C.; Speizer, F.E.; Spiegelman, D.; Stampfer, M.J. Comparison of dietary calcium with supplemental calcium and other nutrients as factors affecting the risk for kidney stones in women. *Ann. Intern. Med.* **1997**, *126*, 497–504. [CrossRef]

109. Lemann, J.; Piering, W.; Lennon, E.J. Possible role of carbohydrate-induced calciuria in calcium oxalate kidney-stone formation. *N. Engl. J. Med.* **1969**, *280*, 232–237. [CrossRef]
110. Barilla, D.E.; Townsend, J.; Pak, C.Y.C. An exaggerated augmentation of renal calcium excretion after oral glucose ingestion in patients with renal hypercalciuria. *Investig. Urol.* **1978**, *15*, 486–488.
111. Nguyen, N.U.; Dumoulin, G.; Henriet, M.T.; Regnard, J. Effects of i.v. insulin bolus on urinary calcium and oxalate excretion in healthy subjects. *Horm. Metab. Res.* **1998**, *30*, 222–226. [CrossRef] [PubMed]
112. Wood, R.J.; Gerhardt, A.; Rosenberg, I.H. Effects of glucose and glucose polymers on calcium absorption in healthy subjects. *Am. J. Clin. Nutr.* **1987**, *46*, 699–701. [CrossRef]
113. Yoon, V.; Adams-Huet, B.; Sakhaee, K.; Maalouf, N.M. Hyperinsulinemia and urinary calcium excretion in calcium stone formers with idiopathic hypercalciuria. *J. Clin. Endocrinol. Metab.* **2013**, *98*, 2589–2594. [CrossRef]
114. Asselman, M.; Verkoelen, C.F. Fructose intake as a risk factor for kidney stone disease. *Kidney Int.* **2008**, *73*, 139–140. [CrossRef]
115. Nguyen, N.U.; Dumoulin, G.; Henriet, M.T.; Regnard, J. Increase in urinary calcium and oxalate after fructose infusion. *Horm. Metab. Res.* **1995**, *27*, 155–158. [CrossRef]
116. Johnson, R.J.; Perez-Pozo, S.E.; Lillo, J.L.; Grases, F.; Schold, J.D.; Kuwabara, M.; Sato, Y.; Hernando, A.A.; Garcia, G.; Jensen, T.; et al. Fructose increases risk for kidney stones: Potential role in metabolic syndrome and heat stress. *BMC Nephrol.* **2018**, *19*, 315. [CrossRef] [PubMed]
117. Emmerson, B.T. Effect of oral fructose on urate production. *Ann. Rheum. Dis.* **1974**, *33*, 276–280. [CrossRef] [PubMed]
118. Fox, I.H.; Kelley, W.N. Studies on the mechanism of fructose-induced hyperuricemia in man. *Metabolism* **1972**, *21*, 713–721. [CrossRef]
119. Rodgers, A.L.; Siener, R. The efficacy of polyunsaturated fatty acids as protectors against calcium oxalate renal stone formation: A review. *Nutrients* **2020**, *12*, 1069. [CrossRef]
120. Baggio, B.; Gambaro, G.; Zambon, S.; Marchini, F.; Bassi, A.; Bordin, L.; Clari, G.; Manzato, E. Anomalous phospholipid n-6 polyunsaturated fatty acid composition in idiopathic calcium nephrolithiasis. *J. Am. Soc. Nephrol.* **1996**, *7*, 613–620. [CrossRef] [PubMed]
121. Lote, C.J.; Haylor, J. Eicosanoids in renal function. *Prostaglandins Leukot. Essent. Fatty Acids* **1989**, *36*, 203–217. [CrossRef]
122. Calder, P.C.; Grimble, R.F. Polyunsaturated fatty acids, inflammation and immunity. *Eur. J. Clin. Nutr.* **2002**, *56*, S14–S19. [CrossRef]
123. Baggio, B.; Budakovic, A.; Nassuato, M.A.; Vezzoli, G.; Manzato, E.; Luisetto, G.; Zaninotto, M. Plasma phospholipid arachidonic acid content and calcium metabolism in idiopathic calcium nephrolithiasis. *Kidney Int.* **2000**, *58*, 1278–1284. [CrossRef]
124. Watkins, B.A.; Li, Y.; Seifert, M.F. Lipids as modulators of bone remodelling. *Curr. Opin. Clin. Nutr. Metab. Care* **2001**, *4*, 105–110. [CrossRef] [PubMed]
125. Buck, A.C.; Lote, C.J.; Sampson, W.F. The influence of renal prostaglandins on urinary calcium excretion in idiopathic urolithiasis. *J. Urol.* **1983**, *129*, 421–426. [CrossRef]
126. Hirayama, H.; Ikegami, K.; Shimomura, T.; Soejima, H.; Yamamoto, T. The possible role of prostaglandin E2 in urinary stone formation. *J. Urol.* **1988**, *139*, 549–551. [CrossRef]
127. Baggio, B.; Priante, G.; Brunati, A.M.; Clari, G.; Bordin, L. Specific modulatory effect of arachidonic acid on human red blood cell oxalate transport: Clinical implications in calcium oxalate nephrolithiasis. *J. Am. Soc. Nephrol.* **1999**, *10*, S381–S384.
128. Cao, J.; Schwichtenberg, K.A.; Hanson, N.Q.; Tsai, M.Y. Incorporation and clearance of omega-3 fatty acids in erythrocyte membranes and plasma phospholipids. *Clin. Chem.* **2006**, *52*, 2265–2272. [CrossRef]
129. Naya, Y.; Ito, H.; Masai, M.; Yamaguchi, K. Association of dietary fatty acids with urinary oxalate excretion in calcium oxalate stone-formers in their fourth decade. *BJU Int.* **2002**, *89*, 842–846. [CrossRef]
130. Siener, R.; Jansen, B.; Watzer, B.; Hesse, A. Effect of n-3 fatty acid supplementation on urinary risk factors for calcium oxalate stone formation. *J. Urol.* **2011**, *185*, 719–724. [CrossRef] [PubMed]
131. Rodgers, A. Aspects of calcium oxalate crystallization: Theory, in vitro studies, and in vivo implementation. *J. Am. Soc. Nephrol.* **1999**, *10*, S351–S354.
132. Pak, C.Y.C.; Adams-Huet, B.; Poindexter, J.R.; Pearle, M.S.; Peterson, R.D.; Moe, O.W. Relative effect of urinary calcium and oxalate on saturation of calcium oxalate. *Kidney Int.* **2004**, *66*, 2032–2037. [CrossRef]
133. Siener, R.; Glatz, S.; Nicolay, C.; Hesse, A. Prospective study on the efficacy of a selective treatment and risk factors for relapse in recurrent calcium oxalate stone patients. *Eur. Urol.* **2003**, *44*, 467–474. [CrossRef]
134. Siener, R.; Ebert, D.; Nicolay, C.; Hesse, A. Dietary risk factors for hyperoxaluria in calcium oxalate stone formers. *Kidney Int.* **2003**, *63*, 1037–1043. [CrossRef] [PubMed]
135. Knight, J.; Jiang, J.; Assimos, D.G.; Holmes, R.P. Hydroxyproline ingestion and urinary oxalate and glycolate excretion. *Kidney Int.* **2006**, *70*, 1929–1934. [CrossRef]
136. Knight, J.; Madduma-Liyanage, K.; Mobley, J.A.; Assimos, D.G.; Holmes, R.P. Ascorbic acid intake and oxalate synthesis. *Urolithiasis* **2016**, *44*, 289–297. [CrossRef] [PubMed]
137. Holmes, R.P.; Goodman, H.O.; Assimos, D.G. Contribution of dietary oxalate to urinary oxalate excretion. *Kidney Int.* **2001**, *59*, 270–276. [CrossRef]
138. Siener, R.; Bade, D.J.; Hesse, A.; Hoppe, B. Dietary hyperoxaluria is not reduced by treatment with lactic acid bacteria. *J. Transl. Med.* **2013**, *11*, 1–7. [CrossRef] [PubMed]
139. Taylor, E.N.; Curhan, G.C. Oxalate intake and the risk for nephrolithiasis. *J. Am. Soc. Nephrol.* **2007**, *18*, 2198–2204. [CrossRef] [PubMed]

140. Crivelli, J.J.; Mitchell, T.; Knight, J.; Wood, K.D.; Assimos, D.G.; Holmes, R.P.; Fargue, S. Contribution of dietary oxalate and oxalate precursors to urinary oxalate excretion. *Nutrients* **2021**, *13*, 62. [CrossRef]
141. Savage, G.P.; Vanhanen, L.; Mason, S.M.; Ross, A.B. Effect of cooking on the soluble and insoluble oxalate content of some New Zealand foods. *J. Food. Compos. Anal.* **2000**, *13*, 201–206. [CrossRef]
142. Chai, W.; Liebman, M. Effect of cooking methods on vegetable oxalate content. *J. Agric. Food Chem.* **2005**, *53*, 3027–3030. [CrossRef]
143. Siener, R.; Hönow, R.; Seidler, A.; Voss, S.; Hesse, A. Oxalate contents of species of the Polygonaceae, Amaranthaceae and Chenopodiaceae families. *Food Chem.* **2006**, *98*, 220–224. [CrossRef]
144. Siener, R.; Hönow, R.; Voss, S.; Seidler, A.; Hesse, A. Oxalate content of cereals and cereal products. *J. Agric. Food Chem.* **2006**, *54*, 3008–3011. [CrossRef] [PubMed]
145. Siener, R.; Seidler, A.; Hönow, R. Oxalate-rich foods. *Food Sci. Technol.* **2021**, *41*, 169–173. [CrossRef]
146. Taylor, E.N.; Curhan, G.C. Determinants of 24-hour urinary oxalate excretion. *Clin. J. Am. Soc. Nephrol.* **2008**, *3*, 1453–1460. [CrossRef] [PubMed]
147. Thomas, L.D.K.; Elinder, C.G.; Tiselius, H.G.; Wolk, A.; Åkesson, A. Ascorbic acid supplements and kidney stone incidence among men: A prospective study. *JAMA Intern. Med.* **2013**, *173*, 386–388. [CrossRef]
148. Traxer, O.; Huet, B.; Poindexter, J.; Pak, C.Y.C.; Pearle, M.S. Effect of ascorbic acid consumption on urinary stone risk factors. *J. Urol.* **2003**, *170*, 397–401. [CrossRef]
149. Siener, R.; Hoppe, B.; Löhr, P.; Müller, S.C.; Latz, S. Metabolic profile and impact of diet in patients with primary hyperoxaluria. *Int. Urol. Nephrol.* **2018**, *50*, 1583–1589. [CrossRef] [PubMed]
150. Fargue, S.; Milliner, D.S.; Knight, J.; Olson, J.B.; Lowther, W.T.; Holmes, R.P. Hydroxyproline metabolism and oxalate synthesis in primary hyperoxaluria. *J. Am. Soc. Nephrol.* **2018**, *29*, 1615–1623. [CrossRef] [PubMed]
151. Voss, S.; Hesse, A.; Zimmermann, D.J.; Sauerbruch, T.; von Unruh, G.E. Intestinal oxalate absorption is higher in idiopathic calcium oxalate stone formers than in healthy controls: Measurements with the [$^{13}C_2$]oxalate absorption test. *J. Urol.* **2006**, *175*, 1711–1715. [CrossRef]
152. Von Unruh, G.E.; Voss, S.; Sauerbruch, T.; Hesse, A. Dependence of oxalate absorption on the daily calcium intake. *J. Am. Soc. Nephrol.* **2004**, *15*, 1567–1573. [CrossRef]
153. Asplin, J.R. The management of patients with enteric hyperoxaluria. *Urolithiasis* **2016**, *44*, 33–43. [CrossRef]
154. Siener, R.; Machaka, I.; Alteheld, B.; Bitterlich, N.; Metzner, C. Effect of fat-soluble vitamins A, D, E and K on vitamin status and metabolic profile in patients with fat malabsorption with and without urolithiasis. *Nutrients* **2020**, *12*, 3110. [CrossRef] [PubMed]
155. Witting, C.; Langman, C.B.; Assimos, D.; Baum, M.A.; Kausz, A.; Milliner, D.; Tasian, G.; Worcester, E.; Allain, M.; West, M.; et al. Pathophysiology and treatment of enteric hyperoxaluria. *Clin. J. Am. Soc. Nephrol.* **2021**, *16*, 487–495. [CrossRef]
156. Earnest, D.L.; Johnson, G.; Williams, H.E.; Admirand, W.H. Hyperoxaluria in patients with ileal resection: An abnormality in dietary oxalate absorption. *Gastroenterology* **1974**, *66*, 1114–1122. [CrossRef]
157. Siener, R.; Petzold, J.; Bitterlich, N.; Alteheld, B.; Metzner, C. Determinants of urolithiasis in patients with intestinal fat malabsorption. *Urology* **2013**, *81*, 17–24. [CrossRef] [PubMed]
158. Siener, R.; Bangen, U.; Sidhu, H.; Hönow, R.; von Unruh, G.; Hesse, A. The role of Oxalobacter formigenes colonization in calcium oxalate stone disease. *Kidney Int.* **2013**, *83*, 1144–1149. [CrossRef]
159. Ticinesi, A.; Nouvenne, A.; Chiussi, G.; Castaldo, G.; Guerra, A.; Meschi, T. Calcium oxalate nephrolithiasis and gut microbiota: Not just a gut-kidney axis. A nutritional perspective. *Nutrients* **2020**, *12*, 548. [CrossRef] [PubMed]
160. Taylor, E.N.; Curhan, G.C. Dietary calcium from dairy and nondairy sources, and risk of symptomatic kidney stones. *J. Urol.* **2013**, *190*, 1255–1259. [CrossRef]
161. Borghi, L.; Schianchi, T.; Meschi, T.; Guerra, A.; Allegri, F.; Maggiore, U.; Novarini, A. Comparison of two diets for the prevention of recurrent stones in idiopathic hypercalciuria. *N. Engl. J. Med.* **2002**, *346*, 77–84. [CrossRef]
162. Muldowney, F.P.; Freaney, R.; Moloney, M.F. Importance of dietary sodium in the hypercalciuria syndrome. *Kidney Int.* **1982**, *22*, 292–296. [CrossRef]
163. Sakhaee, K.; Harvey, J.A.; Padalino, P.K.; Whitson, P.; Pak, C.Y.C. The potential role of salt abuse on the risk for kidney stone formation. *J. Urol.* **1993**, *150*, 310–312. [CrossRef]
164. Massey, L.K.; Whiting, S.J. Dietary salt, urinary calcium and kidney-stone risk. *Nutr. Rev.* **1995**, *53*, 131–134. [CrossRef]
165. Nouvenne, A.; Meschi, T.; Prati, B.; Guerra, A.; Allegri, F.; Vezzoli, G.; Soldati, L.; Gambaro, G.; Maggiore, U.; Borghi, L. Effects of a low-salt diet on idiopathic hypercalciuria in calcium-oxalate stone formers: A 3-mo randomized controlled trial. *Am. J. Clin. Nutr.* **2010**, *91*, 565–570. [CrossRef]

nutrients

MDPI

Review

Diet and Nutrition in Pediatric Inflammatory Bowel Diseases

Ugo Cucinotta, Claudio Romano * and Valeria Dipasquale

Pediatric Gastroenterology and Cystic Fibrosis Unit, Department of Human Pathology in Adulthood and Childhood "G. Barresi", University of Messina, 98124 Messina, Italy; ugocucinotta@gmail.com (U.C.); dipasquale.valeria@libero.it (V.D.)

* Correspondence: romanoc@unime.it; Tel.: +39-090-221-2919

Abstract: Both genetic and environmental factors are involved in the onset of inflammatory bowel disease (IBD). In particular, diet composition is suspected to significantly contribute to IBD risk. In recent years, major interest has raised about the role of nutrition in disease pathogenesis and course, and many studies have shown a clear link between diet composition and intestinal permeability impairment. Moreover, many IBD-related factors, such as poor dietary intake, nutrients loss and drugs interact with nutritional status, thus paving the way for the development of many therapeutic strategies in which nutrition represents the cornerstone, either as first-line therapy or as reversing nutritional deficiencies and malnutrition in IBD patients. Exclusive enteral nutrition (EEN) is the most rigorously supported dietary intervention for the treatment of Crohn's Disease (CD), but is burdened by a low tolerability, especially in pediatric patients. Promising alternative regimens are represented by Crohn's Disease Exclusion Diet (CDED), and other elimination diets, whose use is gradually spreading. The aim of the current paper is to provide a comprehensive and updated overview on the latest evidence about the role of nutrition and diet in pediatric IBD, focusing on the different nutritional interventions available for the management of the disease.

Keywords: diet; enteral nutrition; inflammatory bowel disease; nutrition; nutritional therapy; prevention

Citation: Cucinotta, U.; Romano, C.; Dipasquale, V. Diet and Nutrition in Pediatric Inflammatory Bowel Diseases. *Nutrients* **2021**, *13*, 655. https://doi.org/10.3390/nu13020655

Academic Editor: Ina Bergheim
Received: 1 February 2021
Accepted: 13 February 2021
Published: 17 February 2021

1. Introduction

The term inflammatory bowel disease (IBD) refers to a heterogeneous group of disorders encompassing Crohn's disease (CD), ulcerative colitis (UC) and unclassified inflammatory bowel disease (IBD-U), characterized by a relapsing-remitting behavior, and variably presenting with abdominal pain, diarrhea, rectal bleeding and weight loss. In the last decades, the incidence of IBD has significantly grown in industrialized countries, in children as in adults [1–3]. It is well known that IBD has a multifactorial etiology, but despite the ongoing scientific efforts, pathogenesis and pathophysiology are still unclear [4]. Both genetic and environmental factors are involved in IBD onset. In particular, the prevailing hypothesis accounts a complex interaction between an exaggerated immune response in genetically predisposed individuals and environmental factors along with intestinal flora alterations, eventually sustaining an anomalous chronic inflammation [4,5].

Several studies have been conducted in the past to evaluate the precise role of nutrition (and malnutrition) and diet composition in immune-mediated diseases risk, leading to the development of a new discipline, referred to as "clinical nutrition". Clinical nutrition is defined by European Society for Clinical Nutrition and Metabolism (ESPEN) as the "discipline that deals with the prevention, diagnosis and management of nutritional and metabolic changes related to acute and chronic diseases and conditions caused by a lack or excess of energy and nutrients" [6]. It encompasses all the diseases in which nutrition plays a main role, not only as a promoter, but also as a therapeutic tool [7]. For what IBD is concerned, the relation between diet composition and disease onset, course and management is supported by scientific evidence: i) epidemiological studies found an association between specific dietary habits and nutrients with an increased risk of IBD; ii) some foods and dietary components are potentially capable to either enhance or reduce the severity of inflammation; iii) for some pediatric patients with an established diagnosis of CD, exclusive enteral nutrition (EEN) can be considered a primary induction treatment with efficacy in achieving mucosal healing; iv) exclusion diets could treat or prevent disease flares; v) in IBD children, especially in CD children, malnutrition and nutrients deficiencies are often present at diagnosis; vi) early nutritional strategies can lead to a better disease control, as well as catch-up growth, bone mineral density improvement and adequate pubertal development [8–10]. The most common therapeutic choices for pediatric IBD consist of systemic and topical corticosteroids, amino salicylate and immunosuppressants (such as thiopurines or methotrexate). Starting from the past decade, use of biologic therapies has significantly increased for their clear efficacy also in pediatric setting, although they have been associated with a loss of response over time [11]. Beside this, non-pharmacological management of pediatric IBD has evolved over years and dietary changes are currently considered major therapeutic tools [10,12,13].

The aim of the present study is to provide a comprehensive and updated overview on the latest evidence about the role of nutrition and diet in pediatric IBD, also focusing on the different nutritional interventions available for the management of the disease.

2. Methods

For the purpose of the paper, a comprehensive search of the published literature through the databases PUBMED MEDLINE and GOOGLE SCHOLAR was performed. The following keywords: "diet", "children", "enteral nutrition", "inflammatory bowel disease", "clinical nutrition", "nutritional therapy", "malnutrition" were used. We focused on the most relevant articles published in English between 2005 and December 2020 including both Meta-analysis, Systematic Reviews, Consensus Guidelines, Randomized Controlled Trials and Cohort Studies. The selected papers were then analyzed in order to extrapolate valid and updated evidences about the role of nutrition and diet in pediatric IBD, with a focus on the nutritional strategies available for the management of the disease.

3. Nutrition and Diet and Intestinal Inflammation

Despite the fact the exact pathogenesis is still unclear, IBD development results from a variable interaction between genetic susceptibility, individual immune response, gut microbiota and environmental factors [14].

The role of the diet composition in the intestinal inflammation has long been controversial, due to sometimes limited and conflicting data resulting from retrospective and case-control studies. However, it has been demonstrated that some individuals are more susceptible than others to develop the disease, depending on specific dietary practices. The increasingly "westernization" of lifestyle, mostly characterized by a high consumption of animal proteins and fats along with a poor intake of fruit, vegetables and fibers is associated with a higher risk of IBD [15,16]. Indeed, some epidemiological studies have demonstrated an increased risk of IBD among individuals moving from low-income countries, with low IBD incidence, toward high-income countries, hypothesizing that new environmental factors could increase the risk of the development of the disease [3]. While overall fat intake would not correlate with IBD, there is enough evidence to support that a low intake of omega-3 (n-3) and a high one of omega-6 (n-6) polyunsaturated fatty acids (PUFAs) is associated with an increased risk of CD. N-6 PUFAs (i.e., linoleic acid) are precursors to proinflammatory eicosanoids, whereas dietary n-3 PUFAs inhibits the formation of proinflammatory prostaglandins and leukotrienes through the arachidonic acid pathway, so their chronic imbalanced consumption may eventually lead to a pro- inflammatory state, oxidative stress and impaired intestinal mucosal permeability [17,18]. On the other side, a long-term intake of dietary fiber, particularly derived from fruit, has been associated with lower risk of CD, while no or conflicting association with UC has been found [19]. The most recent ESPEN guidelines recommend a diet rich in fruit and vegetables and low in *n*-6 PUFAs [20]. Western diet increases the risk of disease through food additives like emulsifiers and saturated fats, maltodextrins, carrageenin and carboxymethylcellulose, that have been associated with impaired intestinal permeability [21–24].

Even early life dietary patterns can be considered risk factors. Breast milk is rich in secretory IgA (sIgA), leukocytes and antimicrobial factors (lysozyme, lactoferrin, nucleotides), and promotes immune system maturation and T-lymphocytes pool creation, as well as immune response against infections [25–27]. It has been suggested that an early discontinuation of breastfeeding may facilitate the onset of several chronic conditions later in life, such as metabolic and autoimmune diseases. A recent meta-analysis comparing the exposure to breast milk among patients with CD and UC and controls confirmed an inverse association between breastfeeding and the risk of developing IBD [28]. Moreover, a dose depending effect has been hypothesized, with higher protection for longer-lasting breast milk exposure [28]. Additionally, since microbiome has been proven to differ according to the type of early diet, with relatively higher proportion of *Firmicutes* and *Actinobacteria* in breast-fed infants, it seems reasonable to assume that dysbiosis during the first months of life may negatively impact on immunological functions of the host microbiota [29–31].

Indeed, IBD has been consistently associated with gut dysbiosis, variably determined by specific foods and dietary habits. While commensal microbiome is physiologically abundant in *Firmicutes, Bacteroidetes, Actinobacteria* and *Proteobacteria*, IBD patients typically exhibit a low concentration of *Firmicutes*, such as *Bifidobacterium, Clostridia* and especially *Faecalibacterium prausnitzii*, whose protective role has been theorized for the possibility to stimulate anti-inflammatory cytokines (including IL-10) [12,32,33]. Conversely, increased concentrations of *Escherichia Coli* and other *Enterobacteriacee* has been found [32,33]. Even metabolic pathways are altered, because short chain fatty acids (SCFAs), such as acetate, propionate, and butyrate, normally produced by commensal bacteria through fermentation of food components, have been found reduced in IBD compared to controls. Further evidence includes reduced tryptophan metabolism and disrupted bile acids metabolism, with low bile acids production, classically associated with anti-inflammatory activities and T-cells regulation [33]. All these alterations eventually contribute to a decreased bacterial diversity, altered host barrier, increased permeability and subsequent intestinal inflammation.

4. Nutritional Strategies in Induction of Remission in Pediatric IBD

4.1. Exclusive Enteral Nutrition

Exclusive enteral nutrition (EEN) is the most important nutritional intervention in pediatric IBD, consisting of a complete liquid formula as the unique source of daily energy requirement for a period of 6-8 weeks [34]. Consensus guidelines of the European Society of Pediatric Gastroenterology, Hepatology and Nutrition (ESPGHAN) and European Crohn's Colitis Organization (ECCO), as well as North American guidelines, recommend EEN as the first-line therapy for mild to moderate pediatric CD to induce remission both in the first flare-up and during relapses of symptoms [20,34–37]. The main indication for EEN is represented by active luminal CD, with solely inflammatory behavior (B1 according to Paris classification) and low-to-medium risk at diagnosis, regardless of the disease location (Table 1) [35].

By contrast, there are insufficient data supporting EEN use for extraintestinal manifestations or for perianal disease [36]. EEN advantages include: (i) high rates (up to 80%) of clinical remission; (ii) steroids sparing; (iii) correction of malnutrition and micronutrients deficiency; (iv) lean body mass increase; (v) improvement of growth and height Z scores; (vi) decreased longer term need for steroid and anti-TNF therapies; (vii) improvement of quality of life (QoL) [38–41]. Data supporting EEN in the management of UC is still lacking, and its use is still not recommended in these patients [36].

4.1.1. EEN Efficacy

Many studies, including systematic reviews, meta-analyses and Cochrane reviews, proved EEN to be as effective as steroid therapy in inducing remission over a 6–8 week-period in CD pediatric patients, by inducing both mucosal and transmural healing [42–47]. The significant advantage of sparing steroids, traditionally used in the clinical setting, is represented by the avoidance of their short- and long-term side effects, such as poor growth, increased susceptibility to infections, and metabolic disorders (Cushing syndrome, osteopenia/osteoporosis, impaired glucose metabolism) [35,36,48,49]. Several trials have demonstrated EEN may be superior to steroids in CD children with active disease, for the achievement of mucosal healing with significant decrease of acute phase reactants, such as erythrocyte sedimentation rate (ESR), C reactive protein (CRP) and fecal calprotectin (FC) [50,51]. Particularly, FC is a reliable, non-invasive inflammatory marker and predictor of mucosal healing, and it has been observed to decrease up to 50% after the administration of EEN [52], although FC levels at the end of EEN regimen would not seem to predict the length of time until a future relapse [53]. Furthermore, a study examining FC changes after food reintroduction found that FC rises to pre-treatment levels within 4 months post-EEN, proving that a subclinical inflammation rapidly occurs after food reintroduction, with free-diet foods presumably acting as triggers of inflammation [54].

Table 1. Summary of EEN characteristics.

Indications (according to Paris Classification at Diagnosis)	B1 (Inflammatory)
Advantages	High rates of clinical remission and mucosal healing Increased steroid-free remission Avoidance of steroids related side-effects Treatment of malnutrition and nutrients deficiencies Adequate growth and better QoL
Disadvantages	Low palatability No other food allowed High risk of early withdrawal High cost related to elemental diet Possible side effects (mostly diarrhea and vomiting)
Route of administration	Oral (preferred choice) Nasogastric tube
Duration of treatment	Minimum of 6 weeks up to 12 weeks
Type of formula	Polymeric or elemental

EEN, Exclusive Enteral Nutrition; B1, non-stricturing non-penetrating Behavior; QoL, Quality of Life.

Although no predictive factors of EEN outcome are currently validated, disease severity and luminal disease seem to be the only predictors of EEN success, with mild to moderate disease [weighted Pediatric Crohn Disease Activity Index (wPCDAI) < 57.5, fecal calprotectin < 500 mcg/ g of stool, ileal involvement, CRP > 15 mg/L] and ileal disease having demonstrated a better response to EEN [55]. However, due to insufficient data supporting the association between efficacy of EEN and disease location [56–58], current guidelines do not recommend taking into account the disease phenotype when starting patients on EEN [20,35,36].

4.1.2. Practical Issues of EEN

EEN is delivered through the administration of enteral nutrition (EN) formulas, available in a wide range of commercial formulations. These formulas are classified according to the protein source into intact proteins (polymeric), modified protein (elemental/semi-elemental) and disease specific formulas [39]. A nutritionally balanced polymeric formula should be the preferred choice in patients with intestinal sufficiency and in the absence of other medical conditions (e.g., co-existence of a cow's milk allergy) [36]. Its optimal nutritional properties (approximately 45–60% carbohydrate, 15–20% protein, 30–40% fat), along with better palatability and lower price, makes polymeric feed the most commonly used [59]. However, despite these advantages, no EN formula has been demonstrated to be superior to others, with similar results in terms of achievement of clinical remission [42,58]. Oral route should be the preferred way, switching to nasogastric tube only in the case of inadequate oral intake [36,60]. Volumes are calculated according to the estimated average daily requirement, minding the caloric content of the chosen formula (standard concentration of 0.86^{-1} kcal/mL). Since resting energy expenditure has not shown to be increased in patients with CD during the different phases of disease, equations for estimating energy requirements can be used independently from the disease phase [60].

Despite the absence of evidence about the exact duration of an EEN regimen, the guidelines recommend a minimum of 6 weeks. Indeed, while symptoms may improve after few days and by week 3 in most cases [61], mucosal healing takes several weeks to occur. Food re-introduction is then performed over a period of 1 to 5 weeks, although no protocols for the proper management are currently available [36]. Notably, a a retrospective cohort study evaluating food re-introduction after a EEN cycle during 1 year of follow-up proved that rapid (3 days) food reintroduction after EEN is as effective as standard (5 weeks) food reintroduction, with no significant differences in clinical relapse rates: rapid 50% vs. standard 47% (p = 0.58) [62]. Further studies are warranted in order to assess the best timing in IBD pediatric population.

4.1.3. Mechanism of Action of EEN

The exact mechanism through which EEN is capable of inducing mucosal healing is still unclear, but some evidence exists about the beneficial exclusion of specific dietary components, microbiome modulation, intestinal rest and direct anti-inflammatory effect of EEN [63,64]. Previously published trials have shown the ineffectiveness of partial enteral nutrition (PEN) associated with a free diet in inducing clinical remission, proving that dietary antigens, even in small quantities, are dangerous, thus suggesting that EEN could also work by excluding them [40,65]. After EEN administration microbiome composition has shown some modifications, including paradoxical reduction of microbial diversity and decreased proportion of presumably beneficial bacterial groups, such as *Bifidobacterium spp, Firmicutes* (including *Faecalibacterium prausnitzii*), *Bacteroides/Prevotella* and *Proteobacteriaceae* [64,66,67]. In particular, a significant correlation was found between the decrease of *Bacteroides/Prevotella* group bacteria and the clinical improvement during EEN treatment [68]. Accordingly, regression of major EEN-induced microbiome changes after return to habitual free diet is well documented [67,68].

4.2. Partial Enteral Nutrition

PEN is a nutritional strategy based on the administration of a liquid enteral formula, not covering 100% of total energy requirements, along with whole foods. In different trials evaluating PEN for treatment of active CD or for maintenance of remission, the volume of formula has ranged from 35% to 90% of total energy requirements, while the foods consumed belonged mainly to free diet or to defined restrictive diets [24]. PEN in association with an unrestricted diet has not shown to be a good choice of treatment [65]. It has been found to be significantly less effective for the achievement of clinical remission than EEN or biologics [40], whereby its use is not currently recommended as induction strategy in active pediatric CD [35,36]. Further nutritional strategies combining PEN with specific exclusion diets are warranted.

4.3. Crohn's Disease Exclusion Diet

The Crohn's Disease Exclusion Diet (CDED) is a validated dietary intervention, firstly conceived in 2014 by Sigall-Boneh and colleagues, combining PEN with a specific exclusion diet [69]. The rationale of CDED is the avoidance of certain foods and dietary components (such as additives like emulsifiers or maltodextrins, food preservatives, etc), mostly belonging to western diets, deemed to act as triggers for intestinal inflammation, dysbiosis, altered intestinal mucous layer and impaired barrier function [12,21,22,70,71]. In addition, the use of real foods, even though in a controlled way, seems to meet the request of patients and their parents, often blaming the monotony of EEN as responsible for a low adherence to treatment [72,73].

4.3.1. CDED Efficacy

Sigall-Boneh and colleagues [69] first examined the effectiveness of CDED on 47 CD patients (mean age of 16 ± 5.6 years), through the administration of a polymeric formula covering 50% of their daily energy requirements and selected foods for the remaining 50%. After 6 weeks of treatment 70% of them (33/47) achieved clinical remission, evaluated through the Harvey-Bradshaw Index (HBI) and the PCDAI. Across the following 6 weeks, the quantity of polymeric formula was gradually reduced to 25%, while more selected foods were added to diet. Eighty percent of patients were still in remission at 12 weeks [69]. Another study demonstrated the efficacy of CDED in inducing remission even for children failing biological therapy [74].

A recent randomized controlled trial comparing CDED (50% PEN + CDED) to EEN in patients with luminal mild to moderate active CD demonstrated that CDED is as effective as EEN in inducing clinical remission at 6 weeks of treatment, with similar decrease in PCDAI and acute phase reactants [75]. Moreover, CDED seems to be even superior to EEN in terms of sustained remission rates and tolerability, leading authors to conclude that CDED + PEN is reasonably a good option either as first line treatment in luminal mild-moderate active CD or for maintenance of remission in the long term [75]. However, due to the lack of endoscopic assessment of mucosal healing - a major therapeutic goal - current guidelines still don't recommend CDED + PEN neither as induction therapy nor for maintenance [35].

Another recent trial compared EEN to PEN plus a controlled diet rather similar to CDED to induce mucosal healing in active pediatric CD over a period of 6 weeks [76]. PEN group received a polymeric formula covering 75% of the daily requirement and one meal a day from an anti-inflammatory diet (AID) inspired to CDED (avoidance of processed foods with additives, animal fat, sugar, dairy products, and gluten) (Table 2).

Table 2. Not allowed foods in all phases - adapted from Sigall-Boneh et al. [69].

- Dairy products of any kind, margarine
- Wheat, breakfast cereals, breads and baked goods of any kind, yeast for baking
- Gluten-free products not listed above, Soya products, potato or corn flour
- Processed or smoked meats and fish (sausages, luncheon meats, salamis, fish sticks)
- Sauces, salad dressings, syrups and jams of any kind
- Canned products and dried fruits
- Packaged snacks (potato chips, pretzels, popcorn, nuts, etc)
- All soft drinks, fruit juices and sweetened beverages, alcoholic beverages, coffee
- Candies, chocolates, cakes, cookies and gum

At week 6 clinical and endoscopic remission, defined as PCDAI < 10 and Simple Endoscopic Score for CD (SES-CD) ≤ 2, respectively, were evaluated, showing similar rates of both between EEN and PEN, with higher rates of clinical remission in the PEN group (81.9%) than described so far in other trials evaluating PEN [34,40]. However, some limitations of the study should be taken into account. PEN group received a greater amount of EN formula (75% of the daily requirement) than in other studies on PEN, perhaps explaining the higher remission rates. Moreover, there was the lack of randomization, which may have affected the clinical evaluation. Additionally, a small number of patients ($n = 11$) completed the protocol. Before recommendations can be formulated, further studies are necessary, aimed to replicate these data and to evaluate the real efficacy on a larger sample size of patients.

4.3.2. Practical Characteristics of CDED

As previously mentioned, CDED owes its progressive diffusion both to the high efficacy in terms of clinical remission and to its variability, leading to a better acceptance by patients. CDED is a multi-stage whole food diet, based on the exclusion of animal and saturated fats, gluten, dairy products, or high-processed foods containing emulsifiers, and all packaged products. At the same time, it provides an increased consumption of fruits, vegetables, and resistant starch. The foods are classified into three main groups: mandatory foods, allowed foods and not allowed foods. CDED is then designed in 3 phases: the first two induction phases lasting 6 weeks each, while the third one starting from the 13th week and referred to as "maintenance phase" [69,73]. During the first phase, 50% of nutritional requirement is given in the form of a polymeric formula, poor in lactose and fibers (e.g., Modulen, Nestlè), while the remaining 50% is supplied by mandatory foods (Table 3), source of high-quality proteins, pectin, resistant starch and other beneficial fibers, all necessary for the production of SCFAs [73,77].

Table 3. Example of Crohn Disease Exclusion Diet - adapted from [70,73].

<table>
<tr><th>Phase</th><th>Polymeric formula</th><th>Mandatory foods</th><th>Allowed foods</th></tr>
<tr><td>1 (Week 1–6)</td><td>Modulen ® (50% of daily requirements)</td><td rowspan="2">Fresh Chicken breast (150–200 g/d)
2 Eggs/d
2 Bananas/d
1 Fresh Apple/d
2 Potatoes/d</td><td rowspan="3">Fresh Strawberries
Fresh Melon (1 slice)
Rice flour
White rice and rice noodles (unlimited)
2 Tomatoes (additional allowed for cooking)
2 Cucumbers (medium size)
2 Avocado halves
1 Carrot
Spinach 1 cup uncooked leaves
Lettuce (3 leaves)
Onion
Fresh green herbs (basil, parsley, coriander, rosemary, thyme, mint, dill)
1 glass of squeezed orange juice from fresh oranges
Water, sparkling water
Salt, pepper, paprika, cinnamon, cumin
3 tablespoons honey
4 teaspoons sugar
Fresh ginger and garlic cloves, lemons</td></tr>
<tr><td>2 (Week 7–12)</td><td>Modulen ® (25% of daily requirements)</td></tr>
<tr><td>3 (Maintenance) (Week 13 onwards)</td><td>Modulen ® (25% of daily requirements)</td><td>No mandatory food</td></tr>
</table>

During the second phase the percentage of calories provided by the liquid formula is lowered to 25%, more foods are permitted with higher inclusion of fruits and vegetables and little quantities of bread, red meat and legumes (found to potentially aggravate symptoms) [78]. The third and last phase still consists of 25% polymeric formula and does not have a specific duration, allowing the patient to continue with a controlled diet, rich in healthy foods, with the final aim of better controlling the disease in the long term [73].

4.4. Crohn's Disease Treatment With Eating Diet

A newly devised nutritional intervention is the Crohn's Disease Treatment With Eating Diet (CD-TREAT), an individualized diet that aims to replicate composition, and hence the effect on the microbiome, of EEN using ordinary foods [71,79]. As previously said, EEN is likely to work by exclusion of dietary components thought to be detrimental for gut function and microbiome composition. Hence, similarly to EEN, in CD-TREAT some specific components are excluded (like lactose, gluten, processed meat or some additives) while others are allowed (lean meats, fish eggs, some fruits and vegetables, rich in macronutrients, vitamins and fibers) (Table 4).

CD-TREAT strength is represented by the higher variability and palatability in comparison to EEN. It has shown good percentages of remission after 8 weeks both in adults and in children: 80% (4/5) of adults showed a clinical response (wPCDAI score change > 17.5), while 60% (3/5) of pediatric patients were in remission (wPCDAI score < 12.5) [79]. Despite the exiguous group of patients tested, which make it necessary to verify these on a larger sample of individuals, such a diet seems to be an encouraging step forward in the future approach to CD treatment, especially as a substitute to the long-term use of EEN, whose acceptability could be limited, and as a solid food-based alternative which would be preferable [72].

Table 4. Exclusion diets with related included and excluded food groups - adapted from [71].

Dietary Intervention	Excluded	Included
Crohn's Disease exclusion diet	Dairy, gluten, processed meats, animal fat, canned and packaged foods, coffee, alcohol, emulsifying agents	Selected fruits and vegetables, fish, eggs, lean meats
Crohn's Disease Treatment With Eating Diet	Lactose, gluten, processed meats, animal fat, alcohol, emulsifying agents	Fruits and vegetables, high fiber content, lean meats. Increased macronutrients, vitamins, minerals (aim to mimic EEN diet)
IBD Anti-InflammatoryDiet	Most dairy, gluten-based grains, fruits with seeds, refined sugars, processed foods, trans-fatty acids	Most fruits and vegetables, soluble fibers, flax meal, chia seed, oats, some type of yogurts and cheeses, nut and legume flours, eggs, fish, lean meats

4.5. Further Nutritional Strategies

Further exclusion diets have been proposed, including the Specific Carbohydrate Diet (SCD), the Diet Low in Fermentable Oligo-, Di- and Monosaccharides and Polyol (FODMAPs), the Paleolithic Diet, and the Vegan Diet. Although some of them have shown interestingly results, their efficacy has been frequently evaluated in studies with several limitations. Many of these studies were mostly focused on the assessment of improvement in symptoms without standardized clinical outcomes, and with poor attention to inflammation and mucosal healing achievement [80]. Moreover, most data have been extrapolated from studies on adults, with sometimes conflicting results between those showing a real improvement of symptoms [81,82] and those showing no significative changing or even worsening of symptoms [83,84]. Moreover, some of these diets expose patients to the high risk of malnutrition due to the restrictiveness of foods allowed, so that their use is not routinely recommended in children/adolescents, unless potential benefits outweigh potential risks of the diet [20,36].

5. Nutritional Strategies in Maintenance of Remission in Pediatric IBD

Beyond the achievement of clinical remission, one of the most important goals in the management of IBD is the maintenance of remission, therefore avoiding, or at least delaying, the occurrence of a future relapse. Maintenance therapy of CD has classically consisted of immunosuppressants, such as thiopurines, or biologics in the last years. Despite their well-known efficacy, both of these treatments are burdened by some negative aspects, including risk of side effects for thiopurines (e.g., infections, pancreatitis), loss of response over time and high cost for biologics [35]. For these reasons, trials investigating nutritional strategies for maintenance of remission have been carried out in the last years. To date, EEN is not recommended for maintenance of remission in pediatric patients, since efficacy has not been thoroughly evaluated in children and long-term adherence may be challenging after a first cycle of induction [36,55,80]. Although other nutritional interventions are usually intended for short-period use, it is not infrequent for many patients to continue using them for longer period of time. This behavior derives, on the one hand, by the current absence of a recommended diet for IBD patients in the long term and, on the other hand, by the fear of relapses after food reintroduction. Indeed, many patients with an established diagnosis of IBD tend to avoid a variety of foods (most commonly grains, dairy products, vegetables and fruits) believed to provoke IBD symptoms and flares, or in order to ameliorate the control of disease [85,86]. Patients and their families frequently advocate dietary advices from physicians and are extremely interested about the proper dietary regimen to follow.

According to the recent ESPEN guidelines [20], no specific diet needs to be followed during remission phases of IBD, since none of the proposed alternative diets seems effective in maintenance of remission. It is important to mention that most exclusion diets (e.g., FODMAPs, gluten-free diet, lactose free diet) may improve symptoms in IBD patients but haven't been demonstrated to affect the inflammatory activity in the long term [9,36,71]. Diets should be then customized on the single patient according to individual's preference or in case of specific intolerance (e.g., reducing high-lactose containing products and/or using lactase treated products if lactase deficiency is suspected), keeping in mind the risk of malnourishment or nutritional deficiencies [20].

A recent pediatric study demonstrated that a subgroup of patients, who achieved remission using EEN and not in therapy with any other medications, can successfully continue with PEN supplements for maintenance of remission with lower flares rates at 1 year of follow up [87]. By contrast, no significative differences in relapse rates have been observed between patients using PEN and azathioprine and those only using azathioprine. For this reason, PEN associated with exclusion diets have been proposed in case of patients with mild disease and who have a low risk of flares, despite amount, duration and timing of PEN is still unknown [36,88].

6. Malnutrition

Malnutrition is a quite frequent occurrence in IBD patients, especially in CD patients, since different sites are affected in comparison to UC (small bowel vs colon-rectum). Pathophysiology of malnutrition is heterogeneous, involving multiple aspects, including: i) anorexia, food avoidance and self-imposed hypocaloric diets; ii) increased energy and nutrient losses, as a result of malabsorption and gut losses [89,90].

Data regarding prevalence of malnutrition in IBD children are mostly available among newly diagnosed patients, accounting approximatively 60% of newly diagnosed CD children and 35% of UC [90]. Protein-energy malnutrition is therefore a common finding in CD at the time of diagnosis, with low weight for age and body mass index (BMI). Then, nutritional status often fluctuates over time according to disease control and flares. Noteworthy, different studies have highlighted a progressive decrease in the degree of malnutrition among newly diagnosed IBD patients in recent years, especially in UC patients, in whom even overweight has been observed [90].

Nutritional assessment is a central component in the management of IBD patients, since a poor nutritional status is strictly related to growth failure, poor bone accrual, anemia, disrupted pubertal development over time and short stature in adulthood, as well as increased complications rates and poor prognosis [36,91,92]. Recommended approach to IBD patients should comprise: i) a periodical dietary history through a 3- to 5-day record of qualitative and quantitative intake of foods (macro and micronutrients); ii) evaluation of nutritional status with assessment of weight-for-age, height-for-age, and BMI z-score at every visit, and assessment of height velocity every 6 months [36,91].

7. Micronutrient Deficiency

Micronutrients and vitamins are frequently lacking in IBD, due to malabsorption linked to inflammation and/or to sub-optimal nutritional intake. Micronutrients deficiency can be present at the diagnosis as well as develop during the clinical course of the disease, so that IBD patients should be checked for micronutrients levels on a regular basis, and specific deficits should be appropriately corrected [92–94]. The most frequent vitamin deficiencies observed in IBD-patients are water-soluble B9 and B12 deficiencies, physiologically absorbed in the duodenum, jejunum and ileum, which are frequent site of inflammation [90,95]. About micronutrients, iron and zinc are the most frequently lacking, with iron-deficient anemia (IDA) being the predominant type [90,96,97].

Necessity for single nutrients intervention varies according to type of elements/minerals considered. Zinc, selenium and magnesium deficiency should be not routinely assessed or supplemented [36]. By contrast, IDA should be periodically ruled out, especially during period of active inflammation. Treatment varies according to the disease phase: oral iron supplementation can be the choice of treatment in case of controlled disease and hemoglobin level > 10 g/dl, always keeping in mind the high risk of intolerance for the oral route. In case of active disease or moderate-severe anemia (hemoglobin < 10 g/dl), intra-venous ferric carboxymaltose replacement should be preferred [36]. Both children and adolescents are at risk of vitamin D deficiency during the clinical course of disease and should be always investigated for it. Vitamin D is a risk factor for low bone mineral density, whose additional risk factors are represented by cumulative corticosteroid dose, height-for-age Z-score, and BMI Z-score [20,96]. For this reason, according to current guidelines, all children affected by CD and steroid-treated, serum calcium and 25(OH) vitamin D should be monitored and supplemented, if required, to help preventing low bone mineral density. Vitamin D level below 50 nmol/L or 20 ng/mL suggests vitamin D deficiency, requiring a proper supplementation. If encountered, osteopenia and osteoporosis should be then managed according to current osteoporosis guidelines [20]. For the risk of folate (B9) deficiency, especially in those using sulfasalazine and methotrexate, its measurement is generally recommended annually, with folate supplementation (1 mg/day or 5 mg/week) in case of documented deficiency [36].

8. Conclusions

Dietary factors and malnutrition play a primary role in the onset and management of pediatric IBD. An appropriate diet can decrease the risk of IBD flares, and an age-adequate nutritional status can decrease the risk of complications and surgery in the long-term stage of the disease. Enteral nutrition has a significant impact on mucosal inflammation in CD, and clinical response to oral polymeric diet is associated with down-regulation of mucosal pro-inflammatory cytokines. Exclusive enteral nutrition or CDED diet have shown to be more effective than steroids in inducing clinical remission in pediatric patients.

Author Contributions: U.C., V.D. and C.R. equally contributed to the manuscript. All authors have read and agreed to the published version of the manuscript.

Funding: This research received no external funding.

Conflicts of Interest: The authors declare no conflict of interest.

References

1. Burisch, J.; Jess, T.; Martinato, M.; Lakatos, P.L.; & ECCO -EpiCom. The burden of inflammatory bowel disease in Europe. *J. Crohns colitis* **2013**, *7*, 322–337. [CrossRef] [PubMed]
2. Ng, S.C.; Kaplan, G.G.; Tang, W.; Banerjee, R.; Adigopula, B.; Underwood, F.E.; Tanyingoh, D.; Wei, S.C.; Lin, W.C.; Lin, H.H.; et al. Population density and risk of inflammatory bowel disease: A prospective population-based study in 13 countries or regions in Asia-Pacific. *Am. J. Gastroenterol.* **2019**, *114*, 107–115. [CrossRef]
3. GBD 2017 Inflammatory Bowel Disease Collaborators. The global, regional, and national burden of inflammatory bowel disease in 195 countries and territories, 1990–2017: A systematic analysis for the Global Burden of Disease Study 2017. *Lancet Gastroenterol. Hepatol.* **2020**, *5*, 17–30.
4. Soon, I.S.; Molodecky, N.A.; Rabi, D.M.; Ghali, W.A.; Barkema, H.W.; Kaplan, G.G. The relationship between urban environment and the inflammatory bowel diseases: A systematic review and meta-analysis. *BMC gastroenterol.* **2012**, *12*, 51. [CrossRef] [PubMed]
5. Lucendo, A.J.; De Rezende, L.C. Importance of nutrition in inflammatory bowel disease. *World J. Gastroenterol.* **2009**, *15*, 2081–2088. [CrossRef] [PubMed]
6. Cederholm, T.; Barazzoni, R.; Austin, P.; Ballmer, P.; Biolo, G.; Bischoff, S.C.; Compher, C.; Correia, I.; Higashiguchi, T.; Holst, M.; et al. ESPEN guidelines on definitions and terminology of clinical nutrition. *Clin. Nutr.* **2017**, *36*, 49–64. [CrossRef]
7. Cardenas, D. What is clinical nutrition? Understanding the epistemological foundations of a new discipline. *Clin. Nutr. ESPEN* **2016**, *11*, e63–e66. [CrossRef]

8. Hou, J.K.; Abraham, B.; El-Serag, H. Dietary intake and risk of developing inflammatory bowel disease: A systematic review of the literature. *Am J Gastroenterol.* **2011**, *106*, 563–573. [CrossRef] [PubMed]
9. Schreiner, P.; Martinho-Grueber, M.; Studerus, D.; Vavricka, S.R.; Tilg, H.; Biedermann, L.; On behalf of Swiss IBDnet, an official working group of the Swiss Society of Gastroenterology. Nutrition in Inflammatory Bowel Disease. *Digestion* **2020**, *101*, 120–135. [CrossRef] [PubMed]
10. Forbes, A.; Escher, J.; Hébuterne, X.; Kłęk, S.; Krznaric, Z.; Schneider, S.; Shamir, R.; Stardelova, K.; Wierdsma, N.; Wiskin, A.E.; et al. ESPEN guideline: Clinical nutrition in inflammatory bowel disease. *Clin Nutr.* **2017**, *36*, 321–347. [CrossRef]
11. Roberti, R.; Iannone, L.F.; Palleria, C.; De Sarro, C.; Spagnuolo, R.; Barbieri, M.A.; Vero, A.; Manti, A.; Pisana, V.; Fries, W.; et al. Safety profiles of biologic agents for inflammatory bowel diseases: A prospective pharmacovigilance study in Southern Italy. *Curr. Med. Res.* **2020**, *36*, 1457–1463.
12. Levine, A.; Sigall Boneh, R.; Wine, E. Evolving role of diet in the pathogenesis and treatment of inflammatory bowel diseases. *Gut* **2018**, *67*, 1726–1738. [CrossRef]
13. Khalili, H.; Chan, S.S.; Lochhead, P.; Ananthakrishnan, A.N.; Hart, A.R.; Chan, A.T. The role of diet in the aetiopathogenesis of inflammatory bowel disease. *Nat. Rev. Gastroenterol. Hepatol.* **2018**, *15*, 525–535. [CrossRef]
14. Ng, S.C.; Bernstein, C.N.; Vatn, M.H.; Lakatos, P.L.; Loftus, E.V.; Tysk, C., Jr.; O'Morain, C.; Moum, B.; Colombel, J.F. Epidemiology and Natural History Task Force of the International Organization of Inflammatory Bowel Disease (IOIBD). Geographical variability and environmental risk factors in inflammatory bowel disease. *Gut* **2013**, *62*, 630–649. [CrossRef] [PubMed]
15. Yatsunenko, T.; Rey, F.E.; Manary, M.J.; Trehan, I.; Dominguez-Bello, M.G.; Contreras, M.; Magris, M.; Hidalgo, G.; Baldassano, R.N.; Anokhin, A.P.; et al. Human gut microbiome viewed across age and geography. *Nature* **2012**, *486*, 222–227. [CrossRef]
16. Telle-Hansen, V.H.; Holven, K.B.; Ulven, S.M. Impact of a healthy dietary pattern on gut microbiota and systemic inflammation in humans. *Nutrients* **2018**, *10*, 1783. [CrossRef]
17. Ananthakrishnan, A.N.; Khalili, H.; Konijeti, G.G.; Higuchi, L.M.; de Silva, P.; Fuchs, C.S.; Willett, W.C.; Richter, J.M.; Chan, A.T. Long-term intake of dietary fat and risk of ulcerative colitis and Crohn's disease. *Gut* **2014**, *63*, 776–784. [CrossRef]
18. Li, X.; Wei, X.; Sun, Y.; Du, J.; Li, X.; Xun, Z.; Li, Y.C. High-fat diet promotes experimental colitis by inducing oxidative stress in the colon. *Am. J. Physiol. Gastrointest. Liver Physiol.* **2019**, *317*, G453–G462. [CrossRef]
19. Ananthakrishnan, A.N.; Khalili, H.; Konijeti, G.G.; Higuchi, L.M.; de Silva, P.; Korzenik, J.R.; Fuchs, C.S.; Willett, W.C.; Richter, J.M.; Chan, A.T. A prospective study of long-term intake of dietary fiber and risk of Crohn's disease and ulcerative colitis. *Gastroenterology* **2013**, *145*, 970–977. [CrossRef]
20. Bischoff, S.C.; Escher, J.; Hébuterne, X.; Kłęk, S.; Krznaric, Z.; Schneider, S.; Shamir, R.; Stardelova, K.; Wierdsma, N.; Wiskin, A.E.; et al. ESPEN practical guideline: Clinical nutrition in inflammatory bowel disease. *Clin. Nutr.* **2020**, *39*, 632–653. [CrossRef] [PubMed]
21. Nickerson, K.P.; McDonald, C. Crohn's disease-associated adherent-invasive escherichia coli adhesion is enhanced by exposure to the ubiquitous dietary polysaccharide maltodextrin. *PLoS ONE* **2012**, *7*, e52132.
22. Roberts, C.L.; Rushworth, S.L.; Richman, E.; Rhodes, J.M. Hypothesis: Increased consumption of emulsifiers as an explanation for the rising incidence of Crohn's disease. *J. Crohns Colitis* **2013**, *7*, 338–341. [CrossRef]
23. Martino, J.V.; Limbergen, J.V.; Cahil, L. The role of carrageenan and carboxymethylcellulose in the development of intestinal inflam- mation. *Front. Pediatr.* **2017**, *5*, 96. [CrossRef] [PubMed]
24. Sigall-Boneh, R.; Levine, A.; Lomer, M.; Wierdsma, N.; Allan, P.; Fiorino, G.; Gatti, S.; Jonkers, D.; Kierkus, J.; Katsanos, K.H.; et al. Research gaps in diet and nutrition in inflammatory bowel disease. A topical review by D-ECCO Working Group [Dietitians of ECCO]. *J. Crohns Colitis* **2017**, *11*, 1407–1419. [CrossRef] [PubMed]
25. Parigi, S.M.; Eldh, M.; Larssen, P.; Gabrielsson, S.; Villablanca, E.J. Breast milk and solid food shaping intestinal immunity. *Front Immunol.* **2015**, *6*, 415. [CrossRef]
26. Rogier, E.W.; Frantz, A.L.; Bruno, M.E.; Wedlund, L.; Cohen, D.A.; Stromberg, A.J.; Kaetzel, C.S. Lessons from mother: Long-term impact of antibodies in breast milk on the gut microbiota and intestinal immune system of breastfed offspring. *Gut microbes* **2014**, *5*, 663–668. [CrossRef] [PubMed]
27. Praveen, P.; Jordan, F.; Priami, C.; Morine, M.J. The role of breast-feeding in infant immune system: A systems perspective on the intestinal microbiome. *Microbiome* **2015**, *3*, 41. [CrossRef] [PubMed]
28. Xu, L.; Lochhead, P.; Ko, Y.; Claggett, B.; Leong, R.W.; Ananthakrishnan, A.N. Systematic review with meta-analysis: Breastfeeding and the risk of Crohn's disease and ulcerative colitis. *Aliment. Pharmacol. Ther.* **2017**, *46*, 780–789. [CrossRef] [PubMed]
29. Hill, C.J.; Lynch, D.B.; Murphy, K.; Ulaszewska, M.; Jeffery, I.B.; O'Shea, C.A.; Watkins, C.; Dempsey, E.; Mattivi, F.; Tuohy, K.; et al. Evolution of gut microbiota composition from birth to 24 weeks in the INFANTMET Cohort. *Microbiome* **2017**, *5*, 4. [CrossRef] [PubMed]
30. Kashtanova, D.A.; Popenko, A.S.; Tkacheva, O.N.; Tyakht, A.B.; Alexeev, D.G.; Boytsov, S.A. Association between the gut microbiota and diet: Fetal life, early childhood, and further life. *Nutrition* **2016**, *32*, 620–627. [CrossRef] [PubMed]
31. Madan, J.C.; Hoen, A.G.; Lundgren, S.N.; Farzan, S.F.; Cottingham, K.L.; Morrison, H.G.; Sogin, M.L.; Li, H.; Moore, J.H.; Karagas, M.R. Association of cesarean delivery and formula supplementation with the intestinal microbiome of 6-Week-old infants. *JAMA Pediatrics* **2016**, *170*, 212–219. [CrossRef]

32. Pittayanon, R.; Lau, J.T.; Leontiadis, G.I.; Tse, F.; Yuan, Y.; Surette, M.; Moayyedi, P. Differences in gut microbiota in patients with vs without inflammatory bowel diseases: A systematic review. *Gastroenterology* **2020**, *158*, 930–946.e1. [CrossRef]
33. Liu, S.; Zhao, W.; Lan, P.; Mou, X. The microbiome in inflammatory bowel diseases: From pathogenesis to therapy. *Protein Cell* **2020**. [CrossRef] [PubMed]
34. Ruemmele, F.M.; Veres, G.; Kolho, K.L.; Griffiths, A.; Levine, A.; Escher, J.C.; Amil Dias, J.; Barabino, A.; Braegger, C.P.; Bronsky, J.; et al. Consensus guidelines of ECCO/ESPGHAN on the medical management of pediatric Crohn's disease. *J. Crohn's Colitis* **2014**, *8*, 1179–1207. [CrossRef]
35. Van Rheenen, P.F.; Aloi, M.; Assa, A.; Bronsky, J.; Escher, J.C.; Fagerberg, U.L.; Gasparetto, M.; Gerasimidis, K.; Griffiths, A.; Henderson, P. The medical management of paediatric Crohn's Disease: An ECCO-ESPGHAN guideline update. *J. Crohns Colitis* **2020**, jjaa161. [CrossRef] [PubMed]
36. Miele, E.; Shamir, R.; Aloi, M.; Assa, A.; Braegger, C.; Bronsky, J.; De Ridder, L.; Escher, J.C.; Hojsak, I.; Kolaˇcek, S.; et al. Nutrition in pediatric inflammatory bowel disease: A position paper on behalf of the Porto Inflammatory Bowel Disease Group of the European Society of Pediatric Gastroenterology, Hepatology and Nutrition. *J. Pediatr. Gastroenterol. Nutr.* **2018**, *66*, 687–708. [CrossRef] [PubMed]
37. Mack, D.R.; Benchimol, E.I.; Critch, J.; deBruyn, J.; Tse, F.; Moayyedi, P.; Church, P.; Deslandres, C.; El-Matary, W.; Huynh, H.; et al. Canadian Association of Gastroenterology Clinical Practice Guideline for the medical management of pediatric luminal Crohn's disease. *Gastroenterology* **2019**, *157*, 320–348. [CrossRef] [PubMed]
38. Swaminath, A.; Feathers, A.; Ananthakrishnan, A.N.; Falzon, L.; Li Ferry, S. Systematic review with meta-Analysis: Enteral nutrition therapy for the induction of remission in paediatric Crohn's disease. *Aliment. Pharmacol. Ther.* **2017**, *46*, 645–656. [CrossRef]
39. Hansen, T.; Duerksen, D.R. Enteral nutrition in the management of pediatric and adult Crohn's disease. *Nutrients* **2018**, *10*, 537. [CrossRef]
40. Lee, D.; Baldassano, R.N.; Otley, A.R.; Albenberg, L.; Griffiths, A.M.; Compher, C.; Chen, E.Z.; Li, H.; Gilroy, E.; Nessel, L.; et al. Comparative effectiveness of nutritional and biological therapy in North American children with active Crohn's disease. *Inflamm. Bowel Dis.* **2015**, *21*, 1786–1793. [CrossRef] [PubMed]
41. Shaoul, R.; Brown, S.; Day, A.S. Reasoning beyond the potential use of Exclusive Enteral Nutrition and other specified diets in children with Ulcerative Colitis. *J. Pediatr. Gastroenterol. Nutr.* **2018**, *66*, 378–382. [CrossRef]
42. Narula, N.; Dhillon, A.; Zhang, D.; Sherlock, M.E.; Tondeur, M.; Zachos, M. Enteral nutritional therapy for induction of remission in Crohn's disease. *Cochrane Database Syst. Rev.* **2018**, *4*, CD000542. [CrossRef] [PubMed]
43. Knight, C.; El-Matary, W.; Spray, C.; Sandhu, B.K. Long-term outcome of nutritional therapy in paediatric Crohn's disease. *Clin. Nutr.* **2005**, *24*, 775–779. [CrossRef]
44. Yu, Y.; Chen, K.C.; Chen, J. Exclusive enteral nutrition versus corticosteroids for treatment of pediatric Crohn's disease: A meta-analysis. *World J. Pediatr.* **2019**, *15*, 26–36. [CrossRef]
45. Comeche, J.M.; Caballero, P.; Gutierrez-Hervas, A.; García-Sanjuan, S.; Comino, I.; Altavilla, C.; Tuells, J. Enteral nutrition in patients with inflammatory bowel disease. Systematic review, meta-analysis, and meta-regression. *Nutrients* **2019**, *11*, 2657. [CrossRef] [PubMed]
46. Grover, Z.; Muir, R.; Lewindon, P. Exclusive enteral nutrition induces early clinical, mucosal and transmural remission in paediatric Crohn's disease. *J. Gastroenterol.* **2014**, *49*, 638–645. [CrossRef]
47. Cohen-Dolev, N.; Sladek, M.; Hussey, S.; Turner, D.; Veres, G.; Koletzko, S.; Martin de Carpi, J.; Staiano, A.; Shaoul, R.; Lionetti, P.; et al. Differences in outcomes over time with Exclusive Enteral Nutrition compared with steroids in children with mild to moderate Crohn's disease: Results from the GROWTH CD Study. *J. Crohn colitis* **2018**, *12*, 306–312. [CrossRef]
48. Werkstetter, K.J.; Schatz, S.B.; Alberer, M.; Filipiak-Pittroff, B.; Koletzko, S. Influence of exclusive enteral nutrition therapy on bone density and geometry in newly diagnosed pediatric crohn's disease patients. *Ann. Nutr. Metab.* **2013**, *63*, 10–16. [CrossRef] [PubMed]
49. Connors, J.; Basseri, S.; Grant, A.; Giffin, N.; Mahdi, G.; Noble, A.; Rashid, M.; Otley, A.; Van Limbergen, J. Exclusive enteral nutrition therapy in paediatric Crohn's disease results in long-term avoidance of corticosteroids: Results of a propensity-score matched cohort analysis. *J. Crohns colitis* **2017**, *11*, 1063–1070. [CrossRef]
50. Borrelli, O.; Cordischi, L.; Cirulli, M.; Paganelli, M.; Labalestra, V.; Uccini, S.; Russo, P.M.; Cucchiara, S. Polymeric diet alone versus corticosteroids in the treatment of active pediatric Crohn's disease: A randomized controlled open-label trial. *Clin. Gastroenterol. Hepatol.* **2006**, *4*, 744–753. [CrossRef]
51. Pigneur, B.; Lepage, P.; Mondot, S.; Schmitz, J.; Goulet, O.; Doré, J.; Ruemmele, F.M. Mucosal healing and bacterial composition in response to enteral nutrition vs steroid-based induction therapy-A randomised prospective clinical trial in children with Crohn's disease. *J. Crohns colitis* **2019**, *13*, 846–855. [CrossRef]
52. Logan, M.; Ijaz, U.Z.; Hansen, R.; Gerasimidis, K.; Russell, R.K. Letter: Reproducible evidence shows that exclusive enteral nutrition significantly reduces faecal calprotectin concentrations in children with active Crohn's disease. *Aliment. Pharmacol. Ther.* **2017**, *46*, 1119–1120. [CrossRef]
53. Gerasimidis, K.; Nikolaou, C.K.; Edwards, C.A.; McGrogan, P. Serial fecal calprotectin changes in children with Crohn's disease on treatment with exclusive enteral nutrition: Associations with disease activity, treatment response, and prediction of a clinical relapse. *J. Clin. Gastroenterol.* **2011**, *45*, 234–239. [CrossRef]

54. Logan, M.; Clark, C.M.; Ijaz, U.Z.; Gervais, L.; Duncan, H.; Garrick, V.; Curtis, L.; Buchanan, E.; Cardigan, T.; Armstrong, L.; et al. The reduction of faecal calprotectin during exclusive enteral nutrition is lost rapidly after food re-introduction. *Aliment. Pharmacol. Ther.* **2019**, *50*, 664–674. [CrossRef]
55. Moriczi, M.; Pujol-Muncunill, G.; Martín-Masot, R.; Jiménez Treviño, S.; Segarra Cantón, O.; Ochoa Sangrador, C.; Peña Quintana, L.; González Santana, D.; Rodríguez Martínez, A.; Rosell Camps, A.; et al. Predictors of response to exclusive enteral nutrition in newly diagnosed Crohn's disease in children: PRESENCE Study from SEGHNP. *Nutrients* **2020**, *12*, 1012. [CrossRef] [PubMed]
56. Buchanan, E.; Gaunt, W.W.; Cardigan, T.; Garrick, V.; McGrogan, P.; Russell, R.K. The use of exclusive enteral nutrition for induction of remission in children with Crohn's disease demonstrates that disease phenotype does not influence clinical remission. *Aliment. Pharmacol. Ther.* **2009**, *30*, 501–517. [CrossRef] [PubMed]
57. Rubio, A.; Pigneur, B.; Garnier-Lengliné, H.; Talbotec, C.; Schmitz, J.; Canioni, D.; Goulet, O.; Ruemmele, F.M. The efficacy of exclusive nutritional therapy in paediatric Crohn's disease, comparing fractionated oral vs. continuous enteral feeding. *Aliment. Pharmacol. Ther.* **2011**, *33*, 1332–1339. [CrossRef]
58. Zachos, M.; Tondeur, M.; Griffiths, A.M. Enteral nutritional therapy for induction of remission in Crohn's disease. *Cochrane Database Syst Rev* **2007**, *1*, CD000542. [CrossRef]
59. Whitten, K.E.; Rogers, P.; Ooi, C.Y.; Day, A.S. International survey of enteral nutrition protocols used in children with Crohn's disease. *J. Dig. Dis.* **2012**, *13*, 107e12. [CrossRef] [PubMed]
60. Ashton, J.J.; Gavin, J.; Beattie, R.M. Exclusive enteral nutrition in Crohn's disease: Evidence and practicalities. *Clin Nutr* **2019**, *38*, 80–89. [CrossRef]
61. Sigall Boneh, R.; Van Limbergen, J.; Wine, E.; Assa, A.; Shaoul, R.; Milman, P.; Cohen, S.; Kori, M.; Peleg, S.; On, A.; et al. Dietary therapies induce rapid response and remission in pediatric patients with active Crohn's disease. *Clin. Gastroenterol. Hepatol.* **2020**, S1542-3565(20)30487-0. [CrossRef]
62. Faiman, A.; Mutalib, M.; Moylan, A.; Morgan, N.; Crespi, D.; Furman, M.; Kader, A. Standard versus rapid food reintroduction after exclusive enteral nutritional therapy in paediatric Crohn's disease. *Eur. J. Gastroenterol. Hepatol.* **2014**, *26*, 276–281. [CrossRef]
63. Day, A.S.; Whitten, K.E.; Sidler, M.; Lemberg, D.A. Systematic review: Nutritional therapy in paediatric Crohn's disease. *Aliment. Pharmacol. Ther.* **2008**, *27*, 293–307. [CrossRef]
64. Diederen, K.; Li, J.V.; Donachie, G.E.; de Meij, T.G.; de Waart, D.R.; Hakvoort, T.; Kindermann, A.; Wagner, J.; Auyeung, V.; Te Velde, A.A.; et al. Exclusive enteral nutrition mediates gut microbial and metabolic changes that are associated with remission in children with Crohn's disease. *Sci. Rep.* **2020**, *10*, 18879. [CrossRef]
65. Johnson, T.; Macdonald, S.; Hill, S.M.; Thomas, A.; Murphy, M.S. Treatment of active Crohn's disease in children using partial enteral nutrition with liquid formula: A randomised controlled trial. *Gut* **2006**, *55*, 356–361. [CrossRef] [PubMed]
66. MacLellan, A.; Connors, J.; Grant, S.; Cahill, L.; Langille, M.G.I.; Van Limbergen, J. The impact of exclusive enteral nutrition (een) on the gut microbiome in Crohn's disease: A review. *Nutrients* **2017**, *9*, 447. [CrossRef]
67. Gerasimidis, K.; Bertz, M.; Hanske, L.; Junick, J.; Biskou, O.; Aguilera, M.; Garrick, V.; Russell, R.K.; Blaut, M.; McGrogan, P.; et al. Decline in presumptively protective gut bacterial species and metabolites are paradoxically associated with disease improvement in pediatric Crohn's disease during enteral nutrition. *Inflamm. Bowel Dis.* **2014**, *20*, 861–871. [CrossRef] [PubMed]
68. Leach, S.; Mitchell, H.; Eng, W.; Zhang, L.; Day, A. Sustained modulation of intestinal bacteria by exclusive enteral nutrition used to treat children with Crohn's disease. *Aliment. Pharmacol. Ther.* **2008**, *28*, 724–733. [CrossRef]
69. Sigall Boneh, R.; Pfeffer-Gik, T.; Segal, I.; Zangen, T.; Boaz, M.; Levine, A. Partial enteral nutrition with a Crohn's disease exclusion diet is effective for induction of remission in children and young adults with Crohn's disease. *Inflamm. Bowel. Dis.* **2014**, *20*, 1353–1360. [CrossRef]
70. Corsello, A.; Pugliese, D.; Gasbarrini, A.; Armuzzi, A. Diet and nutrients in gastrointestinal chronic diseases. *Nutrients* **2020**, *12*, 2693. [CrossRef] [PubMed]
71. Sasson, A.N.; Ananthakrishnan, A.N.; Raman, M. Diet in treatment of inflammatory bowel diseases. *Clin Gastroenterol Hepatol.* **2021**, *19*, 425–435.e3. [CrossRef]
72. Svolos, V.; Gerasimidis, K.; Buchanan, E.; Curtis, L.; Garrick, V.; Hay, J.; Laird, S.; Munro, J.; Gaya, D.R.; Russell, R.K.; et al. Dietary treatment of Crohn's disease: Perceptions of families with children treated by exclusive enteral nutrition, a questionnaire survey. *BMC Gastroenterol.* **2017**, *17*, 14. [CrossRef]
73. Herrador-López, M.; Martín-Masot, R.; Navas-López, V.M. EEN Yesterday and Today ... CDED Today and Tomorrow. *Nutrients* **2020**, *12*, E3793. [CrossRef] [PubMed]
74. Sigall Boneh, R.; Sarbagili Shabat, C.; Yanai, H.; Chermesh, I.; Ben Avraham, S.; Boaz, M.; Levine, A. Dietary therapy with the Crohn's disease exclusion diet is a successful strategy for induction of remission in children and adults failing biological therapy. *J. Crohns Colitis* **2017**, *11*, 1205–1212. [CrossRef]
75. Levine, A.; Wine, E.; Assa, A.; Sigall Boneh, R.; Shaoul, R.; Kori, M.; Cohen, S.; Peleg, S.; Shamaly, H.; On, A. Crohn's disease exclusion diet plus partial enteral nutrition induces sustained remission in a randomized controlled trial. *Gastroenterology* **2019**, *157*, 440–450.e8. [CrossRef]
76. Urlep, D.; Benedik, E.; Brecelj, J.; Orel, R. Partial enteral nutrition induces clinical and endoscopic remission in active pediatric Crohn's disease: Results of a prospective cohort study. *Eur. J. Pediatr.* **2020**, *179*, 431–438. [CrossRef] [PubMed]
77. Van Limbergen, J.; Dunn, K.; Wine, E.; Sigall Boneh, R.; Bielawski, J.; Levine, A. OP22 Crohn's disease exclusion diet reduces bacterial dysbiosis towards healthy controls in paediatric Crohn's disease. *J. Crohns Colitis* **2020**, *14*, S019–S020. [CrossRef]

78. Levine, A.; El-Matary, W.; Van Limbergen, J. A case-based approach to new directions in dietary therapy of Crohn's disease: Food for thought. *Nutrients* **2020**, *12*, 880. [CrossRef]
79. Svolos, V.; Hansen, R.; Nichols, B.; Quince, C.; Ijaz, U.Z.; Papadopoulou, R.T.; Edwards, C.A.; Watson, D.; Alghamdi, A.; Brejnrod, A.; et al. Treatment of active Crohn's disease with an ordinary food-based diet that replicates exclusive enteral nutrition. *Gastroenterology* **2019**, *156*, 1354–1367.e6. [CrossRef] [PubMed]
80. Gkikas, K.; Gerasimidis, K.; Milling, S.; Ijaz, U.Z.; Hansen, R.; Russell, R.K. Dietary strategies for maintenance of clinical remission in inflammatory bowel diseases: Are we there yet? *Nutrients* **2020**, *12*, 2018. [CrossRef]
81. Cohen, S.A.; Gold, B.D.; Oliva, S.; Lewis, J.; Stallworth, A.; Koch, B.; Eshee, L.; Mason, D. Clinical and mucosal improvement with specific carbohydrate diet in pediatric Crohn disease. *J. Pediatr. Gastroenterol. Nutr.* **2014**, *59*, 516–521. [CrossRef] [PubMed]
82. Gearry, R.B.; Irving, P.M.; Barrett, J.S.; Nathan, D.M.; Shepherd, S.J.; Gibson, P.R. Reduction of dietary poorly absorbed short-chain carbohydrates (FODMAPs) improves abdominal symptoms in patients with inflammatory bowel disease-a pilot study. *J. Crohn's Colitis* **2009**, *3*, 8–14. [CrossRef]
83. Wahbeh, G.T.; Ward, B.T.; Lee, D.Y.; Giefer, M.J.; Suskind, D.L. Lack of mucosal healing from modified specific carbohydrate diet in pediatric patients with Crohn disease. *J Pediatr Gastroenterol Nutr* **2017**, *65*, 289–292. [CrossRef] [PubMed]
84. Cox, S.R.; Prince, A.C.; Myers, C.E.; Irving, P.M.; Lindsay, J.O.; Lomer, M.C.; Whelan, K. Fermentable carbohydrates [Fodmaps] exacerbate functional gastrointestinal symptoms in patients with inflammatory bowel disease: A randomised, double-blind, placebo-controlled, cross-over, re-challenge trial. *J. Crohn's Colitis* **2017**, *11*, 1420–1429. [CrossRef]
85. Limdi, J.K.; Aggarwal, D.; McLaughlin, J.T. Dietary practices and beliefs in patients with inflammatory bowel disease. *Inflamm. Bowel Dis.* **2016**, *22*, 164–170. [CrossRef] [PubMed]
86. Bergeron, F.; Bouin, M.; D'Aoust, L.; Lemoyne, M.; Presse, N. Food avoidance in patients with inflammatory bowel disease: What, when and who? *Clin. Nutr.* **2018**, *37*, 884–889. [CrossRef]
87. Duncan, H.; Buchanan, E.; Cardigan, T.; Garrick, V.; Curtis, L.; McGrogan, P.; Barclay, A.; Russell, R.K. A retrospective study showing maintenance treatment options for paediatric CD in the first year following diagnosis after induction of remission with EEN: Supplemental enteral nutrition is better than nothing! *BMC Gastroenterol.* **2014**, *14*, 50. [CrossRef]
88. Schulman, J.M.; Pritzker, L.; Shaoul, R. Maintenance of remission with partial enteral nutrition therapy in pediatric Crohn's disease: A retrospective study. *Can. J. Gastroenterol. Hepatol.* **2017**, *2017*. [CrossRef]
89. Gavin, J.; Ashton, J.J.; Heather, N.; Marino, L.V.; Beattie, R.M. Nutritional support in paediatric Crohn's disease: Outcome at 12 months. *Acta Paediatr* **2018**, *107*, 156–162. [CrossRef] [PubMed]
90. Gerasimidis, K.; McGrogan, P.; Edwards, C.A. The aetiology and impact of malnutrition in paediatric inflammatory bowel disease. *J. Hum. Nutr Diet.* **2011**, *24*, 313–326. [CrossRef]
91. Wiskin, A.E.; Wootton, S.A.; Hunt, T.M.; Cornelius, V.R.; Afzal, N.A.; Jackson, A.A.; Beattie, R.M. Body composition in childhood inflammatory bowel disease. *Clin Nutr.* **2011**, *30*, 112–115. [CrossRef] [PubMed]
92. Massironi, S.; Rossi, R.E.; Cavalcoli, F.A.; Della Valle, S.; Fraquelli, M.; Conte, D. Nutritional deficiencies in inflammatory bowel disease: Therapeutic approaches. *Clin. Nutr.* **2013**, *32*, 904–910. [CrossRef]
93. Weisshof, R.; Chermesh, I. Micronutrient deficiencies in inflammatory bowel disease. *Curr. Op. Clin. Nutr. Metab. Care* **2015**, *18*, 576–581. [CrossRef] [PubMed]
94. Fritz, J.; Walia, C.; Elkadri, A.; Pipkorn, R.; Dunn, R.K.; Sieracki, R.; Goday, P.S.; Mariano Cabrera, J. A systematic review of micronutrient deficiencies in pediatric inflammatory bowel disease. *Inflamm. Bowel Dis.* **2019**, *25*, 445–459. [CrossRef]
95. Ehrlich, S.; Mark, A.G.; Rinawi, F.; Shamir, R.; Assa, A. Micronutrient deficiencies in children with inflammatory bowel diseases. *Nutr. Clin. Pract.* **2020**, *35*, 315–322. [CrossRef] [PubMed]
96. Lopes, L.H.; Sdepanian, V.L.; Szejnfeld, V.L.; de Morais, M.B.; Fagundes-Neto, U. Risk factors for low bone mineral density in children and adolescents with inflammatory bowel disease. *Dig. Dis. Sci.* **2008**, *53*, 2746–2753. [CrossRef]
97. Hwang, C.; Ross, V.; Mahadevan, U. Micronutrient deficiencies in inflammatory bowel disease: From a to zinc. *Inflamm. Bowel Dis.* **2012**, *18*, 1961–1981. [CrossRef] [PubMed]

nutrients

MDPI

Review

Diet in Intestinal Fibrosis: A Double-Edged Sword

Rachel Marion-Letellier [1,2,*], Mathilde Leboutte [1,2], Asma Amamou [3], Maitreyi Raman [4,5], Guillaume Savoye [1,2,6] and Subrata Ghosh [3]

1 UNIROUEN, INSERM UMR 1073 Nutrition, Inflammation and Gut-Brain Axis, Normandie University, 76183 Rouen, France; mathilde.leboutte@etu.univ-rouen.fr (M.L.); guillaume.savoye@chu-rouen.fr (G.S.)
2 Institute for Research and Innovation in Biomedicine (IRIB), UNIROUEN, 76183 Rouen, France
3 APC Microbiome Ireland, Biosciences Building, University College Cork, Cork, Ireland; AAmamou@ucc.ie (A.A.); SubrataGhosh@ucc.ie (S.G.)
4 Division of Gastroenterology, University of Calgary, Calgary, AB T2N 4N1, Canada; mkothand@ucalgary.ca
5 Department of Community Health Sciences, University of Calgary, Calgary, AB T2N 4N1, Canada
6 Department of Gastroenterology, Rouen University Hospital, 76031 Rouen, France
* Correspondence: rachel.letellier@univ-rouen.fr

Abstract: The natural history of inflammatory bowel diseases, especially Crohn's disease, is frequently complicated by intestinal fibrosis. Because of the lack of effective treatments for intestinal fibrosis, there is an urgent need to develop new therapies. Factors promoting intestinal fibrosis are currently unclear, but diet is a potential culprit. Diet may influence predisposition to develop intestinal fibrosis or alter its natural history by modification of both the host immune response and intestinal microbial composition. Few studies have documented the effects of dietary factors in modulating IBD-induced intestinal fibrosis. As the mechanisms behind fibrogenesis in the gut are believed to be broadly similar to those from extra-intestinal organs, it may be relevant to investigate which dietary components can inhibit or promote fibrosis factors such as myofibroblasts progenitor activation in other fibrotic diseases.

Keywords: fibrosis; Crohn's disease; diet; gut microenvironment

Citation: Marion-Letellier, R.; Leboutte, M.; Amamou, A.; Raman, M.; Savoye, G.; Ghosh, S. Diet in Intestinal Fibrosis: A Double-Edged Sword. *Nutrients* **2021**, *13*, 3148. https://doi.org/10.3390/nu13093148

Academic Editor: Ina Bergheim

Received: 23 July 2021
Accepted: 7 September 2021
Published: 9 September 2021

1. Intestinal Fibrosis

Inflammatory bowel diseases (IBD) are relapsing systemic inflammatory diseases, mainly affecting the gastrointestinal tract. IBD occurs in people with susceptibility genes triggered by environmental factors. It leads to an exacerbated immune response associated with a gut dysbiosis. The natural history of IBD is frequently complicated by intestinal fibrosis and strictures formation. More than half of patients with Crohn's disease (CD) develop intestinal fibrosis, especially when the ileum is involved (Montreal classification L1) [1,2]. We and others have previously shown that stricturing CD may not respond well to anti-inflammatory therapy such as anti-TNFα, the gold standard in IBD treatment [3]. Eighty percent of CD patients with intestinal fibrosis and strictures undergo resection, but it frequently recurs, leading to repeated surgeries. Chronic inflammation induces remodeling of the intestinal wall by a cascade of events from intestinal epithelial damages to angiogenesis and immune and mesenchymal cells activation [4]. There is currently no specific therapy to prevent or inhibit intestinal fibrosis and this therefore constitutes an unmet need in IBD.

Western Diet

Environmental factors, especially diet, may influence predisposition to develop IBD or alter its course. Diet can target both the host immune response [5,6], and intestinal microbial composition. In addition, diet is a recurrent concern for IBD patients and most of IBD patients believe that diet may be a trigger of disease activity [5,7,8]. The IBD incidence is higher in Western countries [5] and this incidence continues to increase in the newly industrialized countries adopting the Western diet [7]. Diet plays a key role in controlling gut immune homeostasis. Factors promoting intestinal fibrosis are currently unknown, but diet is a potential culprit. Indeed, the Western diet promotes fibrosis in studies from other organs [8,9]. The Western diet is characterized by an insufficient intake of healthy foodstuffs and an excessive amount of saturated fats, sugar, and salt. Even IBD patients in remission have considerably distorted and unhealthy dietary intake leading to an increased risk of nutritional deficiencies [10]. Very recently, a prospective cohort study has demonstrated an association between ultra-processed food consumption and IBD risk [11].

2. Obesity

The incidence of obesity is increasing worldwide and we previously demonstrated increasing body weight over time from 1991 to 2008 in CD as evidenced by baseline data from 40 randomized clinical trials [12]. In addition, CD patients exhibited a higher clinical disease activity and duration over the same time period [12]. Adiposity may thus play a potential role in initiating and perpetuating intestinal inflammation. In addition, obesity is associated with chronic gut inflammation and visceral fat accumulation as observed in CD. Recently, visceral obesity was associated with adverse outcomes in severe CD patients [13] while IBD patients with weight loss after bariatric surgery had fewer complications [14].

High-fat diets (HFD) have a significant impact on gut physiology and mucosal defenses (Figure 1). Thirty-days HFD is sufficient to alter the spatial distribution and composition of the microbiota [15]. Innate immunity is also altered with: (i) decreased antimicrobial peptide with (ii) a reduction in Paneth cells and (iii) a decrease in goblet cell number and mucus secretion [15,16]. In addition, HFD induced greater intestinal permeability [15]. A few studies have also investigated the effects of HFD on experimental colitis. Mice receiving HFD were more susceptible to chemically-induced colitis and exhibited more severe colonic inflammation [16–18]. Several mechanisms have been suggested: (i) a gut barrier dysfunction, (ii) intestinal hyperpermeability, (iii) pathobiont expansion, and (iv) decreased plasma myokine irisin or adipokine levels [16–18]. Nevertheless, effects of HFD are only investigated on acute chemically-induced colitis and their effects on chronic colitis and intestinal fibrosis are still unknown. In fibrosis from extra-intestinal organs, HFD has a deleterious effect. Wnt-β catenin signaling is activated in intestinal fibrosis and can be induced through diet-induced obesity. For example, consumption of HFD induced an upregulation of β catenin and activated epithelial-mesenchymal transition (EMT) in a murine model of colon cancer [19]. This work is in accordance with findings from numerous extra-intestinal models of fibrosis such as hepatic [20] or renal fibrosis [21] where HFD is associated with higher EMT through TGF-β and β catenin signaling.

Environmental factors promoting intestinal fibrosis are currently unknown. High fat diets (HFD) may contribute to promotes fibrosis, as observed in extra-intestinal organs.

HFD have a significant impact on gut barrier function. HFD altered innate immunity response results in: (i) lower production of antimicrobial peptides, (ii) a reduced number of Paneth cells, and (iii) a decrease in goblet cell number and mucus secretion. In addition, HFD induced higher intestinal permeability and HFD-fed mice are more susceptible to colitis. As observed in fibrosis from extra-intestinal organs, HFD may also induce factors that promote intestinal fibrosis such as Wnt-β catenin or epithelial-mesenchymal transition (EMT). A second putative mechanism of HFD on intestinal fibrosis is the involvement of epithelial endoplasmic reticulum stress. Diet is the main modulator of gut microbiota composition. HFD is associated with dysbiosis and may thus impact microbial components. It has been demonstrated that certain bacteria are able to activate ECM production in intestinal fibroblasts. In addition, bacterial ligands are able to promote angiogenesis and it may contribute to fibrosis development. Effects of HFD may be mediated through an increased adiposity. Adiposity may thus play a potential role in initiating and perpetuating intestinal inflammation. Indeed, visceral obesity is associated with higher complications in patients with IBD. Adipocytes are able to secrete adipokines such as leptin or adiponectin. While leptin may exacerbate intestinal fibrosis, adiponectin may exert anti-fibrosis properties through a reduced extracellular matrix (ECM) deposition.

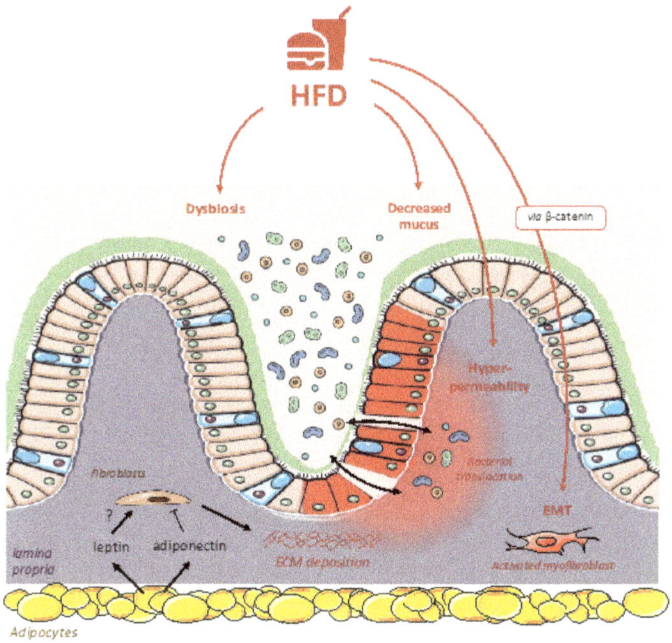

Figure 1. How high fat diets may contribute to intestinal fibrosis?

Involvement of epithelial endoplasmic reticulum stress has been recently suggested in CD fibrosis [22] and more precisely HFD is able to exacerbate endoplasmic reticulum stress in a model of pulmonary fibrosis [23].

Adipocytes secrete adipokines such as leptin or adiponectin in IBD mesenteric adipose tissue and serum. While leptin promote a Th1 profile, adiponectin antagonized TNFα and decreased adhesion molecules [24]. Very recently, Xie et al. investigated the effects of intraperitoneal injection of adiponectin in mice with chronic TNBS-induced colitis and they observed that adiponectin treatment reduced inflammatory markers such as colon myeloperoxidase activity and pro-inflammatory cytokines [25]. Adiponectin treatment also reduced extracellular matrix (ECM) deposition. The authors investigated the effect of adiponectin incubation in TGF-β1-treated primary human intestinal fibroblasts and they found that adiponectin reduced collagen level and the phosphorylation of Smad2 [25]. Diet-induced obesity may thus promote intestinal fibrosis via leptin. Other mechanisms may also be involved through microbial components. It has been recently demonstrated that bacterial ligands are able to promote angiogenesis through interaction with CEACAM1 in human intestinal microvascular endothelial cells (HIMEC) [26].

It has been demonstrated that creeping fat are associated to strictures [27]. More recently, Rieder's team deciphered the mechanisms underlying cross-talks between adipocyte environment from creeping fat and human intestinal muscle cells [1]. Fatty acids derived from creeping fat are able to induce human intestinal muscle cells hyperplasia. In addition, co-culture of HIMEC with whole creeping fat tissue induced their proliferation. They also demonstrated that adipokines from creeping fat of CD patients were able to induce an M2 macrophage subtype and TGF-β, a core cytokine in intestinal fibrosis [28]. Very recently, Devkota's team also demonstrated that there is specific translocation of a subset of viable bacteria such as *C. innocuum* from the gut microbiota to creeping fat [29]. It limits the systemic dissemination of gut bacteria but also leads to fibrosis development [29]. These mechanisms observed in creeping fat may be also observed from adipocytes from visceral adiposity in obese IBD patients and further studies are required to decipher mechanisms underlying the effects of obesity on complications in IBD patients.

3. High Salt

The Western diet is characterized by a high amount of sodium intake >100 mmol/day [30], which is in excess of physiological need (i.e., 10–20 mmol/day). In Western countries processed foods are the main provider of sodium intake (approximately 75% of intake) [30]. A few recent studies have shown the potential of dietary salt to promote intestinal inflammation in colitis models [31–33]. It thus raised the potential for dietary salt to induce a more vulnerable environment to inflammatory insults. We have recently demonstrated that a high-salt diet (4%) exacerbates intestinal fibrosis in a rat model of chronic TNBS-induced colitis and fibrosis [34]. We also demonstrated that high salt fed colitic rats had higher undernutrition compared to standard diet fed colitic rats [34]. We investigated the effect of high salt in TGF-β-induced human colon fibroblasts, and reported that NaCl promoted ECM-associated proteins in fibroblasts. Taken together, our study suggested that dietary salt can activate intestinal fibroblasts, thereby contributing to exacerbation of intestinal fibrosis. Further clinical studies are required to investigate whether dietary salt may be considered as a risk factor for intestinal fibrosis.

4. High Sugar

Few studies have investigated the effect of sweet diets in colitis models [35,36]. Laffin et al. fed mice with a high sugar diet (50% of sucrose) 2 days before chemically-induced colitis induction. These mice had a higher susceptibility to acute colitis, a higher intestinal permeability, a decreased microbial diversity and a reduced production of short chain fatty acids. In addition, macrophages from high sugar fed mice were more responsive to liposaccharides. Interestingly, the authors were able to reduce high sugar-mediated proinflammatory effects such as histological score or epithelial damage by supplementation with short chain fatty acid acetate in the drinking water [35].

Khan et al. used a different approach by studying the effects of simple sugars in mice. Pre-treatment with simple sugars such as glucose, fructose, or sucrose at 10% in drinking water for 7 days upregulated the histological score and worsened colitis development in chemically-induced colitis development. The authors of the study fed IL-10$^{-/-}$ mice with glucose and this was associated with higher colon inflammatory mediators such as lipocalin-2 or pro-inflammatory cytokines [36]. They also observed a gut dysbiosis in high sugar-diet fed mice, in particular higher abundance of the mucus-degrading bacteria *Akkermansia muciniphila* [36].

The effects of a high sugar diet are not yet documented in preclinical models of intestinal fibrosis. Of note is that glucose is able to induce EMT in many extra-intestinal organs [37,38] and this mechanism may be also relevant in IBD-associated intestinal fibrosis.

5. Beneficial Effects of Dietary Components

Contrary to potential deleterious effects mediated by westernized dietary patterns, certain components of diet can prevent intestinal fibrosis development. These nutrients can target several mechanisms involved in intestinal fibrosis. They can act to inhibit or suppress inflammatory processes. They can target specific receptors such as PPARγ or AhR with anti-fibrotic properties. The nutrients can also act at a cellular level by down-regulating EMT processes. Diet is also the main modulator of gut microbiota which may prevent or inhibit fibrogenesis.

5.1. Dietary Modulation of Receptors with Anti-Fibrosis Properties

Some nutrients are able to target specific receptors such as PPARγ, AhR, or VDR with anti-fibrotic properties. PPARγ is a nuclear receptor highly expressed in the colon and these anti-fibrotic properties have been investigated by natural and synthetic ligands in IBD models.

AhR belongs to the basic helix–loop–helix superfamily of transcription factors and nutrients such as curcumin or tryptophan metabolites can act as AhR ligands. The AhR is widely expressed in the gut and its activation has been associated with intestinal homeostasis. The role of Vitamin D receptor (VDR) to regulate intestinal inflammation is well documented in preclinical models of IBD. More recently, its involvement in intestinal fibrosis has been reported and invalidation of VDR promotes intestinal fibrosis development in mice in response to DSS-induced colitis.

5.1.1. Peroxisome Proliferator-Activated Receptor γ (PPARγ)

PPARγ is a nuclear receptor highly expressed in the colon and regulates intestinal inflammation [39]. Its anti-fibrotic properties have been investigated in IBD models. Speca et al. have used GED-0507-34, a novel PPARγ agonist, and have shown that preventive PPARγ agonist treatment reduced chronic colitis-induced intestinal fibrosis in mice and ECM-associated factors in TGF-β-induced intestinal fibroblasts and epithelial cells [40].

Many nutrients can target PPARγ [39,41] (Figure 2). The natural PPARγ agonist curcumin treatment from 2.5 to 10 μM reduced ECM-associated factors in TGFβ-induced intestinal fibroblasts as can the synthetic PPARγ agonist, rosiglitazone [42]. As myofibroblasts can be derived from various cell types in intestinal fibrosis, the authors validated their findings in epithelial cells and found that curcumin treatment also downregulated TGF-β-associated signaling in intestinal epithelial cells [42]. In addition, this effect was reversed by the use of GW9662, a synthetic PPARγ antagonist, showing the involvement of PPARγ in curcumin-induced anti-fibrotic effects [42]. The authors then confirmed their finding in vivo showing that curcumin at 200 mg/kg reduced chronic colitis-induced intestinal fibrosis and ECM-associated proteins such as fibronectin or CTGF [42]. This is also in accordance with a preclinical study using focal irradiation-induced fibrosis model [43]. In this murine model, 100 mg/kg of curcumin by gavage was able to reduce apoptosis in the injured area and intestinal and plasma IL-6 production [43]. As curcumin use is already validated in UC patients, the authors of this preclinical study hypothesized that curcumin may be relevant as a radioprotector. Novel therapeutic forms of curcumin have been developed, such as polycurcumin [44] or nanoparticle curcumin [45]. Both have been tested in preclinical IBD models and both reduced chemically-induced colitis.

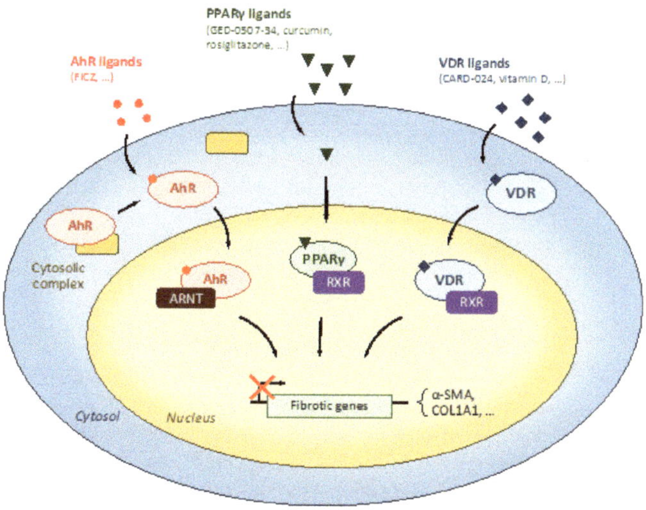

Figure 2. Dietary modulation of receptors involved in the regulation of intestinal fibrosis.

5.1.2. Aryl Hydrocarbon Receptor (AhR)

AhR is a member of the basic helix–loop–helix superfamily of transcription factors, which were first associated with cellular responses to xenobiotics [46,47]. More recently, nutrients such as curcumin or tryptophan metabolites can act as AhR ligands [46,47]. Upon ligand binding, a conformational change leads to AhR translocation into the nucleus and AhR with ARNT heterodimerization to induce target gene expression.

The AhR is widely expressed in the gut and its activation has been associated with intestinal homeostasis regulation [48]. Lamas et al. have shown that treatment with 6-formylindolo(3,2-b) carbazole (FICZ), an AhR agonist, reduced intestinal inflammation in Card9$^{-/-}$ mice [49]. IBD patients exhibited a reduced production of AhR ligands from the gut microbiota [49]. As dietary components can activate AhR to modulate inflammatory responses, Monteleone et al. have investigated whether FICZ exerts anti-fibrotic properties into the gut [50]. From other extra-intestinal fibrotic diseases, dietary ligands of AhR such as 2-(1′H-indole-3′-carbonyl)-thiazole-4-carboxylic acid methyl ester (ITE), L-kynurenin [51] or curcumin [42,52] are able to down-regulate ECM-associated proteins in fibroblasts.

Effects of AhR ligands on IBD-associated intestinal fibrosis are less documented. Treatment with FICZ from 100 to 400 nM decreased ECM-associated gene in stimulated fibroblasts from CD patients [50]. Similarly, 5 ng/mL of TGF-β up-regulated ECM-associated genes such as ACTA2 and COL1A1 in skin fibroblasts while FICZ at 100 nM decreased them [53].

In primary culture of human orbital fibroblasts, 1 ng/mL of TGF-β up-regulated ECM-associated proteins such as fibronectin, collagen I and α-SMA while ITE at 1 μM reduced them [54]. These data are consistent with a study in a liver fibrosis context where ITE treatment at 1 μM for 6 days inhibited ECM-associated proteins such as α-SMA in hepatic stellate cells [55].

5.1.3. Vitamin D Receptor

Epidemiological studies have suggested that low serum vitamin D is associated with an increased IBD risk [56,57]. Similarly, vitamin D and its receptor VDR mediated anti-inflammatory properties in experimental IBD models [58,59]. The role of vitamin D in intestinal fibrosis has been investigated. Johnson et al. have demonstrated that CARD-024, a vitamin D analogue was able to reduce ECM-associated markers in TGF-β-stimulated or stiffness-induced colonic fibroblasts [60]. In addition, down-regulation of colon VDR is observed in chronic CD patients and also in mice with chronic DSS-induced colitis and fibrosis [61]. VDR is also reduced in fibroblasts from CD patients [62]. Mitochondrial dysfunction has been described in patients with IBD and genetic deletion of prohibitin 1, a key protein of the inner mitochondrial membrane decreased in IBD, which can provoke ileitis in mice [63]. VDR is also involved in mitochondrial dysfunction [61] and its specific role in intestinal fibrosis has been recently demonstrated [61]. The authors of this elegant study first demonstrated that VDR expression was lower in intestinal stenotic areas in CD patients [61]. They then induced colitis-induced fibrosis by TNBS or DSS in mice with intestine-specific VDR deletion and they found that VDR deletion exacerbated intestinal fibrosis in both models. To decipher the mechanism underlying these anti-fibrosis effects, they used VDR invalidation in colonic fibroblasts, leading to their activation [61]. VDR invalidation also induced mitochondrial dysfunction mediated epithelial integrity.

5.2. Dietary Modulation of Anti-Fibrosis Signaling

Nuclear factor E2-related factor 2 (Nrf2) is a transcription factor involved in anti-oxidant response through the regulation of gene expression. Nrf2 signaling can regulate intestinal inflammation into the gut and has been recently proposed as a putative target in intestinal fibrosis. Nrf2 signaling can be activated by synthetic agonists and nutrients [64]. Sesamin derived from sesame seeds can counterbalance oxidative stress in intestinal epithelial cell line in response to H_20_2 and Nrf2 knockdown abolished the sesamin effect [65]. The authors of this study also evaluated the effects of sesamin at 100 mg/kg in a chemically-induced colitis model and they observed that sesamin was more effective compared to 5-ASA at 50 mg/kg [65]. Other dietary compounds such as numerous polyphenols have been identified to activate NrF2 signaling. Very interestingly, biotransformation of plants by various *lactobacillus* leads to compounds that are dietary ligands of Nrf2 and the Western diet is also characterized by a low consumption of fermented foods compared to our ancient traditional dietary patterns [66].

Some fatty acids derivatives can be partial agonists of cannabinoids receptors (CB1, CB2) agonists and are defined as endocannabinoids. This is the case of anandamide and 2-arachidonylglycerol. By its dual role in intestinal inflammation and metabolic disorders, the endocannabinoid system may be a relevant target in the context of intestinal fibrosis [67]. Indeed, cannabinoid analogues treatment by palmitoylethanolamide (PEA) for 5 weeks was able to counterbalance an ovariectomy-induced mild obesity model with reduced food intake, body weight, and fat mass [68]. Interestingly, PEA treatment also reduced inflammation in colonic biopsies from UC patients and in mice with DSS-induced colitis and these inflammatory effects were mediated through PPARα [69]. This effect has not yet been evaluated in IBD-associated fibrosis but its effect in extra-intestinal fibrosis has been demonstrated [70,71]. Targeting the endocannabinoid system may be particularly useful in the context of obese IBD patients.

5.3. Inhibition of Pro-Fibrotic Molecules by Amino Acids

Glutamine is a conditionally essential amino acid [72] and is the preferred fuel used by intestinal cells to promote enterocyte proliferation. Glutamine also regulates tight junctions and reduces proinflammatory signaling [73]. Its effect on preclinical models of intestinal fibrosis has been evaluated. Glutamine enemas at 25 mg/kg reduced colon fibrosis, the number of α-SMA stained cells in the submucosa and ECM-associated proteins in rats with TNBS-induced colitis [74]. These results are in accordance with a study evaluating glutamine treatment in a radiation-induced model where glutamine administration at 1 g/kg/day was able to prevent radiation-induced enteropathy in rats [75]. Similarly, glutamine enemas from 4 to 12 weeks after the surgery reduced colonoscopic and histological scores and reduced the number of collagen fibers in tissue in an experimental model of diversion colitis [76]. Nevertheless, a recent meta-analysis performed on seven published articles about glutamine use in IBD found that glutamine supplementation has no effect on disease course and inflammatory markers in patients with IBD [77] but its effect on fibrosis prevention and/or inhibition has never been evaluated in IBD patients.

Arginine is also a conditionally essential amino acid and we previously demonstrated that arginine treatment was able to down-regulate IL-8 production in cultured intestinal biopsies from CD patients [78]. Nitric oxide is a product of the enzymatic conversion of arginine to citrulline. The role of arginine in intestinal fibrosis is not yet documented but it may be protective through NO pathway. Invalidation of iNOS accelerated high-fat-induced liver fibrosis and inflammation development in mice [79] and this effect was mediated through NO- mediated NF-κB activation. Interestingly, we have shown in an intestinal epithelial cell line that arginine treatment down-regulated cytokines-induced inflammation through the NO pathway [80] and Horowitz et al. have demonstrated that L-arginine treatment up-regulated NO production in HIMEC [81]. Targeting iNOS/NO pathway may be relevant in IBD-associated intestinal fibrosis.

5.4. Anti-Fibrosis Properties of n-3 PUFA

We have previously shown anti-inflammatory effects on n-3 PUFA in experimental models of IBD [82] but their effects on intestinal fibrosis are not yet documented. In extra-intestinal organs, n-3 PUFA such as EPA reduced ECM-associated markers and SMAD signaling in TGF-β-induced hepatic stellate cells [83]. These n-3 PUFA effects were reduced by PPARγ knockdown while GW9662, a PPARγ antagonist, did not alter n-3 PUFA effects [83]. In LPS-stimulated dermal fibroblasts, the effects of EPA and DHA were evaluated on fibrosis markers. DHA reduced mRNA levels of α-SMA and collagen

III whereas EPA did not. Interestingly, the DHA effect was reinforced when combined with short chain fatty acid butyrate [84]. It is in accordance with a study from Zeng et al., showing that DHA inhibits TGF-β-induced rat renal fibroblast activation at a dose and time-dependent manner [85]. DHA derivative such as resolvin D1 was also evaluated in extra-intestinal fibrosis model [86]. Resolvin D1 treatment was able to reduce mechanical stretch-induced EMT and SMAD signaling in a murine model of lung fibrosis [86].

5.5. Dietary Manipulation of the Gut Microbiota

While IBD are strongly associated with shifts in the gut microbiome, the role of microbial factors in intestinal fibrosis is largely unexplored. In vitro, bacterial ligands are able to induce proliferation and migration of intestinal endothelial cells [26]. In chronic DSS, epithelial damage contribute to bacterial translocation and a recent study highlighted the role of flagellin to induce ECM components by intestinal fibroblasts [87]. In vivo, intestinal fibrosis can be abrogated in germ free mice and fibrosis severity is associated with specific microbes in mice overexpressing a member of the TNF superfamilly called TL1a [88]. These specific bacterial strains are able to promote in vitro fibrosis [88]. Imai J et al. have infected mice models with CD-associated pathobiont adherent-invasive Escherichia coli (AIEC). While healthy mice were able to gradually eradicate their infection from the intestine, mice from Salmonella- or DSS-induced colitis models, AIEC infection exploited inflammation to persist leading to intestinal fibrosis through IL-33 receptor signaling [89]. As probiotics strains such as Saccharomyces cerevisiae CNCM I-3856 [90], Lactobacillus, or Bifidobacterium [91] can counterbalance AIEC-promoting inflammation, it may open novel therapeutic avenues in the treatment of intestinal fibrosis.

Diet is a strong modulator of gut microbiota by affecting its composition or as a substrate for microbial production of metabolites. For example, curcumin treatment is also associated with gut microbiota changes. Treatment with nanoparticle curcumin reduced colitis development and increased the abundance of butyrate producing-bacteria and fecal butyrate production [45].

Short chain fatty acids such as butyrate has been evaluated on in vitro angiogenesis in primary cultures of HIMEC [92]. The authors of this study found that butyrate treatment reduced VEGF-induced cellular proliferation, transmigration, and tube formation of HIMEC through down-regulation of COX-2.

5.6. Reduction of Myofibroblast Activation

Recently, berberine, an alkaloid extracted from medicinal plants, was able to inhibit EMT [93]. This study used conditioned medium from human intestinal fibroblasts to induce morphological changes and ECM-associated markers in a colonic epithelial cell line and these effects were reversed by berberine treatment at 100 μg/mL for 24 h [93]. The authors demonstrated that berberine reduced EMT by acting on the TGF-β/Smads signaling [93]. Myofibroblasts can be derived from various origins in intestinal fibrosis. For example, endothelial to mesenchymal transition (EndoMT) has been demonstrated in intestinal fibrosis [94]. Nutrients can modify this EndoMT and numerous nutritional approaches have been evaluated. We have previously evaluated DHA, a long chain n-3 PUFA in the primary culture of HIMEC. We have demonstrated that DHA pre-treatment can reduce IL-1β-activated HIMEC pro-inflammatory effects [82] such as decreased adhesion molecule VCAM-1, TLR-4 or cytokine production of IL-6, IL-8. Similarly, curcumin treatment at 10 μM reduced TNF and LPS-induced or irradiation-induced VCAM-1 through NFκB activation in HIMEC [95,96].

5.7. Mucosal Healing

The potential effect of probiotics on wound healing have been evaluated. Conditioned medium with the strain *Bacillus polyfermenticus* had pro-angiogenic properties in HIMEC by increasing cell migration, permeability, and tube formation and this effect is mediated though IL-8 production and NF-κB activation [97]. Results were confirmed in vivo in an acute model of DSS colitis [97].

Very few studies have also investigated more complex nutrients compared to unique nutrients in preclinical models of intestinal fibrosis. We have evaluated the effects of a polymeric diet enriched in TGF-β2 in a model of pre-pubertal rats with chronic TNBS-induced colitis and we failed to reverse the inflammation or intestinal fibrosis in our tested conditions [98]. Very recently, a study investigated the effects of fermented rice bran on post-colitis restoration demonstrating a reduction of ECM-associated markers and TGF-β/Smad signaling [99].

6. Conclusions

Diet may represent an underestimated risk factor for intestinal fibrosis. A better understanding of the crosstalk between nutrients and factors that promote intestinal fibrosis may enable to provide a better rationale for dietary advice to limit complications in IBD patients. As stipulated by the last ESPEN guidelines for clinical nutrition in IBD, all IBD patients should benefit from dietary counseling by a dietician, which will contribute to limit nutrition-related disorders [100]. In particular, restrictive diets are very popular in IBD patients and are being evaluated in clinical trials and these diets may contribute to a poor psychological well-being [101,102] and lead to undernutrition [100] unless closely supervised. Interestingly, two very recent studies highlighted the potential of Mediterranean diet in IBD patients. While the beneficial effect of this diet has been demonstrated on fibrosis in NAFLD patients with reduced cardiovascular or diabetes risk [103], a 6-month Mediterranean diet was able in IBD patients to reduce malnutrition-associated disorders, improved disease activity and inflammatory markers with a concomitant increased of a quality of life score [104]. The need for further nutritional intervention trials with diets that target factors-promoting intestinal fibrosis and/or address the Westernization of food in IBD-associated intestinal fibrosis are urgently required.

Author Contributions: All authors contributed equally. All authors have read and agreed to the published version of the manuscript.

Funding: Mathilde Leboutte was supported by a RIN grant from the Region Normandie and Asma Amamou by an international mobility grant from the Société Française de Nutrition Clinique et Métabolisme (SFNCM). The authors thank the "François Aupetit" Association (AFA Crohn-RCH) for its support.

Institutional Review Board Statement: Not applicable.

Informed Consent Statement: Not applicable.

Data Availability Statement: Not applicable.

Conflicts of Interest: The authors declare no conflict of interest.

Abbreviations

CD, Crohn's disease; ECM, extracellular matrix, EMT, epithelial-mesenchymal transition; IBD, inflammatory bowel disease; Nrf2, Nuclear factor E2-related factor 2, PPARγ, Peroxisome Proliferator-activated Receptor γ, UC, ulcerative colitis.

References

1. Mao, R.; Kurada, S.; Gordon, I.O.; Baker, M.E.; Gandhi, N.; McDonald, C.; Coffey, J.C.; Rieder, F. The Mesenteric Fat and Intestinal Muscle Interface: Creeping Fat Influencing Stricture Formation in Crohn's Disease. *Inflamm. Bowel Dis.* **2019**, *25*, 421–426. [CrossRef]
2. Kugathasan, S.; Denson, L.A.; Walters, T.D.; Kim, M.-O.; Marigorta, U.M.; Schirmer, M.; Mondal, K.; Liu, C.; Griffiths, A.; Noe, J.D.; et al. Prediction of Complicated Disease Course for Children Newly Diagnosed with Crohn's Disease: A Multicentre Inception Cohort Study. *Lancet* **2017**, *389*, 1710–1718. [CrossRef]

3. Moran, G.W.; Dubeau, M.; Kaplan, G.G.; Yang, H.; Seow, C.H.; Fedorak, R.N.; Dieleman, L.A.; Barkema, H.W.; Ghosh, S.; Panaccione, R. Phenotypic Features of Crohn's Disease Associated With Failure of Medical Treatment. *Clin. Gastroenterol. Hepatol.* **2014**, *12*, 434–442.e1. [CrossRef]
4. Rieder, F.; Fiocchi, C.; Rogler, G. Mechanisms, Management, and Treatment of Fibrosis in Patients with Inflammatory Bowel Diseases. *Gastroenterology* **2017**, *152*, 340–350.e6. [CrossRef] [PubMed]
5. Marion-Letellier, R.; Savoye, G.; Ghosh, S. IBD: In Food We Trust. *J. Crohns Colitis* **2016**, *10*, 1351–1361. [CrossRef] [PubMed]
6. Marion-Letellier, R.; Savoye, G.; Beck, P.L.; Panaccione, R.; Ghosh, S. Polyunsaturated Fatty Acids in Inflammatory Bowel Diseases: A Reappraisal of Effects and Therapeutic Approaches. *Inflamm. Bowel Dis.* **2013**, *19*, 650–661. [CrossRef]
7. Ng, S.C.; Shi, H.Y.; Hamidi, N.; Underwood, F.E.; Tang, W.; Benchimol, E.I.; Panaccione, R.; Ghosh, S.; Wu, J.C.Y.; Chan, F.K.L.; et al. Worldwide Incidence and Prevalence of Inflammatory Bowel Disease in the 21st Century: A Systematic Review of Population-Based Studies. *Lancet* **2017**, *390*, 2769–2778. [CrossRef]
8. Guanghong, J.; Habibi, J.; DeMarco, V.G.; Martinez-Lemus, L.A.; Ma, L.; Whaley-Connell, A.T.; Aroor, A.R.; Domeier, T.L.; Zhu, Y.; Meininger, G.A.; et al. Endothelial Mineralocorticoid Receptor Deletion Prevents Diet-Induced Cardiac Diastolic Dysfunction in Females. *Hypertension* **2015**, *66*, 1159–1167. [CrossRef]
9. Aroor, A.R.; Habibi, J.; Nistala, R.; Ramirez-Perez, F.I.; Martinez-Lemus, L.A.; Jaffe, I.Z.; Sowers, J.R.; Guanghong, J.; Whaley-Connell, A. Diet-Induced Obesity Promotes Kidney Endothelial Stiffening and Fibrosis Dependent on the Endothelial Mineralocorticoid Receptor. *Hypertension* **2019**, *73*, 849–858. [CrossRef]
10. Taylor, L.; Almutairdi, A.; Shommu, N.; Fedorak, R.; Ghosh, S.; Reimer, R.A.; Panaccione, R.; Raman, M. Cross-Sectional Analysis of Overall Dietary Intake and Mediterranean Dietary Pattern in Patients with Crohn's Disease. *Nutrients* **2018**, *10*, 1761. [CrossRef]
11. Narula, N.; Wong, E.C.L.; Dehghan, M.; Mente, A.; Rangarajan, S.; Lanas, F.; Lopez-Jaramillo, P.; Rohatgi, P.; Lakshmi, P.V.M.; Varma, R.P.; et al. Association of Ultra-Processed Food Intake with Risk of Inflammatory Bowel Disease: Prospective Cohort Study. *BMJ* **2021**, *374*, n1554. [CrossRef] [PubMed]
12. Moran, G.W.; Dubeau, M.-F.; Kaplan, G.G.; Panaccione, R.; Ghosh, S. The Increasing Weight of Crohn's Disease Subjects in Clinical Trials: A Hypothesis-Generatings Time-Trend Analysis. *Inflamm. Bowel Dis.* **2013**, *19*, 2949–2956. [CrossRef]
13. Grillot, J.; D'Engremont, C.; Parmentier, A.-L.; Lakkis, Z.; Piton, G.; Cazaux, D.; Gay, C.; De Billy, M.; Koch, S.; Borot, S.; et al. Sarcopenia and Visceral Obesity Assessed by Computed Tomography Are Associated with Adverse Outcomes in Patients with Crohn's Disease. *Clin. Nutr.* **2020**, *39*, 3024–3030. [CrossRef]
14. Braga Neto, M.B.; Gregory, M.H.; Ramos, G.P.; Bazerbachi, F.; Bruining, D.H.; Abu Dayyeh, B.K.; Kushnir, V.M.; Raffals, L.E.; Ciorba, M.A.; Loftus, E.V.; et al. Impact of Bariatric Surgery on the Long-Term Disease Course of Inflammatory Bowel Disease. *Inflamm. Bowel Dis.* **2020**, *26*, 1089–1097. [CrossRef] [PubMed]
15. Tomas, J.; Mulet, C.; Saffarian, A.; Cavin, J.-B.; Ducroc, R.; Regnault, B.; Tan, C.K.; Duszka, K.; Burcelin, R.; Wahli, W.; et al. High-Fat Diet Modifies the PPAR-γ Pathway Leading to Disruption of Microbial and Physiological Ecosystem in Murine Small Intestine. *Proc. Natl. Acad. Sci. USA* **2016**, *113*, E5934–E5943. [CrossRef] [PubMed]
16. Lee, J.-C.; Lee, H.-Y.; Kim, T.K.; Kim, M.-S.; Park, Y.M.; Kim, J.; Park, K.; Kweon, M.-N.; Kim, S.-H.; Bae, J.-W.; et al. Obesogenic Diet-Induced Gut Barrier Dysfunction and Pathobiont Expansion Aggravate Experimental Colitis. *PLoS ONE* **2017**, *12*, e0187515. [CrossRef] [PubMed]
17. Mazur-Bialy, A.I.; Bilski, J.; Wojcik, D.; Brzozowski, B.; Surmiak, M.; Hubalewska-Mazgaj, M.; Chmura, A.; Magierowski, M.; Magierowska, K.; Mach, T.; et al. Beneficial Effect of Voluntary Exercise on Experimental Colitis in Mice Fed a High-Fat Diet: The Role of Irisin, Adiponectin and Proinflammatory Biomarkers. *Nutrients* **2017**, *9*, 410. [CrossRef]
18. Ma, X.; Torbenson, M.; Hamad, A.R.A.; Soloski, M.J.; Li, Z. High-Fat Diet Modulates Non-CD1d-Restricted Natural Killer T Cells and Regulatory T Cells in Mouse Colon and Exacerbates Experimental Colitis. *Clin. Exp. Immunol.* **2008**, *151*, 130–138. [CrossRef] [PubMed]
19. Tang, F.-Y.; Pai, M.-H.; Chiang, E.-P.I. Consumption of High-Fat Diet Induces Tumor Progression and Epithelial–Mesenchymal Transition of Colorectal Cancer in a Mouse Xenograft Model. *J. Nutr. Biochem.* **2012**, *23*, 1302–1313. [CrossRef]
20. Kwapisz, O.; Górka, J.; Korlatowicz, A.; Kotlinowski, J.; Waligórska, A.; Marona, P.; Pydyn, N.; Dobrucki, J.W.; Jura, J.; Miekus, K. Fatty Acids and a High-Fat Diet Induce Epithelial-Mesenchymal Transition by Activating TGFβ and β-Catenin in Liver Cells. *Int. J. Mol. Sci.* **2021**, *22*, 1272. [CrossRef]
21. Ha, S.; Kim, M.J.; Kim, D.H.; Kim, B.M.; Chung, K.W.; Chung, H.Y. Short-Term Intake of High Fat Diet Aggravates Renal Fibrosis in Aged Sprague-Dawley Rats. *Exp. Gerontol.* **2020**, *142*, 111108. [CrossRef]
22. Vieujean, S.; Hu, S.; Bequet, E.; Salee, C.; Massot, C.; Bletard, N.; Pierre, N.; Quesada Calvo, F.; Baiwir, D.; Mazzucchelli, G.; et al. Potential Role of Epithelial Endoplasmic Reticulum Stress and Anterior Gradient Protein 2 Homolog in Crohn's Disease Fibrosis. *J. Crohns Colitis* **2021**, jjab061. [CrossRef] [PubMed]
23. Chu, S.G.; Villalba, J.A.; Liang, X.; Xiong, K.; Tsoyi, K.; Ith, B.; Ayaub, E.A.; Tatituri, R.V.; Byers, D.E.; Hsu, F.-F.; et al. Palmitic Acid–Rich High-Fat Diet Exacerbates Experimental Pulmonary Fibrosis by Modulating Endoplasmic Reticulum Stress. *Am. J. Respir. Cell Mol. Biol.* **2019**, *61*, 737–746. [CrossRef]
24. Latella, G.; Rieder, F. Intestinal Fibrosis: Ready to Be Reversed. *Curr. Opin. Gastroenterol.* **2017**, *33*, 239–245. [CrossRef]
25. Xie, M.; Xiong, Z.; Yin, S.; Xiong, J.; Li, X.; Jin, L.; Zhang, F.; Chen, H.; Lan, P.; Lian, L. Adiponectin Alleviates Intestinal Fibrosis by Enhancing AMP-Activated Protein Kinase Phosphorylation. *Dig. Dis. Sci.* **2021**. [CrossRef] [PubMed]

26. Schirbel, A.; Rebert, N.; Sadler, T.; West, G.; Rieder, F.; Wagener, C.; Horst, A.; Sturm, A.; de la Motte, C.; Fiocchi, C. Mutual Regulation of TLR/NLR and CEACAM1 in the Intestinal Microvasculature: Implications for IBD Pathogenesis and Therapy. *Inflamm. Bowel Dis.* **2019**, *25*, 294–305. [CrossRef]
27. Borley, N.R.; Mortensen, N.J.; Jewell, D.P.; Warren, B.F. The Relationship between Inflammatory and Serosal Connective Tissue Changes in Ileal Crohn's Disease: Evidence for a Possible Causative Link. *J. Pathol.* **2000**, *190*, 196–202. [CrossRef]
28. Kredel, L.I.; Batra, A.; Stroh, T.; Kühl, A.A.; Zeitz, M.; Erben, U.; Siegmund, B. Adipokines from Local Fat Cells Shape the Macrophage Compartment of the Creeping Fat in Crohn's Disease. *Gut* **2013**, *62*, 852–862. [CrossRef]
29. Ha, C.W.Y.; Martin, A.; Sepich-Poore, G.D.; Shi, B.; Wang, Y.; Gouin, K.; Humphrey, G.; Sanders, K.; Ratnayake, Y.; Chan, K.S.L.; et al. Translocation of Viable Gut Microbiota to Mesenteric Adipose Drives Formation of Creeping Fat in Humans. *Cell* **2020**, *183*, 666–683.e17. [CrossRef] [PubMed]
30. Brown, I.J.; Tzoulaki, I.; Candeias, V.; Elliott, P. Salt Intakes around the World: Implications for Public Health. *Int. J. Epidemiol.* **2009**, *38*, 791–813. [CrossRef]
31. Monteleone, I.; Marafini, I.; Dinallo, V.; Di Fusco, D.; Troncone, E.; Zorzi, F.; Laudisi, F.; Monteleone, G. Sodium Chloride-Enriched Diet Enhanced Inflammatory Cytokine Production and Exacerbated Experimental Colitis in Mice. *J. Crohns Colitis* **2017**, *11*, 237–245. [CrossRef]
32. Aguiar, S.L.F.; Miranda, M.C.G.; Guimarães, M.A.F.; Santiago, H.C.; Queiroz, C.P.; Cunha, P.D.S.; Cara, D.C.; Foureaux, G.; Ferreira, A.J.; Cardoso, V.N.; et al. High-Salt Diet Induces IL-17-Dependent Gut Inflammation and Exacerbates Colitis in Mice. *Front. Immunol.* **2018**, *8*. [CrossRef] [PubMed]
33. Tubbs, A.L.; Liu, B.; Rogers, T.D.; Sartor, R.B.; Miao, E.A. Dietary Salt Exacerbates Experimental Colitis. *J. Immunol. Baltim. Md 1950* **2017**, *199*, 1051–1059. [CrossRef]
34. Amamou, A.; Rouland, M.; Yaker, L.; Goichon, A.; Guérin, C.; Aziz, M.; Savoye, G.; Marion-Letellier, R. Dietary Salt Exacerbates Intestinal Fibrosis in Chronic TNBS Colitis via Fibroblasts Activation. *Sci. Rep.* **2021**, *11*, 15055. [CrossRef]
35. Laffin, M.; Fedorak, R.; Zalasky, A.; Park, H.; Gill, A.; Agrawal, A.; Keshteli, A.; Hotte, N.; Madsen, K.L. A High-Sugar Diet Rapidly Enhances Susceptibility to Colitis via Depletion of Luminal Short-Chain Fatty Acids in Mice. *Sci. Rep.* **2019**, *9*, 12294. [CrossRef]
36. Khan, S.; Waliullah, S.; Godfrey, V.; Khan, M.A.W.; Ramachandran, R.A.; Cantarel, B.L.; Behrendt, C.; Peng, L.; Hooper, L.V.; Zaki, H. Dietary Simple Sugars Alter Microbial Ecology in the Gut and Promote Colitis in Mice. *Sci. Transl. Med.* **2020**, *12*. [CrossRef]
37. Gao, L.; Fan, Y.; Zhang, X.; Yang, L.; Huang, W.; Hang, T.; Li, M.; Du, S.; Ma, J. Zinc Supplementation Inhibits the High Glucose-induced EMT of Peritoneal Mesothelial Cells by Activating the Nrf2 Antioxidant Pathway. *Mol. Med. Rep.* **2019**, *20*, 655–663. [CrossRef] [PubMed]
38. Das, N.A.; Carpenter, A.J.; Belenchia, A.; Aroor, A.R.; Noda, M.; Siebenlist, U.; Chandrasekar, B.; DeMarco, V.G. Empagliflozin Reduces High Glucose-Induced Oxidative Stress and MiR-21-Dependent TRAF3IP2 Induction and RECK Suppression, and Inhibits Human Renal Proximal Tubular Epithelial Cell Migration and Epithelial-to-Mesenchymal Transition. *Cell. Signal.* **2020**, *68*, 109506. [CrossRef]
39. Marion-Letellier, R.; Déchelotte, P.; Iacucci, M.; Ghosh, S. Dietary Modulation of Peroxisome Proliferator-Activated Receptor Gamma. *Gut* **2009**, *58*, 586–593. [CrossRef] [PubMed]
40. Speca, S.; Rousseaux, C.; Dubuquoy, C.; Rieder, F.; Vetuschi, A.; Sferra, R.; Giusti, I.; Bertin, B.; Dubuquoy, L.; Gaudio, E.; et al. Novel PPARγ Modulator GED-0507-34 Levo Ameliorates Inflammation-Driven Intestinal Fibrosis. *Inflamm. Bowel Dis.* **2016**, *22*, 279–292. [CrossRef] [PubMed]
41. Marion-Letellier, R.; Butler, M.; Déchelotte, P.; Playford, R.J.; Ghosh, S. Comparison of Cytokine Modulation by Natural Peroxisome Proliferator–Activated Receptor γ Ligands with Synthetic Ligands in Intestinal-like Caco-2 Cells and Human Dendritic Cells—Potential for Dietary Modulation of Peroxisome Proliferator–Activated Receptor γ in Intestinal Inflammation. *Am. J. Clin. Nutr.* **2008**, *87*, 939–948. [CrossRef]
42. Xu, S.; Jiang, B.; Wang, H.; Shen, C.; Chen, H.; Zeng, L. Curcumin Suppresses Intestinal Fibrosis by Inhibition of PPARγ-Mediated Epithelial-Mesenchymal Transition. *Evid.-Based Complement. Altern. Med. ECAM* **2017**, *2017*, 7876064. [CrossRef] [PubMed]
43. Verginadis, I.I.; Kanade, R.; Bell, B.; Koduri, S.; Ben-Josef, E.; Koumenis, C. A Novel Mouse Model to Study Image-Guided, Radiation-Induced Intestinal Injury and Preclinical Screening of Radioprotectors. *Cancer Res.* **2017**, *77*, 908–917. [CrossRef]
44. Qiao, H.; Fang, D.; Chen, J.; Sun, Y.; Kang, C.; Di, L.; Li, J.; Chen, Z.; Chen, J.; Gao, Y. Orally Delivered Polycurcumin Responsive to Bacterial Reduction for Targeted Therapy of Inflammatory Bowel Disease. *Drug Deliv.* **2017**, *24*, 233–242. [CrossRef] [PubMed]
45. Ohno, M.; Nishida, A.; Sugitani, Y.; Nishino, K.; Inatomi, O.; Sugimoto, M.; Kawahara, M.; Andoh, A. Nanoparticle Curcumin Ameliorates Experimental Colitis via Modulation of Gut Microbiota and Induction of Regulatory T Cells. *PLoS ONE* **2017**, *12*, e0185999. [CrossRef]
46. Larigot, L.; Juricek, L.; Dairou, J.; Coumoul, X. AhR Signaling Pathways and Regulatory Functions. *Biochim. Open* **2018**, *7*, 1–9. [CrossRef]
47. Rothhammer, V.; Quintana, F.J. The Aryl Hydrocarbon Receptor: An Environmental Sensor Integrating Immune Responses in Health and Disease. *Nat. Rev. Immunol.* **2019**, *19*, 184–197. [CrossRef]
48. Lamas, B.; Natividad, J.M.; Sokol, H. Aryl Hydrocarbon Receptor and Intestinal Immunity. *Mucosal Immunol.* **2018**, *11*, 1024–1038. [CrossRef] [PubMed]

49. Lamas, B.; Richard, M.L.; Leducq, V.; Pham, H.-P.; Michel, M.-L.; Da Costa, G.; Bridonneau, C.; Jegou, S.; Hoffmann, T.W.; Natividad, J.M.; et al. CARD9 Impacts Colitis by Altering Gut Microbiota Metabolism of Tryptophan into Aryl Hydrocarbon Receptor Ligands. *Nat. Med.* **2016**, *22*, 598–605. [CrossRef]
50. Monteleone, I.; Zorzi, F.; Marafini, I.; Di Fusco, D.; Dinallo, V.; Caruso, R.; Izzo, R.; Franzè, E.; Colantoni, A.; Pallone, F.; et al. Aryl Hydrocarbon Receptor-Driven Signals Inhibit Collagen Synthesis in the Gut. *Eur. J. Immunol.* **2016**, *46*, 1047–1057. [CrossRef]
51. Dolivo, D.M.; Larson, S.A.; Dominko, T. Tryptophan Metabolites Kynurenine and Serotonin Regulate Fibroblast Activation and Fibrosis. *Cell. Mol. Life Sci. CMLS* **2018**, *75*, 3663–3681. [CrossRef]
52. Nakai, R.; Fukuda, S.; Kawase, M.; Yamashita, Y.; Ashida, H. Curcumin and Its Derivatives Inhibit 2,3,7,8,-Tetrachloro-Dibenzo-p-Dioxin-Induced Expression of Drug Metabolizing Enzymes through Aryl Hydrocarbon Receptor-Mediated Pathway. *Biosci. Biotechnol. Biochem.* **2018**, *82*, 616–628. [CrossRef]
53. Murai, M.; Tsuji, G.; Hashimoto-Hachiya, A.; Kawakami, Y.; Furue, M.; Mitoma, C. An Endogenous Tryptophan Photo-Product, FICZ, Is Potentially Involved in Photo-Aging by Reducing TGF-β-Regulated Collagen Homeostasis. *J. Dermatol. Sci.* **2018**, *89*, 19–26. [CrossRef] [PubMed]
54. Lehmann, G.M.; Xi, X.; Kulkarni, A.A.; Olsen, K.C.; Pollock, S.J.; Baglole, C.J.; Gupta, S.; Casey, A.E.; Huxlin, K.R.; Sime, P.J.; et al. The Aryl Hydrocarbon Receptor Ligand ITE Inhibits TGFβ1-Induced Human Myofibroblast Differentiation. *Am. J. Pathol.* **2011**, *178*, 1556–1567. [CrossRef]
55. Yan, J.; Tung, H.-C.; Li, S.; Niu, Y.; Garbacz, W.G.; Lu, P.; Bi, Y.; Li, Y.; He, J.; Xu, M.; et al. Aryl Hydrocarbon Receptor Signaling Prevents Activation of Hepatic Stellate Cells and Liver Fibrogenesis in Mice. *Gastroenterology* **2019**, *157*, 793–806.e14. [CrossRef] [PubMed]
56. Suibhne, T.N.; Cox, G.; Healy, M.; O'Morain, C.; O'Sullivan, M. Vitamin D Deficiency in Crohn's Disease: Prevalence, Risk Factors and Supplement Use in an Outpatient Setting. *J. Crohns Colitis* **2012**, *6*, 182–188. [CrossRef] [PubMed]
57. Raftery, T.; Merrick, M.; Healy, M.; Mahmud, N.; O'Morain, C.; Smith, S.; McNamara, D.; O'Sullivan, M. Vitamin D Status Is Associated with Intestinal Inflammation as Measured by Fecal Calprotectin in Crohn's Disease in Clinical Remission. *Dig. Dis. Sci.* **2015**, *60*, 2427–2435. [CrossRef]
58. Kong, J.; Zhang, Z.; Musch, M.W.; Ning, G.; Sun, J.; Hart, J.; Bissonnette, M.; Li, Y.C. Novel Role of the Vitamin D Receptor in Maintaining the Integrity of the Intestinal Mucosal Barrier. *Am. J. Physiol. Gastrointest. Liver Physiol.* **2008**, *294*, G208–G216. [CrossRef]
59. Wu, S.; Zhang, Y.-G.; Lu, R.; Xia, Y.; Zhou, D.; Petrof, E.O.; Claud, E.C.; Chen, D.; Chang, E.B.; Carmeliet, G.; et al. Intestinal Epithelial Vitamin D Receptor Deletion Leads to Defective Autophagy in Colitis. *Gut* **2015**, *64*, 1082–1094. [CrossRef]
60. Johnson, L.A.; Sauder, K.L.; Rodansky, E.S.; Simpson, R.U.; Higgins, P.D.R. CARD-024, a Vitamin D Analog, Attenuates the pro-Fibrotic Response to Substrate Stiffness in Colonic Myofibroblasts. *Exp. Mol. Pathol.* **2012**, *93*, 91–98. [CrossRef] [PubMed]
61. Yu, M.; Wu, H.; Wang, J.; Chen, X.; Pan, J.; Liu, P.; Zhang, J.; Chen, Y.; Zhu, W.; Tang, C.; et al. Vitamin D Receptor Inhibits EMT via Regulation of the Epithelial Mitochondrial Function in Intestinal Fibrosis. *J. Biol. Chem.* **2021**, *296*, 100531. [CrossRef] [PubMed]
62. Gisbert-Ferrándiz, L.; Cosín-Roger, J.; Hernández, C.; Macias-Ceja, D.C.; Ortiz-Masiá, D.; Salvador, P.; Esplugues, J.V.; Hinojosa, J.; Navarro, F.; Calatayud, S.; et al. Diminished Vitamin D Receptor Protein Levels in Crohn's Disease Fibroblasts: Effects of Vitamin D. *Nutrients* **2020**, *12*, 973. [CrossRef] [PubMed]
63. Jackson, D.N.; Panopoulos, M.; Neumann, W.L.; Turner, K.; Cantarel, B.L.; Thompson-Snipes, L.; Dassopoulos, T.; Feagins, L.A.; Souza, R.F.; Mills, J.C.; et al. Mitochondrial Dysfunction during Loss of Prohibitin 1 Triggers Paneth Cell Defects and Ileitis. *Gut* **2020**, *69*, 1928–1938. [CrossRef] [PubMed]
64. Pompili, S.; Sferra, R.; Gaudio, E.; Viscido, A.; Frieri, G.; Vetuschi, A.; Latella, G. Can Nrf2 Modulate the Development of Intestinal Fibrosis and Cancer in Inflammatory Bowel Disease? *Int. J. Mol. Sci.* **2019**, *20*, 4061. [CrossRef]
65. Bai, X.; Gou, X.; Cai, P.; Xu, C.; Cao, L.; Zhao, Z.; Huang, M.; Jin, J. Sesamin Enhances Nrf2-Mediated Protective Defense against Oxidative Stress and Inflammation in Colitis via AKT and ERK Activation. *Oxid. Med. Cell. Longev.* **2019**, *2019*, e2432416. [CrossRef]
66. Senger, D.R.; Li, D.; Jaminet, S.-C.; Cao, S. Activation of the Nrf2 Cell Defense Pathway by Ancient Foods: Disease Prevention by Important Molecules and Microbes Lost from the Modern Western Diet. *PLoS ONE* **2016**, *11*, e0148042. [CrossRef]
67. Cani, P.D.; Plovier, H.; Van Hul, M.; Geurts, L.; Delzenne, N.M.; Druart, C.; Everard, A. Endocannabinoids–at the Crossroads between the Gut Microbiota and Host Metabolism. *Nat. Rev. Endocrinol.* **2016**, *12*, 133–143. [CrossRef]
68. Mattace Raso, G.; Santoro, A.; Russo, R.; Simeoli, R.; Paciello, O.; Di Carlo, C.; Diano, S.; Calignano, A.; Meli, R. Palmitoylethanolamide Prevents Metabolic Alterations and Restores Leptin Sensitivity in Ovariectomized Rats. *Endocrinology* **2014**, *155*, 1291–1301. [CrossRef]
69. Esposito, G.; Capoccia, E.; Turco, F.; Palumbo, I.; Lu, J.; Steardo, A.; Cuomo, R.; Sarnelli, G.; Steardo, L. Palmitoylethanolamide Improves Colon Inflammation through an Enteric Glia/Toll like Receptor 4-Dependent PPAR-α Activation. *Gut* **2014**, *63*, 1300–1312. [CrossRef]
70. Ohara, M.; Ohnishi, S.; Hosono, H.; Yamamoto, K.; Fu, Q.; Maehara, O.; Suda, G.; Sakamoto, N. Palmitoylethanolamide Ameliorates Carbon Tetrachloride-Induced Liver Fibrosis in Rats. *Front. Pharmacol.* **2018**, *9*, 709. [CrossRef]
71. Ye, S.; Chen, Q.; Jiang, N.; Liang, X.; Li, J.; Zong, R.; Huang, C.; Qiu, Y.; Ma, J.-X.; Liu, Z. PPARα-Dependent Effects of Palmitoylethanolamide Against Retinal Neovascularization and Fibrosis. *Invest. Ophthalmol. Vis. Sci.* **2020**, *61*, 15. [CrossRef]

72. Coëffier, M.; Marion-Letellier, R.; Déchelotte, P. Potential for Amino Acids Supplementation during Inflammatory Bowel Diseases. *Inflamm. Bowel Dis.* **2010**, *16*, 518–524. [CrossRef]
73. Bertrand, J.; Marion-Letellier, R.; Azhar, S.; Chan, P.; Legrand, R.; Goichon, A.; Ghouzali, I.; Aziz, M.; Vaudry, D.; Savoye, G.; et al. Glutamine Enema Regulates Colonic Ubiquitinated Proteins but Not Proteasome Activities during TNBS-Induced Colitis Leading to Increased Mitochondrial Activity. *Proteomics* **2015**, *15*, 2198–2210. [CrossRef] [PubMed]
74. San-Miguel, B.; Crespo, I.; Kretzmann, N.A.; Mauriz, J.L.; Marroni, N.; Tuñón, M.J.; González-Gallego, J. Glutamine Prevents Fibrosis Development in Rats with Colitis Induced by 2,4,6-Trinitrobenzene Sulfonic Acid. *J. Nutr.* **2010**, *140*, 1065–1071. [CrossRef]
75. Jensen, J.C.; Schaefer, R.; Nwokedi, E.; Bevans, D.W.; Baker, M.L.; Pappas, A.A.; Westbrook, K.C.; Klimberg, V.S. Prevention of Chronic Radiation Enteropathy by Dietary Glutamine. *Ann. Surg. Oncol.* **1994**, *1*, 157–163. [CrossRef]
76. Rodrigo Goulart Pacheco, C.C.E.; Ller, M.T.C.-B. Use of Butyrate or Glutamine in Enema Solution Reduces Inflammation and Fibrosis in Experimental Diversion Colitis. *World J. Gastroenterol.* **2012**, *18*, 4278–4287. [CrossRef]
77. Severo, J.S.; da Silva Barros, V.J.; Alves da Silva, A.C.; Luz Parente, J.M.; Lima, M.M.; Moreira Lima, A.Â.; dos Santos, A.A.; Matos Neto, E.M.; Tolentino, M. Effects of Glutamine Supplementation on Inflammatory Bowel Disease: A Systematic Review of Clinical Trials. *Clin. Nutr. ESPEN* **2021**, *42*, 53–60. [CrossRef]
78. Lecleire, S.; Hassan, A.; Marion-Letellier, R.; Antonietti, M.; Savoye, G.; Bôle-Feysot, C.; Lerebours, E.; Ducrotté, P.; Déchelotte, P.; Coëffier, M. Combined Glutamine and Arginine Decrease Proinflammatory Cytokine Production by Biopsies from Crohn's Patients in Association with Changes in Nuclear Factor-KappaB and P38 Mitogen-Activated Protein Kinase Pathways. *J. Nutr.* **2008**, *138*, 2481–2486. [CrossRef]
79. Nozaki, Y.; Fujita, K.; Wada, K.; Yoneda, M.; Kessoku, T.; Shinohara, Y.; Imajo, K.; Ogawa, Y.; Nakamuta, M.; Saito, S.; et al. Deficiency of INOS-Derived NO Accelerates Lipid Accumulation-Independent Liver Fibrosis in Non-Alcoholic Steatohepatitis Mouse Model. *BMC Gastroenterol.* **2015**, *15*, 42. [CrossRef] [PubMed]
80. Marion, R.; Coëffier, M.; Lemoulan, S.; Gargala, G.; Ducrotté, P.; Déchelotte, P. L-Arginine Modulates CXC Chemokines in the Human Intestinal Epithelial Cell Line HCT-8 by the NO Pathway. *Biochimie* **2005**, *87*, 1048–1055. [CrossRef] [PubMed]
81. Horowitz, S.; Binion, D.G.; Nelson, V.M.; Kanaa, Y.; Javadi, P.; Lazarova, Z.; Andrekopoulos, C.; Kalyanaraman, B.; Otterson, M.F.; Rafiee, P. Increased Arginase Activity and Endothelial Dysfunction in Human Inflammatory Bowel Disease. *Am. J. Physiol. Gastrointest. Liver Physiol.* **2007**, *292*, G1323–G1336. [CrossRef]
82. Ibrahim, A.; Mbodji, K.; Hassan, A.; Aziz, M.; Boukhettala, N.; Coëffier, M.; Savoye, G.; Déchelotte, P.; Marion-Letellier, R. Anti-Inflammatory and Anti-Angiogenic Effect of Long Chain n-3 Polyunsaturated Fatty Acids in Intestinal Microvascular Endothelium. *Clin. Nutr. Edinb. Scotl.* **2011**, *30*, 678–687. [CrossRef]
83. Hu, S.; Bae, M.; Park, Y.-K.; Lee, J.-Y. N-3 PUFAs Inhibit TGFβ1-Induced Profibrogenic Gene Expression by Ameliorating the Repression of PPARγ in Hepatic Stellate Cells. *J. Nutr. Biochem.* **2020**, *85*, 108452. [CrossRef]
84. Maeshige, N.; Torii, K.; Tabuchi, H.; Imai, M.; Koga, Y.; Uemura, M.; Aoyama-Ishikawa, M.; Miyoshi, M.; Fujino, H.; Terashi, H.; et al. Inhibitory Effects of Short-Chain Fatty Acids and ω-3 Polyunsaturated Fatty Acids on Profibrotic Factors in Dermal Fibroblasts. *Eplasty* **2019**, *19*, e4. [PubMed]
85. Zeng, Z.; Yang, H.; Wang, Y.; Ren, J.; Dai, Y.; Dai, C. Omega-3 Polyunsaturated Fatty Acids Attenuate Fibroblast Activation and Kidney Fibrosis Involving MTORC2 Signaling Suppression. *Sci. Rep.* **2017**, *7*, 46146. [CrossRef] [PubMed]
86. Yang, Y.; Hu, L.; Xia, H.; Chen, L.; Cui, S.; Wang, Y.; Zhou, T.; Xiong, W.; Song, L.; Li, S.; et al. Resolvin D1 Attenuates Mechanical Stretch-Induced Pulmonary Fibrosis via Epithelial-Mesenchymal Transition. *Am. J. Physiol.-Lung Cell. Mol. Physiol.* **2019**, *316*, L1013–L1024. [CrossRef] [PubMed]
87. Zhao, S.; Dejanovic, D.; Yao, P.; Bhilocha, S.; Sadler, T.; Schirbel, A.; West, G.; Doyon, G.; Lopez, R.; Mao, R.; et al. Selective Deletion of MyD88 Signaling in α-SMA Positive Cells Ameliorates Experimental Intestinal Fibrosis via Post-Transcriptional Regulation. *Mucosal Immunol.* **2020**. [CrossRef]
88. Jacob, N.; Jacobs, J.P.; Kumagai, K.; Ha, C.W.Y.; Kanazawa, Y.; Lagishetty, V.; Altmayer, K.; Hamill, A.M.; Von Arx, A.; Sartor, R.B.; et al. Inflammation-Independent TL1A-Mediated Intestinal Fibrosis Is Dependent on the Gut Microbiome. *Mucosal Immunol.* **2018**, *11*, 1466–1476. [CrossRef]
89. Imai, J.; Kitamoto, S.; Sugihara, K.; Nagao-Kitamoto, H.; Hayashi, A.; Morhardt, T.L.; Kuffa, P.; Higgins, P.D.R.; Barnich, N.; Kamada, N. Flagellin-Mediated Activation of IL-33-ST2 Signaling by a Pathobiont Promotes Intestinal Fibrosis. *Mucosal Immunol.* **2019**, *12*, 632–643. [CrossRef]
90. Sivignon, A.; de Vallée, A.; Barnich, N.; Denizot, J.; Darcha, C.; Pignède, G.; Vandekerckove, P.; Darfeuille-Michaud, A. Saccharomyces Cerevisiae CNCM I-3856 Prevents Colitis Induced by AIEC Bacteria in the Transgenic Mouse Model Mimicking Crohn's Disease. *Inflamm. Bowel Dis.* **2015**, *21*, 276–286. [CrossRef]
91. Leccese, G.; Bibi, A.; Mazza, S.; Facciotti, F.; Caprioli, F.; Landini, P.; Paroni, M. Probiotic Lactobacillus and Bifidobacterium Strains Counteract Adherent-Invasive Escherichia Coli (AIEC) Virulence and Hamper IL-23/Th17 Axis in Ulcerative Colitis, but Not in Crohn's Disease. *Cells* **2020**, *9*, 1824. [CrossRef]
92. Ogawa, H.; Rafiee, P.; Fisher, P.J.; Johnson, N.A.; Otterson, M.F.; Binion, D.G. Sodium Butyrate Inhibits Angiogenesis of Human Intestinal Microvascular Endothelial Cells through COX-2 Inhibition. *FEBS Lett.* **2003**, *554*, 88–94. [CrossRef]
93. Huang, C.; Wang, X.-L.; Qi, F.-F.; Pang, Z.-L. Berberine Inhibits Epithelial-Mesenchymal Transition and Promotes Apoptosis of Tumour-Associated Fibroblast-Induced Colonic Epithelial Cells through Regulation of TGF-β Signalling. *J. Cell Commun. Signal.* **2020**, *14*, 53–66. [CrossRef] [PubMed]

94. Rieder, F.; Kessler, S.P.; West, G.A.; Bhilocha, S.; de la Motte, C.; Sadler, T.M.; Gopalan, B.; Stylianou, E.; Fiocchi, C. Inflammation-Induced Endothelial-to-Mesenchymal Transition: A Novel Mechanism of Intestinal Fibrosis. *Am. J. Pathol.* **2011**, *179*, 2660–2673. [CrossRef] [PubMed]
95. Binion, D.G.; Heidemann, J.; Li, M.S.; Nelson, V.M.; Otterson, M.F.; Rafiee, P. Vascular Cell Adhesion Molecule-1 Expression in Human Intestinal Microvascular Endothelial Cells Is Regulated by PI 3-Kinase/Akt/MAPK/NF-KB: Inhibitory Role of Curcumin. *Am. J. Physiol. Gastrointest. Liver Physiol.* **2009**, *297*, G259–G268. [CrossRef]
96. Rafiee, P.; Binion, D.G.; Wellner, M.; Behmaram, B.; Floer, M.; Mitton, E.; Nie, L.; Zhang, Z.; Otterson, M.F. Modulatory Effect of Curcumin on Survival of Irradiated Human Intestinal Microvascular Endothelial Cells: Role of Akt/MTOR and NF-{kappa}B. *Am. J. Physiol. Gastrointest. Liver Physiol.* **2010**, *298*, G865–G877. [CrossRef] [PubMed]
97. Im, E.; Choi, Y.J.; Kim, C.H.; Fiocchi, C.; Pothoulakis, C.; Rhee, S.H. The Angiogenic Effect of Probiotic Bacillus Polyfermenticus on Human Intestinal Microvascular Endothelial Cells Is Mediated by IL-8. *Am. J. Physiol. Gastrointest. Liver Physiol.* **2009**, *297*, G999–G1008. [CrossRef] [PubMed]
98. Dupont-Lucas, C.; Marion-Letellier, R.; Pala, M.; Guerin, C.; Amamou, A.; Jarbeau, M.; Bôle-Feysot, C.; Nicol, L.; David, A.; Aziz, M.; et al. A Polymeric Diet Rich in Transforming Growth Factor Beta 2 Does Not Reduce Inflammation in Chronic 2,4,6-Trinitrobenzene Sulfonic Acid Colitis in Pre-Pubertal Rats. *BMC Gastroenterol.* **2020**, *20*, 1–11. [CrossRef]
99. Agista, A.Z.; Rusbana, T.B.; Islam, J.; Ohsaki, Y.; Sultana, H.; Hirakawa, R.; Watanabe, K.; Nochi, T.; Ardiansyah; Budijanto, S.; et al. Fermented Rice Bran Supplementation Prevents the Development of Intestinal Fibrosis Due to DSS-Induced Inflammation in Mice. *Nutrients* **2021**, *13*, 1869. [CrossRef]
100. Bischoff, S.C.; Escher, J.; Hébuterne, X.; Kłęk, S.; Krznaric, Z.; Schneider, S.; Shamir, R.; Stardelova, K.; Wierdsma, N.; Wiskin, A.E.; et al. ESPEN Practical Guideline: Clinical Nutrition in Inflammatory Bowel Disease. *Clin. Nutr.* **2020**, *39*, 632–653. [CrossRef]
101. Borren, N.Z.; van der Woude, C.J.; Ananthakrishnan, A.N. Fatigue in IBD: Epidemiology, Pathophysiology and Management. *Nat. Rev. Gastroenterol. Hepatol.* **2019**, *16*, 247–259. [CrossRef] [PubMed]
102. Schreiner, P.; Yilmaz, B.; Rossel, J.-B.; Franc, Y.; Misselwitz, B.; Scharl, M.; Zeitz, J.; Frei, P.; Greuter, T.; Vavricka, S.R.; et al. Vegetarian or Gluten-Free Diets in Patients with Inflammatory Bowel Disease Are Associated with Lower Psychological Well-Being and a Different Gut Microbiota, but No Beneficial Effects on the Course of the Disease. *United Eur. Gastroenterol. J.* **2019**, *7*, 767–781. [CrossRef] [PubMed]
103. Kouvari, M.; Boutari, C.; Chrysohoou, C.; Fragkopoulou, E.; Antonopoulou, S.; Tousoulis, D.; Pitsavos, C.; Panagiotakos, D.B.; Mantzoros, C.S.; ATTICA study Investigators. Mediterranean Diet Is Inversely Associated with Steatosis and Fibrosis and Decreases Ten-Year Diabetes and Cardiovascular Risk in NAFLD Subjects: Results from the ATTICA Prospective Cohort Study. *Clin. Nutr. Edinb. Scotl.* **2021**, *40*, 3314–3324. [CrossRef] [PubMed]
104. Chicco, F.; Magrì, S.; Cingolani, A.; Paduano, D.; Pesenti, M.; Zara, F.; Tumbarello, F.; Urru, E.; Melis, A.; Casula, L.; et al. Multidimensional Impact of Mediterranean Diet on IBD Patients. *Inflamm. Bowel Dis.* **2021**, *27*, 1–9. [CrossRef] [PubMed]

nutrients

MDPI

Review

The Gluten-Free Diet for Celiac Disease and Beyond

Bara Aljada, Ahmed Zohni and Wael El-Matary *

Section of Pediatric Gastroenterology, Department of Pediatrics and Child Health, Max Rady College of Medicine, University of Manitoba, Winnipeg, MB R3A 1S1, Canada; aljadab@myumanitoba.ca (B.A.); zohniah@gmail.com (A.Z.)
* Correspondence: welmatary@hsc.mb.ca

Abstract: The gluten-free diet (GFD) has gained popularity beyond its main medical indication as the treatment for gluten-induced immune-mediated disorders such as celiac disease (CD), dermatitis herpetiformis, gluten ataxia, wheat allergy, and non-celiac gluten sensitivity. However, the diet carries some disadvantages such as elevated costs, nutritional deficiencies, and social and psychological barriers. The present work aims to review indications, proven benefits, and adverse events of a gluten-free diet. Close follow-up with patients following the diet is recommended. More data is needed to assess the effectiveness of the diet in managing mental and cognitive disorders and to establish a connection between the brain and gluten.

Keywords: celiac disease; gluten; gluten-free diet

Citation: Aljada, B.; Zohni, A.; El-Matary, W. The Gluten-Free Diet for Celiac Disease and Beyond. *Nutrients* **2021**, *13*, 3993. https://doi.org/10.3390/nu13113993

Academic Editor: Ina Bergheim

Received: 14 October 2021
Accepted: 3 November 2021
Published: 9 November 2021

1. Introduction

Wheat is responsible for 20% of global caloric consumption, making it amongst the most valuable crops worldwide. Due to its versatility, wheat can be incorporated into various foods such as bread, pasta, cereals, and baked goods, which has propelled this crop into a staple food across the temperate world [1]. Despite its traditional view as a nutritious source containing proteins, vitamins, and minerals, concerns have been raised towards a specific component of wheat called gluten. As an ingredient, gluten consumption dates back to 6th-century Chinese cuisine, where its popularity grew amongst Buddhists who used gluten as a substitute for meat. Jia Sixie's *Qimin Yaoshu*, a Chinese agricultural encyclopedia written in 544 CE, mentions the use of gluten in noodles called bótuō. References of gluten in Western literature appear much later. Bartolomeo Beccari authored *De Frumento*, an Italian treatise on wheat, in 1745, which documented the extraction of gluten from wheat flour. In 1803, John Imson defined gluten in the English language in *Elements of Science and Art* [2]. The industrial revolution played a prominent role in the rising popularity of wheat as a staple food in the Western diet. Over this time, wheat was inexpensively milled in large quantities and quickly distributed using the developing railroad systems [3,4]. Western popularity of wheat also rose during the Great Depression and World War II, when wheat-containing products, such as bread and pasta, served as cheaper substitutes of rationed foods such as dairy and meat [5,6]. Today, global wheat consumption has increased at a faster rate than all other cereals [7]. As a result, there is increasing attention towards the health effects of gluten.

2. Gluten and Celiac Disease

Gluten is a mixture of water-insoluble prolamin proteins. The prolamins, a complex group of alcohol-soluble lectins, constitute the significant seed proteins in cereals. They comprise about 80% of the starch endosperm storage proteins in mature cereal grains [8] and are yet to be found in other parts of the grain [9]. The most abundant gluten prolamins (called gliadin and glutenin) are predominantly found in wheat. However, prolamins can be found in different cereal species under specific names, such as in barley (called hordeins), rye (secalins), oats (avenins), and other closely related grains although each has different molecular properties [10]. Gliadins comprise four significant alcohol-soluble monomers that collectively allow the gluten to elongate while providing intermolecular binding sites. The α-helices and β-sheets of α/β- and γ-gliadins allow for hydrogen and disulfide bonding, whereas ω-gliadins are composed of β-turns and have no α-helices or β-sheets [11]. In contrast, glutenins are alcohol-insoluble polymers that contribute to the flexibility and stability of gluten. When flour and water are mixed, a thiol group from glutenin interacts with disulfide bonds in gliadin, resulting in a shift towards intermolecular disulfide bonds [12]. The high concentration of glutamine amino acids results in many inter-chain hydrogen bonds that collectively provide strength [11,12]. In addition, gluten's high proline content alters the protein structure to provide elasticity [11].

Gluten is infamous for its role in celiac disease (CD). This autoimmune condition affects 1% of the population and leads to a reversible inflammatory process in small bowel mucosa with acute repercussions such as diarrhea, constipation, bloating, nausea, and vomiting [13–15]. Long-term consequences of mucosal damage and inflammation include malabsorption of nutrients such as calcium, vitamin D [16], iron [17], vitamin B12, folic acid, and zinc [18], leading to debilitating consequences such as osteoporosis, anemia, and stunted growth [19]. The clinical presentation of CD can vary depending on age. The classic presentation in pediatric patients includes malnutrition, failure to thrive, abdominal pain, and distension. In contrast, adults commonly present with gastrointestinal symptoms but with less severity [20], with most patients experiencing severe diarrhea [21].

Calcium and vitamin D absorption is of particular concern in the growth and development of pediatric patients with CD. Several factors influence bone mineral density, including inflammation from chronic disease, diet, absorption in the duodenum, and metabolism [22,23]. In patients with CD, mucosal damage of the small bowel impairs calcium and vitamin D absorption, leading to impaired bone health. Whereas vitamin D is involved in the hormonal regulation of bone remodeling and calcium absorption [24], calcium serves a structural role in bones as a component of hydroxyapatite [25]. Pediatric patients with CD are at risk of short stature and constitutional delay of puberty. One study [26] found CD in 2–8% of children with short stature and no gastrointestinal symptoms. After ruling out endocrine causes for short stature, the same study found that the proportion of CD increased to 19–59%. When using a growth chart, pediatric patients with CD typically demonstrate a decline in both weight and stature velocity, crossing several percentile lines in both categories [27]. In addition, Ludvigsson et al. [28] found that patients with CD are at increased risk of subsequent hip fracture and fracture of any kind, independent of age or sex. A lower bone mineral density is one theory for the observed fracture risk, specifically in the femoral neck region, which Melton et al. [29] determined to be the strongest predictor of future hip fracture. Kemppainen et al. [30] supported this finding after they determined that patients had significantly lower bone mineral density at the lumbar spine and femoral neck, with over 64% of men and 71% of female patients presenting with low calcifediol, a form of vitamin D produced in the liver.

The pathophysiology of CD involves a complex interplay between patients' genetics and environment [31,32] that leads to an inappropriate immune response. In turn, the maladaptive response can cause enterocyte destruction and subsequent villous atrophy [20]. Once consumed, gluten's glutamine and proline components prevent complete hydrolyzation of the immunoreactive epitope, producing peptides longer than ten amino acids in length [33]. Most notably, 13-, 19-, and 33-mer peptides are associated with the inflammatory reaction seen in CD [34,35]. In addition, gliadin prolamin upregulates the production of the intestinal peptide zonulin, which increases the permeability of tight junctions in the intestines. Several studies have shown increased levels of zonulin in patients with CD, making it a leading culprit in the pathogenesis of the disease [36,37]. In turn, these changes allow increased paracellular and transcellular peptide transport into the lamina propria [38]. Once in the gut mucosa, tissue transglutaminase (tTG) recognizes the glutamine and proline components, resulting in a series of deamidation and transamidation reactions that increases peptide affinity to antigen-presenting major histocompatibility complex class two (MHC II) molecules [20,39]. One study found human leukocyte antigens (HLA)-DQ2 and DQ8 present in 98.4% of patients with CD and a presence of 89.6% in their families, suggesting a genetic component to the disease [40]. Antigen-presenting cells, such as dendritic cells, present the peptides to gluten-specific T cells, triggering both the innate and adaptive immune response. The innate response releases interleukin (IL)-15, leading to the destruction of gut epithelial cells by CD8+ (cytotoxic) T-lymphocytes [41]. The role of IL-17 in the pathogenesis of CD is still under investigation. Scaleia et al. [42] found lower levels of IL-17-producing T cells in the intra-epithelial lymphocyte (IEL) compartment of CD patients. They speculate that these changes negatively affect the homeostasis of the mucosal barrier while contributing to the altered permeability of the gut mucosa. In addition, the adaptive response generates inflammatory cytokines, activating either interferon-gamma (IFN-γ) producing T helper (Th)1 cells or Th2 cells that promote B-lymphocyte development into plasma cells. In turn, plasma cells produce anti-gliadin and anti-tissue-transglutaminase antibodies [43]. The effects of gluten on the gut mucosa of susceptible individuals vary but can include gut inflammation, villous atrophy, crypt hyperplasia, and CD4+ and CD8+ T-cell lymphocytic invasion of the intraepithelial tissue [44]. When studying the histopathological effects of CD and response to treatment, clinicians have traditionally used the Marsh-Oberhuber classification system (Table 1), which grades biopsies of the intestinal mucosa into four categories. A diagnosis of CD is reserved for Marsh 2&3 biopsies, which show increased IELs, crypt hyperplasia, and villous atrophy. Marsh 3 can be divided into three subgroups based on the degree of villous atrophy [45].

Table 1. Modified Marsh-Oberhuber classification system.

Type	Intraepithelial Lymphocytes/100 Enterocytes (Duodenum)	Crypt	Villous Architecture	Lesion
0	<30	Normal	Normal	Pre-infiltrative
1	>30	Normal	Normal	Infiltrative
2	>30	Hyperplasia	Normal	Infiltrative-hyperplastic
3a	>30	Hyperplasia	Mild atrophy	Flat destructive
3b	>30	Hyperplasia	Marked atrophy	Flat destructive
3c	>30	Hyperplasia	Complete atrophy	Flat destructive

References [45–47].

3. Gluten-Free Diet for Celiac Disease

A lifetime gluten-free diet (GFD) is the treatment for individuals with CD [48]. Continuing to ingest gluten can exacerbate clinical symptoms, further intestinal damage, and increase the risk of future cancers, including small intestinal adenocarcinoma, esophageal cancer, melanoma, and non-Hodgkin's lymphoma [49]. For best results, this diet involves complete removal of gluten-containing foods from one's diet, including gluten proteins in wheat (gliadin), barley (hordeins), rye (secalins), oats (avenins), and other closely related grains. Due to such dietary cutbacks, individuals on a GFD are encouraged to incorporate other nutritious food sources such as fruits, vegetables, fish, meat, and gluten-free products. Over the years, scientific discovery, aggressive marketing, and media coverage of the benefits of a GFD have pushed food companies to produce more gluten-free options. As a result, 2016 saw over $15.5 billion in retail sales of gluten-free foods, more than double 2011 figures [50]. The marked increase in gluten-free substitutes allows CD patients to reproduce the dietary habits and patterns of the general population [51]. To support consumers following a GFD, the Food and Drug Administration (FDA) passed a gluten-free labeling rule that outlined the legal requirements for labeling a product "gluten-free", "free of gluten" ,"without gluten", or "no gluten". A gluten-free product is defined as having <20 ppm of gluten while considering possible contamination during product creation [52]. In addition, local organic food stores commonly sell gluten-free products such as bread and pasta, albeit at a slightly higher cost and with a different taste than their gluten-containing counterparts.

3.1. Efficacy of Gluten-Free Diet in Celiac Disease

There has been extensive research on the efficacy of the GFD. A strict GFD can restore the histology of the small bowel architecture in 95% of children within two years [53], whereas 34% and 66% of adult patients experience mucosal recovery after two and five years, respectively [54]. However, some data show incomplete recovery in older patients (between 30 and 60 years) and no statistically significant recovery in individuals older than 60 years [55]. With small bowel recovery, a GFD can also improve symptoms of malabsorption, including diarrhea, steatorrhea, and weight loss. In addition, several studies have demonstrated significant improvement in bone mineral density after one year of the diet [56–58], although complete reversal of osteopenia could not be observed [59]. Soliman et al. [60] found that pediatric patients on a GFD for two years demonstrate average growth in height and weight compared to age-matched controls, with significant catch-up growth (increase in percentile position on a growth curve) in some patients. When comparing the efficacy of GFD between patients with mild enteropathy and those with villous atrophy, Kurppa et al. established that the GFD has similar outcomes in mucosal architecture recovery, reduction of intestinal mucosal inflammation, antibody concentrations, and symptom improvement [61]. Another study examining the GFD in patients with borderline enteropathy that does not meet the criteria of CD demonstrated restoration of mucosal structure and marked improvement in clinical symptoms within 8–12 months of adhering to the diet compared to controls [62].

3.2. Skepticism of the Gluten-Free Diet

Despite the extensive literature on the GFD, questions and skepticism remain. Even with careful preparation and storage of gluten-free food, the likelihood of cross-contamination has raised questions about the effects of chronic low-dose gluten exposure [63]. Therefore, the focus towards the GFD has shifted from the absolute removal of gluten from one's diet to limiting gluten intake below a specific threshold yet to be determined [64,65]. To identify the levels of safe gluten exposure, Akobeng and Thomas [66] reviewed thirty-five studies and found that gluten tolerability differed across studies and among study participants. While some patients had no histological abnormalities on a diet containing an average of 36 mg of gluten per day, others developed mucosal changes after only consuming 10 mg per day. They concluded that a daily intake of less than 10 mg is "unlikely to cause significant histological abnormalities." In comparison, definite mucosal changes were observed with daily intakes of 100 mg and 500 mg, respectively [67]. Taken all together, one may conclude that achieving a conclusive threshold that could result in mucosal changes in 100% of patients with CD is unlikely to occur, although a daily intake of less than 10 mg is likely to produce the safest results. Skepticism has also been raised about the GFD's ability to completely reverse abnormal changes in the gut mucosa. Gluten activation of the immune system has been shown to produce changes in the intra-epithelial lymphocyte compartment (IEL) and is associated with increased γ/δ IELs. Recent cell sequencing work found high levels of γ/δ IELs in histologically normal-appearing tissue, suggesting that some changes persist following a GFD [68].

3.3. Challenges of a Gluten-Free Diet

Given that gluten-containing food represents staple dietary components in many households worldwide, a GFD represents a dramatic lifestyle change that can pose many challenges. The threat of cross-contamination is a daily issue for individuals on a GFD. Sharing cupboards, countertops, and kitchen appliances with individuals who do not follow a GFD present possible contamination opportunities that impair the diet's success. For increased safety, meals should be prepared and stored away from non-gluten-free food. A similar concern extends to eating at restaurants, food courts, and food stands. Although individuals may face difficulty finding gluten-free options, restaurants are increasing their gluten-free options due to the rising popularity of the GFD. Of note, one study found that 32% of gluten-free labeled restaurant food tested positive for gluten, with gluten-free pizza and pasta being the most likely culprits [69]. Processed foods made from gluten-containing ingredients represent another area of concern. Potentially hidden sources of gluten include certain soups, processed meat, French fries, seasonings, and beer. Although the degree of susceptibility to gluten-containing food varies between individuals, one study suggests a safe gluten contamination cutoff of 100 ppm (1/4 mg/kg) in gluten-free foods [70]. Therefore, eating foods that have a gluten-free label is generally a safe option for avoiding gluten-contaminated food. Finally, adhering to a GFD can be costly. In one study, gluten-free products were 242% more expensive than their gluten-containing counterparts in the same food group [71]. Several studies echoed this sentiment, demonstrating lower availability and higher cost of gluten-free foods [72,73]. However, despite these challenges, a prospective study by Mustalahti et al. [74] found declining symptoms and a significantly improved quality of life in patients with CD on a GFD, suggesting that the diet was not particularly distressing for the majority of patients.

3.4. How to Monitor a Gluten-Free Diet for Celiac Disease

Strict adherence to a gluten-free diet is the only recommended treatment for CD [75]. As such, one may suggest that newly diagnosed and symptomatic patients require more frequent assessment, especially as the gut mucosa is undergoing repair and clinical symptoms are improving. Several studies have investigated when patients should be followed up after initiating a GFD and with whom, given that there is no clear consensus. In a study examining patient preferences towards follow-up, most preferred to be seen by a dietician (with a physician available if needed), and 67% of respondents preferring annual appointments [76]. Kurppa et al. [77] found that follow-up by primary care physicians was just as successful as a follow-up in tertiary centers, with average GFD adherence rates at 88%. Current guidelines recommend routine blood tests at each follow-up visit, including checking for intestinal absorption with a complete blood count, serum calcium, ferritin, vitamin B12, and alkaline phosphate. In addition, thyroid function tests such as thyroid-stimulating hormone and thyroid hormone should be checked to screen for other autoimmune conditions, alongside liver function tests such as aspartate aminotransferase and alanine aminotransferase levels to monitor for autoimmune liver disease [44]. While there are no strong recommendations towards a particular monitoring tool, there are several methods for monitoring gluten-free diet adherence and efficacy in CD, including symptom assessment, dietetic interview, serology, stool and urine markers, and small bowel biopsy (Figure 1).

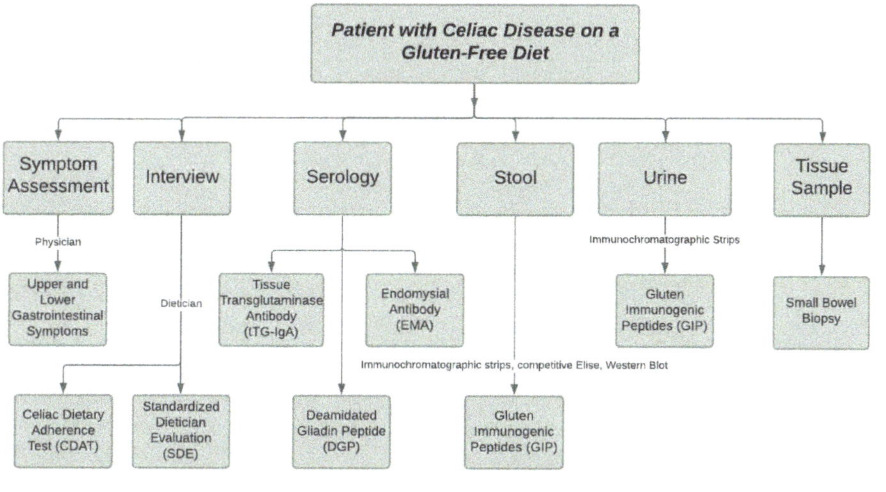

Figure 1. Methods for monitoring adherence to gluten-free diet.

3.4.1. Symptom Assessment

The first step of monitoring the GFD in patients with CD is to identify ongoing symptoms and their severity. One study [78] found that upper gastrointestinal symptoms disappear first, while lower gastrointestinal tract symptoms, such as constipation, remain unchanged when re-evaluated 12–28 months after beginning a GFD. Abdominal bloating (51.3%), abdominal pain (45.9%), and constipation (29.7%) represent the most common symptoms at follow-up. The strongest positive predictors for ongoing symptoms at re-evaluation include experiencing symptoms for five years or more before diagnosis (OR 5.3, 95% CI 1.3 to 21.8) and having constipation at the time of diagnosis (OR 7.4, 95% CI 1.3 to 42). However, Rubio-Tapia et al. [54] found that clinical response to a GFD was an inaccurate marker for mucosal repair. Additionally, 62% of patients who experienced a clinical response to a GFD have continued mucosal damage at their follow-up biopsy, although symptomatic patients do not present with more severe histological lesions than asymptomatic patients [78]. Lahdeaho et al. [70] supported the limited utility of clinical response as a monitoring tool when they found that 22% of patients with significant small bowel damage had no symptoms. Nonetheless, symptomatic improvement is a potential motivator for the continued adherence to a GFD and serves as a limited tool for monitoring the disease.

3.4.2. Dietetic Interview

A second option for monitoring the GFD is a dietetic interview conducted by a trained dietician or physician. There are various questionnaires available in a variety of languages that assess self-reported compliance with the GFD. The results from these surveys are often combined with visual analog scales that contain unmarked lines with anchor statements such as 'I never follow my diet' and 'I always follow my diet' at the boundaries [44,79]. Currently, the Standardized Dietician Evaluation (SDE) is the gold-standard interview format for assessing adherence to the GFD. A trained dietician conducts this interview, consisting of three main parts from which answers are graded according to a 6-point Likert scale. First, the dietician analyzes the patient's diet over twenty-four hours or three days. The patient then participates in a food-label quiz to determine which ingredients and additives are likely to contain gluten from a list of twenty-eight. Finally, the patient is assessed on their ability to check the labels of medicines, supplements, and cosmetics for gluten [80]. The Celiac Dietary Adherence Test (CDAT) is another popular screening tool. Developed by gastroenterologists, dieticians, psychologists, and celiac patients, this tool grades participants' answers to seven questions regarding their knowledge, opinions, and adherence to a GFD on a 5-point Likert scale [81]. Although the CDAT is highly correlated with the SDE [80], the SDE shows a stronger correlation with serological titers and duodenal biopsies [81]. Subjectivity, fear of judgment, and under-reporting of gluten consumption represent significant limitations to the interview format [82].

3.4.3. Serology

Serological testing for antibodies associated with CD is another option for monitoring the GFD. Elevated levels of tissue transglutaminase antibodies (tTG-IgA), endomysial antibodies (EMA), and deamidated gliadin peptide (DGP) antibodies can indicate poor adherence to or efficacy of a GFD. Testing for tTG-IgA is a first-line diagnostic tool in the workup of CD with sensitivity and specificity levels above 95% [83,84]. Relative to other markers, the combination of high sensitivity, functionality, and cheaper costs of tTG-IgA make it a preferred choice for initial serological testing. Positive tTG-IgA results are often followed up by confirmatory EMA testing, which shows a higher specificity (99.0–100%) for CD [85,86]. Some studies have shown rising serum EMA levels before the appearance of villous atrophy, making it a potential early marker in CD [61,87]. Limitations of this marker include higher operating costs and less objective results due to the use of labor-intensive, resource-demanding, and operator-dependent immunofluorescence [88]. Testing for DGP is a newer technique for CD, although it poses a lower sensitivity (88%) and specificity

(94%) in the general population than the markers mentioned earlier [89]. Nonetheless, new evidence suggests that testing for DGP is of better use in pediatric patients for diagnosing CD and monitoring a GFD. In an investigation of forty children less than two years of age with features of chronic enteropathy, Barbato et al. found eleven patients with normal tTG, and EMA titers with elevated DGP and endoscopic changes consistent with CD [90]. In addition, Monzani et al. demonstrated that testing for DGP IgA and IgG in children had a sensitivity of 100% for screening for CD and was 52% more sensitive than tTG for monitoring GFD adherence [91]. A study by Liu et al. found that DGP levels normalized faster than tTg in children following initiation of a GFD, making DGP a possible early marker of a response to a GFD [92]. However, despite these antibodies' reported high specificity and sensitivity for diagnosing CD, serology has several limitations in its use as a marker for GFD efficacy, especially as it relates to mucosal repair. Serological markers represent the body's immune response to the disease and are not directly correlated with intestinal damage. A meta-analysis on the sensitivity and specificity of tTG IgA and EMA IgA assays determined that both serological markers had a poor correlation with mucosal damage in celiac patients undergoing a follow-up biopsy while on a GFD. In patients with villous atrophy (Marsh 3), tTG IgA had a sensitivity of 0.50 (95% CI 0.41–0.60) and a specificity of 0.83 (95% CI 0.79–0.87), while EMA IgA had a sensitivity of 0.45 (95% CI 0.34–0.57) and specificity of 0.91 (95% CI 0.87–0.94). Although a positive test result is a good indicator of persistent villous atrophy, most patients with mucosal damage will have normal antibody titers while on the GFD, making serology an unreliable marker for following mucosal repair and monitoring adherence [93].

3.4.4. Stool and Urine Markers

Clinicians can also use stool and urine markers for monitoring GFD. Specific gluten peptides, such as the immunotoxic 33-mer peptide, are resistant to gastrointestinal degradation. In one study, over 30% of 33-mer peptides resisted hydrolysis in vitro simulated gastrointestinal digestion [94]. The degree of immunotoxic peptide absorption and excretion varies among individuals and can be influenced by differences in the gut microbiome and diet [95]. Some peptides are subsequently excreted in feces and can be detected by immunochromatographic strips, competitive ELISA, and Western blot [94,96]. In turn, detection of gliadin peptides in the stool can be used as evidence of gluten consumption and as a non-invasive marker of compliance with a GFD [96–98]. Whereas immunochromatographic strips are more likely to be used as clinical standard assays in point-of-care settings, ELISA is more likely to be used for more detailed quantification of gluten exposure when monitoring the efficacy of a GFD [99,100]. Comino et al. [94] found that ingestion of 50 mg of gluten was enough for detection in stool samples and that levels of gluten consumption were "roughly" correlated with gluten excretion 2–4 days after ingestion. The study concluded that the non-invasive nature of the immunologic tests could be used to monitor short-term adherence to GFD, involuntary gluten consumption from contaminated food, and for assessing the effectiveness of novel treatments for CD such as enzymatic therapies designed to destroy toxic gluten peptides. Another multicenter study [100] examined the use of ELISA to detect immunogenic gluten peptides (GIP) in patients on a GFD for at least one year and to compare the assay to other GFD monitoring tools. Researchers found that 30% of patients on a GFD had detectable GIP in their stools, suggesting they were either non-compliant with the GFD or involuntarily consuming contaminated food. The presence of GIP was strongly associated with symptoms associated with gluten exposure, with up to two-thirds of patients unresponsive to a GFD having detectable GIP on ELISA. In contrast, stand-alone use of dietary questionnaires and serum anti-tTG antibody levels revealed non-compliance in 18% of the same patients. The same study found no significant association between stool GIP and dietary questionnaires or serum anti-tTG antibody levels.

Researchers have also explored the use of urine samples as a monitoring tool for compliance and efficacy of a GFD. Moreno et al. [101] found that ingestion of greater than

25 mg of gluten results in urine GIP that are detectable on immunochromatographic strips as early as four hours after ingestion and remain detectable in the urine for up to 48 h. In addition, GIP levels were positively correlated with the level of gluten intake. Moreover, 89% of patients with CD and no intestinal mucosa damage on duodenal biopsy had no detectable GIP in their urine. Consequently, all patients with incomplete recovery of the mucosa had quantifiable GIP. Various factors can influence GIP concentration in the urine, including diet, daily liquid intake, weight, and gut microbiota.

3.4.5. Small Bowel Biopsy and Pathology

At present, assessing small bowel pathology is the most accurate method for monitoring mucosal recovery in patients on a GFD. Several studies have shown mucosal damage on biopsy in patients with normal serology and clinical response to a GFD [54,102]. When performing a biopsy, multiple small bowel samples are collected given the patchy nature of histological abnormalities in CD and the declining sensitivity of biopsies for CD when less than four samples are taken [103,104]. Current guidelines give a strong recommendation, backed by a high level of evidence, for at least one duodenal bulb biopsy and at least four biopsies of the distal duodenum [104,105]. Pathologists look for the presence of crypt elongation and villous atrophy, the density of IELs, and the crypt-villous ratio while classifying the specimen according to the Marsh-Oberhuber scale [105]. Rubio-Tapia et al. [54] examined the rate of mucosal recovery, defined as a villous to crypt ratio of 3 to 1, at the first follow-up biopsy for adult patients on a GFD. Moreover, 35% of patients on a GFD receiving a biopsy within two years of starting the diet showed mucosal recovery, whereas 43% showed mucosal recovery when the first biopsy was taken between two to five years after initial diagnosis. Histological improvement, characterized by an increase of villous to crypt ratio $\geq$2.0 points relative to baseline, was observed in 45% of patients at the first follow-up biopsy. In addition, the average recovery time for mucosal repair was determined to be three years after starting a GFD. Although complete histological recovery is not universally achieved on a GFD, various studies suggest that mucosal healing can be seen in 57–76% of patients [44]. Compared to adults, pediatric patients s89howed a better response to the GFD, with up to 95% showing mucosal recovery within two years [53]. However, intestinal biopsies are more invasive, expensive, and unreasonable for monitoring every patient with CD than other monitoring tools. For this reason, the American College of Gastroenterology [104] gives a strong recommendation for long-term follow-up of a GFD based on history and serology alone. They further suggest that biopsies should be reserved for patients showing inadequate clinical response or relapse in symptoms while on a GFD.

4. Gluten-Free Diet for Other Health Problems

The gluten-free diet is recognized as the standard protocol for patients diagnosed with CD. However, the diet has recently gone mainstream, and individuals excluded from the CD diagnosis now make up most adherents. Chuong and colleagues [106] found that between 2009 and 2014, the prevalence of CD in the American population remained constant (0.7%) while the demographic of people who avoid gluten (PWAG) grew from 0.5% to 1.7%. Since the gluten-free diet is no longer a niche treatment for a select diagnosis and is now utilized more broadly by the general population, many studies have analyzed the benefits of the diet. Beyond patients with CD, the gluten-free diet is also recognized in the treatment of gluten ataxia, dermatitis herpetiformis, cognitive impairment, inflammatory bowel disease and irritable bowel syndrome, dermatitis herpetiformis, and non-celiac gluten sensitivity (Figure 2).

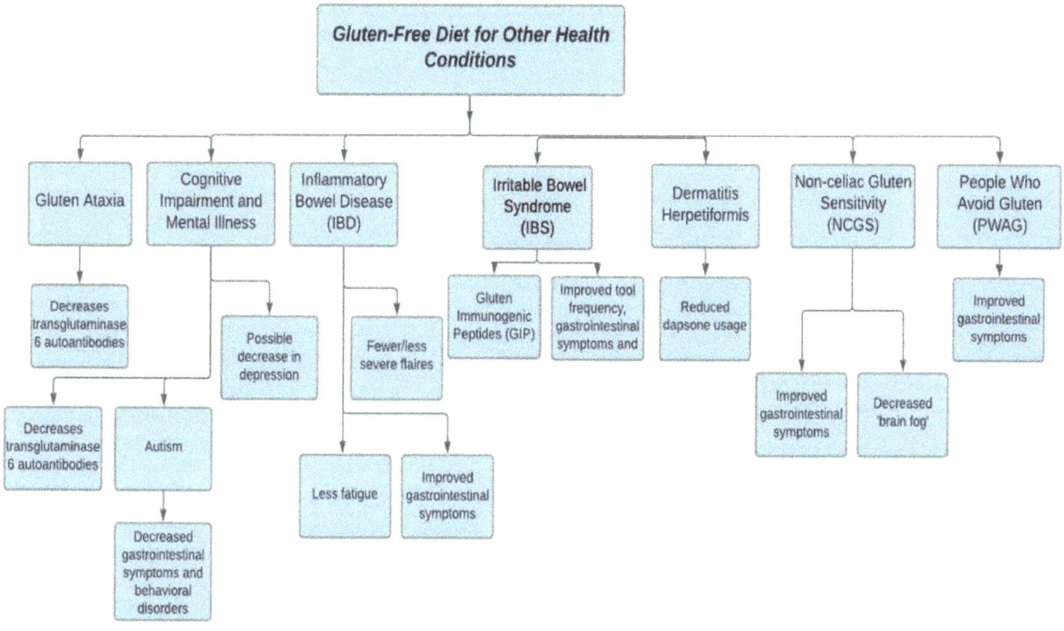

Figure 2. The GFD for health conditions other than celiac disease.

4.1. Gluten Ataxia

Gluten ataxia is an immune-mediated disease wherein ingestion of gluten causes the body's immune system to attack the nervous system tissue, specifically the cerebellum. Transglutaminase 6 (TG6) autoantibodies are more abundant in patients with gluten ataxia and have become an efficient marker for diagnosing the condition, as demonstrated by Hadjivassiliou and colleagues [107]. These antibodies are suspected to be the primary mechanism through which neurological diseases develop in individuals with gluten sensitivities. A study by Dipper and colleagues demonstrated that patients placed on a GFD experienced a decrease in TG6 autoantibodies and a sustained normalization in those who continued to follow the diet [108], suggesting that a GFD can be used to contain symptoms of gluten ataxia. While the GFD has proven its efficacy in treating gluten ataxia and CD, much of its perceived benefits towards other health problems remain questionable.

4.2. Cognitive Impairment and Neurological and Mental Illnesses

Recent studies have shown that there may be a correlation between gluten sensitivity and neurological diseases. Since TG6 autoantibodies are known to attack the nervous system as an immune-mediated reaction to gluten ingestion, a link may exist between this mechanism and other neurological illnesses beyond gluten ataxia. A study conducted by Hadjivassiliou and colleagues [109] analyzed the serum levels of antigliadin antibodies in 147 neurological patients, of which 53 (25 ataxia, 20 peripheral neuropathy, 5 mononeuritis multiplex, 4 myopathy, 3 motor neuropathy, 2 myelopathy) had no known cause for their diagnosis despite full investigation. They were compared alongside a second group of 94 patients that had known causes for their diagnosis. Finally, 50 healthy blood donors were used as the third group. Results demonstrated that the first group had significantly higher positive serum anti-gliadin antibodies than the other groups (57%, 5%, and 12%, respectively). These data establish a strong correlation between gluten sensitivity and neurological illnesses. Finally, another neurological illness of concern as it pertains to the GFD is autism. Autism diagnosis has started to increase, with a diagnosis of 1 in 88 children [110]. Patients with autism have a higher prevalence of IgG antibodies to gliadin, the same antibodies associated with CD and gluten ataxia [111]. Since many children with autism have gastrointestinal symptoms, there seems to be a link between autism and gluten sensitivity. In a study conducted by Ghalichi and colleagues [112], 80 children with autism spectrum disorders (ASD) received either a GFD treatment (n = 40) or a regular diet treatment (n = 40); 53.9% of the children reported having gastrointestinal abnormalities. The ROME III questionnaire for evaluating gastrointestinal symptoms and Gilliam Autism Rating Scale 2 questionnaire (GARS-2) for assessing psychometric properties were used to evaluate the effects of the GFD versus the regular diet. Results demonstrated that children placed on the GFD experienced a significant decrease in both gastrointestinal symptoms (40.57% vs. 17.10%, $p < 0.05$) and behavioral disorders (80.03 $\pm$ 14.07 vs. 75.82 $\pm$ 15.37, $p < 0.05$). The children placed on the regular diet experienced an insignificant increase in both metrics. The research, however, is somewhat conflicting on this topic. Piwowarczyk and colleagues demonstrated that a GFD did not influence autistic symptoms, maladaptive behaviors, or intellectual abilities [113]. The relief in gastrointestinal symptoms in children with autism placed on a GFD is in accord with most of the literature, given that patients with elevated levels of IgG antibodies to gliadin tend to experience similar effects. However, the influence of the GFD on autistic symptoms and intellectual abilities is not well established.

Some evidence has emerged on the potential benefits of the gluten-free diet for depressive disorders, although the studies on this topic are scarce, and further investigations may be needed. Peters and colleagues [114] conducted a study that tested 22 patients with irritable bowel syndrome who had a negative CD diagnosis. The authors utilized a double-blind cross-over method which consisted of 3 days of one of 3 dietary challenges (diet supplemented with gluten, whey, and no supplement (placebo)) followed by a 3-day washout period before crossing-over. The mental state was assessed using the Spielberger State-Trait Personality Inventory (STPI), and results demonstrated that depression scores in the gluten group were higher than the placebo group ($M = 2.03$, 95% CI (0.55–3.51), $p = 0.010$). The whey ingestion group did not show significant differences in depression rates, cortisol secretion, or gastrointestinal symptoms. These results prompted the conclusion that a correlation could exist between depressive disorders and gluten ingestion. Another study conducted by Zylberberg and colleagues [115] found similar results in people who avoided gluten. Data from 22,274 participants of the 2009–2014 National Health and Nutrition Examination Survey compared depression, insomnia, quality of life variables, and psychotropic medication use in CD patients and people who avoid gluten to controls. The results obtained showed no increased odds of depression or sleep difficulty among CD patients. People who avoid gluten, however, had lower odds of depression compared to control after adjustments. The study calls for further investigation into the correlation between gluten exposure and depression. Since people who avoid gluten do so out of their conviction, they could be more health-conscious than the CD patients and control group without any formal diagnosis. Given that physical health is closely associated with mental health, there could be some confounding effects [116,117]. Moreover, schizophrenia is a particular mental health disease of interest when discussing the gluten-free diet. Some studies have shown that schizophrenic patients tend to have elevated anti-gliadin antibodies and transglutaminase 6 antibodies [115] despite not having a CD diagnosis. A review of articles conducted by Ergün, Urhan, and Ayer [118] found that symptoms of schizophrenia improved following the elimination of gluten from the diet. Another systematic review, conducted by Levinta and colleagues [119], searched different databases and found 9 studies relevant to gluten and schizophrenia; 6 of the studies demonstrated beneficial effects, namely decreased severity in symptoms and improved functioning. However, they found that only one of the studies was a randomized controlled trial, while seven were cross-over studies and one was an open-label pilot study. For this reason, the conclusions of the systematic review are limited. Nonetheless, there seems to be a connection between the consumption of gluten and schizophrenic disorders.

4.3. Inflammatory Bowel Disease and Irritable Bowel Syndrome

The GFD has also been utilized as a potential treatment for irritable bowel syndrome (IBS). Diarrhea-dominant irritable bowel syndrome (d-IBS) patients tend to experience symptom relief following the introduction of a gluten-free diet. In a study conducted by Wahnschaffe and colleagues, 60% of d-IBS patients positive for human leukocyte antigen (HLA)-DQ2 T-cell haplotypes, and CD-associated serum IgG had improved stool frequency. Moreover, gastrointestinal symptom scores returned to normal after 6 months of a gluten-free diet compared to 12% negative for these biomarkers [120]. While the patients with d-IBS were positive for CD biomarkers, these antibodies were not always collected. Therefore, the patients would be classified as having non-celiac gluten sensitivity. Another study conducted by Aziz et al. analyzed the effect of a 6-week gluten-free diet on patients with d-IBS (20 HLA-DQ2/8-positive and 21 HLA-DQ2/8-negative). Twenty-nine patients (71%) reported having their symptoms relieved following the completion of the trial [121]. These two studies demonstrate the potential benefits of the gluten-free diet for patients with d-IBS. Patients with inflammatory bowel disease (IBD) also appear to benefit from a GFD. Patients with CD are also more likely to have an IBD diagnosis than the general population [122]. Herfarth and colleagues [123] conducted a study analyzing the effects of a GFD on 1647 patients with IBD. CD and non-celiac gluten sensitivity were reported by 10 (0.6%) and 81 (4.9%) respondents, respectively; 314 participants reported having previously tried a GFD, and 135 reported current use of GFD (19.1% and 8.2%, respectively). Overall, 65.6% of all patients who attempted a GFD described improving their gastrointestinal symptoms, and 38.3% reported fewer or less severe IBD flares. Patients who were strict in GFD adherence also reported less fatigue. In addition, Lindberg and colleagues [124] compared the levels of IgG, IgA, and IgM antibodies to baker's yeast (saccharomyces cerevisiae), yeast mannan, gliadin, ovalbumin, and beta-lactoglobulin in twins with IBD versus those of healthy controls. Results demonstrated that the twins with ulcerative colitis had elevated IgA antibodies to gliadin levels than the other twins and healthy controls. For these reasons, the GFD may be an effective symptom managing diet in patients with ulcerative colitis IBD.

4.4. Dermatitis Herpetiformis

Treatment of patients with dermatitis herpetiformis (DH) with a GFD has been demonstrated to be highly effective [125,126]. In a study conducted by Reunala and colleagues [125], 81 patients with DH were treated with a GFD and a standard diet (control); 93% of patients placed on a GFD were able to reduce their dosage of dapsone, an antibiotic used in the treatment of DH, versus 16% in the control group. In addition, 28% of the GFD group were able to eliminate the antibiotic without experiencing any symptom aggravation. Another study conducted by Lionel et al. [126] demonstrated similar results. Twenty-four patients with DH were treated with a GFD and 16 (80%) were able to reduce their dapsone usage. Ten of the patients were able to eliminate the antibiotic and were free of any skin lesions. These two studies provide satisfactory evidence demonstrating the efficacy of a GFD in the treatment of DH.

4.5. Non-Celiac Gluten Sensitivity (NCGS) and People Who Avoid Gluten

While the benefits of the GFD in treating CD and gluten ataxia are established in the literature, many studies have sought to investigate whether the diet is viable in treating other conditions. A biopsy is generally needed to diagnose a patient with CD, requiring a gluten-free diet for treatment. In recent times, however, patients who were excluded from a CD diagnosis but had IBS-like symptoms when exposed to gluten have been put under the non-celiac gluten sensitivity (NCGS) umbrella. Patients with NCGS tend to have normal small intestinal permeability and will experience IBS-like symptoms such as bloating, stomach pain, fatigue, rash, and discomfort upon consuming gluten. The scientific literature is not always clear when establishing a diagnosis for this condition as the overlap with irritable bowel syndrome is strong. Patients with NCGS do not express CD-related antibodies and are generally harder to diagnose as they do not have well-defined biomarkers. Still, there is strong evidence that supports the existence of this condition [127–129]. Theories have proposed that patients with NCGS may be sensitive to another component of wheat besides gluten, namely the amylase-trypsin inhibitors, which trigger a similar immune response as gluten [130]. Wheat germ agglutinin is another plant protein found in wheat that has been shown to trigger similar immune responses [131]. Thus, nonceliac gluten sensitivity patients may be more sensitive to wheat in general instead of specifically gluten, and the term non-celiac wheat sensitivity may describe the condition better. To complicate matters, a study conducted by Skodje and colleagues [132] found that NCGS patients experienced worsened symptoms following consumption of fructan but not gluten. Fructan is oligo, di, and monosaccharides that are often found in foods that also contain gluten. The double-blind cross-over challenge found that 59 self-diagnosed NCGS individuals following a gluten-free diet experienced more symptoms based on the Gastrointestinal Symptom Rating Scale Irritable Bowel Syndrome (GSRS-IBS) following ingestion of fructan than following ingestion of gluten. No significant differences were found between gluten and placebo or fructan and placebo. While NCGS patients may not be specifically sensitive to gluten, they could be sensitive to other factors that are generally found alongside gluten. Impairments in cognitive health have been observed in some patients with gluten sensitivity before treatment. Brain fog, which is in the spectrum of non-celiac gluten sensitivity (NCGS), refers to problems involving memory, attention, executive function, and cognitive processing speed. Patients with NCGS often report this condition, and a gluten-free diet has been observed to improve some of these symptoms after one year of adherence [133].

People who avoid gluten (PWAG) are a broader term that describes GFD adherents excluded from CD and non-celiac gluten sensitivity through rechallenge tests. PWAG make up the largest demographic of gluten adherents. PWAG generally tend to do so due to its perceived benefits. However, as mentioned below, the gluten-free diet does not come without its adverse outcomes [50]. Therefore, it is important to educate individuals who adhere to the GFD without any diagnosis about the potential risks, given that these individuals do not immediately require the diet.

5. Adverse Events of GFD

While the benefits of a GFD seem alluring, it is important to consider the risks associated with the regiment. Much of the studies conducted on its health complications appear inconclusive and even conflicting. One of the main concerns of the GFD is the lack of beneficiary whole grains consumed by adherents, which can be a factor in coronary heart disease [134–137]. Assessing this hypothesis, Lebwohl et al. [138] studied the development of coronary heart disease in 64,714 women in the Nurses' Health Study and 45,303 men in the Health Professionals Follow-up Study. Food diaries were updated every 4 years from 1986 through 2010 and were used to assess the amount of gluten consumed. Results demonstrated an inverse relation between gluten intake and coronary heart disease risks. On the other hand, a systematic review conducted by Potter and colleagues [139] analyzed 27 articles on patients who adopted the GFD. Findings included increases in high-density lipoproteins, fasting glycemia, total cholesterol, and body mass index, although the increases in metrics were within a healthy range. The review did not find any increase in triglycerides, low-density lipoprotein, or blood pressure, prompting the conclusion that the GFD is not associated with coronary heart disease. Of note, only one of the articles had a control group and was limited by several confounders, so proper analysis is limited. Another analysis, conducted by Heikkilä and colleagues [140], found some support for the association of coronary heart disease with the GFD; however, they state that the evidence base was weak and had limitations. Finally, Kim and colleagues [141] demonstrated that the GFD was beneficial in waist circumference reduction and lowered BMI while maintaining that the diet was not associated with elevated cardiovascular disease risks. GFD followers, who were primarily women and were health-conscious, were found to have lower metabolic syndrome and lower cardiovascular disease risks, although the difference was not statistically significant. Overall, most studies have called for more research to examine this hypothesis, as no conclusive findings have been made. Many studies lean towards excluding the GFD as a factor in cardiovascular disease risk. While the literature seems inconclusive regarding the GFD and coronary heart disease, other adversities associated with the regiment are clearer. Recent evidence suggests that the diet may worsen the gut microbiota while having nutritional deficiencies in iron, calcium, and fiber [142–145]. The diet is also associated with a high cost due to the further processing required for gluten-free alternatives [146]. Finally, some research has raised concerns about the negative social and psychological impacts that many GFD adherents experience, mainly due to the diet's restrictive nature [147,148].

5.1. Gluten and the Gut Microbiome

The importance of healthy gut microbiota in maintaining good health is becoming increasingly evident in the literature. The human gut contains two main phyla of bacteria, Bacteroidetes and Firmicutes. The role of these bacteria is highly diverse and includes the metabolism of nutrients consumed by the host, xenobiotic and drug metabolism, maintenance of structural integrity of the gut mucosal barrier, protection against pathogens, and immunomodulation [149]. Diet can also affect the health of gut flora, along with other factors such as birthing method (vaginal canal vs. cesarean) [150] and use of antibiotics [151]. In their study, David and colleagues [152] demonstrated that changes in diet rapidly influence the composition of the gut flora. Thus, it is important to consider the effects of gluten consumption and restriction on the health of the bacteria found in the host's gut. Golfetto and colleagues [153] conducted a study on 42 healthy subjects and 14 patients with CD to analyze the health of their gut bacteria. The study found that patients with CD had an imbalance in their intestinal microbiota despite being on a gluten-free diet. It is unknown whether the patients with CD had this imbalance before adhering to a GFD or whether they developed it later. Regardless, the diet does seem to cause this imbalance to persist. It is important to address these imbalances as they can cause the gastrointestinal symptoms patients may experience when consuming gluten [154]. In another study conducted by Palma and colleagues [155], 10 healthy subjects were introduced to a GFD, and their gut microbiota was monitored for a month. Results demonstrated a reduction in beneficial gut bacteria, raising concerns over the potential risks of the GFD. If gluten exclusion from the diet results in imbalances and a reduction in healthy gut flora, it is important to address those issues by providing support. Probiotic supplements are of particular interest as they can balance the gut flora and provide it with the nutrients it needs to remain healthy [145,156,157].

5.2. Nutritional Deficiencies

Concerns have been raised about the nutritional quality of GFD. As the diet has gained popularity from media coverage and celebrity promotion, many people have adopted the regiment despite having no diagnosed CD. For these individuals, gluten avoidance may cause nutritional deficiencies which could otherwise be prevented. For example, abnormal intake of vitamin D has been linked to the GFD. In a study conducted by Deora and colleagues [142], the medical records of 140 children with CD were assessed, and 70% of these children had vitamin D deficiency at the time of diagnosis. After 6 months of GFD adherence, these children found a slight improvement in their vitamin D uptake, although levels remained abnormal. Given that vitamin D is crucial to intestinal uptake of minerals, it is important to address this issue through supplementation and dietary adjustments. The diet also presents other deficiency concerns beyond vitamin D. In 2005, a survey conducted by Thompson and colleagues [143] in patients with CD found that women had a mean average intake of 46%, 44%, and 31% of their daily fiber, iron, and calcium intake requirements. In men, the values were 88%, 100%, and 63%, respectively. These results demonstrated that women who adhered to the GFD might be at risk of nutritional deficiencies, even more so than men. In addition, a systematic review conducted by Di

Nardo et al. [158] found that all children, regardless of whether they were diagnosed with CD or not, were at risk of nutritional deficiencies (insufficient fiber, iron, vitamin D, and calcium). Moreover, children with CD following a GFD had inadequate folate, magnesium, and zinc consumption, and higher consumption of high glycemic index foods. The paper suggested the need for therapeutic protocols to include education about these deficiencies so patients can ensure their diet is complete. Another disadvantage of the GFD is the potentially elevated level of lipid and protein consumption. Mariani and colleagues [159], in a survey analyzing the 3-day alimentary intake of 47 adolescents with CD, found that strict adherents to the GFD had increased intakes in protein and lipid, as well as a more significant prevalence of obesity (72% vs. 47% in control). These results are expected, as gluten tends to occur in carbohydrate-rich foods naturally and not protein- or lipid-rich foods. It is important to note that the quality of the lipids and proteins should be of concern and not necessarily the amount consumed. An analysis of gluten-free biscuits by Caponio et al. [160] found that they contained a sizeable mean amount of low-quality oleic trans-isomer fats (9.39%). Much of the literature suggests mitigating this negative side of the GFD by consuming more naturally gluten-free products and avoiding processed gluten-free alternatives as they do not seem to provide many nutritional benefits. It is important to note that many of the studies conducted on the deficiencies of the GFD have studied CD patients who suffer from gut inflammation and lack proper nutrient uptake. This may be a confounding factor as the results pertain to individuals affected by the disease and may not apply to those without CD. With that in mind, the GFD seems to have some nutritional disadvantages, namely deficiencies in vitamin D, iron, calcium, folate, and dietary fibers, and a higher amount of low-quality lipids found in some gluten-free alternatives [142–144]. Whether afflicted with CD or not, adherents of the diet should ensure that they reach daily recommended requirements for all minerals listed above. Avoiding processed gluten-free alternatives and eating naturally-occurring gluten-free foods high in iron, such as meats, fish, and green vegetables, is a recommended solution to this dietary problem associated with the diet [158].

5.3. Cost

Cost is another challenge associated with the GFD. Most products that naturally contain gluten, such as pasta and bread, require little to no processing to produce. Gluten-containing foods have been around for thousands of years and are found in many popular recipes. Bread, for example, is a staple in many dishes and diets across the world. Grains generally tend to be cheap to produce and grow in a wide range of climates, making them ideal for consumption. As these tend to contain gluten naturally, further processing is required to remove the protein while maintaining palatability. Significant price disparities are found across most gluten-free alternatives of gluten-containing foods due to this further processing requirement. In a study conducted by Missbach and colleagues [146], 63 gluten-free products and 126 of their gluten-containing counterparts were analyzed in 12 different Austrian supermarkets. The products included a broad range of items: bread, cereals, baking mixes, pasta, cookies, cakes, and snacks. Results showed that on average, gluten-free foods were 205% (cereals) to 267% (bread and bakery goods) more expensive than their gluten-containing counterparts. Whether this large price gap is because of overpricing due to high demand or processing costs is unclear. A 2-fold price gap between the two counterparts creates a tremendous burden on strict followers and may have detrimental financial effects. Other studies have confirmed this significant price difference as well [71,161]. Another study, conducted by Singh and Whelan [120], found that gluten-free products were more expensive (wheat-based products were 76–518% more expensive) and had limited availability in stores. Regular supermarkets had almost all the gluten-free alternative products (18/20, 90%); however, corner stores and budget supermarkets had limited gluten-free alternative products (1.8/20, 9%). Limited availability in convenience stores can further increase the cost of adherence, perhaps due to the time spent traveling to a regular supermarket that may be further away. One solution to circumvent this problem, provided by Di Nardo and colleagues [158], is to build the diet around naturally-occurring gluten-free foods and avoid the processed gluten-free alternatives altogether. This strategy can mitigate the price difference between the two counterparts and increase the number of stores one can buy from.

5.4. Social and Psychological Impact

Strict adherence to a gluten-free diet has been shown to cause some social and psychological adversities. Food is deeply embedded in cultures worldwide and can be found at the center of many social constructs. People gather and enjoy different foods to celebrate career accomplishments, weddings, religious rituals, and birthdays. Given that food exerts a significant influence on daily life, strict restrictions on dietary options can be a source of social isolation and unhappiness. In a study conducted by Zarkadas and colleagues [147], questionnaires were sent to members of the Canadian Celiac Association (5240 members) and 2681 adults (aged 16 or older) who had biopsy-proven CD. The questionnaire aimed to assess the recipients' quality of life based on celiac-associated questions and the "SF-12," a self-reported outcome measure assessing the impact of health on an individual's everyday life. It was discovered that 44% of respondents reported having difficulties following the diet for various reasons, including determining if foods were gluten-free (85%), finding gluten-free foods in stores (83%), avoiding restaurants (79%), and avoiding travel (38%). However, due to the rising popularity of the gluten-free diet, many restaurants now include labels on the menu identifying any gluten-free items. Another study, conducted by Silvester et al. [148], further demonstrated the social isolation associated with the diet. The study found that non-CD responders to the questionnaire were less likely to adhere to the diet strictly and would sometimes ingest gluten intentionally. This group was associated with more pleasure and less anger and depression than CD responders who were stricter in adherence. The study also found that social isolation was more pronounced in CD responders, and eating was mainly at home instead of in public spaces. These results further demonstrate the challenges with adhering to the diet at the psychological and sociological levels. MacCulloch and Rashid [161] conducted a survey and found that improved labeling, government support through income, and education for schools and restaurants greatly help adherents of the diet. The social frustrations associated with a GFD can also be seen in type 1 diabetes, another autoimmune disease requiring restrictive dieting [162–164]. Patients with CD have shown a higher prevalence of type 1 diabetes mellitus than the general population (4.4–11.1% versus 0.5%), and Camarca and colleagues [165] found that 50% of patients with CD and type 1 diabetes comply with the GFD compared to the higher rate of 73% in patients with only CD. In adolescents, significantly, strict compliance has been associated with a worsened quality of life. Although not recognized by the Diagnostic and Statistical Manual of Mental Disorders, a phenomenon involving restrictive eating called orthorexia nervosa represents another cause for concern in adherents of the GFD. Orthorexia nervosa describes the behavior of healthy individuals who pursue increasingly restrictive diets despite not needing to do so (patients have a healthy weight and no diagnosed condition), and can experience a decrease in quality of life and overall health [166]. A study conducted by Wolf and colleagues [167] found that highly vigilant GFD adherents had lower quality of life due to anxiety, putting them at risk of orthorexia nervosa as they vigorously pursue their gluten-free lifestyle. It is crucial to address these socio-psychological issues as they tend to be harder to quantify. Close follow-up of quality of life, level of adherence with a GFD, and patient education on possible risks in CD patients following the diet is essential.

6. Conclusions

The GFD remains the primary treatment for celiac disease and may work in other health conditions. Patients with celiac disease must adhere to a lifelong GFD as it is currently the best-known treatment. Treatment of patients with celiac disease should be done at an early age, as younger individuals tend to show more significant reversal of gastrointestinal symptoms and healing from damage to the gut mucosa. While the diet is recognized in treating gluten ataxia, little is known about its other benefits. Patients with d-IBS and IBD experience relief in gastrointestinal symptoms following treatment with a GFD. Patients with NCSG experience similar improvements following the diet. Maintaining a strict gluten-free lifestyle has many challenges, including nutritional deficiencies, high costs due to adherence, and social and psychological barriers. These issues should be addressed when recommending the diet for any individual. More research is required to assess the benefits of the diet in treating mental, neurological, and cognitive diseases (depressive disorders, autism spectrum disorder, and "brain fog", respectively). Large sample size studies can significantly help the current effort to assess the diet's risks and benefits, which is needed to educate individuals who follow the diet without any diagnosis. This cohort of people makes up the most prominent GFD adherents who usually follow the diet because of the reported benefits. Studies that provide strong evidence are needed in order to aid individuals in making well-educated decisions on whether to follow the diet.

Author Contributions: All authors: data curation, B.A. and A.Z.; writing—original draft preparation, W.E.-M.; writing—review and editing, W.E.-M.; visualization, W.E.-M.; supervision, W.E.-M.; project administration. All authors have read and agreed to the published version of the manuscript.

Funding: This review received no external funding.

Institutional Review Board Statement: Not applicable.

Informed Consent Statement: Not applicable.

Data Availability Statement: Not applicable.

Conflicts of Interest: The authors declare no conflict of interests.

References

1. Shiferaw, B.; Smale, M.; Braun, H.-J.; Duveiller, E.; Reynolds, M.; Muricho, G. Crops that feed the world 10. Past successes and future challenges to the role played by wheat in global food security. *Food Secur* **2013**, *5*, 291–317. [CrossRef]
2. Shurtleff, W.; Huang, H.T.; Aoyagi, A. *History of Soybeans and Soyfoods in China and Taiwan, and in Chinese Cookbooks, Restaurants, and Chinese Work with Soyfoods outside China (1024 BCE TO 2014)*; Soyinfo Center: Lafayette, CA, USA, 2014.
3. Nature or Nurture? Explaining English Wheat Yields in the Industrial Revolution, c.1770 on JSTOR. Available online: https://www.jstor.org/stable/3874947 (accessed on 26 October 2021).
4. Atack, J.; Margo, R.A. The Impact of Access to Rail Transportation on Agricultural Improvement: The American Midwest as a Test Case, 1850–1860. *J. Transp. Land Use* **2011**, *4*, 5–18. [CrossRef]
5. Fitzgerald, D. World War II and the Quest for Time-Insensitive Foods. *Osiris* **2020**, *35*, 291–309. [CrossRef]
6. Theien, I. Food rationing during World War two: A special case of sustainable consumption? *Anthr. Food* **2009**. [CrossRef]
7. Kearney, J. Food consumption trends and drivers. *Philos. Trans. R. Soc. B Biol. Sci.* **2010**, *365*, 2793–2807. [CrossRef]
8. Wen, S.; Wen, N.; Pang, J.; Langen, G.; Brew-Appiah, R.; Mejías, J.; Osorio, C.E.; Yang, M.; Gemini, R.; Moehs, C.; et al. Structural genes of wheat and barley 5-methylcytosine DNA glycosylases and their potential applications for human health. *Proc. Natl. Acad. Sci. USA* **2012**, *109*, 20543–20548. [CrossRef]
9. Shewry, P. What Is Gluten—Why Is It Special? *Front. Nutr.* **2019**, *6*, 101. [CrossRef]
10. Schalk, K.; Lexhaller, B.; Koehler, P.; Scherf, K.A. Isolation and characterization of gluten protein types from wheat, rye, barley and oats for use as reference materials. *PLoS ONE* **2017**, *12*, e0172819. [CrossRef] [PubMed]
11. Balakireva, A.V.; Zamyatnin, A.A. Properties of Gluten Intolerance: Gluten Structure, Evolution, Pathogenicity and Detoxification Capabilities. *Nutrients* **2016**, *8*, 644. [CrossRef] [PubMed]
12. Li, H.; Wang, J.; Pan, L.; Lu, Q. Effect of amino and thiol groups of wheat gluten on the quality characteristics of Chinese noodles. *J. Food Sci. Technol.* **2019**, *56*, 2825–2835. [CrossRef]
13. Moreno, M.D.L.; Rodríguez-Herrera, A.; Sousa, C.; Comino, I. Biomarkers to Monitor Gluten-Free Diet Compliance in Celiac Patients. *Nutrients* **2017**, *9*, 46. [CrossRef]
14. Gujral, N. Celiac disease: Prevalence, diagnosis, pathogenesis and treatment. *World J. Gastroenterol.* **2012**, *18*, 6036–6059. [CrossRef]
15. Caio, G.; Volta, U.; Sapone, A.; Leffler, D.A.; de Giorgio, R.; Catassi, C.; Fasano, A. Celiac disease: A comprehensive current review. *BMC Med.* **2019**, *17*, 1–20. [CrossRef]
16. Stazi, A.V.; Trecca, A.; Trinti, B. Osteoporosis in celiac disease and in endocrine and reproductive disorders. *World J. Gastroenterol.* **2008**, *14*, 498–505. [CrossRef]
17. Ndez-Bañares, H.M.; Monzón, H.; Forné, M. A short review of malabsorption and anemia. *World J. Gastroenterol.* **2009**, *15*, 4644–4652. [CrossRef] [PubMed]

18. Rondanelli, M.; Faliva, M.A.; Gasparri, C.; Peroni, G.; Naso, M.; Picciotto, G.; Riva, A.; Nichetti, M.; Infantino, V.; Alalwan, T.; et al. Micronutrients Dietary Supplementation Advices for Celiac Patients on Long-Term Gluten-Free Diet with Good Compliance: A Review. *Medicina* **2019**, *55*, 337. [CrossRef]
19. Kreutz, J.M.; Adriaanse, M.P.M.; Van Der Ploeg, E.M.C.; Vreugdenhil, A.C.E. Narrative Review: Nutrient Deficiencies in Adults and Children with Treated and Untreated Celiac Disease. *Nutrients* **2020**, *12*, 500. [CrossRef] [PubMed]
20. Barker, J.M.; Liu, E. Celiac Disease: Pathophysiology, Clinical Manifestations, and Associated Autoimmune Conditions. *Adv. Pediatr.* **2008**, *55*, 349–365. [CrossRef] [PubMed]
21. Balaban, D.V.; Dima, A.; Jurcut, C.; Popp, A.; Jinga, M. Celiac crisis, a rare occurrence in adult celiac disease: A systematic review. *World J. Clin. Cases* **2019**, *7*, 311–319. [CrossRef]
22. De Pablo, P.; Cooper, M.S.; Buckley, C.D. Association between bone mineral density and C-reactive protein in a large population-based sample. *Arthritis Rheum.* **2012**, *64*, 2624–2631. [CrossRef]
23. Hardy, R.; Cooper, M.S. Bone loss in inflammatory disorders. *J. Endocrinol.* **2009**, *201*, 309–320. [CrossRef] [PubMed]
24. Khundmiri, S.J.; Murray, R.D.; Lederer, E. PTH and Vitamin D. *Compr. Physiol.* **2016**, *6*, 561–601. [CrossRef] [PubMed]
25. Flynn, A. The role of dietary calcium in bone health. *Proc. Nutr. Soc.* **2003**, *62*, 851–858. [CrossRef]
26. Van Rijn, J.C.W.; Grote, F.K.; Oostdijk, W.; Wit, J.M. Short stature and the probability of coeliac disease, in the absence of gastrointestinal symptoms. *Arch. Dis. Child.* **2004**, *89*, 882–883. [CrossRef] [PubMed]
27. Garganta, M.D.; Bremer, A.A. Clinical Dilemmas in Evaluating the Short Child. *Pediatr. Ann.* **2014**, *43*, 321–327. [CrossRef] [PubMed]
28. Ludvigsson, J.F.; Michaelsson, K.; Ekbom, A.; Montgomery, S.M. Coeliac disease and the risk of fractures—A general population-based cohort study. *Aliment. Pharmacol. Ther.* **2006**, *25*, 273–285. [CrossRef]
29. Melton, L.J.; Beck, T.J.; Amin, S.; Khosla, S.; Achenbach, S.J.; Oberg, A.L.; Riggs, B.L. Contributions of bone density and structure to fracture risk assessment in men and women. *Osteoporos. Int.* **2005**, *16*, 460–467. [CrossRef] [PubMed]
30. Kemppainen, T.; Kröger, H.; Janatuinen, E.; Arnala, I.; Kosma, V.-M.; Pikkarainen, P.; Julkunen, R.; Jurvelin, J.; Alhava, E.; Uusitupa, M. Osteoporosis in adult patients with celiac disease. *Bone* **1999**, *24*, 249–255. [CrossRef]
31. Sollid, L.M.; McAdam, S.N.; Molberg, Á.; Quarsten, H.; Arentz-Hansen, H.; Louka, A.S.; Lundin, K.E.A. Genes and envi-ronment in celiac disease. *Acta Odontol. Scand.* **2001**, *59*, 183–186. [CrossRef]
32. Lionetti, E.; Catassi, C. The Role of Environmental Factors in the Development of Celiac Disease: What Is New? *Diseases* **2015**, *3*, 282. [CrossRef]
33. Tian, N.; Leffler, D.A.; Kelly, C.P.; Hansen, J.; Marietta, E.V.; Murray, J.A.; Schuppan, D.; Helmerhorst, E.J. Despite sequence homologies to gluten, salivary proline-rich proteins do not elicit immune responses central to the pathogenesis of celiac disease. *Am. J. Physiol. Liver Physiol.* **2015**, *309*, G910–G917. [CrossRef] [PubMed]
34. Fernández-Pérez, S.; Pérez-Andrés, J.; Gutiérrez, S.; Navasa, N.; Martínez-Blanco, H.; Ferrero, M.; Vivas, S.; Vaquero, L.; Iglesias, C.; Casqueiro, J.; et al. The Human Digestive Tract Is Capable of Degrading Gluten from Birth. *Int. J. Mol. Sci.* **2020**, *21*, 7696. [CrossRef] [PubMed]
35. Camarca, A.; Anderson, R.P.; Mamone, G.; Fierro, O.; Facchiano, A.; Costantini, S.; Zanzi, D.; Sidney, J.; Auricchio, S.; Sette, A.; et al. Intestinal T Cell Responses to Gluten Peptides Are Largely Heterogeneous: Implications for a Peptide-Based Therapy in Celiac Disease. *J. Immunol.* **2009**, *182*, 4158–4166. [CrossRef] [PubMed]
36. Drago, S.; El Asmar, R.; Di Pierro, M.; Clemente, M.G.; Sapone, A.T.A.; Thakar, M.; Iacono, G.; Carroccio, A.; D'Agate, C.; Not, T.; et al. Gliadin, zonulin and gut permeability: Effects on celiac and non-celiac intestinal mucosa and intestinal cell lines. *Scand. J. Gastroenterol.* **2006**, *41*, 408–419. [CrossRef] [PubMed]
37. Clemente, M.G.; De Virgiliis, S.; Kang, J.S.; Macatagney, R.; Musu, M.P.; Di Pierro, M.R.; Drago, S.; Congia, M.; Fasano, A. Early effects of gliadin on enterocyte intracellular signalling involved in intestinal barrier function. *Gut* **2003**, *52*, 218–223. [CrossRef]
38. Fasano, A. Zonulin, regulation of tight junctions, and autoimmune diseases. *Ann. N. Y. Acad. Sci.* **2012**, *1258*, 25–33. [CrossRef] [PubMed]
39. Ráki, M.; Schjetne, K.W.; Stamnaes, J.; Molberg, Ø.; Jahnsen, F.L.; Issekutz, T.B.; Bogen, B.; Sollid, L.M. Surface Expression of Transglutaminase 2 by Dendritic Cells and its Potential Role for Uptake and Presentation of Gluten Peptides to T Cells. *Scand. J. Immunol.* **2007**, *65*, 213–220. [CrossRef]
40. Cecilio, L.A.; Bonatto, M.W. The Prevalence Of Hla Dq2 And Dq8 In Patients With Celiac Disease, In Family and in General Population. *Arq. Bras. Cir. Dig.* **2015**, *28*, 183–185. [CrossRef]
41. Mazzarella, G. Effector and suppressor T cells in celiac disease. *World J. Gastroenterol.* **2015**, *21*, 7349–7356. [CrossRef]
42. La Scaleia, R.; Barba, M.; Di Nardo, G.; Bonamico, M.; Oliva, S.; Nenna, R.; Valitutti, F.; Mennini, M.; Barbato, M.; Montuori, M.; et al. Size and dynamics of mucosal and peripheral IL-17A+ T-cell pools in pediatric age, and their disturbance in celiac disease. *Mucosal Immunol.* **2012**, *5*, 513–523. [CrossRef]
43. Parzanese, I.; Qehajaj, D.; Patrinicola, F.; Aralica, M.; Chiriva-Internati, M.; Stifter, S.; Elli, L.; Grizzi, F. Celiac disease: From pathophysiology to treatment. *World J. Gastrointest. Pathophysiol.* **2017**, *8*, 27–38. [CrossRef] [PubMed]
44. Ludvigsson, J.F.; Bai, J.C.; Biagi, F.; Card, T.; Ciacci, C.; Ciclitira, P.J.; Green, P.H.R.; Hadjivassiliou, M.; Holdoway, A.; van Heel, D.; et al. Diagnosis and management of adult coeliac disease: Guidelines from the British Society of Gastroenterology. *Gut* **2014**, *63*, 1210–1228. [CrossRef]
45. Corazza, G.R. Coeliac disease. *J. Clin. Pathol.* **2005**, *58*, 573–574. [CrossRef] [PubMed]

46. Bañares, F.F.; Mariné, M.; Rosinach, M.; Carrasco, A.; Esteve, M.; Rodrigo, L. Type 1 Marsh Celiac Disease: Diagnosis and Response. *OmniaScience Monogr.* **2014**, *5*, 289–302. [CrossRef]
47. Kamboj, A.K.; Oxentenko, A.S. Clinical and Histologic Mimickers of Celiac Disease. *Clin. Transl. Gastroenterol.* **2017**, *8*, e114. [CrossRef] [PubMed]
48. Itzlinger, A.; Branchi, F.; Elli, L.; Schumann, M. Gluten-Free Diet in Celiac Disease—Forever and for All? *Nutrients* **2018**, *10*, 1796. [CrossRef]
49. Green, P.H.; Fleischauer, A.T.; Bhagat, G.; Goyal, R.; Jabri, B.; Neugut, A.I. Risk of malignancy in patients with celiac disease. *Am. J. Med.* **2003**, *115*, 191–195. [CrossRef]
50. Niland, B.; Cash, B.D. Health Benefits and Adverse Effects of a Gluten-Free Diet in Non–Celiac Disease Patients. *Gastroenterol. Hepatol.* **2018**, *14*, 82–91.
51. Valitutti, F.; Iorfida, D.; Anania, C.; Trovato, C.M.; Montuori, M.; Cucchiara, S.; Catassi, C. Cereal Consumption among Subjects with Celiac Disease: A Snapshot for Nutritional Considerations. *Nutrients* **2017**, *9*, 396. [CrossRef]
52. Jones, A.L. The Gluten-Free Diet: Fad or Necessity? *Diabetes Spectr.* **2017**, *30*, 118–123. [CrossRef]
53. Wahab, P.J.; Meijer, J.W.; Mulder, C.J. Histologic Follow-up of People With Celiac Disease on a Gluten-Free Diet. *Am. J. Clin. Pathol.* **2002**, *118*, 459–463. [CrossRef] [PubMed]
54. Rubio-Tapia, A.; Rahim, M.W.; See, J.A.; Lahr, B.D.; Wu, T.-T.; Murray, J.A. Mucosal Recovery and Mortality in Adults With Celiac Disease After Treatment With a Gluten-Free Diet. *Am. J. Gastroenterol.* **2010**, *105*, 1412–1420. [CrossRef]
55. Tursi, A.; Brandimarte, G.; Giorgetti, G.; Elisei, W.; Inchingolo, C.; Monardo, E.; Aiello, F. Endoscopic and histological findings in the duodenum of adults with celiac disease before and after changing to a gluten-free diet: A 2-year prospective study. *Endoscopy* **2006**, *38*, 702–707. [CrossRef] [PubMed]
56. Kavak, U.S.; Yüce, A.; Koçak, N.; Demir, H.; Saltik, I.N.; Gürakan, F.; Özen, H. Bone Mineral Density in Children With Untreated and Treated Celiac Disease. *J. Pediatr. Gastroenterol. Nutr.* **2003**, *37*, 434–436. [CrossRef]
57. Nicoletta, M.; Maria Pia, C.; Maria Teresa, B.; Sergio, O.; Giorgio Giambattista, G.; Paolo, B. Bone Mineral Density in Adult Celiac Patients and the Effect of Gluten-Free Diet from Childhood. *Am. J. Gastroenterol.* **1990**, *85*, 51–53.
58. Barera, G.; Mora, S.; Brambilla, P.; Ricotti, A.; Menni, L.; Beccio, S.; Bianchi, C. Body composition in children with celiac disease and the effects of a gluten-free diet: A prospective case-control study. *Am. J. Clin. Nutr.* **2000**, *72*, 71–75. [CrossRef] [PubMed]
59. Kalayci, A.G.; Kansu, A.; Girgin, N.; Kucuk, N.O.; Aras, G. Bone Mineral Density and Importance of a Gluten-Free Diet in Patients With Celiac Disease in Childhood. *Pediatrics* **2001**, *108*, 89. [CrossRef]
60. Soliman, A.T.; Laham, M.; Jour, C.; Shaat, M.; Souikey, F.; Itani, M.; Al-Safi, A.; Karmallah, A.; Qudaisat, A.; Alarabi, Z.; et al. Linear growth of children with celiac disease after the first two years on gluten- free diet: A controlled study. *Acta Biomed* **2019**, *90*, 20–27. [CrossRef] [PubMed]
61. Kurppa, K.; Collin, P.; Viljamaa, M.; Haimila, K.; Saavalainen, P.; Partanen, J.; Laurila, K.; Huhtala, H.; Paasikivi, K.; Mäki, M.; et al. Diagnosing Mild Enteropathy Celiac Disease: A Randomized, Controlled Clinical Study. *Gastroenterology* **2009**, *136*, 816–823. [CrossRef]
62. Tursi, A.; Brandimarte, G. The Symptomatic and Histologic Response to a Gluten-Free Die in Patients With Borderline Enteropathy. *J. Clin. Gastroenterol.* **2003**, *36*, 13–17.
63. Wieser, H.; Segura, V.; Ruiz-Carnicer, A.; Sousa, C.; Comino, I. Food Safety and Cross-Contamination of Gluten-Free Products: A Narrative Review. *Nutrients* **2021**, *13*, 2244. [CrossRef]
64. Hischenhuber, C.; Crevel, R.; Jarry, B.; Maki, M.; Moneret-Vautrin, D.A.; Romano, A.; Troncone, R.; Ward, R. Review article: Safe amounts of gluten for patients with wheat allergy or coeliac disease. *Aliment. Pharmacol. Ther.* **2006**, *23*, 559–575. [CrossRef] [PubMed]
65. Lähdeaho, M.-L.; Mäki, M.; Laurila, K.; Huhtala, H.; Kaukinen, K. Small- bowel mucosal changes and antibody responses after low- and moderate-dose gluten challenge in celiac disease. *BMC Gastroenterol.* **2011**, *11*, 129. [CrossRef] [PubMed]
66. Akobeng, A.K.; Thomas, A.G. Systematic review: Tolerable amount of gluten for people with coeliac disease. *Aliment. Pharmacol. Ther.* **2008**, *27*, 1044–1052. [CrossRef] [PubMed]
67. Catassi, C.; Rossini, M.; Ratsch, I.M.; Bearzi, I.; Santinelli, A.; Castagnani, R.; Pisani, E.; Coppa, G.V.; Giorgi, P.L. Dose dependent effects of protracted ingestion of small amounts of gliadin in coeliac disease children: A clinical and jejunal morphometric study. *Gut* **1993**, *34*, 1515–1519. [CrossRef] [PubMed]
68. Dunne, M.; Byrne, G.; Chirdo, F.G.; Feighery, C. Coeliac Disease Pathogenesis: The Uncertainties of a Well-Known Immune Mediated Disorder. *Front. Immunol.* **2020**, *11*, 1374. [CrossRef] [PubMed]
69. Lerner, B.A.; Vo, L.T.P.; Yates, S.; Rundle, A.G.; Green, P.H.; Lebwohl, B. Detection of Gluten in Gluten-Free Labeled Restaurant Food: Analysis of Crowd-Sourced Data. *Am. J. Gastroenterol.* **2019**, *114*, 792–797. [CrossRef] [PubMed]
70. Collin, P.; Thorell, L.; Kaukinen, K.; Mäki, M. The safe threshold for gluten contamination in gluten-free products. Can trace amounts be accepted in the treatment of coeliac disease? *Aliment. Pharmacol. Ther.* **2004**, *19*, 1277–1283. [CrossRef]
71. Stevens, L.; Rashid, M. Gluten-Free and Regular Foods: A Cost Comparison. *Can. J. Diet. Pr. Res.* **2008**, *69*, 147–150. [CrossRef]
72. Lee, A.R.; Ng, D.L.; Zivin, J.; Green, P.H.R. Economic burden of a gluten-free diet. *J. Hum. Nutr. Diet.* **2007**, *20*, 423–430. [CrossRef]
73. Singh, J.; Whelan, K. Limited availability and higher cost of gluten-free foods. *J. Hum. Nutr. Diet.* **2011**, *24*, 479–486. [CrossRef]
74. Mustalahti, K.; Lohiniemi, S.; Collin, P.; Vuolteenaho, N.; Laippala, P.; Mäki, M. Gluten-free diet and quality of life in patients with screen-detected celiac disease. *Eff. Clin. Pract.* **2002**, *5*, 105–113.

75. Plugis, N.M.; Khosla, C. Therapeutic approaches for celiac disease. *Best Pr. Res. Clin. Gastroenterol.* **2015**, *29*, 503–521. [CrossRef] [PubMed]
76. Bebb, J.R.; Lawson, A.; Knight, T.; Long, R.G. Long-term follow-up of coeliac disease—What do coeliac patients want? *Aliment. Pharmacol. Ther.* **2006**, *23*, 827–831. [CrossRef] [PubMed]
77. Kurppa, K.; Lauronen, O.; Collin, P.; Ukkola, A.; Laurila, K.; Huhtala, H.; Mäki, M.; Kaukinen, K. Factors Associated with Dietary Adherence in Celiac Disease: A Nationwide Study. *Digestion* **2012**, *86*, 309–314. [CrossRef]
78. Galli, G.; Carabotti, M.; Pilozzi, E.; Lahner, E.; Annibale, B.; Conti, L. Relationship between Persistent Gastrointestinal Symptoms and Duodenal Histological Findings after Adequate Gluten-Free Diet: A Gray Area of Celiac Disease Management in Adult Patients. *Nutrients* **2021**, *13*, 600. [CrossRef]
79. Leffler, D.A.; Dennis, M.; George, J.E.; Jamma, S.; Cook, E.F.; Schuppan, D.; Kelly, C.P. A Validated Disease-Specific Symptom Index for Adults With Celiac Disease. *Clin. Gastroenterol. Hepatol.* **2009**, *7*, 1328–1334.e3. [CrossRef] [PubMed]
80. Leffler, D.A.; Dennis, M.; George, J.B.E.; Jamma, S.; Magge, S.; Cook, E.F.; Schuppan, D.; Kelly, C.P. A Simple Validated Gluten-Free Diet Adherence Survey for Adults With Celiac Disease. *Clin. Gastroenterol. Hepatol.* **2009**, *7*, 530–536.e2. [CrossRef] [PubMed]
81. Gładyś, K.; Dardzińska, J.; Guzek, M.; Adrych, K.; Małgorzewicz, S. Celiac Dietary Adherence Test and Standardized Dietician Evaluation in Assessment of Adherence to a Gluten-Free Diet in Patients with Celiac Disease. *Nutrients* **2020**, *12*, 2300. [CrossRef]
82. Wieser, H.; Ruiz-Carnicer, Á.; Segura, V.; Comino, I.; Sousa, C. Challenges of Monitoring the Gluten-Free Diet Adherence in the Management and Follow-Up of Patients with Celiac Disease. *Nutrients* **2021**, *13*, 2274. [CrossRef] [PubMed]
83. Hill, I.D. What are the sensitivity and specificity of serologic tests for celiac disease? Do sensitivity and specificity vary in different populations? *Gastroenterology* **2005**, *128*, S25–S32. [CrossRef]
84. Rostom, A.; Dubé, C.; Cranney, A.; Saloojee, N.; Sy, R.; Garritty, C.; Sampson, M.; Zhang, L.; Yazdi, F.; Mamaladze, V.; et al. The diagnostic accuracy of serologic tests for celiac disease: A systematic review. *Gastroenterology* **2005**, *128*, S38–S46. [CrossRef] [PubMed]
85. Maglione, M.A.; Okunogbe, A.; Ewing, B.; Grant, S.; Newberry, S.J.; Motala, A.; Shanman, R.; Mejia, N.; Arifkhanova, A.; Shekelle, P.; et al. Diagnosis of Celiac Disease. *Diagnosis Celiac Dis.* **2016**, *162*, 1.
86. Vitoria, J.C.; Arrieta, A.; Arranz, C.; Ayesta, A.; Sojo, A.; Maruri, N.; García-Masdevall, M.D. Antibodies to Gliadin, Endomysium, and Tissue Transglutaminase for the Diagnosis of Celiac Disease. *J. Pediatr. Gastroenterol. Nutr.* **1999**, *29*, 571–574. [CrossRef] [PubMed]
87. Maki, M.; Holm, K.; Hallstrom, O.; Collin, P.; Viander, M.; Savilahti, E.; Lipsanen, V.; Koskimies, S. Serological markers and HLA genes among healthy first-degree relatives of patients with coeliac disease. *Lancet* **1991**, *338*, 1350–1353. [CrossRef]
88. Adriaanse, M.; Leffler, D.A. Serum Markers in the Clinical Management of Celiac Disease. *Dig. Dis.* **2015**, *33*, 236–243. [CrossRef]
89. Lewis, N.R.; Scott, B.B. Meta-analysis: Deamidated gliadin peptide antibody and tissue transglutaminase antibody compared as screening tests for coeliac disease. *Aliment. Pharmacol. Ther.* **2010**, *31*, 73–81. [CrossRef]
90. Barbato, M.; Maiella, G.; Di Camillo, C.; Guida, S.; Valitutti, F.; Lastrucci, G.; Mainiero, F.; Cucchiara, S. The anti-deamidated gliadin peptide antibodies unmask celiac disease in small children with chronic diarrhoea. *Dig. Liver Dis.* **2011**, *43*, 465–469. [CrossRef] [PubMed]
91. Monzani, A.; Rapa, A.; Fonio, P.; Tognato, E.; Panigati, L.; Oderda, G. Use of Deamidated Gliadin Peptide Antibodies to Monitor Diet Compliance in Childhood Celiac Disease. *J. Pediatr. Gastroenterol. Nutr.* **2011**, *53*, 55–60. [CrossRef]
92. Liu, E.; Li, M.; Emery, L.; Taki, I.; Barriga, K.; Tiberti, C.; Eisenbarth, G.S.; Rewers, M.J.; Hoffenberg, E.J. Natural History of Antibodies to Deamidated Gliadin Peptides and Transglutaminase in Early Childhood Celiac Disease. *J. Pediatr. Gastroenterol. Nutr.* **2007**, *45*, 293–300. [CrossRef]
93. Silvester, J.A.; Kurada, S.; Szwajcer, A.; Kelly, C.P.; Leffler, D.A.; Duerksen, D.R. Tests for Serum Transglutaminase and Endomysial Antibodies Do Not Detect Most Patients With Celiac Disease and Persistent Villous Atrophy on Gluten-free Diets: A Meta-analysis. *Gastroenterology* **2017**, *153*, 689–701. [CrossRef]
94. Comino, I.; Real, A.; Vivas, S.; Síglez, M.; Caminero, A.; Nistal, E.; Casqueiro, J.; Rodríguez-Herrera, A.; Cebolla, Á.; Sousa, C. Monitoring of gluten-free diet compliance in celiac patients by assessment of gliadin 33-mer equivalent epitopes in feces. *Am. J. Clin. Nutr.* **2012**, *95*, 670–677. [CrossRef] [PubMed]
95. Sonnenburg, J.L.; Fischbach, M.A. Community Health Care: Therapeutic Opportunities in the Human Microbiome. *Sci. Transl. Med.* **2011**, *3*, 78ps12. [CrossRef]
96. Morón, B.; Cebolla, Á.; Manyani, H.; Álvarez-Maqueda, M.; Megías, M.; Thomas, M.D.C.; López, M.C.; Sousa, C. Sensitive detection of cereal fractions that are toxic to celiac disease patients by using monoclonal antibodies to a main immunogenic wheat peptide. *Am. J. Clin. Nutr.* **2008**, *87*, 405–414. [CrossRef] [PubMed]
97. Comino, I.; Segura, V.; Ortigosa, L.; Espín, B.; Castillejo, G.; Garrote, J.A.; Sierra, C.; Millán-Jiménez, A.; Ribes-Koninckx, C.; Román, E.; et al. Prospective longitudinal study: Use of faecal gluten immunogenic peptides to monitor children diagnosed with coeliac disease during transition to a gluten-free diet. *Aliment. Pharmacol. Ther.* **2019**, *49*, 1484–1492. [CrossRef] [PubMed]
98. Cebolla, Á.; Moreno, M.D.L.; Coto, L.; Sousa, C. Gluten Immunogenic Peptides as Standard for the Evaluation of Potential Harmful Prolamin Content in Food and Human Specimen. *Nutrients* **2018**, *10*, 1927. [CrossRef] [PubMed]
99. Baviera, L.C.B.; Aliaga, E.D.; Ortigosa, L.; Litwin, N.; Peña-Quintana, L.; Méndez, V.; González, M.V.; López-Manzanares, J.M.; Méndez, E.; Koninckx, C.R. Celiac Disease Screening by Immunochromatographic Visual Assays: Results of a Multicenter Study. *J. Pediatr. Gastroenterol. Nutr.* **2007**, *45*, 546–550. [CrossRef] [PubMed]

100. Comino, I.; Fernández-Bañares, F.; Esteve, M.; Ortigosa, L.; Castillejo, G.; Fambuena, B.; Ribes-Koninckx, C.; Sierra, C.; Rodríguez-Herrera, A.; Salazar, J.C.; et al. Fecal Gluten Peptides Reveal Limitations of Serological Tests and Food Questionnaires for Monitoring Gluten-Free Diet in Celiac Disease Patients. *Am. J. Gastroenterol.* **2016**, *111*, 1456–1465. [CrossRef] [PubMed]
101. Moreno, M.D.L.; Cebolla, Á.; Muñoz-Suano, A.; Carrillo-Carrion, C.; Comino, I.; Pizarro, Á.; León, F.; Rodríguez-Herrera, A.; Sousa, C. Detection of gluten immunogenic peptides in the urine of patients with coeliac disease reveals transgressions in the gluten-free diet and incomplete mucosal healing. *Gut* **2015**, *66*, 250–257. [CrossRef]
102. Leonard, M.M.; Weir, D.C.; DeGroote, M.; Mitchell, P.D.; Singh, P.; Silvester, J.; Leichtner, A.M.; Fasano, A. Value of IgA tTG in Predicting Mucosal Recovery in Children With Celiac Disease on a Gluten-Free Diet. *J. Pediatr. Gastroenterol. Nutr.* **2017**, *64*, 286–291. [CrossRef]
103. Pais, W.P.; Duerksen, D.R.; Pettigrew, N.M.; Bernstein, C.N. How many duodenal biopsy specimens are required to make a diagnosis of celiac disease? *Gastrointest. Endosc.* **2008**, *67*, 1082–1087. [CrossRef]
104. Rubio-Tapia, A.; Hill, I.D.; Kelly, C.P.; Calderwood, A.; Murray, J.A. ACG Clinical Guidelines: Diagnosis and Management of Celiac Disease. *Am. J. Gastroenterol.* **2013**, *108*, 656–676. [CrossRef] [PubMed]
105. Husby, S.; Koletzko, S.; Korponay-Szabó, I.; Mearin, M.; Phillips, A.; Shamir, R.; Troncone, R.; Giersiepen, K.; Branski, D.; Catassi, C.; et al. European Society for Pediatric Gastroenterology, Hepatology, and Nutrition Guidelines for the Diagnosis of Coeliac Disease. *J. Pediatr. Gastroenterol. Nutr.* **2012**, *54*, 136–160. [CrossRef] [PubMed]
106. Choung, R.S.; Unalp-Arida, A.; Ruhl, C.E.; Brantner, T.L.; Everhart, J.E.; Murray, J.A. Less Hidden Celiac Disease But Increased Gluten Avoidance Without a Diagnosis in the United States. *Mayo Clin. Proc.* **2016**, *92*, 30–38. [CrossRef]
107. Hadjivassiliou, M.; Aeschlimann, P.; Sanders, D.S.; Mäki, M.; Kaukinen, K.; Grunewald, R.A.; Bandmann, O.; Woodroofe, N.; Haddock, G.; Aeschlimann, D. Transglutaminase 6 antibodies in the diagnosis of gluten ataxia. *Neurology* **2013**, *80*, 1740–1745. [CrossRef]
108. Dipper, C.R.; Maitra, S.; Thomas, R.; Lamb, C.A.; Mclean-Tooke, A.P.C.; Ward, R.; Smith, D.; Spickett, G.; Mansfield, J.C. Anti-tissue transglutaminase antibodies in the follow-up of adult coeliac disease. *Aliment. Pharmacol. Ther.* **2009**, *30*, 236–244. [CrossRef] [PubMed]
109. Hadjivassiliou, M.; Gibson, A.; Davies-Jones, G.; Lobo, A.; Stephenson, T.; Milford-Ward, A. Does cryptic gluten sensitivity play a part in neurological illness? *Lancet* **1996**, *347*, 369–371. [CrossRef]
110. Buie, T. The Relationship of Autism and Gluten. *Clin. Ther.* **2013**, *35*, 578–583. [CrossRef]
111. Lau, N.M.; Green, P.H.R.; Taylor, A.K.; Hellberg, D.; Ajamian, M.; Tan, C.Z.; Kosofsky, B.E.; Higgins, J.J.; Rajadhyaksha, A.M.; Alaedini, A. Markers of Celiac Disease and Gluten Sensitivity in Children with Autism. *PLoS ONE* **2013**, *8*, e66155. [CrossRef]
112. Ghalichi, F.; Ghaemmaghami, J.; Malek, A.; Ostadrahimi, A. Effect of gluten free diet on gastrointestinal and behavioral indices for children with autism spectrum disorders: A randomized clinical trial. *World J. Pediatr.* **2016**, *12*, 436–442. [CrossRef] [PubMed]
113. Piwowarczyk, A.; Horvath, A.; Pisula, E.; Kawa, R.; Szajewska, H. Gluten-Free Diet in Children with Autism Spectrum Disorders: A Randomized, Controlled, Single-Blinded Trial. *J. Autism Dev. Disord.* **2019**, *50*, 482–490. [CrossRef]
114. Peters, S.L.; Biesiekierski, J.; Yelland, G.; Muir, J.G.; Gibson, P.R. Randomised clinical trial: Gluten may cause depression in subjects with non-coeliac gluten sensitivity—An exploratory clinical study. *Aliment. Pharmacol. Ther.* **2014**, *39*, 1104–1112. [CrossRef] [PubMed]
115. Zylberberg, H.M.; Demmer, R.T.; Murray, J.A.; Green, P.H.; Lebwohl, B. Depression and insomnia among individuals with celiac disease or on a gluten-free diet in the USA. *Eur. J. Gastroenterol. Hepatol.* **2017**, *29*, 1091–1096. [CrossRef] [PubMed]
116. Ohrnberger, J.; Fichera, E.; Sutton, M. The relationship between physical and mental health: A mediation analysis. *Soc. Sci. Med.* **2017**, *195*, 42–49. [CrossRef] [PubMed]
117. Blanchet, S.; Chikhi, S.; Maltais, D. The benefits of physical activities on cognitive and mental health in healthy and pathological aging. *Geriatrie et psychologie neuropsychiatrie du vieillissement* **2018**, *16*, 197–205. [CrossRef] [PubMed]
118. Ergün, C.; Urhan, M.; Ayer, A. A review on the relationship between gluten and schizophrenia: Is gluten the cause? *Nutr. Neurosci.* **2017**, *21*, 455–466. [CrossRef] [PubMed]
119. Levinta, A.; Mukovozov, I.; Tsoutsoulas, C. Use of a Gluten-Free Diet in Schizophrenia: A Systematic Review. *Adv. Nutr.* **2018**, *9*, 824–832. [CrossRef] [PubMed]
120. Wahnschaffe, U.; Ullrich, R.; Riecken, E.; Schulzke, J. Celiac disease–like abnormalities in a subgroup of patients with irritable bowel syndrome. *Gastroenterology* **2001**, *121*, 1329–1338. [CrossRef] [PubMed]
121. Aziz, I.; Trott, N.; Briggs, R.; North, J.R.; Hadjivassiliou, M.; Sanders, D.S. Efficacy of a Gluten-Free Diet in Subjects With Irritable Bowel Syndrome-Diarrhea Unaware of Their HLA-DQ2/8 Genotype. *Clin. Gastroenterol. Hepatol.* **2015**, *14*, 696–703.e1. [CrossRef]
122. Celiac disease, non-celiac gluten sensitivity and inflammatory bowel disease. *Minerva Gastroenterol. Dietol.* **2015**, *61*, 267–271. Available online: https://www.minervamedica.it/en/journals/gastroenterology/article.php?cod=R08Y2015N04A0267 (accessed on 8 September 2021).
123. Herfarth, H.H.; Martin, C.F.; Sandler, R.S.; Kappelman, M.D.; Long, M.D. Prevalence of a Gluten-free Diet and Improvement of Clinical Symptoms in Patients with Inflammatory Bowel Diseases. *Inflamm. Bowel Dis.* **2014**, *20*, 1194–1197. [CrossRef]
124. Lindberg, E.; Magnusson, K.-E.; Tysk, C.; Jarnerot, G. Antibody (IgG, IgA, and IgM) to baker's yeast (Saccharomyces cerevisiae), yeast mannan, gliadin, ovalbumin and betalactoglobulin in monozygotic twins with inflammatory bowel disease. *Gut* **1992**, *33*, 909–913. [CrossRef]

125. Reunala, T.; Blomqvist, K.; Tarpila, S.; Halme, H.; Kangas, K. Gluten-free diet in dermatitis herpetiformis. *Br. J. Dermatol.* **1977**, *97*, 473–480. [CrossRef] [PubMed]
126. Fry, L.; Riches, D.; Seah, P.; Hoffbrand, A. Clearance of skin lesions in dermatitis herpetiformis after gluten withdrawal. *Lancet* **1973**, *301*, 288–291. [CrossRef]
127. Catassi, C.; Elli, L.; Bonaz, B.; Bouma, G.; Carroccio, A.; Castillejo, G.; Cellier, C.; Cristofori, F.; De Magistris, L.; Dolinsek, J.; et al. Diagnosis of Non-Celiac Gluten Sensitivity (NCGS): The Salerno Experts' Criteria. *Nutrients* **2015**, *7*, 4966–4977. [CrossRef] [PubMed]
128. Kabbani, T.A.; Vanga, R.R.; Leffler, D.A.; Villafuerte-Galvez, J.; Pallav, K.; Hansen, J.; Mukherjee, R.; Dennis, M.; Kelly, C.P. Celiac Disease or Non-Celiac Gluten Sensitivity? An Approach to Clinical Differential Diagnosis. *Am. J. Gastroenterol.* **2014**, *109*, 741–746. [CrossRef] [PubMed]
129. Roszkowska, A.; Pawlicka, M.; Mroczek, A.; Bałabuszek, K.; Nieradko-Iwanicka, B. Non-Celiac Gluten Sensitivity: A Review. *Medicina* **2019**, *55*, 222. [CrossRef] [PubMed]
130. Fasano, A.; Sapone, A.; Zevallos, V.; Schuppan, D. Nonceliac Gluten Sensitivity. *Gastroenterology* **2015**, *148*, 1195–1204. [CrossRef]
131. Pellegrina, C.D.; Perbellini, O.; Scupoli, M.; Tomelleri, C.; Zanetti, C.; Zoccatelli, G.; Fusi, M.; Peruffo, A.; Rizzi, C.; Chignola, R. Effects of wheat germ agglutinin on human gastrointestinal epithelium: Insights from an experimental model of immune/epithelial cell interaction. *Toxicol. Appl. Pharmacol.* **2009**, *237*, 146–153. [CrossRef] [PubMed]
132. Skodje, G.I.; Sarna, V.K.; Minelle, I.H.; Rolfsen, K.L.; Muir, J.G.; Gibson, P.R.; Veierød, M.B.; Henriksen, C.; Lundin, K.E. Fructan, Rather Than Gluten, Induces Symptoms in Patients With Self-Reported Non-Celiac Gluten Sensitivity. *Gastroenterology* **2018**, *154*, 529–539.e2. [CrossRef]
133. Yelland, G.W. Gluten-induced cognitive impairment ("brain fog") in coeliac disease. *J. Gastroenterol. Hepatol.* **2017**, *32*, 90–93. [CrossRef]
134. Truswell, A.S. Cereal grains and coronary heart disease. *Eur. J. Clin. Nutr.* **2002**, *56*, 1–14. [CrossRef] [PubMed]
135. Flight, I.; Clifton, P. Cereal grains and legumes in the prevention of coronary heart disease and stroke: A review of the literature. *Eur. J. Clin. Nutr.* **2006**, *60*, 1145–1159. [CrossRef] [PubMed]
136. Jensen, M.K.; Koh-Banerjee, P.; Hu, F.B.; Franz, M.; Sampson, L.; Grønbaek, M.; Rimm, E.B. Intakes of whole grains, bran, and germ and the risk of coronary heart disease in men. *Am. J. Clin. Nutr.* **2004**, *80*, 1492–1499. [CrossRef]
137. Tang, G.; Wang, D.; Long, J.; Yang, F.; Si, L. Meta-Analysis of the Association Between Whole Grain Intake and Coronary Heart Disease Risk. *Am. J. Cardiol.* **2015**, *115*, 625–629. [CrossRef] [PubMed]
138. Lebwohl, B.; Cao, Y.; Zong, G.; Hu, F.B.; Green, P.H.R.; Neugut, A.I.; Rimm, E.B.; Sampson, L.; Dougherty, L.W.; Giovannucci, E.; et al. Long term gluten consumption in adults without celiac disease and risk of coronary heart disease: Prospective cohort study. *BMJ* **2017**, *357*, j1892. [CrossRef] [PubMed]
139. Potter, M.D.E.; Brienesse, S.C.; Walker, M.M.; Boyle, A.; Talley, N.J. Effect of the gluten-free diet on cardiovascular risk factors in patients with coeliac disease: A systematic review. *J. Gastroenterol. Hepatol.* **2017**, *33*, 781–791. [CrossRef]
140. Heikkilä, K.; Koskinen, O.; Agarwal, A.; Tikkinen, K.; Mäki, M.; Kaukinen, K. Associations of coeliac disease with coronary heart disease and cerebrovascular disease: A systematic review and meta-analysis. *Nutr. Metab. Cardiovasc. Dis.* **2015**, *25*, 816–831. [CrossRef]
141. Kim, H.-S.; Demyen, M.F.; Mathew, J.; Kothari, N.; Feurdean, M.; Ahlawat, S.K. Obesity, Metabolic Syndrome, and Cardiovascular Risk in Gluten-Free Followers Without Celiac Disease in the United States: Results from the National Health and Nutrition Examination Survey 2009–2014. *Dig. Dis. Sci.* **2017**, *62*, 2440–2448. [CrossRef]
142. Deora, V.; Aylward, N.; Sokoro, A.; El-Matary, W. Serum Vitamins and Minerals at Diagnosis and Follow-up in Children With Celiac Disease. *J. Pediatr. Gastroenterol. Nutr.* **2017**, *65*, 185–189. [CrossRef]
143. Thompson, T.; Dennis, M.; Higgins, L.A.; Lee, A.R.; Sharrett, M.K. Gluten-free diet survey: Are Americans with coeliac disease consuming recommended amounts of fibre, iron, calcium and grain foods? *J. Hum. Nutr. Diet.* **2005**, *18*, 163–169. [CrossRef]
144. Sue, A.; Dehlsen, K.; Ooi, C.Y. Paediatric Patients with Coeliac Disease on a Gluten-Free Diet: Nutritional Adequacy and Macro- and Micronutrient Imbalances. *Curr. Gastroenterol. Rep.* **2018**, *20*, 2. [CrossRef]
145. Hemarajata, P.; Versalovic, J. Effects of probiotics on gut microbiota: Mechanisms of intestinal immunomodulation and neuromodulation. *Ther. Adv. Gastroenterol.* **2012**, *6*, 39–51. [CrossRef]
146. Missbach, B.; Schwingshackl, L.; Billmann, A.; Mystek, A.; Hickelsberger, M.; Bauer, G.; Koenig, J. Gluten-free food database: The nutritional quality and cost of packaged gluten-free foods. *PeerJ* **2015**, *3*, e1337. [CrossRef] [PubMed]
147. Zarkadas, M.; Cranney, A.; Case, S.; Molloy, M.; Switzer, C.; Graham, I.D.; Butzner, J.D.; Rashid, M.; Warren, R.E.; Burrows, V. The impact of a gluten-free diet on adults with coeliac disease: Results of a national survey. *J. Hum. Nutr. Diet.* **2006**, *19*, 41–49. [CrossRef] [PubMed]
148. Silvester, J.A.; Weiten, D.; Graff, L.A.; Walker, J.R.; Duerksen, D.R. Living gluten-free: Adherence, knowledge, lifestyle adaptations and feelings towards a gluten-free diet. *J. Hum. Nutr. Diet.* **2015**, *29*, 374–382. [CrossRef] [PubMed]
149. Jandhyala, S.M. Role of the normal gut microbiota. *World J. Gastroenterol.* **2015**, *21*, 8787–8803. [CrossRef]
150. Dominguez-Bello, M.G.; Costello, E.K.; Contreras, M.; Magris, M.; Hidalgo, G.; Fierer, N.; Knight, R. Delivery mode shapes the acquisition and structure of the initial microbiota across multiple body habitats in newborns. *Proc. Natl. Acad. Sci. USA* **2010**, *107*, 11971–11975. [CrossRef]

151. Bohnhoff, M.; Miller, C.P. Enhanced Susceptibility to Salmonella Infection in Streptomycin-Treated Mice. *J. Infect. Dis.* **1962**, *111*, 117–127. [CrossRef]
152. David, L.A.; Maurice, C.F.; Carmody, R.N.; Gootenberg, D.; Button, J.E.; Wolfe, B.E.; Ling, A.V.; Devlin, A.S.; Varma, Y.; Fischbach, M.; et al. Diet rapidly and reproducibly alters the human gut microbiome. *Nature* **2013**, *505*, 559–563. [CrossRef]
153. Golfetto, L.; De Senna, F.D.; Hermes, J.; Beserra, B.T.S.; França, F.D.S.; Martinello, F. Lower bifidobacteria counts in adult patients with celiac disease on a gluten-free diet. *Arq. Gastroenterol.* **2014**, *51*, 139–143. [CrossRef]
154. Wacklin, P.; Laurikka, P.; Lindfors, K.; Collin, P.; Salmi, T.; Lähdeaho, M.-L.; Saavalainen, P.; Mäki, M.; Mättö, J.; Kurppa, K.; et al. Altered Duodenal Microbiota Composition in Celiac Disease Patients Suffering From Persistent Symptoms on a Long-Term Gluten-Free Diet. *Am. J. Gastroenterol.* **2014**, *109*, 1933–1941. [CrossRef]
155. De Palma, G.; Nadal, I.; Collado, M.C.; Sanz, Y. Effects of a gluten-free diet on gut microbiota and immune function in healthy adult human subjects. *Br. J. Nutr.* **2009**, *102*, 1154–1160. [CrossRef] [PubMed]
156. Yang, B.; Xiao, L.; Liu, S.; Liu, X.; Luo, Y.; Ji, Q.; Yang, P.; Liu, Z. Exploration of the effect of probiotics supplementation on intestinal microbiota of food allergic mice. *Am. J. Transl. Res.* **2017**, *9*, 376–385. [PubMed]
157. Sánchez, B.; Delgado, S.; Blanco-Míguez, A.; Lourenço, A.; Gueimonde, M.; Margolles, A. Probiotics, gut microbiota, and their influence on host health and disease. *Mol. Nutr. Food Res.* **2016**, *61*, 1600240. [CrossRef] [PubMed]
158. Di Nardo, G.; Villa, M.P.; Conti, L.; Ranucci, G.; Pacchiarotti, C.; Principessa, L.; Raucci, U.; Parisi, P. Nutritional Deficiencies in Children with Celiac Disease Resulting from a Gluten-Free Diet: A Systematic Review. *Nutrients* **2019**, *11*, 1588. [CrossRef] [PubMed]
159. Mariani, P.; Viti, M.G.; Montouri, M.; La Vecchia, A.; Cipolletta, E.; Calvani, L.; Bonamico, M. The Gluten-Free Diet: A Nutritional Risk Factor for Adolescents with Celiac Disease? *J. Pediatr. Gastroenterol. Nutr.* **1998**, *27*, 519–523. [CrossRef] [PubMed]
160. Caponio, F.; Summo, C.; Clodoveo, M.L.; Pasqualone, A. Evaluation of the nutritional quality of the lipid fraction of gluten-free biscuits. *Eur. Food Res. Technol.* **2007**, *227*, 135–139. [CrossRef]
161. MacCulloch, K.; Rashid, M. Factors affecting adherence to a gluten-free diet in children with celiac disease. *Paediatr. Child Heal.* **2014**, *19*, 305–309. [CrossRef] [PubMed]
162. Lennerz, B.S.; Barton, A.; Bernstein, R.K.; Dikeman, R.D.; Diulus, C.; Hallberg, S.; Rhodes, E.T.; Ebbeling, C.B.; Westman, E.C.; Yancy, W.S.; et al. Management of Type 1 Diabetes With a Very Low–Carbohydrate Diet. *Pediatrics* **2018**, *141*, e20173349. [CrossRef]
163. Rosenfalck, A.M.; Almdal, T.; Viggers, L.; Madsbad, S.; Hilsted, J. A low-fat diet improves peripheral insulin sensitivity in patients with Type 1 diabetes. *Diabet. Med.* **2006**, *23*, 384–392. [CrossRef]
164. Patton, S.R. Adherence to Diet in Youth with Type 1 Diabetes. *J. Am. Diet. Assoc.* **2011**, *111*, 550–555. [CrossRef]
165. Camarca, M.E.; Mozzillo, E.; Nugnes, R.; Zito, E.; Falco, M.; Fattorusso, V.; Mobilia, S.; Buono, P.; Valerio, G.; Troncone, R.; et al. Celiac disease in type 1 diabetes mellitus. *Ital. J. Pediatrics* **2012**, *38*, 10. [CrossRef] [PubMed]
166. Scarff, J.R. Orthorexia Nervosa: An Obsession With Healthy Eating. *Fed. Pract.* **2017**, *34*, 36–39. [PubMed]
167. Wolf, R.L.; Lebwohl, B.; Lee, A.R.; Zybert, P.; Reilly, N.R.; Cadenhead, J.; Amengual, C.; Green, P.H.R. Hypervigilance to a Gluten-Free Diet and Decreased Quality of Life in Teenagers and Adults with Celiac Disease. *Dig. Dis. Sci.* **2018**, *63*, 1438–1448. [CrossRef] [PubMed]

MDPI
St. Alban-Anlage 66
4052 Basel
Switzerland
Tel. +41 61 683 77 34
Fax +41 61 302 89 18
www.mdpi.com

Nutrients Editorial Office
E-mail: nutrients@mdpi.com
www.mdpi.com/journal/nutrients

www.ingramcontent.com/pod-product-compliance
Ingram Content Group UK Ltd.
Pitfield, Milton Keynes, MK11 3LW, UK
UKHW060123010826
14089UKWH00006B/115

9 783036 548395